Lecture Notes in Computer Science   11172

Commenced Publication in 1973
Founding and Former Series Editors:
Gerhard Goos, Juris Hartmanis, and Jan van Leeuwen

More information about this series at http://www.springer.com/series/7407

Marco Dorigo · Mauro Birattari
Christian Blum · Anders L. Christensen
Andreagiovanni Reina · Vito Trianni (Eds.)

Swarm Intelligence

11th International Conference, ANTS 2018
Rome, Italy, October 29–31, 2018
Proceedings

 Springer

Editors
Marco Dorigo ⓘD
Université Libre de Bruxelles
Brussels
Belgium

Mauro Birattari ⓘD
Université Libre de Bruxelles
Brussels
Belgium

Christian Blum ⓘD
Artificial Intelligence Research Institute
Bellaterra
Spain

Anders L. Christensen ⓘD
University of Southern Denmark
Odense
Denmark

Andreagiovanni Reina ⓘD
University of Sheffield
Sheffield
UK

Vito Trianni ⓘD
National Research Council
Rome
Italy

ISSN 0302-9743   ISSN 1611-3349   (electronic)
Lecture Notes in Computer Science
ISBN 978-3-030-00532-0   ISBN 978-3-030-00533-7   (eBook)
https://doi.org/10.1007/978-3-030-00533-7

Library of Congress Control Number: 2018954072

LNCS Sublibrary: SL1 – Theoretical Computer Science and General Issues

This Springer imprint is published by the registered company Springer Nature Switzerland AG
The registered company address is: Gewerbestrasse 11, 6330 Cham, Switzerland

Preface

These proceedings contain the papers presented at ANTS 2018, the 11th International Conference on Swarm Intelligence, held at the National Research Council (CNR) in Rome, Italy, during October 29–31, 2018. The ANTS series started in 1998 with the First International Workshop on Ant Colony Optimization (ANTS 1998). Since then, ANTS, which is held bi-annually, has gradually become an international forum for researchers in the wider field of swarm intelligence. In 2004, this development was acknowledged by the inclusion of the term "Swarm Intelligence" (next to "Ant Colony Optimization") in the conference title. Since 2010, the ANTS conference is officially devoted to the field of swarm intelligence as a whole, without any bias towards specific research directions. This is reflected in the title of the conference: "International Conference on Swarm Intelligence."

This volume contains the best papers selected out of 69 submissions. Of these, 24 were accepted as full-length papers, while 12 were accepted as short papers. This corresponds to an overall acceptance rate of 52%. Also included in this volume are 7 extended abstracts.

All the contributions were presented as posters. The full-length papers were also presented orally in a plenary session. Extended versions of the best papers presented at the conference will be published in a special issue of the *Swarm Intelligence* journal.

We would like to take this opportunity to thank the large number of people that were involved in making this conference a success. We would like to express our gratitude to the authors who contributed their work, to the members of the International Program Committee, to the additional referees for their qualified and detailed reviews, and to the staff at the Institute of Cognitive Sciences and Technologies (ISTC) of the CNR for helping with organizational matters.

We hope the reader will find this volume useful both as a reference to current research in swarm intelligence and as a starting point for future work.

July 2018

Marco Dorigo
Mauro Birattari
Christian Blum
Anders L. Christensen
Andreagiovanni Reina
Vito Trianni

Organization

Organizing Committee

General Chair

Marco Dorigo   Université Libre de Bruxelles, Belgium

Vice-General Chair

Mauro Birattari   Université Libre de Bruxelles, Belgium

Local Organizing and Publicity Chair

Vito Trianni   Italian National Research Council, Italy

Technical Program Chairs

Christian Blum   Spanish National Research Council, Spain
Anders L. Christensen   University of Southern Denmark, Denmark,
        and Instituto Universitário de Lisboa (ISCTE-IUL),
        Portugal

Publication Chair

Andreagiovanni Reina   University of Sheffield, UK

Paper Submission Chair

Volker Strobel   Université Libre de Bruxelles, Belgium

Program Committee

Michael Allwright   Université Libre de Bruxelles, Belgium
Prasanna Balaprakash   Argonne National Laboratory, USA
Jacob Beal   BBN Technologies, USA
Giovanni Beltrame   Polytechnique Montréal, Canada
Tim Blackwell   Goldsmiths, University of London, UK
Mohammad Reza Bonyadi   The University of Adelaide, Australia
Darko Bozhinoski   Université Libre de Bruxelles, Belgium
Alexandre Campo   Université Libre de Bruxelles, Belgium
Marco Chiarandini   University of Southern Denmark, Denmark
Maurice Clerc   Independent Consultant on Optimisation, France
Carlos Coello Coello   CINVESTAV-IPN, Mexico
Oscar Cordon   University of Granada, Spain
Nikolaus Correll   University of Colorado Boulder, USA
Guido De Croon   Delft University of Technology, The Netherlands

Gianni Di Caro	Carnegie Mellon University, USA
Luca Maria Gambardella	Istituto Dalle Molle di Studi sull'Intelligenza Artificiale, Switzerland
Melvin Gauci	Harvard University, USA
Luca Di Gaspero	University of Udine, Italy
Haibin Duan	Beihang University, China
Andries Engelbrecht	University of Pretoria, South Africa
Eliseo Ferrante	University of Birmingham, Dubai, UAE
Gianpiero Francesca	Toyota Motor Europe, Belgium
José García-Nieto	University of Málaga, Spain
Simon Garnier	New Jersey Institute of Technology, USA
Jorge Gomes	University of Lisbon, Portugal
Morten Goodwin	University of Agder, Norway
Roderich Gross	University of Sheffield, UK
Frédéric Guinand	University of Le Havre, France
Heiko Hamann	University of Lübeck, Germany
Julia Handl	University of Manchester, UK
J. Michael Herrmann	University of Edinburgh, UK
Yara Khaluf	Ghent University, Belgium
Xiaodong Li	RMIT University, Australia
Simone Ludwig	North Dakota State University, USA
Manuel López-Ibáñez	University of Manchester, UK
Vittorio Maniezzo	University of Bologna, Italy
Alcherio Martinoli	Ecole Polytechnique Fédérale de Lausanne, Switzerland
Massimo Mastrangeli	Delft University of Technology, The Netherlands
Nithin Mathews	Netcetera, Switzerland
Michalis Mavrovouniotis	University of Cyprus, Cyprus
Yi Mei	Victoria University of Wellington, New Zealand
Ronaldo Menezes	Florida Institute of Technology, USA
Bernd Meyer	Monash University, Australia
Martin Middendorf	University of Leipzig, Germany
Alan Millard	University of York, UK
Nicolas Monmarché	University of Tours, France
Roberto Montemanni	Istituto Dalle Molle di Studi sull'Intelligenza Artificiale, Switzerland
Marco Montes de Oca	Northeastern University, USA
Sanaz Mostaghim	Otto von Guericke University Magdeburg, Germany
Konstantinos Parsopoulos	University of Ioannina, Greece
Paola Pellegrini	IFSTTAR, France
Carlo Pinciroli	Worcester Polytechnic Institute, USA
Lenka Pitonakova	University of Bristol, UK
Günther Raidl	Vienna University of Technology, Austria
Katya Rodriguez-Vazquez	National Autonomous University of Mexico, Mexico
Mike Rubenstein	Northwestern University, USA
Erol Sahin	Middle East Technical University, Turkey

Roberto Santana	University of the Basque Country, Spain
Thomas Schmickl	University of Graz, Austria
Kevin Seppi	Brigham Young University, USA
Christine Solnon	LIRIS, CNRS, France
Thomas Stützle	Université Libre de Bruxelles, Belgium
Dirk Sudholt	University of Sheffield, UK
Yasumasa Tamura	Tokyo Institute of Technology, Japan
Danesh Tarapore	University of Southampton, UK
Guy Theraulaz	Paul Sabatier University, France
Dhananjay Thiruvady	Monash University, Australia
Jon Timmis	University of York, UK
Elio Tuci	Middlesex University, UK
Ali Emre Turgut	Katholieke Universiteit Leuven, Belgium
Gabriele Valentini	Arizona State University, USA
Michael Vrahatis	University of Patras, Greece
Justin Werfel	Harvard University, USA
Alan Winfield	University of the West of England, UK
Masahito Yamamoto	Hokkaido University, Japan

Additional Reviewers

Nicolas Cambier	Université de Technologie de Compiègne, France
Yue Gu	University of Sheffield, UK
Bahar Haghighat	Ecole Polytechnique Fédérale de Lausanne, Switzerland
Matthew Hall	University of Sheffield, UK
Marcos Oliveira	Leibniz Institute for the Social Sciences, Germany
Anil Ozdemir	University of Sheffield, UK
Diego Pinheiro	Florida Institute of Technology, USA
Judhi Prasetyo	Middlesex University Dubai, UAE
Leonardo Stella	University of Sheffield, UK

Contents

Full Papers

A Study on Force-Based Collaboration in Flying Swarms 3
Chiara Gabellieri, Marco Tognon, Lucia Pallottino,
and Antonio Franchi

Automatic Design of Communication-Based Behaviors for Robot Swarms . . . 16
Ken Hasselmann, Frédéric Robert, and Mauro Birattari

Behavior Trees as a Control Architecture in the Automatic Modular
Design of Robot Swarms. 30
Jonas Kuckling, Antoine Ligot, Darko Bozhinoski, and Mauro Birattari

Guidance of Swarms with Agents Having Bearing Only
and Limited Visibility Sensors 44
Rotem Manor and Alfred M. Bruckstein

Hybrid Control of Swarms for Resource Selection. 57
Marco Trabattoni, Gabriele Valentini, and Marco Dorigo

Local Communication Protocols for Learning Complex Swarm Behaviors
with Deep Reinforcement Learning 71
Maximilian Hüttenrauch, Adrian Šošić, and Gerhard Neumann

Morphogenesis as a Collective Decision of Agents Competing
for Limited Resource: A Plants Approach 84
Payam Zahadat, Daniel Nicolas Hofstadler, and Thomas Schmickl

Negative Updating Combined with Opinion Pooling in the
Best-of-n Problem in Swarm Robotics. 97
Chanelle Lee, Jonathan Lawry, and Alan Winfield

On Mimicking the Effects of the Reality Gap with
Simulation-Only Experiments. 109
Antoine Ligot and Mauro Birattari

Optimization of Swarm Behavior Assisted by an Automatic Local
Proof for a Pattern Formation Task 123
Mario Coppola and Guido C. H. E. de Croon

Quality-Sensitive Foraging by a Robot Swarm Through Virtual
Pheromone Trails   135
  Anna Font Llenas, Mohamed S. Talamali, Xu Xu, James A. R. Marshall,
  and Andreagiovanni Reina

Search in a Maze-Like Environment with Ant Algorithms: Complexity,
Size and Energy Study.   150
  Zainab Husain, Dymitr Ruta, Fabrice Saffre, Yousof Al-Hammadi,
  and Abdel F. Isakovic

Self-adaptive Quantum Particle Swarm Optimization
for Dynamic Environments.   163
  Gary Pamparà and Andries P. Engelbrecht

Simulating Kilobots Within ARGoS: Models and Experimental Validation. . .   176
  Carlo Pinciroli, Mohamed S. Talamali, Andreagiovanni Reina,
  James A. R. Marshall, and Vito Trianni

Simulating Multi-robot Construction in ARGoS   188
  Michael Allwright, Navneet Bhalla, Carlo Pinciroli, and Marco Dorigo

Stability Analysis of the Multi-objective Multi-guided Particle
Swarm Optimizer   201
  Christopher W. Cleghorn, Christiaan Scheepers,
  and Andries P. Engelbrecht

Swarm Attack: A Self-organized Model to Recover from Malicious
Communication Manipulation in a Swarm of Simple Simulated Agents   213
  Giuseppe Primiero, Elio Tuci, Jacopo Tagliabue, and Eliseo Ferrante

Task-Agnostic Evolution of Diverse Repertoires of Swarm Behaviours   225
  Jorge Gomes and Anders Lyhne Christensen

The Best-of-*n* Problem with Dynamic Site Qualities: Achieving
Adaptability with Stubborn Individuals   239
  Judhi Prasetyo, Giulia De Masi, Pallavi Ranjan, and Eliseo Ferrante

The Impact of Interaction Models on the Coherence of Collective
Decision-Making: A Case Study with Simulated Locusts   252
  Yara Khaluf, Ilja Rausch, and Pieter Simoens

The Importance of Component-Wise Stochasticity in Particle
Swarm Optimization   264
  Elre T. Oldewage, Andries P. Engelbrecht,
  and Christopher W. Cleghorn

The Importance of Information Flow Regulation in Preferentially
Foraging Robot Swarms.. 277
   Lenka Pitonakova, Richard Crowder, and Seth Bullock

The Role of Largest Connected Components in Collective Motion 290
   Heiko Hamann

Why the Intelligent Water Drops Cannot Be Considered
as a Novel Algorithm 302
   Christian Leonardo Camacho-Villalón, Marco Dorigo,
   and Thomas Stützle

Short Papers

A Cooperative Opposite-Inspired Learning Strategy
for Ant-Based Algorithms 317
   Nicolás Rojas-Morales, María-Cristina Riff, Carlos A. Coello Coello,
   and Elizabeth Montero

A Solution for the Team Selection Problem Using ACO 325
   Lázaro Lugo, Marilyn Bello, Ann Nowe, and Rafael Bello

Boundary Constraint Handling Techniques for Particle Swarm
Optimization in High Dimensional Problem Spaces.................. 333
   Elre T. Oldewage, Andries P. Engelbrecht,
   and Christopher W. Cleghorn

Does the $ACO_{\mathbb{R}}$ Algorithm Benefit from the Use of Crossover? 342
   Ashraf M. Abdelbar and Khalid M. Salama

Embodied Evolution of Self-organised Aggregation
by Cultural Propagation...................................... 351
   Nicolas Cambier, Vincent Frémont, Vito Trianni, and Eliseo Ferrante

Experimental Evaluation of ACO for Continuous Domains
to Solve Function Optimization Problems........................ 360
   Ryouei Takahashi, Yukihiro Nakamura, and Toshihide Ibaraki

Gaussian-Valued Particle Swarm Optimization 368
   Kyle Robert Harrison, Beatrice M. Ombuki-Berman,
   and Andries P. Engelbrecht

Individual Activity Level and Mobility Patterns of Ants Within Nest Site ... 378
   Kazutaka Shoji

Learning Based Leadership in Swarm Navigation 385
   Ovunc Tuzel, Gilberto Marcon dos Santos, Chloë Fleming,
   and Julie A. Adams

Maintaining Diversity in Robot Swarms with Distributed
Embodied Evolution   395
    Iñaki Fernández Pérez, Amine Boumaza, and François Charpillet

On Steering Swarms   403
    Ariel Barel, Rotem Manor, and Alfred M. Bruckstein

Vector Field Benchmark for Collective Search in Unknown
Dynamic Environments   411
    Palina Bartashevich, Welf Knors, and Sanaz Mostaghim

Extended Abstracts

A Honey Bees Mating Optimization Algorithm with Path Relinking
for the Vehicle Routing Problem with Stochastic Demands   423
    Yannis Marinakis and Magdalene Marinaki

Blockchain Technology for Robot Swarms: A Shared Knowledge
and Reputation Management System for Collective Estimation   425
    Volker Strobel and Marco Dorigo

Declarative Physicomimetics for Tangible Swarm
Application Development.   427
    *Ayberk Özgür, Wafa Johal, Arzu Guneysu Ozgur,
    Francesco Mondada, and Pierre Dillenbourg*

Influence of Leaders and Predators on Steering a Large-Scale
Robot Swarm   429
    John D. Lewis, Himanshi Jain, and Sujit P. Baliyarasimhuni

Movement-Based Localisation for PSO-Inspired Search Behaviour
of Robotic Swarms   431
    Sebastian Mai, Christoph Steup, and Sanaz Mostaghim

Of Bees and Botnets   433
    Vijay Sarvepalli

Using Particle Swarms to Build Strategies for Market Timing:
A Comparative Study   435
    Ismail Mohamed and Fernando E. B. Otero

Author Index   437

Full Papers

A Study on Force-Based Collaboration in Flying Swarms

Chiara Gabellieri[2] , Marco Tognon[1] , Lucia Pallottino[2] ,
and Antonio Franchi[1]([✉])

[1] LAAS-CNRS, Université de Toulouse, CNRS, Toulouse, France
{marco.tognon,antonio.franchi}@laas.fr
[2] Centro di Ricerca "E. Piaggio", Dipartimento di Ingegneria dell'Informazione,
Università di Pisa, Pisa, Italy
lucia.pallottino@unipi.it

Abstract. This work investigates collaborative aerial transportation by swarms of agents based only on implicit information, enabled by the physical interaction among the agents and the environment. Such a coordinating mechanism in collaborative transportation is a basic skill in groups of social animals. We consider cable-suspended objects transported by a swarm of flying robots and we formulate several hypothesis on the behavior of the overall system which are validated thorough numerical study. In particular, we show that a nonzero internal force reduces to one the number of asymptotically stable equilibria and that the internal force intensity is directly connected to the convergence rate. As such, the internal force represents the cornerstone of a communication-less cooperative manipulation paradigm in swarms of flying robots. We also show how a swarm can achieve a stable transportation despite the imprecise knowledge of the system parameters.

1 Introduction

Cooperative transportation without explicit communication, but based only on on the indirect exchange of information through the physical interaction with the environment is a very important feature for social animals. From a scientific point of view, the problem has been addressed as *Stigmery theory* [8]; later, it has been regarded as the main coordinating mechanism in groups of ants for object transportation [9] (see Fig. 1 on the left) and has been indeed observed and studied in [3,5,6,13].

Such skills observed in nature have inspired researchers to transfer them to swarms of simple robotic agents. In fact, avoiding explicit communication would reduce hardware and software complexity, and overcome possible communication failure issues. So far, the interest has been mainly focused on terrestrial systems [3,9,17,22], where the possibility of decentralized transportation based on physical interactions has been proved. Instead, in this work, we are interested in the communication-less *aerial* transportation of objects by swarms. This is particularly interesting not only from a scientific point of view, thanks to the

© Springer Nature Switzerland AG 2018
M. Dorigo et al. (Eds.): ANTS 2018, LNCS 11172, pp. 3–15, 2018.
https://doi.org/10.1007/978-3-030-00533-7_1

Fig. 1. Left: red ants in cooperative transportation. Right: the tension in spider webs influences their natural frequency [2].

higher complexity of the problem, but also because a simple, robust and scalable object aerial manipulation technique could meet the requirements demanded by many real applications. Aerial transportation can benefit from a larger workspace and the independence from uneven terrains. However, aerial robots, e.g., multi-rotors, though agile and low cost, are typically characterized by a limited payload. Hence, a cooperative approach is a very suitable solution. Some examples of applications can be found in industrial contexts, in agriculture, and in search and rescue missions to carry necessary equipment or first aid.

Considering the lack of results for aerial systems compared to grounded ones, our purpose is to start filling this gap. In particular, we investigate the possibility of a communication-less approach for cooperative aerial manipulation with a swarm of flying agents. Furthermore, we investigate if and how the load and the cable physically connecting the robots may play the role of an implicit communication channel exploiting the forces exchanged. To the best of our knowledge, this is the first work proposing a bio-inspired algorithm for communication-less aerial manipulation by flying swarms, going beyond the two-robot-'only' scenario, recently considered in [7,20,21].

The algorithm that we propose exploits a leader-follower paradigm where the leader agent knows where to go and hence steers the object toward the desired position. On the other hand, the follower agents follow the leader and help to sustain the weight and manipulate the load, exploiting only the implicit information contained in the force received from the load itself. It has been observed that also groups of army ants *Eciton burchelli* [5] and *Dorylus wilverthi* [6] adopt a distinct caste distribution in transportation groups, in the sense that groups have a significant tendency to contain only one submajor, i.e., a particular type of ant. These species of army ants have proved to be very efficient in transporting objects together. Additionally, in the same works, it has been noticed that it is usually a single ant, the submajor, that starts the motion of the object, and then the rest of the group moves accordingly. Such a behavior is actually replicated in the leader-follower paradigm proposed in this paper.

While animals usually deploy items by directly touching them (direct manipulation), in our framework we have chosen an indirect manipulation technique of the load through cables. This choice is motivated by different reasons and it has been proved to be a very effective solution for cooperative aerial transportation [10–12,19,23]. First of all, a cable attached to the agent center of mass allows to minimize the coupling between the rotational dynamics of the agent itself from the rest of the system dynamics. This is particularly useful for aerial agents that are underactuated – the most common case – since they need to change their attitude to be able to apply forces in any direction. Furthermore, compared to other possible decoupling gripping mechanisms, as the ones in [14], cables are simpler, low cost, and in general lighter. In this paper we demonstrate through several numerical simulations that a swarm of N flying agents is capable of collaborative manipulation skills based only on implicit communication. We show that a twofold major role is played by the *internal force* applied to the transported object. Internal forces are forces applied at the contact points on the object that stretch or compress it without producing any movement, since they counterbalance each other. The condition of zero internal force corresponds to the case in which the agents transport the object while keeping the cables vertical and applying only a force that compensates for the gravity. Firstly, we have found that nonzero internal forces allow the swarm to univocally set the attitude of the commonly transported object, and secondly, that larger internal forces reduce the convergence time of the overall system to such unique equilibrium. This creates an interesting analogy with the role of tension in spider webs, see Fig. 1-right. The breadth of analysis covers also the thorough investigation of the leader forces depending on the swarm parameters and the analysis of the benefits of a saturated nonlinear law for the leader force in order to tradeoff compliance/safety and transportation accuracy.

The paper structure is the following: Sect. 2 illustrates the dynamic model of the system. Then, we formulate the hypotheses regarding the properties of the swarm, supported by the numerical results presented in Sect. 3. A thorough discussion follows in Sect. 4. Final conclusions and future developments are presented in Sect. 5.

2 Model

The system is composed by a set of N flying agents attached to a cable suspended load that must be deployed to a particular configuration. In our framework, each agent interacts with the environment, hence we aim at a soft response similar to the behavior of human or of animal during everyday interaction tasks. We model the commonly transported object (the load) as a rigid body. The attitude of the load is parametrized by Euler angles yaw, pitch and roll, indicated with ψ, θ, and ϕ, respectively. Each agent is attached to the load by means of a cable by means of which it can transfer forces. The cables are attached to the load at the points L_i, with $i = 1, \ldots, N$, placed on the same plane, denoted by $\mathcal{I}$. The object center of mass (CoM) is indicated with G.

Each cable is modeled as a linear unidirectional spring, with a dissipative term that damps its longitudinal oscillations. We assume that each flying agent is endowed with a position controller. If the latter is sufficiently precise, we can model the closed loop system as a simple double integrator. In this way we can consider each agent as an actuated point mass capable of exchanging a force with the external world.

By doing so, the proposed method can be applied to different aerial robots. If we consider multidirectional-thrust platforms capable of controlling position and orientation independently (popular in the field of aerial physical interaction) [4,15,16,18], the double integrator is an exact model of the position-controlled closed loop system. In the case of underactuated unidirectional-thrust vehicles using a standard position controller, the double integrator is instead a very good approximation. Furthermore, the time-scale separation between the translational and rotational dynamics has been exploited in other works on aerial manipulation like [12,14] where the robots are considered as point masses and modeled as double integrators. Denoting by the vector $\boldsymbol{p}_{Ri} \in \mathbb{R}^3$ the position of the i-th agent, by $\boldsymbol{M}_i = m_{Ai}\boldsymbol{I}_3$ its inertia matrix (with $m_{Ai} \in \mathbb{R}_{>0}$) and by $\boldsymbol{f}_i = \begin{bmatrix} f_{i,x} \ f_{i,y} \ f_{i,z} \end{bmatrix}^\top$ the force that the i-th cable exerts on the object (so that $-\boldsymbol{f}_i$ is the force exerted by the cable on the agent), the dynamics of the i-th agent is:

$$\boldsymbol{M}_i \ddot{\boldsymbol{p}}_{Ri} = -\boldsymbol{f}_i + \boldsymbol{f}_{Ci}, \tag{1}$$

$$\text{where} \quad \boldsymbol{f}_{Ci} = \boldsymbol{B}_i(\dot{\boldsymbol{p}}_{Ri}^d - \dot{\boldsymbol{p}}_{Ri}) + \boldsymbol{K}_i(\boldsymbol{p}_{Ri}^d - \boldsymbol{p}_{Ri}) + \boldsymbol{\pi}_i \tag{2}$$

is the 'control' force of each agent. To better understand how to control a multirotor aerial robot in such a way, the reader can refer to [21]. Such control force models three simple actions. First, the agent implements a spring like action to move towards a desired position or follow a desired path (the apex d indicates the 'desired' quantities). Secondly, the agent implements a dissipative derivative term proportional to the velocity error. This action damps the oscillations induced by the spring action. Finally, there is a force bias indicated with $\boldsymbol{\pi}_i = \begin{bmatrix} \pi_{i,x} \ \pi_{i,y} \ \pi_{i,z} \end{bmatrix}^\top$. This bias is essential to make the flying agents sustain the weight of the load. We shall show that it plays an important role also for shaping the system equilibria. Through $\boldsymbol{\pi}_i$ it is possible to set reference internal forces (forces that do not result in a motion of the object).

The static equilibrium equation of the object subject to the gravity vector $\boldsymbol{g} \in \mathbb{R}^3$ is given by $[m_L\boldsymbol{g}^\top \ 0_{1\times3}]^\top = W\bar{\boldsymbol{f}}$, where $W \in \mathbb{R}^{6\times3N}$ is the grasp matrix, that maps the forces at the contact points to a wrench applied at the object center of mass and $\bar{\boldsymbol{f}} \in \mathbb{R}^{3N\times1}$ collects the equilibrium cable forces. Resolving the equation for $\bar{\boldsymbol{f}}$ we obtain:

$$\bar{\boldsymbol{f}} = W^\dagger[m_L\boldsymbol{g}^\top \ 0_{1\times3}]^\top + \boldsymbol{t}, \tag{3}$$

where † indicates a right (pseudo)inverse, and $\boldsymbol{t} \in \text{null}(W)$ is the internal force, which neither influences the object dynamics nor balances any external wrench.

Generally, in leader follower approaches it may occur that only one agent, i.e., the leader, is aware of the desired trajectory ($\boldsymbol{p}_{Ri}^d, \dot{\boldsymbol{p}}_{Ri}^d$). On the other hand, slave agents tend to stay where they are if no external action intervenes. We model such agents by setting $\boldsymbol{K}_i = \boldsymbol{0}\,\mathrm{Nm}^{-1}$ and $\dot{\boldsymbol{p}}_{Ri}^d = \boldsymbol{0}\,\mathrm{ms}^{-1}$ in (2). When the leader starts moving, the followers will perceive a modification of the environment through a change in their cable force. This, in turn, will make the followers move toward the leader agent trying to bring back the force to the initial equilibrium value. It is worthy to note that, once each agent stops, possible load oscillations are damped thanks to the dissipative action modeled in (2).

3 Numerical Study

For a system of only *two* aerial agents and a beam/like load, in [21] we formally proved, using a Lyapunov-based approach, the stability or instability of all the possible equilibria and the passivity of the overall controlled system. However, it is not trivial to extend those theoretical results to $N > 2$ and to a more generally shaped object. In particular, it is not straightforward to solve the so called *equilibria inverse problem* in [21], namely to find all the possible positions of the agents and forces in the cables for each stable pose of the load. The authors in [1] showed that such problem is very complex to solve even if just less than six cables of assigned length are used. Analytical answers to the problem are therefore difficult to reach. However, a numerical study of whether some of the properties discovered in [21] for the two-agent system apply also to the swarm case, where the number of follower agents may be arbitrary large and the object not only a bar, is equally interesting. Thus, in this section we extrapolate some conjectures on the expected behavior of the swarm system and validate these hypotheses through a wide numerical study.

Table 1. The different simulations setups (scenarios).

Goal of the study	Scenario	Internal forces	CoM	Points L_i	Unknown parameters
Internal force role	S_a	$t = 0$ [N]	$G \in \mathcal{I}$	On a circle	None
	S_b	$t = 0$ [N]	$G \notin \mathcal{I}$	On a circle	None
	S_c	$t \neq 0$ [N]	$G \notin \mathcal{I}$	On a circle	None
Parametric uncertainty	S_d	$t = 0$ [N]	$G \notin \mathcal{I}$	Random	m_L, N (only bounds)

Table 1 contains the description of the simulated conditions. The load has mass $m_L = 5\,\mathrm{kg}$ and inertial matrix equal to $\boldsymbol{J}_L = \boldsymbol{I}_3\,\mathrm{kgm}^2$. The leader agent, set as the agent 1, is fed with a sufficiently smooth reference trajectory constituted by a 5th order polynomial in 3D (rest to rest trajectory with zero acceleration at start and end points), which lasts 10 s and covers 2 m. $\boldsymbol{M}_1 = 0.5\boldsymbol{I}_3\,\mathrm{kg}$, $\boldsymbol{M}_i = 0.01\boldsymbol{I}_{3(N-1)}\,\mathrm{kg}$, $\boldsymbol{B}_1 = 100\boldsymbol{I}_3\,\mathrm{Nsm}^{-1}$ $\boldsymbol{B}_i = 1.5\boldsymbol{I}_{3(N-1)}\mathrm{Nsm}^{-1}$, $\boldsymbol{K}_1 = 1000\boldsymbol{I}_3\,\mathrm{Nsm}^{-1}$, and $\boldsymbol{K}_i = 0\boldsymbol{I}_{3(N-1)}\,\mathrm{Nm}^{-1}$ for $i \neq 1$. When $N > 30$, we changed the apparent mass $\boldsymbol{M}_i$, and the damping $\boldsymbol{B}_i$ of the followers, i.e., for $i = 2, ..., N$ reducing both by 90%. This allows the leader agent to drag the system without

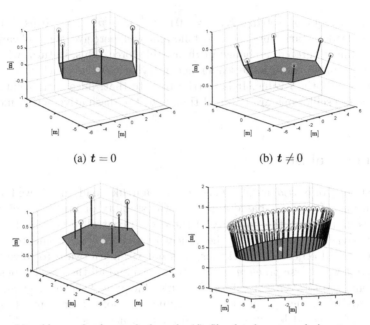

(a) $t = 0$

(b) $t \neq 0$

(c) cables randomly attached on the surface of the object

(d) Simulated system during transportation

Fig. 2. Simulated system for the case $N = 6$ ((a), (b), (c)) and $N = 60$ (d)). The object is depicted as a grey surface, and the light spot on it coincide with its center of mass G. In case $G \notin \mathcal{I}$, the light spot is instead the projection of G on $\mathcal{I}$. The cables are the black lines, while the circles represent the robots (the darker one is the leader).

applying too large forces, similarly to what happens in biological systems as described in Sect. 4.

Scenarios S_a, S_b and S_c refer to the cases in which the parameters of the system are perfectly known, and so is N. Based on what we demonstrated in [21] for the two agents system, we decided to investigate the role of the internal force on the object for the equilibria of the system. We considered the case in which the points L_i lie on a circle centered around the projection of G on $\mathcal{I}$ (see Figs. 2(a), (b), and (d)). We ran different simulations where N is a random number between 2 and 100, and the initial pose of the object has a random value of yaw between 0 rad and $\pi/4$ rad.

In the scenarios S_a and S_b the force bias in (2) is set to sustain the weight of the load without internal forces, namely $\boldsymbol{\pi}_i = [0 \ 0 \ m_L g/N]^\top \ \forall i = 1, \ldots, N$, where g is the intensity of $\boldsymbol{g}$. This implies vertical cables. Once the leader agent stops, the system converges to an equilibrium in which the final attitude of the transported object is not univocally determined but depends on the trajectory resulting from the initial condition and the leader desired trajectory. In particular, the results in the first line of plots of Fig. 3 shows a completely arbitrary

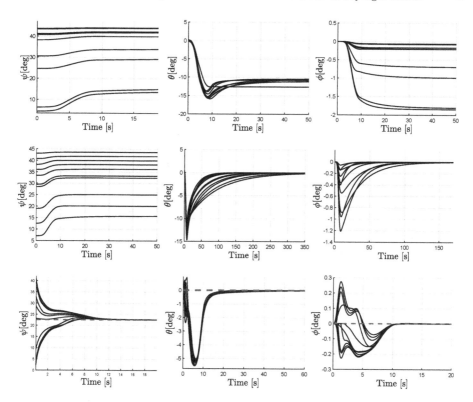

Fig. 3. Object attitude during transportation. First row: scenario S_a: $f_{int} = 0\,\mathrm{N}$ and $G \in \mathcal{I}$. Second row: scenario S_b: $f_{int} = 0\,\mathrm{N}$ and G below $\mathcal{I}$. Third row: scenario S_c: $f_{int} = 0.8\mathrm{N}$ and G placed below $\mathcal{I}$. The dotted line is the desired value. The first, second, and third columns refer to the trajectories of the yaw (ψ), pitch (θ), and roll (ϕ) angles of the object, respectively.

final orientation of the load for the scenario S_a, and the second line of Fig. 3 an arbitrary final yaw for the scenario S_b.

In the scenario S_c we set the force bias $\boldsymbol{\pi}_i$ in (2) so that the system reaches an equilibrium with the cables forces applying non zero horizontal forces. More in detail, indicating with $\bar{l}_i$ the vector that connects the projection of G on $\mathcal{I}$ to the position of point L_i at the final attitude, we set $[\pi_{i,x} \;\; \pi_{i,y}]^{\top} = f_{int}\bar{l}_i/\|\bar{l}_i\|$, where f_{int} is thus the intensity of each agent's planar force bias. In other words, the desired horizontal forces in the cables are oriented radially and outward the object in the final configuration, similarly to what is depicted in Fig. 2(b). Since the object reaches an equilibrium, and the external wrench on the object at the equilibrium is only the vertical force due to the gravity, as in (3), such non vertical components of the cable forces, which do not cause any motion of the object and do not compensate any external wrench, generate an internal force $t \neq 0$. In this way, the object reaches always the same attitude: zero pitch, zero roll, and the same yaw. The third line of Fig. 3 shows the results of such simulations. Notice that the same results have been obtained even when $G \in \mathcal{I}$.

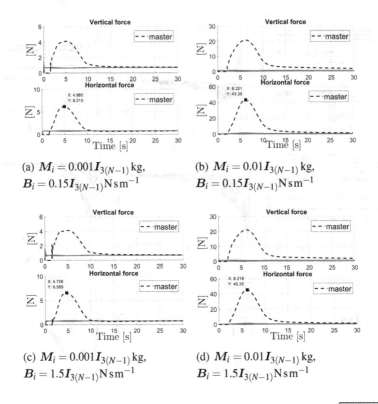

(a) $M_i = 0.001 I_{3(N-1)}$ kg,
$B_i = 0.15 I_{3(N-1)} \mathrm{N\,s\,m}^{-1}$

(b) $M_i = 0.01 I_{3(N-1)}$ kg,
$B_i = 0.15 I_{3(N-1)} \mathrm{N\,s\,m}^{-1}$

(c) $M_i = 0.001 I_{3(N-1)}$ kg,
$B_i = 1.5 I_{3(N-1)} \mathrm{N\,s\,m}^{-1}$

(d) $M_i = 0.01 I_{3(N-1)}$ kg,
$B_i = 1.5 I_{3(N-1)} \mathrm{N\,s\,m}^{-1}$

Fig. 4. Intensities of the vertical force $f_{i,z}$ and of the horizontal force $\sqrt{f_{i,x}^2 + f_{i,y}^2}$ that the agents apply to the object during the transportation for four different parameters of the followers in scenario S_c with $N = 70$. The dotted line refers to the leader agent, while the solid lines refer to the followers. In the subcaptions $i = 2, \ldots, N$.

Figure 4 shows the evolution of the forces that the agents apply to the load in four different cases belonging to scenario S_c.

We also decided to see whether the value of f_{int} influences the speed of convergence with which the swarm stabilizes the object in scenario S_c. Figure 5 shows the evolution of the load attitude when transported by a group of five agents for different values of f_{int}.

We present now the results concerning scenario S_d, where neither the exact real mass of the load, m_L, nor the exact total number of agents, N, are known by the agents. Only some upper and lower bounds are given, indicated with m_{max}, m_{min}, N_{max} and N_{min} respectively. Therefore, we introduced a particular choice of π_i.

$$
\pi_i =
\begin{bmatrix}
0 \\
0 \\
f_{i,z} + U_i^z
\end{bmatrix}
\quad \text{where} \quad
U_i^z =
\begin{cases}
-K_Z & \text{if } |f_{i,z}| > f_{max} \\
K_Z & \text{if } |f_{i,z}| < f_{min} \\
0 & \text{otherwise}
\end{cases}
\quad . \tag{4}
$$

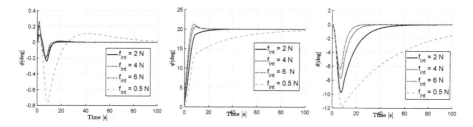

Fig. 5. Evolution of the object attitude (yaw, pitch and roll, respectively) during transportation for different values of the intensity of the internal force in scenario S_c with $N = 5$.

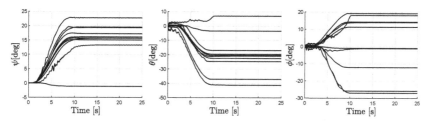

Fig. 6. Attitude of the load (yaw, pitch and roll, respectively) during the transportation task in scenario S_d with random cables attachment points and uncertain load mass and agents number.

Namely, we are defining a dead-zone in the sensed vertical force such that if the robots perceive a vertical force in the cable that is inside a certain range defined by f_{max} and f_{min}, they ignore it; otherwise, they apply an upward or downward force trying to restore a vertical force inside the predefined range. K_Z is a constant that determines the responsiveness of the robots in trying to maintain the vertical force inside $[-f_{max}, -f_{min}]$. We choose $f_{max} = \frac{m_{max}g}{N_{min}}$ and $f_{min} = \frac{m_{min}g}{N_{max}}$. In this way, we guarantee that the overall force exerted by the agents will be enough to sustain the object weight. In particular, we are not choosing a precise reference force distribution. The actual final force in each cable is induced by the choice of the bias (which does not depend on the exact values of N and m_L) and is not the same for all the agents. Such an implementation allows the robots to successfully cope with a variation of the parameters of the swarm and of the object. Simulations results for this scenario, see Fig. 6, show that the system stabilizes after the transportation. However, due to the uncertain conditions, the final pose of the object cannot be known a priori. The upper and lower bounds have been set to $m_{max} = 7\,\text{kg}$, $m_{min} = 3\,\text{kg}$, $N_{max} = 50$ and $N_{min} = 7$.

We conclude this section proposing and testing an alternative version of (2) for the leader agent. Thanks to the model (1) and (2), the leader agent does not blindly follow the desired trajectory but is aware of the outer world. In the choice of the parameters of (2) one has to face a clear trade-off between compliance

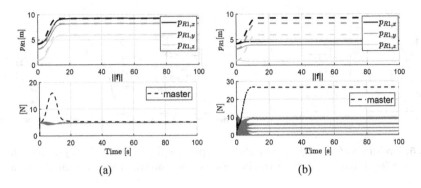

Fig. 7. The three components of $\boldsymbol{p}_{R1}$ and $\boldsymbol{p}_{R1}^d$ (dotted lines of the same color) and the intensity of the cable forces. By modeling $\boldsymbol{f}_{C1}$ as in (5) it is possible, with the same parameters, to have the leader follow its desired trajectory under normal conditions (a), and to limit the forces exerted in case the swarm does not follow it (b).

(i.e., energy consumption and safety) and tracking error (i.e., performance). A behavior that is too compliant can compromise the reach of the final position, a behavior that is too stiff can require a large amount of force from the leader side and thus a lot of energy consumption and an increased risk of ruptures. Due to space limitations we omit the performed numerical results showing such intuitive trade-off. A possible solution to better deal with such trade-off is to introduce a nonlinear saturation model like, e.g., the following one:

$$\boldsymbol{f}_{C1} = \boldsymbol{B}_1 \tan^{-1}(\dot{\boldsymbol{p}}_{R1}^d - \dot{\boldsymbol{p}}_{R1}) + \boldsymbol{K}_i \tan^{-1}(\boldsymbol{p}_{R1}^d - \boldsymbol{p}_{R1}) + \boldsymbol{\pi}_i. \tag{5}$$

The relative results are shown in Fig. 7.

4   Discussion

We first simulated symmetric cables attachment points, lying on a circle around the object center of mass (or its projection on $\mathcal{I}$). Applying a non-zero internal force that stretches the object the agents are capable of controlling univocally the final orientation of the object without the need of communication. This result actually reflects what we had discovered and formally proven for the two-agent system and a beam-like load in [21]. It is possible to assume that also biological agents, transporting an object, may be able to pull the edge of the object towards themselves, applying a force orthogonal to the edge, and thus generating a resulting internal force altogether. For example, applying a constant horizontal force bias that equilibrates the object at the initial configuration, the swarm may be capable of transporting the load and repositioning it with the same attitude without the need of explicit communication. This suggests that the local sensing allows a sort of collective memory, as if the swarm of agents *remembered* the initial orientation of the object and were capable of recreating it.

We were also interested in understanding the role played by the internal forces in the convergence rate. In fact, observing some biological structures, it seems reasonable to assume that a higher internal force in the object might lead to a faster convergence of the system to the equilibrium. For example, in [2] the authors observe the dynamic response of different spider webs. It came out that an initial tension in the net changes the natural frequency in the sense that the net is faster in recovering its equilibrium after being perturbed with respect to the case without initial tension. Similarly, we found that increasing the intensity of the horizontal bias force in f_C, and hence in turn increasing the internal force in the object, leads to faster a convergence rate, as shown in Fig. 5.

The apparent mass of the leader agent has been set to a larger value than the one of the followers. Actually, it has been observed in groups of different army ants (*Eciton burchelli, Dorylus wilverthi*) that the front ant in prey retrieval groups is larger than the others, and it is characterized by a larger dry weight. Additionally, in our simulations, during the dynamic part, the leader agent has to apply greater forces to the object than the other ones in the group. An example is shown in Fig. 4. However, this seems to be true also in super efficient prey retrieval groups of army ants [5], where a single submajor initiates the motion of the item by itself, and only after the rest of the group helps in the transportation.

The damping parameter of the followers B_i and their masses M_i have been tuned depending on the number of robots. This has been done to reduce the force that the leader agent needs to apply in order to let each follower start moving. This force increases with the increasing of the damping parameters of the followers that are related to energy dissipation of the overall system. The master force in case of a small N with a large damping is comparable to the force that the master applies with a large N and small damping, a behavior which is reported in Fig. 4(a). To equilibrate as much as possible the force exerted by the leader for different values of N, one solution is to modify the damping parameter of the followers based on N. Another way to modify the effect of the follower inertia on the leader it is to modify their mass. Compare Figs. 4(d) and (b) to see the effect of the follower inertia on the master applied force, and Figs. 4(d) and (b) to see the effect of the damping parameter. Finally, comparing Figs. 4(d) and (a) the benefits of decreasing both the followers mass and damping emerges. It is not immediately clear which is the possible biological meaning of tuning the mass of the agents though. However, it is interesting to highlight that *Eciton burchelli* ants, very efficient in transportation tasks, tend to maintain a constant ratio between the total weight of porters and the weight of the carried load [5]. f Concerning the leader spring parameter, it is not trivial to find the right balance in order to have a leader agent capable of both following the desired trajectory not being too much perturbed by small forces (for instance the ones caused by the followers initial resistance to the motion) and being sensitive to large forces on the cable indicating that something is not going as expected. However, that is exactly what an intelligent biological system would do. We were able to mimic this behavior by using a nonlinear control action on the leader agent, see (5). In this way, the leader follows the desired trajectory accurately for small deviations, but it changes its behavior consistently if the followers stay still and not follow as expected (see Fig. 7).

Finally, the results in Fig. 6, which refer to scenario S_d, show the potential of the algorithm also in conditions were the swarm is not completely aware of the parameters of the system. Even with a limited and imperfect knowledge the agents are capable of commonly carrying an object in a stable way.

5   Conclusions

This work is a simulative study on communication-less cooperative object transportation by swarms of aerial agents. The main focus concerns the important role of the internal force to make asymptotically stable certain equilibria of the system and to enhance the manipulation capabilities of the swarm. An imperfect knowledge of the system parameters has been also treated. Several parallelisms with biological examples are discussed as well. Validation of the proposed algorithm on real platforms and theoretical proofs is left as an important future work. Of course, in order to realize a practical implementation, additional aspects, such as a collision avoidance technique, are required, especially for very large number of agents. Furthermore, the relaxation of some hypothesis, e.g., on the cables attachment points or on the rigidity of the load, may represent an interesting enlargement of the application domain.

Acknowledgments. This research was partially supported by the ANR, Project ANR-17-CE33-0007 MuRoPhen.

References

1. Abbasnejad, G., Carricato, M.: Direct geometrico-static problem of underconstrained cable-driven parallel robots with n cables. IEEE Trans. Robot. **31**(2), 468–478 (2015)
2. Alam, M.S., Wahab, M.A., Jenkins, C.H.: Mechanics in naturally compliant structures. Mech. Mater. **39**(2), 145–160 (2007)
3. Berman, S., Lindsey, Q., Sakar, M.S., Kumar, V., Pratt, S.C.: Experimental study and modeling of group retrieval in ants as an approach to collective transport in swarm robotic systems. Proc. IEEE **99**(9), 1470–1481 (2011)
4. Brescianini, D., D'Andrea, R.: Design, modeling and control of an omni-directional aerial vehicle. In: 2016 IEEE International Conference on Robotics and Automation, Stockholm, Sweden, pp. 3261–3266, May 2016
5. Franks, N.R.: Teams in social insects: group retrieval of prey by army ants (eciton burchelli, hymenoptera: Formicidae). Behav. Ecol. Sociobiol. **18**(6), 425–429 (1986)
6. Franks, N.R., Sendova-Franks, A.B., Anderson, C.: Division of labour within teams of new world and old world army ants. Anim. Behav. **62**(4), 635–642 (2001)
7. Gassner, M., Cieslewski, T., Scaramuzza, D.: Dynamic collaboration without communication: vision-based cable-suspended load transport with two quadrotors. In: 2017 IEEE International Conference on Robotics and Automation, Singapore, pp. 5196–5202, May 2017

8. Grassé, P.P.: La reconstruction du nid et les coordinations interindividuelles chezbellicositermes natalensis etcubitermes sp. la théorie de la stigmergie: Essai d'interprétation du comportement des termites constructeurs. Insectes Sociaux **6**(1), 41–80 (1959)
9. Kube, C.R., Bonabeau, E.: Cooperative transport by ants and robots. Robot. Auton. Syst. **30**(1–2), 85–101 (2000)
10. Manubens Ferriol, M., Devaurs, D., Ros, G.L., Cortés, J.: A motion planning approach to 6-D manipulation with aerial towed-cable systems. In: Proceedings of the 2013 International Micro Air Vehicle Conference and Flight Competition, Toulouse, France, pp. 1–7 (2013)
11. Masone, C., Bülthoff, H.H., Stegagno, P.: Cooperative transportation of a payload using quadrotors: a reconfigurable cable-driven parallel robot. In: 2016 IEEE/RSJ International Conference on Intelligent Robots and Systems, pp. 1623–1630, October 2016
12. Michael, N., Fink, J., Kumar, V.: Cooperative manipulation and transportation with aerial robots. Auton. Robot. **30**(1), 73–86 (2011)
13. Moffett, M.W.: Cooperative food transport by an asiatic ant. Natl. Geogr. Res. **4**(3), 386–394 (1988)
14. Nguyen, H.N., Park, S., Lee, D.J.: Aerial tool operation system using quadrotors as rotating thrust generators. In: 2015 IEEE/RSJ International Conference on Intelligent Robots and Systems, Hamburg, Germany, pp. 1285–1291, October 2015
15. Park, S., Her, J., Kim, J., Lee, D.: Design, modeling and control of omni-directional aerial robot. In: 2016 IEEE/RSJ International Conference on Intelligent Robots and Systems, Daejeon, South Korea, pp. 1570–1575 (2016)
16. Rajappa, S., Ryll, M., Bülthoff, H.H., Franchi, A.: Modeling, control and design optimization for a fully-actuated hexarotor aerial vehicle with tilted propellers. In: 2015 IEEE International Conference on Robotics and Automation, Seattle, WA, pp. 4006–4013, May 2015
17. Rubenstein, M., Cabrera, A., Werfel, J., Habibi, G., McLurkin, J., Nagpal, R.: Collective transport of complex objects by simple robots: theory and experiments. In: 2013 International Conference on Autonomous Agents and Multi-agent Systems, pp. 47–54 (2013)
18. Ryll, M., et al.: 6D physical interaction with a fully actuated aerial robot. In: 2017 IEEE International Conference on Robotics and Automation, Singapore, pp. 5190–5195, May 2017
19. Sreenath, K., Kumar, V.: Dynamics, control and planning for cooperative manipulation of payloads suspended by cables from multiple quadrotor robots. In: Robotics: Science and Systems, Berlin, Germany, June 2013
20. Tagliabue, A., Kamel, M., Verling, S., Siegwart, R., Nieto, J.: Collaborative transportation using MAVs via passive force control. In: 2017 IEEE International Conference on Robotics and Automation, Singapore, pp. 5766–5773 (2016)
21. Tognon, M., Gabellieri, C., Pallottino, L., Franchi, A.: Aerial co-manipulation with cables: the role of internal force for equilibria, stability, and passivity. IEEE Robot. Autom. Lett. Spec. Issue Aer. Manip. **3**(3), 2577–2583 (2018). https://doi.org/10.1109/LRA.2018.2803811
22. Wang, Z., Schwager, M.: Force-amplifying n-robot transport system (force-ANTS) for cooperative planar manipulation without communication. Int. J. Robot. Res. **35**(13), 1564–1586 (2016)
23. Wu, G., Sreenath, K.: Geometric control of multiple quadrotors transporting a rigid-body load. In: 53rd IEEE Conference on Decision and Control, Los Angeles, CA, pp. 6141–6148, December 2014

Automatic Design
of Communication-Based Behaviors
for Robot Swarms

Ken Hasselmann[1], Frédéric Robert[2], and Mauro Birattari[1]([✉])

[1] IRIDIA, Université Libre de Bruxelles, Brussels, Belgium
{khasselm,mbiro}@ulb.ac.be
[2] BEAMS, Université Libre de Bruxelles, Brussels, Belgium
frrobert@ulb.ac.be

Abstract. We introduce Gianduja, an automatic design method that generates communication-based behaviors for robot swarms. Gianduja extends Chocolate, a previously published design method. It does so by providing the robots with the capability to communicate using one message. The semantics of the message is not a priori fixed. It is the automatic design process that implicitly defines it, on a per-mission basis, by prescribing the conditions under which the message is sent by a robot and how the receiving peers react to it. We empirically study Gianduja on three missions and we compare it with the aforementioned Chocolate and with EvoCom, a rather standard evolutionary robotics method that generates communication-based behaviors. We evaluate the behaviors produced by the three automatic design methods on a swarm of 20 e-puck robots. The results show that Gianduja uses communication meaningfully and effectively in all the three missions considered. The aggregate results indicate that, on the three missions considered, Gianduja performs significantly better than the two other methods under analysis.

1   Introduction

In swarm robotics, communication plays a central role and can significantly enhance collective performance [3]. Designing effective communication mechanisms is challenging and design choices can have an important impact on the effectiveness, complexity, and cost of a swarm [2]. Notwithstanding the advancements achieved in the last decade [4,7,24,29,34,43,51], the design of robot swarms is still at dawn and no generally applicable methodology has been proposed so far [8,11,21]. Automatic design methods are a promising way of approaching the issue [6,15]. In automatic methods, the design problem is cast into an optimization problem: a space of solutions is searched via an optimization algorithm, with the goal of maximizing a performance measure. Most of the

---

The proposed method was implemented and tested by KH. The experiments were designed by the three authors. This paper was drafted by KH, refined by MB, and revised by the three authors. The research was conceived and directed by MB.

M. Dorigo et al. (Eds.): ANTS 2018, LNCS 11172, pp. 16–29, 2018.
https://doi.org/10.1007/978-3-030-00533-7_2

research on the automatic design of robot swarms has been inspired by neuro-evolution [37,47]. In this approach, robots are controlled by a neural network, whose parameters are obtained via artificial evolution [12,27,31,38,44,45,47,50]. Other methods have been proposed that are based on different control archi-tectures and/or different optimization algorithms [16,18,22,30]. Among them, Chocolate [16] produces probabilistic finite state machines by using the irace optimization algorithm [35] to assemble preexisting low-level behaviors and con-ditions, and to fine-tune their parameters. The low-level behaviors, which define the actions that individual robots can perform, are: exploration, stop, phototaxis, anti-phototaxis, attraction to neighbors, repulsion from neighbors. The condi-tions, which define events that cause a transition between low-level behaviors, are: black-floor, white-floor, gray-floor, neighbor-count, inverted-neighbor-count, fixed-probability.

In this paper, we study the automatic design of collective behaviors that rely on communication. In particular, we are interested in exploring the case in which messages exchanged by the robots do not have an a priori defined semantics. We wish to develop an automatic design process that, on a per-mission basis, defines (i) the conditions under which a robot broadcasts a message and (ii) the effects that this message has on the behavior of the receiving peers.

We introduce Gianduja, a new instance of AutoMoDe [18]. Gianduja extends Chocolate by adding the capability of locally broadcasting a single message and reacting to it. We test Gianduja on three missions that we shall call AGGREGA-TION, STOP, and DECISION. We present results of experiments performed with a swarm of 20 e-puck robots [36].

Within the evolutionary robotics approach, it has already been shown that an automatic design process can (implicitly) give a semantics to an a priori mean-ingless message. Nonetheless, this has been demonstrated only on teams of two robots [1,49]. The novel elements that we propose in this paper are that: (1) we study the emergence of a message semantics in swarm robotics and we demon-strate it with a swarm of 20 robots; (2) we show that a message semantics can emerge also when robots are controlled by a finite state machine; and (3) we consider three different missions in which the emerging semantics is different.

2   Related Work

Communication—be it direct or indirect, explicit or implicit—is an integral part of most robot swarms demonstrated so far. As a result, the literature on com-munication in swarm robotics is extremely large and covering it goes beyond the scope of this paper. In particular, we will not cover studies in which communi-cation has been a priori defined by the designer—e.g., [2,3,9,14,28]. Instead, we will focus on studies in which communication has been automatically designed.

The vast majority of studies in which communication emerged from an auto-matic design process belong within evolutionary robotics [37,47,48]. Quinn et al. [41,42] were the first to study the emergence of communication between agents. In their studies, robots move in an arbitrary direction while staying close to each other. Robots do not have dedicated communication devices. Nonetheless, they evolved a simple form of implicit communication: using their proximity sensors, robots detect motion in their peers and establish a social interaction. In particular, they coordinate to assume the roles of leader or follower. Nolfi [39] evolved a behavior for solving a collective navigation problem. Robots are controlled by neural networks and can communicate using four different signals. Although the evolutionary process did not explicitly reward the use of communication, it produced a behavior in which the robots effectively use communication to coordinate. The behavior obtained was tested in simulation on a swarm of four robots. Floreano et al. [13] studied the evolution of robots that can produce visual signals to provide information on food location. The authors evolved behaviors for a swarm of ten robots that were eventually able to reliably find the food source. Communication increases the performance of the swarm compared to the case in which robot cannot communicate. The behavior was then tested with real robots. Ampatzis et al. [1] evolved the behavior of two robots to recognize features of the environment and react accordingly. The robots are controlled by neural networks and can use their on-board speakers and microphones to send/receive a sound message. Although communication is not strictly needed to solve the task and was not explicitly rewarded in the evolutionary process, it emerged as it improves performance. The behavior obtained was tested both in simulation and in reality with two s-bot robots. Tuci [49] studied the origin of communication from an evolutionary perspective. The author considered a setting in which two robots, which might communicate via a sound message, need to categorize the environment and act accordingly. Also in this case, although communication was not explicitly rewarded, the evolutionary process produced behaviors that effectively use the available communication capabilities to perform the mission. Experiments were conducted in simulation only.

Among all the studies highlighted above, the research we present in this paper is most closely related to [1,49]. Indeed, as in those studies, we consider the case in which the semantics of the message exchanged by the robots is not a priori defined but is the result of the automatic design process.

3   AutoMoDe-Gianduja

By introducing Gianduja, we address one of the limitations of AutoMoDe: the instances of AutoMoDe defined so far, Vanilla and Chocolate, are unable to design behaviors that exploit explicit communication. The behaviors automatically generated by Gianduja can rely on sending and receiving a single message whose semantics is not fixed a priori. Gianduja is a proper extension of Chocolate [16] that adds the ability to (i) locally broadcast a message, (ii) change state when the message is received (or is not received), and (iii) approach (or retract from) neighboring peers that broadcast the message.

As Chocolate and Vanilla, Gianduja designs control software for the e-puck platform. Nonetheless, it considers a reference model that is an extension of the one considered by Chocolate and Vanilla—RM 1.1 [25]. Precisely, the

Table 1. Reference model RM 2: novelties with respect to RM 1.1 are highlighted.

Input	Value	Description
$prox_{i \in \{1,\ldots,8\}}$	[0,1]	reading of proximity sensor i
$light_{i \in \{1,\ldots,8\}}$	[0,1]	reading of light sensor i
$gnd_{j \in \{1,2,3\}}$	{black, gray, white}	reading of ground sensor j
n	[0,20]	number of neighboring robots perceived
V	$([0, 0.70]\,\mathrm{m}, [0, 2\pi]\,\mathrm{rad})$	direction of attraction to them
b	[0,20]	number of messaging neighbors perceived
V_b	$([0, 0.70]\,\mathrm{m}, [0, 2\pi]\,\mathrm{rad})$	direction of attraction to them

Output	Value	Description
$v_{k \in \{l,r\}}$	$[-0.12, 0.12]\,\mathrm{m\,s^{-1}}$	target linear wheel velocity
s	{on, off}	broadcast message

Period of the control cycle: 100 ms

extension concerns the ability to (a) locally broadcast the message and (b) sense the broadcasting peers that are within the perception range. The new reference model, which we shall call RM 2, is given in Table 1. The variables highlighted are the elements of novelty with respect to RM 1.1: b, V_b, and s. Before we explain these variables, it is convenient that we first recall the mechanism that allows robots to perceive their neighbors—both in RM 1.1 and in RM 2. Using their range-and-bearing module [23], all robots continuously broadcast a "heartbeat" signal whose payload encodes their unique ID. At every time step, every robot receives the heartbeat signal of the peers that are within its perception range, which is of about 0.70 m. It can therefore infer the number of neighboring peers and their relative positions: range and bearing. This information is made available to the control software via the variables r and V. The former is the number of neighboring peers and the latter is a vector indicating the direction of attraction to these neighboring peers, which is computed based on the framework on virtual potential fields [46].

In RM 2, every robot locally broadcasts the message by setting a specific bit of its heartbeat's payload. Due to this extension, at every time step, a robot can infer the number and relative position of the neighboring peers that are broadcasting the message. The information that is made available to the control software is stored in the variables b and V_b. The former is the number of neighboring peers that broadcast the message and the latter is a vector indicating the direction of attraction to these neighboring peers, which also in this case is computed following the framework on virtual potential fields [46]. Formally,

$$V_b = \begin{cases} \sum_{m=1}^{b}(\alpha/r_m^2, \angle b_m), & \text{if } b > 0 \text{ broadcasting robots are perceived;} \\ (1, \angle 0), & \text{otherwise.} \end{cases}$$

Here, r_m and $\angle b_m$ are the range and bearing of the m-th neighboring peer that is broadcasting the message and α a real value parameter. The variable s can be set by the control software and indicates whether, during the following control cycle, the robot should broadcast the message or not. It can take two values: *on* or *off*.

Gianduja produces control software in the form of probabilistic finite state machines, as Chocolate does. It does so by combining and fine-tuning (a) the original transition conditions of Chocolate (and Vanilla) [16,18]; (b) an extended version of the low-level behaviors of Chocolate (and Vanilla) [16,18]; (c) four additional modules: two low-level behaviors and two transition conditions. We extend the preexisting low-level behaviors of Chocolate (and Vanilla) by adding a binary parameter: if the parameter is set, the robot continuously broadcasts the message while performing the low-level behavior; otherwise, it does not. We conceived the four additional modules specifically for exploiting the extended functionalities provided by RM 2. The two additional low-level behaviors are: **attraction to message** – the robot moves in the direction indicated by V_b; **repulsion from message** – the robot moves in the opposite direction. Also these additional behaviors have the aforementioned binary parameter that specifies whether the message should be broadcast or not. The two additional conditions are: **message count** – a state transition occurs if the number of neighboring peers broadcasting the message is larger than the value of a parameter; **inverted message count** – a state transition occurs if the number of neighboring peers broadcasting the message is smaller than the value of a parameter. The additional modules are modeled after the original attraction, repulsion, neighbor-count, and inverted-neighbor-count of Chocolate (and Vanilla) [16,18]. The optimization algorithm used to search the space of the possible probabilistic finite state machines that can be obtained by assembling the available modules and fine-tuning their parameters is irace [35]—the same algorithm used in Chocolate. As in Chocolate (and Vanilla), valid probabilistic finite state machines have at most four states and each state has at most four outgoing transitions. Finally, as in Chocolate (and Vanilla), the design process is performed in simulation using ARGoS [20,40].

4   Experimental Setting

We test Gianduja on three missions and we compare it with two other methods.

4.1   Missions

In all three missions, the robots operate in a dodecagonal area of $4.91\,\mathrm{m}^2$. The arena is surrounded by walls. Its floor is gray, apart from some specific areas that, on a per-mission basis, could be white or black, as detailed in the following. The time available to the robots for performing a mission is $T = 120\,\mathrm{s}$. The three missions considered are AGGREGATION, STOP, and DECISION; they are described in the following. We have selected them because, according to our a priori expectations, communication should play a different role in them. Indeed, we expect that AGGREGATION can be solved without using communication. On the other hand, we expect that STOP and DECISION require communication for being solved effectively. We also expect that the semantics implicitly attached to the message by the automatic design process will be different in STOP and DECISION. We will detail this in the following, on a per-mission basis.

AGGREGATION. The arena's floor is marked by two circular spots, with diameter of 0.6 m: one is white and the other black. They are positioned on the left-hand side of the arena, separated by a gap of 0.25 m. At the beginning of each run, the robots are randomly positioned in the right-hand half of the arena, so that no robot is already on the spots—see Fig. 1(*right*). The mission prescribes that the robots quickly aggregate on the white spot. The black spot is not supposed to play any role and simply acts as a disturbance to the automatic design process. The performance of the robots is measured via the following objective function—the higher, the better:

$$C_{\mathrm{A}} = 24000 - \sum_{t=1}^{T} \sum_{i=1}^{N} I_i(t); \qquad I_i(t) = \begin{cases} 0, & \text{if robot } i \text{ is on the white spot;} \\ 1, & \text{otherwise.} \end{cases}$$

Here, i is an index that spans over all the robots of the swarm, N is the total number of robots, and $T = 120$ s is duration of the experiment. 24 000 is the maximum theoretical score that the robots could achieve. It is included in the definition of the objective function to guarantee that its value is non-negative and ranges from 0 to its theoretical maximum.

As already mentioned, we think communication is not needed in this mission.

STOP. The arena's floor is marked by a circular white spot, with diameter of 0.2 m, positioned near the walls, on the top-left quadrant. At the beginning of each experimental run, all robots are randomly positioned in the right-hand half of the arena: none of them is on the white spot—see Fig. 1(*center*). The mission prescribes that the robots search for the spot and, as one of them finds it, all stop quickly. The performance measure—the higher, the better—is:

$$C_{\mathrm{S}} = 48000 - \left(\bar{t}N + \sum_{t=1}^{\bar{t}} \sum_{i=1}^{N} \bar{I}_i(t) + \sum_{t=\bar{t}+1}^{T} \sum_{i=1}^{N} I_i(t) \right);$$

$$I_i(t) = \begin{cases} 1, & \text{if robot } i \text{ is moving;} \\ 0, & \text{otherwise;} \end{cases} \qquad \bar{I}_i(t) = 1 - I_i(t).$$

Here, i, N, and T are defined as above; $\bar{t}$ is the time at which a robot steps on the white spot for the first time. The performance measure ranges from 0 to its maximum of 48 000. In the definition of I_i (and $\bar{I}_i$), a robot is considered to be moving if its center has traveled more than 5 mm in the last time step.

We expect that communication is needed in this mission and that Gianduja produces behaviors in which (i) robots broadcast the message if they step on the white spot; (ii) upon receiving the message, robots stop and possibly relay it.

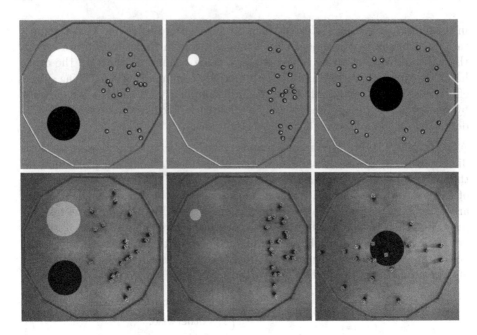

Fig. 1. Arenas for the three missions: AGGREGATION, STOP, and DECISION (*from left to right*); simulation (*top*) and real setup (*bottom*).

DECISION. The arena's floor is marked by a circular spot, with diameter of 0.6m, located in the center of the arena. The spot can be either white or black, with a probability of 0.5. A light source is placed outside the arena, on the right-hand side. At the beginning of each run, robots are randomly positioned—see Fig. 1(*right*). The mission prescribes that the robots quickly relocate into the right-hand half of the arena, when the spot is black; and into the left-hand half, when the spot is white. The performance measure—the higher, the better—is:

$$C_{\mathrm{D}} = 24000 - \sum_{t=1}^{T} \sum_{i=1}^{N} I_i(t);$$

$$I_i(t) = \begin{cases} 0, & \text{if robot } i \text{ is in the correct half of the arena;} \\ 1, & \text{otherwise.} \end{cases}$$

Here, i, N, and T are defined as above. The performance measure ranges between 0 and its theoretical maximum of 24 000.

Also in this case, we expect that communication is needed. A straightforward solution would require two distinct messages: one per spot color/half of the arena in which robots should relocate. As the robots have only one message available, the solution we foresee is that they go in one direction by default and revert to the opposite one in case they receive a message sent by a robot that steps on the spot, should its color indicate that the correct direction is not the default one.

4.2   Protocol

We compare Gianduja with Chocolate [16,32] and EvoCom. Chocolate was originally defined in [16] and is used here unmodified. EvoCom is an evolutionary method that we introduce for this study. It is an extension of EvoStick—a design method that, via an evolutionary process, tunes the parameters of a neural networks to control the e-puck platform, as it is modeled by RM 1.1. EvoStick was formally defined in [18] to serve as a yardstick in the study of Vanilla, but had been previously analyzed in [19]. It was subsequently included in other empirical studies [5,16,17,33]. EvoCom targets the e-puck platform, as it is modeled by RM 2—see Table 1. With respect to EvoStick, it has the further capability of locally broadcasting a message and reacting to it. It features (i) one extra output node for s; and (ii) five extra input nodes: one for b and four for the projections of V_b on the four unit vectors pointing at 45°, 135°, 225°, and 315° with respect to the head of the robot. The neural network is optimized using a standard evolutionary algorithm, the same adopted in EvoStick—see [18,32] for the details. Artificial evolution is based on simulations performed with ARGoS [20,40]—under the same conditions that hold for Gianduja and Chocolate.

We consider a swarm of 20 e-puck robots. For each of the three missions, each of the three methods under analysis is executed 15 times to obtain 15 instances of control software. Each design process can rely on a maximum of 200 000 simulated runs. The simulator adopted in the study is ARGoS3, beta 48. We evaluate each instance of control software obtained by the three design methods: once in simulation and once on the physical robots. The initial positions of the robots and the order of the experimental runs are randomized to avoid any bias. In robot experiments, the value of the objective function is computed automatically using a tracking system that extracts information from images taken with an overhead camera every 100 ms.

Statistics. We report per-mission boxplots of the performance registered in simulation and reality. When appropriate, we report also the outcome of a Wilcoxon rank-sum test, at 95% confidence [10]. Eventually, we aggregate all the results of the robot experiments by ranking across each mission the performance obtained by the instances of control software generated by each method. We present the outcome of a Friedman test [10] in a plot that displays the average rank of each method and its 95% interval of confidence. If two intervals do not overlap, the results we registered for the corresponding methods are significantly different. In the following, statements like "A performs significantly better that B" imply that an appropriate statistical test—either a Wilcoxon or a Friedman test—has been employed and has detected significance with confidence of at least 95%.

5   Results

We present the results on a per-mission basis and then we aggregate them across the three missions. Numerical results, videos, code, and finite state machines generated by Gianduja and Chocolate are available in [26].

AGGREGATION. Results are reported in Fig. 2(*left*). Both `Gianduja` and `Chocolate` perform significantly better than `EvoCom`. Although `Gianduja` performs significantly better than `Chocolate` in simulation, the results of the two methods on the robots are similar.

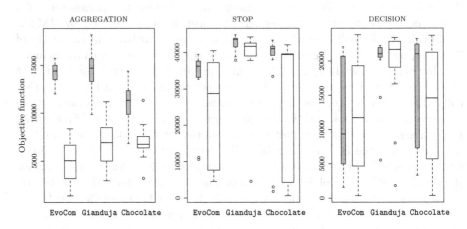

Fig. 2. AGGREGATION, STOP, and DECISION (*from left to right*). Thick white boxes represent the results of robot experiments; thin gray ones, those of simulations.

At visual inspection, `EvoCom` seems to be unable to use communication effectively. The robots randomly explore the arena—sometimes forming moving clusters. If they enter the white spot, they spin in place. Also in `Chocolate` and `Gianduja`, robots navigate randomly, but they stop upon reaching the white spot. In `Gianduja`, when on the white spot, robots typically broadcast the message; the receiving peers converge towards them eventually reaching the white spot. Although elegant, `Gianduja`'s solution does not significantly improve over `Chocolate`'s one. It can be observed that `Gianduja` suffers the reality gap more than `Chocolate`. This could be due to the fact that the ground sensor of the e-puck robot is quite prone to report false positives in the detection of white/black floor. As it can be seen in simulation, the behaviors produced by `Gianduja` rely on communication to attract peers once the white spot is detected. In the presence of false positives, this feature could hinder performance. `Chocolate`, which does not rely on communication, is apparently less affected by false positives.

STOP. Results are reported in Fig. 2(*center*). `Gianduja` performs significantly better than `EvoCom` and `Chocolate`, both in simulation and reality.

At visual inspection, `EvoCom` seems to be unable to use its communication capabilities effectively, whereas `Gianduja` does. In `Gianduja`, robots move randomly until one reaches the white spot and stop. This robot broadcasts the message. The receiving peers relay it and stop. In `EvoCom` and `Chocolate` (which is not endowed with communication capabilities), robot move in random directions until being stopped by the walls. Although trivial, this behavior often scores

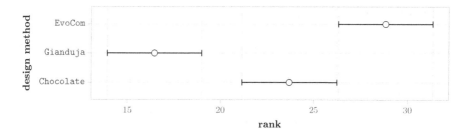

Fig. 3. Friedman test on the aggregated results of the three missions. The plot represents the average rank of the three methods and their 95% confidence interval.

better than one would expect because there is some relatively high chance that, before robots stop against a wall, at least one of them has reached the white spot. In this mission, the behaviors generated by Gianduja cross the reality gap better than those of EvoCom and Chocolate.

DECISION. Results are reported in Fig. 2(*right*). Gianduja performs significantly better than EvoCom both in simulation and reality. Concerning the comparison between Gianduja and Chocolate, although the difference is not significant in simulation, on the robots Gianduja performs significantly better.

Gianduja uses communication effectively. By default, robots go towards one side of the arena. If one robot steps on the central spot and its color indicates that the correct side of the arena is not the default one, the robot itself broadcasts the message. Receiving peers relay the message and all robots head to the correct direction. In some instances of control software designed by Gianduja, the selected default side is the right-hand one, and in others is the left-hand. Accordingly, robots start by performing phototaxis or anti-phototaxis and then possibly switch depending on the color of the central spot. In Chocolate, the behavior is similar but, as the robots are not endowed with communication capabilities, only the robots that individually step on the central spot are able to revert their default choice, should it be needed. EvoCom failed to produce any consistently meaningful behavior. The score has a very large variability and appears to be determined by chance. In simulation, the performance observed is even worse than random behavior, which should produce an expected score of 12 000—half of the maximum. In robot experiments, the score observed matches the profile of a random behavior. Gianduja's behaviors cross the reality gap nicely while those of Chocolate appear to experience a large performance drop. As the performance of EvoCom is particularly poor, any consideration on how the method handles the reality gap would be meaningless.

Aggregate Results. The aggregate results are presented in Fig. 3. The plot confirms that, across the three missions considered, Gianduja performs significantly better that both Chocolate and EvoCom.

6 Conclusions

We have studied the problem of the automatic design of collective behaviors that rely on communication. We have focused on the case in which robots are able to locally broadcast a single message whose semantics is not fixed a priori: the automatic design method can re-define it on a per-mission basis, as needed.

We have introduced Gianduja, an automatic design method based on the previously published Chocolate. Gianduja generates control software by assembling preexisting software modules into a probabilistic finite state machine. We tested Gianduja on three missions, showing that the way in which the message is used by the robots is different—and meaningful—in each of them. As desired, the (implicit) semantics of the message is automatically defined on a per-mission basis by the design process. On all three missions, Gianduja performs significantly better that EvoCom, a rather standard evolutionary robotics methods for robots that are able to broadcast and receive a message. On two of the three missions, Gianduja performs also significantly better than Chocolate, which is not endowed with communication capabilities. The only mission on which the performance of Gianduja and Chocolate is comparable is one in which we a priori expected that communication is not needed. When aggregated, the results of the robot experiments indicate that, across the three missions considered, Gianduja performs significantly better that both EvoCom and Chocolate.

On the missions considered, Gianduja has also shown a weakness: It appears to be more sensitive than Chocolate to noisy readings from the ground sensor. We observed this issue in AGGREGATION but it might have had an impact also in the other two missions. The reason why this issue has a relative lower impact in the other missions is possibly that communication is strictly needed to accomplish them. This clearly gives a major advantage to Gianduja over Chocolate and greatly compensates the increased sensitivity to sensor noise.

Future work will focus on testing Gianduja on further missions. We will also study the possibility of extending Gianduja so that it can handle multiple messages and therefore generate more complex collective behaviors. Finally, we will address also the sensitivity of Gianduja to sensor noise. A possible way to handle the issue is to improve the noise models used in simulation so as to produce behaviors that are more robust to false positives in the detection of white/black ground. We are considering also to adopt ideas from game theory to prevent that malicious (or simply fallacious, erroneous, unintended) messages propagate across the swarm and negatively impact its collective behavior.

Acknowledgements. The project has received funding from the European Research Council (ERC) under the European Union's Horizon 2020 research and innovation programme (grant agreement No 681872). Mauro Birattari acknowledges support from the Belgian *Fonds de la Recherche Scientifique* – FNRS.

References

1. Ampatzis, C., Tuci, E., Trianni, V., Dorigo, M.: Evolution of signaling in a multi-robot system: categorization and communication. Adapt. Behav. **16**(1), 5–26 (2008)
2. Balch, T.: Communication, diversity and learning: cornerstones of swarm behavior. In: Şahin, E., Spears, W.M. (eds.) SR 2004. LNCS, vol. 3342, pp. 21–30. Springer, Heidelberg (2005). https://doi.org/10.1007/978-3-540-30552-1_3
3. Balch, T., Arkin, R.C.: Communication in reactive multiagent robotic systems. Auton. Robot. **1**(1), 27–52 (1994)
4. Berman, S., Kumar, V., Nagpal, R.: Design of control policies for spatially inhomogeneous robot swarms with application to commercial pollination. In: Zexiang, L. (ed.) IEEE International Conference Robotics and Automation, ICRA, pp. 378–385. IEEE Press, Piscataway (2011)
5. Birattari, M., Delhaisse, B., Francesca, G., Kerdoncuff, Y.: Observing the effects of overdesign in the automatic design of control software for robot swarms. In: Dorigo, M. (ed.) ANTS 2016. LNCS, vol. 9882, pp. 149–160. Springer, Cham (2016). https://doi.org/10.1007/978-3-319-44427-7_13
6. Bozhinoski, D., Birattari, M.: Designing control software for robot swarms: software engineering for the development of automatic design methods. In: ACM/IEEE 1st International Workshop on Robotics Software Engineering. RoSE, pp. 33–35. ACM, New York (2018)
7. Brambilla, M., Brutschy, A., Dorigo, M., Birattari, M.: Property-driven design for swarm robotics: a design method based on prescriptive modeling and model checking. ACM Trans. Auton. Adapt. Syst. **9**(4), 17.1–17.28 (2015)
8. Brambilla, M., Ferrante, E., Birattari, M., Dorigo, M.: Swarm robotics: a review from the swarm engineering perspective. Swarm Intell. **7**(1), 1–41 (2013)
9. Cao, Y., Fukunaga, A., Kahng, A., Meng, F.: Cooperative mobile robotics: antecedents and directions. In: IEEE/RSJ International Conference on Intelligent Robots and Systems. Human Robot Interaction and Cooperative Robots, vol. 1, pp. 226–234. IEEE Press, Piscataway (1997)
10. Conover, W.J.: Practical Nonparametric Statistics, 3rd edn. Wiley, New York (1999)
11. Dorigo, M., Birattari, M., Brambilla, M.: Swarm robotics. Scholarpedia **9**(1), 1463 (2014)
12. Floreano, D., Husbands, P., Nolfi, S.: Evolutionary robotics. In: Handbook of Robotics, pp. 1423–1451 (2008)
13. Floreano, D., Mitri, S., Magnenat, S., Keller, L.: Evolutionary conditions for the emergence of communication in robots. Curr. Biol. **17**(6), 514–519 (2007)
14. Fong, T., Nourbakhsh, I.: Socially interactive robots. Robot. Auton. Syst. **42**(3–4), 139–141 (2009)
15. Francesca, G., Birattari, M.: Automatic design of robot swarms: achievements and challenges. Front. Robot. AI **3**(29), 1–9 (2016)
16. Francesca, G., et al.: AutoMoDe-chocolate: automatic design of control software for robot swarms. Swarm Intell. **9**(2/3), 125–152 (2015)
17. Francesca, G., et al.: An experiment in automatic design of robot swarms. In: Dorigo, M. (ed.) ANTS 2014. LNCS, vol. 8667, pp. 25–37. Springer, Cham (2014). https://doi.org/10.1007/978-3-319-09952-1_3
18. Francesca, G., Brambilla, M., Brutschy, A., Trianni, V., Birattari, M.: AutoMoDe: a novel approach to the automatic design of control software for robot swarms. Swarm Intell. **8**(2), 89–112 (2014)

19. Francesca, G., Brambilla, M., Trianni, V., Dorigo, M., Birattari, M.: Analysing an evolved robotic behaviour using a biological model of collegial decision making. In: Ziemke, T., Balkenius, C., Hallam, J. (eds.) SAB 2012. LNCS (LNAI), vol. 7426, pp. 381–390. Springer, Heidelberg (2012). https://doi.org/10.1007/978-3-642-33093-3_38

20. Garattoni, L., Francesca, G., Brutschy, A., Pinciroli, C., Birattari, M.: Software infrastructure for e-puck (and TAM). Technical report TR/IRIDIA/2015-004, IRIDIA, Université libre de Bruxelles, Belgium (2015)

21. Garattoni, L., Birattari, M.: Swarm robotics. In: Webster, J. (ed.) Wiley Encyclopedia of Electrical and Electronics Engineering. Wiley, Hoboken (2016)

22. Gauci, M., Chen, J., Li, W., Dodd, T.J., Groß, R.: Self-organized aggregation without computation. Int. J. Robot. Res. **33**(8), 1145–1161 (2014)

23. Gutiérrez, Á., Campo, A., Dorigo, M., Donate, J., Monasterio-Huelin, F., Magdalena, L.: Open e-puck range & bearing miniaturized board for local communication in swarm robotics. In: Kosuge, K. (ed.) IEEE International Conference on Robotics and Automation, ICRA, pp. 3111–3116. IEEE Press, Piscataway (2009)

24. Hamann, H., Wörn, H.: A framework of space-time continuous models for algorithm design in swarm robotics. Swarm Intell. **2**(2), 209–239 (2008)

25. Hasselmann, K., Ligot, A., Francesca, G., Birattari, M.: Reference models for AutoMoDe. Technical report TR/IRIDIA/2018-002, IRIDIA, Université libre de Bruxelles, Belgium (2018)

26. Hasselmann, K., Robert, F., Birattari, M.: Automatic design of communication-based behaviors for robot swarms: supplementary material. http://iridia.ulb.ac.be/supp/IridiaSupp2018-003/ (2018)

27. Jakobi, N., Husbands, P., Harvey, I.: Noise and the reality gap: the use of simulation in evolutionary robotics. In: Morán, F., Moreno, A., Merelo, J.J., Chacón, P. (eds.) ECAL 1995. LNCS, vol. 929, pp. 704–720. Springer, Heidelberg (1995). https://doi.org/10.1007/3-540-59496-5_337

28. Jones, C., Mataric, M.J.: Automatic synthesis of communication-based coordinated multi-robot systems. In: International Conference on Intelligent Robots and Systems, IROS, vol. 1, pp. 381–387. IEEE Press, Piscataway (2004)

29. Kazadi, S., Lee, J.R., Lee, J.: Model independence in swarm robotics. Int. J. Intell. Comput. Cybern. **2**(4), 672–694 (2009)

30. König, L., Mostaghim, S.: Decentralized evolution of robotic behavior using finite state machines. Int. J. Intell. Comput. Cybern. **2**(4), 695–723 (2009)

31. Koos, S., Mouret, J.B., Doncieux, S.: The transferability approach: crossing the reality gap in evolutionary robotics. IEEE Trans. Evol. Comput. **17**(1), 122–145 (2013)

32. Ligot, A., Hasselmann, K., Delhaisse, B., Garattoni, L., Francesca, G., Birattari, M.: AutoMoDe, NEAT, and EvoStick: implementations for the e-puck robot in ARGoS3. Technical report TR/IRIDIA/2017-002, IRIDIA, Université libre de Bruxelles, Belgium (2017)

33. Ligot, A., Birattari, M.: On mimicking the effects of the reality gap with simulation only experiments. In: Dorigo, M. (ed.) ANTS 2018. LNCS, vol. 11172, pp. 109–122. Springer, Berlin (2018)

34. Lopes, Y.K., Trenkwalder, S.M., Leal, A.B., Dodd, T.J., Groß, R.: Supervisory control theory applied to swarm robotics. Swarm Intell. **10**(1), 65–97 (2016)

35. López-Ibáñez, M., Dubois-Lacoste, J., Pérez Cáceres, L., Birattari, M., Stützle, T.: The irace package: iterated racing for automatic algorithm configuration. Oper. Res. Perspect. **3**, 43–58 (2016)

36. Mondada, F., et al.: The e-puck, a robot designed for education in engineering. In: Gonçalves, P., Torres, P., Alves, C. (eds.) Proceedings of the 9th Conference on Autonomous Robot Systems and Competitions, pp. 59–65. Instituto Politécnico de Castelo Branco, Castelo Branco (2009)
37. Nolfi, S., Floreano, D.: Evolutionary Robotics. MIT Press, Cambridge (2000)
38. Nolfi, S., Floreano, D., Miglino, G., Mondada, F.: How to evolve autonomous robots: different approaches in evolutionary robotics. In: Brooks, R.A., Maes, P. (eds.) Artificial Life IV: Proceedings of the Workshop on the Synthesis and Simulation of Living Systems. pp. 190–197. MIT Press, Cambridge (1994)
39. Nolfi, S.: Emergence of communication in embodied agents: co-adapting communicative and non-communicative behaviours. Connect. Sci. **17**(3–4), 231–248 (2005)
40. Pinciroli, C., et al.: ARGoS: a modular, parallel, multi-engine simulator for multi-robot systems. Swarm Intell. **6**(4), 271–295 (2012)
41. Quinn, M.: Evolving communication without dedicated communication channels. In: Kelemen, J., Sosík, P. (eds.) ECAL 2001. LNCS (LNAI), vol. 2159, pp. 357–366. Springer, Heidelberg (2001). https://doi.org/10.1007/3-540-44811-X_38
42. Quinn, M., Smith, L., Mayley, G., Husbands, P.: Evolving controllers for a homogeneous system of physical robots: structured cooperation with minimal sensors. Philos. Trans. R. Soc. Lond. A: Math. Phys. Eng. Sci. **361**(1811), 2321–2343 (2003)
43. Reina, A., Valentini, G., Fernàndez-Oto, C., Dorigo, M., Trianni, V.: A design pattern for decentralised decision making. PLoS One **10**(10), e0140950 (2015)
44. Silva, F., Duarte, M., Correia, L., Oliveira, S., Christensen, A.: Open issues in evolutionary robotics. Evol. Comput. **24**(2), 205–236 (2016)
45. Silva, F., Urbano, P., Correia, L., Christensen, A.L.: odNEAT: an algorithm for decentralised online evolution of robotic controllers. Evol. Comput. **23**(3), 421–449 (2015)
46. Spears, W.M., Spears, D., Hamann, J.C., Heil, R.: Distributed, physics-based control of swarms of vehicles. Auton. Robot. **17**, 137–162 (2004)
47. Trianni, V.: Evolutionary Swarm Robotics. Springer, Berlin (2008)
48. Trianni, V.: Evolutionary robotics: model or design? Front. Robot. AI **1**(13), 1–6 (2014)
49. Tuci, E.: An investigation of the evolutionary origin of reciprocal communication using simulated autonomous agents. Biol. Cybern. **101**(3), 183–199 (2009)
50. Urzelai, J., Floreano, D.: Evolutionary robotics: coping with environmental change. In: Whitney, L.D., et al. (eds.) Proceedings of Conference on the Genetic and Evolutionary Computation Conference, GECCO, pp. 941–948. Morgan Kaufmann, San Francisco (2000)
51. Werfel, J., Petersen, K., Nagpal, R.: Designing collective behavior in a termite-inspired robot construction team. Science **343**(6172), 754–758 (2014)

Behavior Trees as a Control Architecture in the Automatic Modular Design of Robot Swarms

Jonas Kuckling⬤, Antoine Ligot⬤, Darko Bozhinoski⬤,
and Mauro Birattari^(✉)⬤

IRIDIA, Université Libre de Bruxelles, Brussels, Belgium
mbiro@ulb.ac.be

Abstract. Previous research has shown that automatically combining low-level behaviors into a probabilistic finite state machine produces control software that crosses the reality gap satisfactorily. In this paper, we explore the possibility of adopting behavior trees as an architecture for the control software of robot swarms. We introduce `Maple`: an automatic design method that combines preexisting modules into behavior trees. To highlight the potential of this control architecture, we present robot experiments in which we compare `Maple` with `Chocolate` and `EvoStick` on two missions: FORAGING and AGGREGATION. `Chocolate` and `EvoStick` are two previously published automatic design methods. `Chocolate` is a modular method that generates probabilistic finite state machines and `EvoStick` is a traditional evolutionary robotics method. The results of the experiments indicate that behavior trees are a viable and promising architecture to automatically generate control software for robot swarms.

Keywords: Swarm robotics · Automatic design · Behavior trees

1 Introduction

In swarm robotics, a group of simple robots works together to achieve a common goal that is beyond the capabilities of a single robot [2,4,5,11,19,34]. The collective behavior of the swarm is the result of the local interactions that each robot has with its neighboring peers and with the environment. One of the biggest challenges is to conceive the control software of the individual robots [11]. Often control software is designed manually in a trial-and-error process [5]. This approach is time-consuming, prone to error and bias and difficult to replicate [4,14].

J. Kuckling and A. Ligot contributed equally to the research and should be considered co–first authors. Behavior trees were originally brought to the attention of the authors by DB. The proposed method was conceived by the four authors. It was implemented and tested by JK and AL. The initial draft of the manuscript was written by JK and AL and then revised by DB and MB. The research was directed by MB.

M. Dorigo et al. (Eds.): ANTS 2018, LNCS 11172, pp. 30–43, 2018.
https://doi.org/10.1007/978-3-030-00533-7_3

A promising alternative is automatic design. In automatic design, the design problem is transformed into an optimization problem. The design space of the possible instances of control software is mapped into a solution space on which an optimization algorithm searches a solution that maximizes a mission-dependent performance measure. Due to numerous constraints (of which time and hardware properties are the most notable ones), automatic design is often performed in simulation. As simulation is unavoidably only an approximation of reality, the so-called reality gap has to be faced by control software developed in simulation. It has been observed that different design methods might be more or less robust to the reality gap [16]. When assessing an automatic design method, it is therefore fundamental to perform tests with real robots to study its ability to cross the reality gap satisfactorily.

A popular approach to the automatic design of robot control software is evolutionary robotics [13]. Evolutionary swarm robotics is the application of evolutionary algorithms to generate control software for swarm robotics [37]. In this approach, robots are controlled by artificial neural networks that map sensor readings to commands that are fed to the actuators. Other approaches have been proposed that generate control software by assembling predefined modules. For example, Duarte et al. [12] generated a set of neural networks to perform low-level actions. These neural networks were then combined into a finite state machine. The benefit of the approach is that it is easier generate multiple neural networks that perform low-level actions rather than a single one that performs the whole mission. The limitation is that the designer still needs to decompose the task into suitable subtasks. Francesca et al. [15,16] defined AutoMoDe, a method in which a set of preexisting mission-agnostic *constituent behaviors* and *conditions* are assembled into a finite state-machine by an optimization process that maximizes a mission-specific performance measure. The authors developed two instances of AutoMoDe: `Vanilla` [16] and `Chocolate` [15], which differ in the optimization algorithm adopted. They compared them with a standard evolutionary method they called `EvoStick` [16]. While `EvoStick` performs better in simulation, `Vanilla` and `Chocolate` proved to be more robust to the reality gap and obtain better results in reality.

In this paper, we explore the possibility of automatically assembling preexisting modules into a behavior tree. Behavior trees are a control architecture that was initially developed as an alternative to finite state machines for specifying the behavior of non-player characters in video games [8,23]. Behavior trees gained popularity in the video game industry mainly because of their inherent modularity [8]. Subsequently, they attracted the attention of the academia, mostly in the robotics domain [9]. Compared to finite state machines, behavior trees promote increased readability, maintainabilty and code reuse [10].

Behavior trees are a promising control architecture to be adopted in swarm robotics. Indeed, they can be seen as generalizations of three classical architecture already studied in the literature: the subsumption architecture [6], sequential behavior compositions [7], and decision trees [30].

In this paper, we show that behavior trees can be used as a control architecture in the automatic design of robot swarms. We propose `Maple`, an automatic design method that fine-tunes and assembles preexisting modules (constituent behaviors and conditions) into a behavior tree. We present the results of experiments in which we automatically design control software for two missions: FORAGING and AGGREGATION. In our experiments, `Maple` outperforms `EvoStick` and obtains results that are comparable with those of `Chocolate`.

The research presented in this article prompts us to reconsider the original definition of AutoMoDe, which arbitrarily restricts AutoMoDe to the generation of probabilistic finite state machines [16]. The defining purpose of AutoMoDe– *automatic modular design* – is to generate control software for robot swarms by assembling and fine-tuning preexisting modules. The architecture into which modules are assembled is a secondary issue which we find it should not limit the methods defined as automatic modular design. In the following, we will consider as instances of AutoMoDe all methods that assemble and fine-tune preexisting modules, irrespective of the architecture into which they are cast and of the optimization algorithm used to generate the solutions. In this precise sense, we consider `Maple` to be an instance of AutoMoDe.

The paper is structured as follows: Sect. 2 provides an overview of the behavior tree architecture. Section 3 introduces `Maple`. Section 4 describe the experimental setup and Sect. 5 presents the results. Section 6 discussed related research and Sect. 7 concludes the paper and sketches future developments.

2   Behavior Trees

Behavior trees have been used as an alternative to finite state machines [27]. In this paper, we follow the definition given by Marzinotto et al. [27]. A behavior tree is a tree structure that contains one root node, control nodes, and execution nodes (actions or conditions). Execution is controlled by a tick generated by the root and propagated through the tree. When ticked by its parent, a node is activated. After execution, it returns one of three possible values: *success*, *running*, or *failure*. Condition nodes that are ticked observe the world state and return *success*, if their condition is fulfilled; and *failure*, otherwise. Action nodes that are ticked returns *success*, if their action is completed; *failure*, if their action cannot be completed; and *running*, if their action is still in progress. Control nodes distribute the tick to their children. Their return value depends on those returned by the children. There are six different types of control nodes: selector, selector*, sequence, sequence*, parallel and decorator—see Table 1.

Additionally, behavior trees implement the principle of two-way control transfers [31]. Not only can control be passed from a parent node to its child node, but the child can return execution to its parent, along with information about the state of execution. In a finite state machine, the control flow is only one-directional, that is, a state cannot return control to the predecessor.

Perhaps the most important property of behavior trees is their inherent modularity [10]. Each subtree of a behavior tree is, by definition, a valid behavior

Table 1. Overview of possible control nodes in a behavior tree.

Name	Symbol	Description
Selector	?	Ticks children sequentially as long as they return *failure*
Selector*	?*	Ticks children sequentially as long as they return *failure* Resumes ticking at last ticked node, if it returned *running*
Sequence	$\rightarrow$	Ticks children sequentially as long as they return *success*
Sequence*	$\rightarrow$*	Ticks children sequentially as long as they return *success*. Resumes ticking at last ticked node, if it returned *running*
Parallel	$\rightrightarrows$	Ticks all children simultaneously. Returns *success* (or *failure*), if a majority of the children return *success* (or *failure*). Otherwise it returns *running*
Decorator	δ	Executes a custom function on its only child. The function can either manipulate the number of ticks given to the child, or the value returned to the parent

tree as well. Thanks to the modularity, it is possible to adjust, remove, or add subtrees without having to account for new or missing interactions [31]. Combining subtree modules in a behavior tree leads to a hierarchical structure, which can simplify the analysis, for both humans and computers [10].

The aforementioned properties are appealing in the automatic design of control software for robot swarms. The enhanced expressiveness and the two-way control transfers could allow the representation of behaviors that cannot be easily implemented using finite state machines. The structural modularity could greatly simplify the implementation of optimization algorithms based on local manipulations. It could also allow pruning unused parts to increment readability. Finally, subtrees could be optimized independently of each other and used afterwards as building blocks to generate more complex behaviors.

3 AutoMoDe-Maple

Maple is an automatic design method that, by combining and fine-tuning pre-existing modules, generates control software in the form of behavior trees. In defining Maple, our goal was to explore the possibility of using behavior trees in the modular design of control software for robot swarms. We wished to define a method that we could then compare with Chocolate, the existing state-of-the-art in modular design, which generates finite state machines. We thought that, at this stage of our research, the comparison would have been the most informative if we reduced the differences between Maple and Chocolate as much as possible, so as to isolate the element we wished to study: the architecture. Therefore, we conceived Maple so that it shares with Chocolate the modules to be assembled and the optimization algorithm. The only difference between Maple and Chocolate is the architecture: behavior trees for the former, finite state machines for the latter.

Table 2. Reference model RM 1.1 [21]. Sensors and actuators of the extended version of the e-puck robot. Period of control cycle: 100 ms.

Sensor/actuator	Variables	Values
Proximity	$prox_i$, with $i \in \{0, ..., 7\}$	$[0, 1]$
Light	$light_i$, with $i \in \{0, ..., 7\}$	$[0, 1]$
Ground	$ground_i$, with $i \in \{0, ..., 2\}$	$\{black, gray, white\}$
Range-and-bearing	n	$\{0, ..., 19\}$
	V_d	$([0, 0.7]\,\mathrm{m}, [0, 2\pi]\,\mathrm{radian})$
Wheels	v_l, v_r	$[-0.12, 0.12]\,\mathrm{m/s}$

The modules assembled by Maple are those used by both Vanilla [16] and Chocolate [15]. To use these modules within a behavior tree, we included in Maple only a subset of the control nodes described in Sect. 2.

3.1 Robotic Platform

Maple generates control software for an extended version of e-puck [18,29]. Formally, the subset of sensors and actuators that are used by Maple, along with the corresponding variables, are defined by the reference model RM 1.1 [21], which we reproduce in Table 2 for the convenience of the reader.

The e-puck is a two wheeled robot. The control software can adjust the velocity of the motors of each wheel (v_r and v_l). The e-puck can detect the presence of nearby obstacles ($prox_i$), measure ambient light ($light_i$), and tell whether the floor situated directly beneath itself is white, gray, or black ($ground_i$). Finally, thanks to its range-and-bearing board [20] the e-puck is aware of the presence of its peers in a range of up to 0.7 m: it knows their number (n) and a vector V_d indicating the direction of attraction to the neighboring peers, following the framework of virtual physics [35].

3.2 Set of Modules

Maple uses the set of preexisting modules originally defined for Vanilla [16]. The set is composed of six low-level behaviors (i.e., activities performed by the robot) and six conditions (i.e., assessments of particular situations experienced by the robot). In a behavior tree, a leaf node is either an action or a condition. Maple selects the action nodes among the set of Vanilla's low-level behaviors, and the condition nodes among the set of Vanilla's conditions. In this section, we briefly describe Vanilla's low-level behaviors and conditions. We refer the reader to the work of Francesca et al. [16] for more details.

Low-Level Behaviors. *Exploration* is a random walk strategy. The robot goes straight until an obstacle is perceived by the front proximity sensors. Then, the robot turns on the spot for a random number of control cycles drawn in $\{0, ..., \tau\}$, where τ is an integer parameter $\in \{0, ..., 100\}$. *Stop* orders the robot to stay still. *Phototaxis* moves the robot towards a light source. If no light source is perceived, the robot goes straight. *Anti-phototaxis* moves the robot away from the light source[1]. If no light source is perceived, the robot goes straight. *Attraction* moves the robot in the direction of the neighboring peers (V_d). The speed of convergence towards the detected peers is controlled by a real parameter $\alpha \in [1, 5]$. If no peer is detected, the robot goes straight. *Repulsion* moves the robot away from the neighboring peers $(-V_d)$. The real parameter $\alpha \in [1, 5]$ controls the speed of divergence. Obstacle avoidance is embedded in all low-level behaviors, with the exception of stop. As stated earlier, this is not a design choice we made for `Maple` but rather an earlier decision from `Vanilla` [16] that is kept to allow comparison with previously obtained results. The parameters τ and α must be tuned by the automatic design process.

Conditions. *Black-*, *gray-* and *white-floor* are true with probability $\beta \in [0, 1]$ if the ground sensor perceives the floor as black, gray, or white, respectively. *Neighbor-count* is true with a probability computed as a function $z(n) \in [0, 1]$ of the number of robots detected via the range-and-bearing board. A real parameter $\eta \in [0, 20]$ and an integer parameter $\xi \in \{0, ..., 10\}$ control the steepness and the inflection point of the function, respectively. *Inverted-neighbor-count* is true with probability $1 - z(n)$. *Fixed-probability* is true with probability $\beta \in [0, 1]$. The parameters β, η and ξ must be tuned by the automatic design process.

3.3 Control Software Architecture

We use the preexisting low-level behaviors of `Vanilla` [16] without any modification. In the traditional implementation of behavior trees, an action node is able to tell whether the system it controls (i) successfully executed,(ii) is still executing, or (iii) failed to execute the required activity. The action node then returns the corresponding state variable (i.e., *success*, *running*, or *failure*).

The low-level behaviors of `Vanilla` were designed to be used as states of probabilistic finite state machines, and were meant to be executed until an external condition was enabled. Because of their implementations, when used as action nodes within `Maple`, the low-level behaviors can only return *running*. As a consequence, part of the control-flow nodes of behavior trees do not work as intended. For example, a sequence node with two `Vanilla`'s behaviors as children would always directly return *running* after the first behavior is executed once, and would never execute the second one—see Table 1.

To use `Vanilla`'s behaviors as action nodes, `Maple` instantiates behavior trees that have a restricted topology and use only a subset of all available control nodes. The root node must be of the type sequence* and can only have selector nodes as children. Within `Maple`, each subtree defined by a selector node is forced to have two children: a condition node as the left child, and an action node as the right child. In order to stay close to `Vanilla`'s restriction of a maximum of

---

[1] In biology this behavior is known as *negative phototaxis* [28].

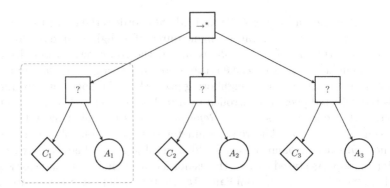

Fig. 1. Illustration of a behavior tree that can be generated by `Maple`. `Maple` determines the number of selector subtrees (highlighted by the dashed box) and specifies the condition and action nodes for each of them. The type of the root node is predefined.

four states in the finite state machine, the behavior tree is allowed to contain a maximum of four selector subtrees. Figure 1 illustrates an example (with only three out of four possible subtrees) of the restricted topology of the behavior trees that `Maple` can produce. In this example, action node A_1 is executed as long as condition node C_1 returns *failure*. When condition node C_1 returns *success*, the sequence* node ticks the next selector subtree, and so forth. Similarly to `Chocolate` [15], `Maple` uses Iterated F-race [26] as the optimization algorithm to search for the best possible instance of behavior tree among all the possible ones.

4   Experimental Setup

In this section, we describe the automatic design methods under analysis, the missions on which we test them, and the protocol we follow.

4.1   Automatic Design Methods

We compare `Maple` with `Chocolate` [15] and `EvoStick` [15,16]. As `Maple`, `Chocolate` and `EvoStick` are based on reference model RM1.1. We briefly describe these methods and we refer the reader to Francesca et al. [15,16] for the details.

Chocolate selects, fine-tunes, and combines preexisting modules into probabilistic finite state machines. It uses the same twelve modules as `Vanilla` and `Maple`. `Chocolate` is restricted to create probabilistic finite state machines comprising up to four states and up to four outgoing edges per state. Similarly to `Maple`, `Chocolate` uses Iterated F-race [26] as optimization algorithm.

EvoStick is an implementation of the evolutionary robotics approach: the topology of a neural network is fixed, and an evolutionary algorithm is used to optimize the weights of the connections. The network considered in `EvoStick` is fully connected, feed-forward and does not contain hidden neurons. It comprises 24 input nodes for the readings of the sensors described in the reference model

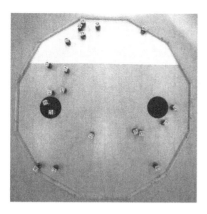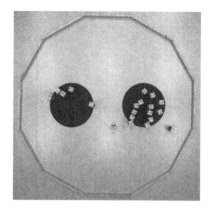

Fig. 2. FORAGING (*left*) and AGGREGATION (*right*).

RM 1.1: 8 for the proximity sensors, 8 for the light sensors, 3 for the ground sensors, and 5 for the range-and-bearing board. Out of the 5 input nodes dedicated to the range-and-bearing board, one is allocated to the number of neighbors, and the four others to the scalar projections of the vector pointing to the center of mass of these neighbors on four unit vectors. The neural network comprises 2 output nodes for the velocities of the left and right wheels.

4.2 Missions

The missions considered are FORAGING and AGGREGATION. They have already been studied in [16]. We refer the reader to the original article for the details. In the two missions, the robots operate in a dodecagonal arena delimited by walls and covering an area of 4.91 m². We limit the duration of the missions to 120 s.

FORAGING. The arena contains two source areas (black circles) and a nest (white area). A light is placed behind the nest to help the robots to navigate (Fig. 2, left). In this idealized version of foraging, a robot is deemed to retrieve an object when it enters a source and then the nest. The goal of the swarm is to retrieve as many objects as possible. The objective function is $F_{\mathrm{F}} = N_i$, where N_i is the number of objects retrieved.

AGGREGATION. The swarm must select one of the two black areas and aggregate there [16,17] (Fig. 2, right). The objective function is $F_{\mathrm{A}} = \max(N_l, N_r)/N$, where N_l and N_r are the number of robots located on the left and right area, respectively; and N is the total number of robots. The objective function is computed at the end of the run, and is maximized when all robots are either on the left or the right area.

4.3 Protocol

We considered a robot swarm composed of 20 e-pucks. The three automatic design methods—Maple, Chocolate and EvoStick—produce control software

FORAGING

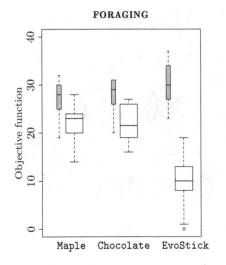

AGGREGATION

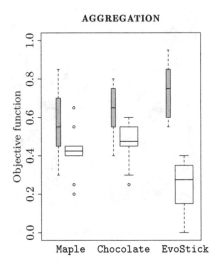

Fig. 3. Results of the experiments. The gray boxes represent the performance assessed in simulation; the white boxes represent the performance assessed in reality.

for two missions—FORAGING and AGGREGATION. Since all the design methods are stochastic, for each mission, each design method is executed 10 times and produces 10 instances of control software. The design budget allocated to each method for each mission is 50000 simulation runs: this is the maximum number of simulation runs allowed during the design process. To study the generalization capabilities of the design methods, we assess the performance of each instance of control software once in simulation, and once in reality [3]. All simulations are performed using ARGoS3, beta 48. [18,33].

In reality, to automatically measure the performance of the swarm, we use a system composed of an overhead camera and markers on the robots to track their position in real time [36]. Experimental runs start from 10 different initial positions/orientations of the robots. We use the tracking system to automatically guide the robots to the initial position/orientation of each run. During a run, we interfere with the robots only if they tip over due to a collision. In this case, we intervene and put them upright to avoid damages.

We present the results in the form of box-and-whiskers boxplots. For each method and each mission, we report two boxplots: one for simulation and one for reality. In the following, statements like "method A is significantly better than B," always imply we performed a Wilcoxon rank-sum test that detected significance with confidence of at least 95%.

5   Results

In this section, we report the results for each mission considered. The instances of control software produced, the details of their performances both in simulation and in reality, and videos of their execution on the robots are available as online supplementary material [25].

FORAGING. Although the performance of control software produced by the three automatic design methods is similar in simulation, `Maple` and `Chocolate` are significantly better than `EvoStick` in reality. The performance of all three methods drops significantly when passing from simulation to reality, but `EvoStick` suffers from the reality gap the most. See Fig. 3(*left*).

`Maple` and `Chocolate` generate control software that displays expected and similar strategies: the robots explore the environment and once a source of food (i.e., a black area) is found, they navigate towards the nest (i.e., the white area) guided by the light. The performance drop that affects `Maple` and `Chocolate` when porting the control software from simulation to reality is probably due to the fact that simulation does not properly reproduce the frictions experienced by the robots. In reality, due to friction, robots become sometimes unable to move and therefore do not contribute to the foraging process. Contrarily to `Maple` and `Chocolate`, and with the exception of a few cases, `EvoStick` was unable to generate instances of control software that display an effective foraging behavior in reality. Indeed, in most cases, the robots seem unable to navigate efficiently.

AGGREGATION. In simulation, `Maple` and `Chocolate` show similar performance, but `EvoStick` performs significantly better than `Maple`. Also in reality, `Maple` and `Chocolate` perform similarly, but they are both significantly better than `EvoStick`. Indeed, the performance of `EvoStick` drops considerably from simulation to reality, whereas the performance drop of `Maple` and `Chocolate` is smaller. See Fig. 3(*right*).

The instances of control software produced by `Maple` and `Chocolate` are able to find the black areas and stop there. Contrarily, the instances of control software produced by `EvoStick` do not efficiently search the space. When a black area is found, the robots tend to leave it quickly. Neither of the three methods produced control software that displayed effective collective decision making.

6 Related Work

Most of the early research on behavior trees has concentrated on their use for game development [1,32]. Subsequently, research has been devoted to the application of behavior trees in robotics. For example, Marzinotto et al. [27] manually designed a behavior tree for manipulation on the NAO robot. Hu et al. [22] described an application of behavior trees to semi-autonomous, simulated surgery.

Jones et al. [24] proposed an automatic design method for robot swarms in which the control architecture of robots is a behavior tree. To the best of our knowledge, that is the first and only application of behavior trees in swarm robotics. The authors used genetic programming to generate control software for kilobots in a foraging mission. The action nodes are atomic commands, such as setting motor state, storing information, or broadcasting a signal. The results show that behavior trees can be effectively used to control the robots of a swarm and that the control software generated is human-readable. Our approach differs both conceptually and methodically from method proposed by Jones et al. [24]. Methodically, we used Iterated F-Race [26] as an optimization algorithm and a more restricted architecture for the behavior trees. Conceptually, we focussed on showing that automatic modular design can cross the reality gap in a satisfactory way, even when using different architectures. Furthermore, for the leaves of the behavior tree we used complex low-level behaviors instead of atomic actions.

7   Conclusions

AutoMoDe, automatic modular design, is an approach in which control software for robot swarms is automatically generated by assembling and fine-tuning preexisting modules. In previous articles, the control software architecture on which AutoMoDe operates was arbitrarily restricted to probabilistic finite state machine. In this article, we went beyond this restriction and we investigated the possibility of adopting behavior trees as a control software architecture. Behavior trees are appealing for a number of reasons. Compared to finite state machines, behavior trees offer greater expressiveness, implement the principle of two-way control transfers, and posses inherent modularity which allows the creation of a hierarchical structure, code reuse, and separation of concerns. Behavior trees are also easier to manipulate without compromising their integrity. This fact could be extremely useful when designing optimization algorithms based on iterative improvement.

We proposed a new instance of AutoMoDe called `Maple`, which fine-tunes and assembles preexisting modules into a behavior tree. To highlight its potential, we performed experiments in simulation and reality for two different missions: FORAGING and AGGREGATION. The results show that `Maple` performs similar to `Chocolate`—the state-of-the-art AutoMoDe method, which generates probabilistic finite state machines. They both cross the reality gap in a satisfactory way. `EvoStick`, which is an evolutionary robotics method, performs better then `Maple` and `Chocolate` in simulation, but significantly worse in reality.

Future work will focus on fully exploiting the potentials of behavior trees. This implies defining modules that are natively conceived to operate within a behavior tree—e.g., modules that properly return their state value (*success*, *running*, or *failure*) and therefore interact correctly with all possible control nodes. Moreover, we will define an ad-hoc optimization algorithm, possibly relying also on iterative improvement, that fully exploits the inherent modularity and hierarchical structure of behavior trees.

Acknowledgements. The project has received funding from the European Research Council (ERC) under the European Union's Horizon 2020 research and innovation programme (grant agreement No 681872). Mauro Birattari acknowledges support from the Belgian *Fonds de la Recherche Scientifique* – FNRS.

References

1. Becroft, D., Bassett, J., Mejía, A., Rich, C., Sidner, C.L.: AIPaint: a sketch-based behavior tree authoring tool. In: Bulitko, V., Riedl, M.O. (eds.) Proceedings of the Seventh AAAI Conference on Artificial Intelligence and Interactive Digital Entertainment, AIIDE-11. AAAI Press, Stanford (2011)
2. Beni, G.: From swarm intelligence to swarm robotics. In: Şahin, E., Spears, W.M. (eds.) SR 2004. LNCS, vol. 3342, pp. 1–9. Springer, Heidelberg (2005). https://doi.org/10.1007/978-3-540-30552-1_1
3. Birattari, M.: On the estimation of the expected performance of a metaheuristic on a class of instances. How many instances, how many runs? Technical report TR/IRIDIA/2004-01, IRIDIA, Université libre de Bruxelles, Belgium (2004)
4. Bozhinoski, D., Birattari, M.: Designing control software for robot swarms: software engineering for the development of automatic design methods. In: ACM/IEEE 1st International Workshop on Robotics Software Engineering, RoSE, pp. 33–35. ACM, New York (2018). https://doi.org/10.1145/3196558.3196564
5. Brambilla, M., Ferrante, E., Birattari, M., Dorigo, M.: Swarm robotics: a review from the swarm engineering perspective. Swarm Intell. **7**(1), 1–41 (2013)
6. Brooks, R.: A robust layered control system for a mobile robot. IEEE J. Robot. Autom. **2**(1), 14–23 (1986)
7. Burridge, R.R., Rizzi, A.A., Koditschek, D.E.: Sequential composition of dynamically dexterous robot behaviors. Int. J. Robot. Res. **18**(6), 534–555 (1999)
8. Champandard, A.J.: Understanding behavior trees (2007). http://aigamedev.com/open/articles/bt-overview/
9. Colledanchise, M., Ögren, P.: How behavior trees modularize hybrid control systems and generalize sequential behavior compositions, the subsumption architecture, and decision trees. IEEE Trans. Robot. **33**(2), 372–389 (2017)
10. Colledanchise, M., Ögren, P.: Behavior trees in robotics and AI: an introduction (2018). https://arxiv.org/abs/1709.00084
11. Dorigo, M., Birattari, M., Brambilla, M.: Swarm robotics. Scholarpedia **9**(1), 1463 (2014)
12. Duarte, M., Gomes, J., Costa, V., Oliveira, S.M., Christensen, A.L.: Hybrid control for a real swarm robotics system in an intruder detection task. In: Squillero, G., Burelli, P. (eds.) EvoApplications 2016. LNCS, vol. 9598, pp. 213–230. Springer, Cham (2016). https://doi.org/10.1007/978-3-319-31153-1_15
13. Floreano, D., Husbands, P., Nolfi, S.: Evolutionary robotics. In: Siciliano, B., Khatib, O. (eds.) Handbook of Robotics, pp. 1423–1451. Springer, Heidelberg (2008)
14. Francesca, G., Birattari, M.: Automatic design of robot swarms: achievements and challenges. Front. Robot. AI **3**(29), 1–9 (2016)
15. Francesca, G., et al.: AutoMoDe-chocolate: automatic design of control software for robot swarms. Swarm Intell. **9**(2/3), 125–152 (2015)
16. Francesca, G., Brambilla, M., Brutschy, A., Trianni, V., Birattari, M.: AutoMoDe: a novel approach to the automatic design of control software for robot swarms. Swarm Intell. **8**(2), 89–112 (2014)
17. Francesca, G., Brambilla, M., Trianni, V., Dorigo, M., Birattari, M.: Analysing an evolved robotic behaviour using a biological model of collegial decision making. In: Ziemke, T., Balkenius, C., Hallam, J. (eds.) SAB 2012. LNCS (LNAI), vol. 7426, pp. 381–390. Springer, Heidelberg (2012). https://doi.org/10.1007/978-3-642-33093-3_38

18. Garattoni, L., Francesca, G., Brutschy, A., Pinciroli, C., Birattari, M.: Software infrastructure for e-puck (and TAM). Technical report TR/IRIDIA/2015-004, IRIDIA, Université libre de Bruxelles, Belgium (2015)
19. Garattoni, L., Birattari, M.: Swarm robotics. In: Webster, J. (ed.) Wiley Encyclopedia of Electrical and Electronics Engineering. Wiley, Hoboken (2016). https://doi.org/10.1002/047134608X.W8312
20. Gutiérrez, Á., Campo, A., Dorigo, M., Donate, J., Monasterio-Huelin, F., Magdalena, L.: Open E-puck range & bearing miniaturized board for local communication in swarm robotics. In: Kosuge, K. (ed.) IEEE International Conference on Robotics and Automation, ICRA, pp. 3111–3116. IEEE Press, Piscataway (2009)
21. Hasselmann, K., Ligot, A., Francesca, G., Birattari, M.: Reference models for AutoMoDe. Technical report TR/IRIDIA/2018-002, IRIDIA, Université libre de Bruxelles, Belgium (2018)
22. Hu, D., Gong, Y., Hannaford, B., Seibel, E.J.: Semi-autonomous simulated brain tumor ablation with Raven II surgical robot using behavior tree. In: Parker, L., et al. (eds.) IEEE International Conference on Robotics and Automation, ICRA, pp. 3868–3875. IEEE Press, Piscataway (2015)
23. Isla, D.: Handling complexity in the Halo 2 AI. In: GDC Proceeding (2005)
24. Jones, S., Studley, M., Hauert, S., Winfield, A.: Evolving behaviour trees for swarm robotics. In: 13th International Symposium on Distributed Autonomous Robotic Systems (DARS) (2016)
25. Kuckling, J., Ligot, A., Bozhinoski, D., Birattari, M.: Behavior trees as a control architecture in the automatic design of robot swarms: Supplementary material (2018). http://iridia.ulb.ac.be/supp/IridiaSupp2018-004/index.html
26. López-Ibáñez, M., Dubois-Lacoste, J., Pérez Cáceres, L., Birattari, M., Stützle, T.: The irace package: iterated racing for automatic algorithm configuration. Oper. Res. Perspect. **3**, 43–58 (2016)
27. Marzinotto, A., Colledanchise, M., Smith, C., Ögren, P.: Towards a unified behavior trees framework for robot control. In: Xi, N., et al. (eds.) IEEE International Conference on Robotics and Automation, ICRA, pp. 5420–5427. IEEE Press, Piscataway (2014)
28. Menzel, R.: Spectral sensitivity and color vision in invertebrates. In: Autrum, H. (ed.) Comparative Physiology and Evolution of Vision in Invertebrates, pp. 503–580. Springer, Heidelberg (1979). https://doi.org/10.1007/978-3-642-66999-6_9
29. Mondada, F., et al.: The e-puck, a robot designed for education in engineering. In: Gonçalves, P., Torres, P., Alves, C. (eds.) Proceedings of the 9th Conference on Autonomous Robot Systems and Competitions, pp. 59–65. Instituto Politécnico de Castelo Branco, Portugal (2009)
30. Nehaniv, C.L., Dautenhahn, K.: Imitation in Animals and Artifacts. MIT Press, Cambridge (2002)
31. Ögren, P.: Increasing modularity of UAV control systems using computer game behavior trees. In: Thienel, J., et al. (eds.) AIAA Guidance, Navigation, and Control Conference 2012, pp. 358–393. AIAA Meeting Papers (2012)
32. Perez, D., Nicolau, M., O'Neill, M., Brabazon, A.: Evolving behaviour trees for the mario AI competition using grammatical evolution. In: Di Chio, C., et al. (eds.) EvoApplications 2011. LNCS, vol. 6624, pp. 123–132. Springer, Heidelberg (2011). https://doi.org/10.1007/978-3-642-20525-5_13
33. Pinciroli, C., et al.: ARGoS: a modular, parallel, multi-engine simulator for multirobot systems. Swarm Intell. **6**(4), 271–295 (2012)

34. Şahin, E.: Swarm robotics: from sources of inspiration to domains of application. In: Şahin, E., Spears, W.M. (eds.) SR 2004. LNCS, vol. 3342, pp. 10–20. Springer, Heidelberg (2005). https://doi.org/10.1007/978-3-540-30552-1_2

35. Spears, W.M., Spears, D., Hamann, J.C., Heil, R.: Distributed, physics-based control of swarms of vehicles. Auton. Robot. **17**, 137–162 (2004)

36. Stranieri, A., et al.: IRIDIA's arena tracking system. Technical report TR/IRIDIA/2013-013, IRIDIA, Université libre de Bruxelles, Belgium (2013)

37. Trianni, V.: Evolutionary Swarm Robotics. Springer, Berlin (2008). https://doi.org/10.1007/978-3-540-77612-3

Guidance of Swarms with Agents Having Bearing Only and Limited Visibility Sensors

Rotem Manor[⊠] and Alfred M. Bruckstein

Technion - Israel Institute of Technology, Technion City, Haifa, Israel
manorrotem@gmail.com

Abstract. We suggest a mechanism for leading a team of mobile, oblivious, identical and indistinguishable agents in desired directions. The agents are assumed to have a compass, i.e. a common North direction, and bearing only sensing within a limited visibility range, and may receive a direction-control broadcast with some given probability. We prove that, under the suggested guidance rule, the swarm of agents gathers to a small disk in the plane and moves in the desired direction with an expected velocity dependent on the probability of receiving the control signal.

1 Introduction

Directing a swarm of simple and low-cost agents, toward a given location in the environment using a global broadcast signal that may be "heard" by only some of the agents, is an interesting challenge. Earlier works already done on systems of simple agents, showed that even inferior capabilities agents can perform task of gathering [1,8,9,11–13]. Here, we intend to control the movement of a system of inferior capabilities agents.

Prior studies on swarm motion control were either based on using global potential fields or external forces [10,14–16], or on using "informed" agents flocking, leader following processes, and either linear or non-linear interaction dynamics, [5–7,17–19].

We here address the problem of controlling a swarm of oblivious, anonymous (identical and indistinguishable) agents without explicit inter-agent communication that are capable of sensing only their neighbours' bearing within a limited visibility range. We assume that all agents have a common compass direction (North), hence the exogenous guidance signal is an azimuth angle that specifies the desired direction of motion in the plane.

We propose and analyse a semi-synchronised discrete time model, where each agent of the swarm has a probability $\gamma > 0$ to "hear" the control signal at each time step. As stated before, the control signal specifies a unit vector $\hat{C}$ in the desired direction of motion.

We start by presenting the agents' local motion law, an extension of the interaction rules presented in [3]. Under this model, each agent jumps inside an

© Springer Nature Switzerland AG 2018
M. Dorigo et al. (Eds.): ANTS 2018, LNCS 11172, pp. 44–56, 2018.
https://doi.org/10.1007/978-3-030-00533-7_4

allowable region, designed for maintaining visibility with its neighbours. Without the presence of the control, this motion law implements a gathering algorithm for swarms of agents with the given capabilities. The proposed law is first shown to gather swarms with an initial connected visibility net to a small region (with a complete visibility net) even in the presence of the control signal. Then, once the swarm's constellation has a complete visibility graph, it moves with a constant expected velocity in the desired direction, broadcast by a controlling entity. We conclude this paper by presenting simulation results, and discuss possible future directions for research.

2   The Motion Law

We consider a system of n identical, anonymous, and oblivious agents in the $\mathbb{R}^2$-plane specified by their time varying locations $\{p_i(k)\}_{i=1,2,\ldots,n}$. We assume that the agents sense only the direction to their neighbours (i.e. bearing only sensing), hence their information about neighbours is partial. Their steps in time are determined by the set of unit vectors pointing to their current neighbours and the control signal, if received. The neighbours of each agent i, at any time step k are defined as the set of agents located within a given visibility range V form the position $p_i(k)$, and are denoted by the set $N_i(k)$. The neighbourhood relation between all the agents is conveniently described by a time dependent visibility-graph.

Our agents move according to the following motion law, first presented in [3]. At time step k, an agent may be active with a strictly positive probability δ. Each active agent not "surrounded" by neighbours, jumps to a random point choosen according to a uniform distribution in an allowable region defined by the geometry of the vectors pointing to its neighbours and the control. The allowable region of an agent is defined as the area the agent can move into, without losing visibility with any of its neighbours. Gordon and Bruckstein proved that in the absence of control such a motion law gathers the swarm from a constellation with a connected visiblity net to a constellation having a complete visibility graph within a finite expected number of time steps [11], and recently in [3] Manor and Bruckstein proved that a small modification of the motion law can further be shown to gather the swarm to a disk of a radius equal to σ (assumed to be $\sigma \leq V/2$), where σ is the agents' maximal allowed step size. This process too happens within a finite expected number of time steps.

In order to adjust this gathering law to a guidance law, we assume that each agent currently "hearing" a control signal acts as if it has yet another neighbouring agent located in the direction provided by the control signal. This approach to a controlled scenario considerably facilitates in the analysis of the dynamics of the controlled swarm.

Let us next formally describe the rule of gathering and guidance. Let $\psi_i(k)$ be the angle of the minimal sector anchored at agent i's position at time step k that contains all its neighbours, and let $\hat{\psi}_i(k)$ be a unit vector in the direction of the bisector of the angle $\psi_i(k)$. If an agent i at time step k has $\psi_i(k) \geq \pi$, it is considered "surrounded" by neighbours.

Let us denote a disk of radius r centered at a point c by $D_r(c)$. An agent i is **active** at each time step k with a strictly positive probability δ, and receives the control signal with probability γ. If an **active** agent i does not currently "hear" the control signal, and is not surrounded by its neighbours, i.e. $\psi_i(k) < \pi$, it jumps to a uniformly selected random point inside its current "allowable region", $ar_i(k)$, defined as follows:

$$\bigcap_{j \in N_i(k)} D_{\frac{\sigma}{2}}\left(p_i(k) + \frac{\sigma}{2}U_{ij}(k)\right) = D_{\frac{\sigma}{2}}\left(p_i(k) + \frac{\sigma}{2}U_i^-(k)\right) \cap D_{\frac{\sigma}{2}}\left(p_i(k) + \frac{\sigma}{2}U_i^+(k)\right)$$

(1)

where recall that $\sigma < V/2$ and $U_{ij}(k)$ is a unit vector pointing from $p_i(k)$ to $p_j(k)$. In (1), $U_i^-(k)$ and $U_i^+(k)$ are unit vectors pointing form agent i to its extremal right and left neighbours, i.e. those defining the sector $\psi_i(k)$ (see Fig. 1).

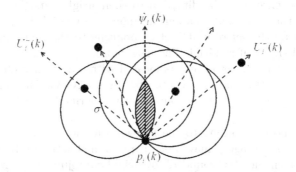

Fig. 1. The allowable region of an agent (dashed area) is the intersection of all disks $D_{\frac{\sigma}{2}}\left(p_i(k) + \frac{\sigma}{2}U_{ij}(k)\right)$ where $j \in N_i(k)$. Notice that it is given by the intersection of the two disks associated with the "extremal" neighbours.

If agent i is **active** and receives the control signal, it adds another virtual agent, v, to its neighbour set. The virtual agent is assumed to be in the guidance direction $\hat{C}$ relative to i's position. Then, agent i proceeds to determine its next position considering the extended neighbour set $v \cup N_i(k)$. We denote the allowable region resulting from the extended set of neighbours by $ar_i^v(k)$. Formally, the local control law for the motion of agent i is therefore the following:

$$p_i(k+1) = \begin{cases} p_i(k) & \text{if } \chi_i(k) = 0 \text{ or } \psi_i(k) \geq \pi \\ \text{a point in } ar_i(k) & \begin{array}{l} \text{if } \chi_i(k) = 1 \text{ and } \xi_i(k) = 0 \\ \text{and } \psi_i(k) < \pi \end{array} \\ \text{a point in } ar_i^v(k) & \begin{array}{l} \text{if } \chi_i(k) = 1 \text{ and } \xi_i(k) = 1 \\ \text{and } \psi_i(k) < \pi \end{array} \end{cases}$$

(2)

where $\chi_i(k)$ is a binary variable equal to 1 or 0 with probability δ or $(1 - \delta)$ determining whether the agent i is active at time step k, and $\xi_i(k)$ is a binary variable equal to 1 or 0 with probability γ or $(1 - \gamma)$ modelling the probabilistic "reception" of the guidance signal by agent i at time step k.

Under the above proposed rule of motion, we have the following straightforward initial result:

Lemma 1. *If all agents move inside their allowable regions, none of them will lose visibility with their neighbours.*

Proof. Consider a pair of neighbour agents i and j. Each one of these agents may jump into its own allowable region, which by (1) is contained in the $\sigma/2$-disk which centered in the direction of its pair, i.e.

$$ar_i(k) \subset D_{\frac{\sigma}{2}}\left(p_i(k) + \frac{\sigma}{2}U_{ij}(k)\right) \quad \text{and} \quad ar_j(k) \subset D_{\frac{\sigma}{2}}\left(p_j(k) + \frac{\sigma}{2}U_{ji}(k)\right)$$

Since $\sigma \leq V/2$, we have that both disks $D_{\sigma/2}\left(p_i(k) + \frac{\sigma}{2}U_{ij}(k)\right)$ and $D_{\sigma/2}\left(p_j(k) + \frac{\sigma}{2}U_{ji}(k)\right)$ are contained in a disk of radius $V/2$ centered at the agents' average location, $(p_i(t) + p_j(t))/2$. Hence, the next possible location for agents i and j, inside their allowable regions, will again result in a less than V distance between them.

Since the allowable region of an agent i is the intersection between all the $\sigma/2$-disks associated with its neighbours, i.e.

$$ar_i^v(k) = \bigcap_{j \in N_i^v(k)} D_{\frac{\sigma}{2}}\left(p_i(k) + \frac{\sigma}{2}U_{ij}(k)\right) \subseteq \bigcap_{j \in N_i(k)} D_{\frac{\sigma}{2}}\left(p_i(k) + \frac{\sigma}{2}U_{ij}(k)\right)$$

Hence, we have that $ar_i^v(k) \subseteq ar_i(k)$, and even in a guided scenario, it will remain, at time $(k+1)$, at a distance less than V from all its current neighbours.

3   Gathering to a Small Region

Before dealing with the movement of the system under the broadcast control, we shall show that, even in a controlled scenario, a swarm of agents acting according to the dynamics given by (2) gathers to within a disk of radius equal to σ in a finite expected number of time steps. To do so, we first recall the essence of the gathering proof given in [3]. The proof states that a swarm of agents acting according to motion law (2) gathers to a disk of diameter V, and then to a disk of radius σ, without any guidance (the case of $\gamma = 0$).

The gathering proof was done in two steps. Theorem 2 states that any constellation having a connected visibility graph reaches a complete visibility graph within a finite expected number of time-steps. Henceforth, all agents remain confined to a disk of diameter V. Then, Theorem 3 states that a constellation having a complete visibility graph further shrinks to within a disk of radius σ in an additional finite expected number of time-steps.

The outline of the proof of Theorem 2 is that at each time-step there is a strictly positive probability of $\delta(1 - \delta)^{n-1}$ for the agent located at the sharpest corner of the constellation's convex-hull to be the only active agent. This agent, when jumping, has a probability bounded away from zero by a constant to reduce its distance from $\bar{p}(k)$, the centroid at time step k of the agents' constellation, by at least a positive constant quantity s^*. As a consequence of this jump, $\mathcal{L}(P(k))$, the sum of all agents' squared distances of from $\bar{p}(k)$, is reduced by at least s^{*2}/n.

We always assume initial constellations with a connected visibility graph. Hence, by Lemma 1 the visibility graphs remain connected, and $\mathcal{L}(P(k))$ is bounded from above. Therefore, $\mathcal{L}(P(k))$ may reach a value of less than $V^2/4$ within a finite number of time steps with strictly positive (though very small!) probability. This happens when a long sequence of decreasing steps occurs. However, once $\mathcal{L}(P(k))$ reaches $V^2/4$, the visibility graph of the constellation is necessarily complete, and by Lemma 1, it remains complete, i.e. all the agents will henceforth be confined to a disc of diameter V.

The proof of Theorem 3 in [3], analyses the dynamics of the system considering the minimal enclosing circle of the agents positions, its radius and center being denoted by $R(k)$ and $C(k)$. In the proof we show that, any agent located at a distance greater than $\sigma/2$ from $C(k)$ can not increase its distance from $C(k)$, and if an agent is located at a distance smaller than or equal to $\sigma/2$ from $C(k)$, it can not jump to a distance greater than σ from $C(k)$. Therefore, if $R(k) > \sigma$, it cannot increase. Furthermore we show that, there are at least two agents located on the circumference of the minimal enclosing circle, or within a close proximity to it, and located at corners of the constellation's convex-hull with angles bounded below π by a constant as well. These agents have strictly positive probabilities, δ, to be **active**, and therefore will jump to locations closer to $C(k)$ with strictly positive probability. Hence, if $R(k) > \sigma$, the radius of the smallest enclosing circle drops significantly with a strictly positive probability within a batch of $\lceil n/2 \rceil$ time-steps. Once $R(k)$ reaches σ it cannot exceed it. For details, please see the technical report [3].

Notice that these proofs are based on agents being in special situations, occurring at some time steps. These agents have strictly positive probabilities to be **active** (δ), and in fact to be the only **active** agent at a time step ($\delta(1-\delta)^{n-1}$). If we add the broadcast control to the model ($\gamma > 0$), we still may use the same proofs assuming further that these agents, in the special situations, do not "hear" the broadcast. Therefore, the above mentioned probabilities should be updated to $\delta(1-\gamma)$ and $\delta(1-\delta)^{n-1}(1-\gamma)$ respectively, leaving them strictly positive, by the assumptions of the model we discuss in this paper, and, as a consequence, we have that our system still gathers within a finite expected number of time steps, even in the presence of broadcast control. The gathering dynamics may be seen in Fig. 2

4   Random Dynamics Analysis

In this section we analyse the dynamics of a constellation of agents with complete visibility graph. We start with the dynamics of the swarm without the presence of a control broadcast.

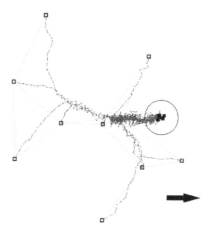

Fig. 2. Gathering under broadcast control according to the motion rule given by (2). Initial configuration is represented by the empty squares (agents' initial locations) which connected by the dotted lines (initial visibility connections between agents), and the trajectories to the current locations (full squares) are drawn. The agents gather into a disk of a radius equal to σ (the black circle), within finite number of time step, and then remain confined to a disk of this size that drifts in the direction given by the broadcast control (black arrow).

In order to characterize the dynamics of the system, we rely on a Theorem from [2] on the convergence in probability of random variables that are sums of uniformly bounded random increments. Given a random variable X_k so that

$$X_k = \sum_{i=1}^{k} Y_i, \text{ i.e. } Y_k = X_k - X_{k-1}.$$

If Y_k satisfies $\mathbb{E}\left[Y_k | \mathcal{F}_{k-1}\right] = 0$ and $\sum_{i=1}^{k} Var\{Y_i\} \overset{a.s.}{\to} \infty$. where $\mathcal{F}_{k-1}$ is the sigma filed generated all prior realization of the process. then $\frac{X_{\tilde{k}}}{\sqrt{\nu}} \overset{p}{\to} N(0,1)$. where $\tilde{k} = \min\{k : \sum_{k'=1}^{k} \mathbb{E}\{\|Y_{k'}\|^2\} > \nu\}$ is a stopping time defined by ν.

We use the linearity of expectation to prove that once the gathering process shrank the constellation of agents to within a disk of radius $V/2$ the expected centroid will either remain in place when no broadcast is heard or it will drift in the direction of the control vector with a velocity equal to $\delta\gamma\sigma/(2n)$.

The above Theorem will further prove that the distribution of the swarm centroid location about the expected trajectory converges to a Gaussian with variance increasing linearly with k.

4.1   No Broadcast Control ($\gamma = 0$)

We first prove that the expected centroid of the agents' constellation does not move in time (see also [3]).

Let us analyse the long term behaviour of the random variable vectors $\bar{p}(k) = 1/n \sum_{i=1}^{n} p_i(k)$ in time. Let $\Delta p_i(k) = p_i(k+1) - p_i(k)$ be the step of agent i at time step k. Then, the vector $\bar{p}(k+1)$ obeys

$$\bar{p}(k+1) = \frac{1}{n} \sum_{i=1}^{n} p_i(k+1) = \frac{1}{n} \sum_{i=1}^{n} (p_i(k) + \Delta p_i(k)) = \bar{p}(k) + \frac{1}{n} \sum_{i=1}^{n} \Delta p_i(k)$$

Therefore, we have to consider the sum of the jumps the agents make at each time-step.

Let $\bar{a}r_i(k)$ be the "mean" location of the current allowable region of agent i. If agent i is not located at a corner of the system's convex-hull, i.e. we have that $\psi_i(k) \geq \pi$, it cannot jump, hence $p_i(k+1) = p_i(k)$. Otherwise, $\psi_i(k)$ is equal to $\varphi_i(k) < \pi$, the inner-angle of the convex-hull corner defined by agent i. By (1), $\bar{a}r_i(k)$ is located at the center of $ar_i(k)$.

$$\bar{a}r_i(k) = \iint_{v \in ar_i(k)} v dv = p_i(k) + \frac{\sigma}{2} \cos(\frac{\varphi_i(k)}{2}) \hat{\psi}_i(k) = p_i(k) + \frac{\sigma}{4} \left(U_i^-(k) + U_i^+(k) \right)$$

An agent i located at a corner of the convex-hull stays put with probability $1 - \delta$ and jumps with probability δ to a uniformly distributed random point in $ar_i(k)$, therefore its expected position at the next time-step is

$$\mathbb{E}(p_i(k+1)) = p_i(k)(1-\delta) + \bar{a}r_i(k)\delta = p_i(k) + \delta \frac{\sigma}{4} \left(U_i^-(k) + U_i^+(k) \right) \quad (3)$$

Assuming the indices of the agents on $\partial CH(P(k))$, the set of agents defining the convex-hull of the constellation $P(k)$, are ordered by the sequence of corners in the convex-hull. We have that the extremal left neighbour of agent $i \in \partial CH(P(k))$ is $i+1 \in \partial CH(P(k))$, and the extremal right neighbour of $i+1$ is i, i.e.

$$U_i^+(k) = -U_{i+1}^-(k)$$

Therefore, the expected position of the agent's centroid at the next time step coincides with its current position, since

$$\mathbb{E}(\bar{p}(k+1)) = \sum_i \mathbb{E}(p_i(k+1)) = \bar{p}(k) + 0 \quad (4)$$

Hence by the linearity, the expected position of the constellation's centroid is stationary once the visibility graph is complete.

Next, we rely on the Theorem from [2] described above to prove that the constellation's centroid distribution converges in probability to a distribution with projections on the x and y axes (and in fact on any direction) of normal distributions. We refer the reader to [3] for a detailed proof.

4.2   Behavior of the Swarm Under Broadcast Control ($\gamma > 0$)

We next show that, in the presence of broadcast control, the centroid of a swarm of agents acting by motion law (2) moves in the desired direction with a constant expected velocity (i.e. expected displacement per time step). Furthermore, we prove that the distribution of the system centroid at time step k, when projected on any arbitrary direction, convergences in probability to a distribution similar to that of a biased 1D random-walk variable.

Let $\hat{C}$ be the desired direction broadcast to the swarm. We next calculate the centroid of the allowable regions of the agents of the set $\partial CH(k)$ when the broadcast is received. We denote the centroid of the allowable region $ar_i^v(k)$ by $\bar{a}r_i^v(k)$.

Let $\hat{C}^\perp$ be a unit vector orthogonal to $\hat{C}$. Let S^u/S^d be the set of agents located at positions with the maximal/minimal projection on $\hat{C}^\perp$, and let u/d be an agent of the set S^u/S^d located at a position with the minimal projection on $\hat{C}$. Furthermore, let the agents from the right and left sides of the segment $[p_d(k), p_u(k)]$ be the sets R and L, see Fig. 3.

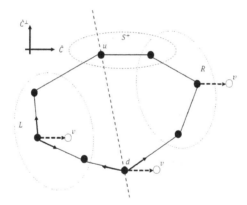

Fig. 3. The agents of the convex-hull divided into subsets based on the influence they may have when receiving the control signal. The allowable regions of agents u and d may be influenced by adding to their neighbour sets a virtual agents in the direction $\hat{C}$. The allowable regions of the agents in the set L is not affected, and while those of the agents in the set R vanish.

The allowable region of an agent i hearing the broadcast is the intersection of all the disks $D_{\sigma/2}(p_i + \frac{\sigma}{2}U_i^j(k))$, where $j \in N_i^v(k)$ (including the virtual agent v), which is equal to the intersection between the two disks associated with the extremal agents (in case $\psi_i(k) < \pi$). Hence, the allowable regions of the agents of the set L are not affected by the virtual agent v, and the allowable regions of the agents of the set R vanish because of it. Furthermore, the virtual agent is the extremal right agent of u, and extremal left agent of d.

We have already seen in (3) that the centroid of a non-surrounded agent's allowable region is located at the current position of the agent plus the sum of the two unit vectors pointing to its extremal neighbours multiplied by $\sigma/4$. Hence, assuming that an agent $i \in \partial CH(k)$ "hears" the signal, we have that the centroid of its allowable region is as follows:

$$\bar{ar}_i^v(k) = p_i(k) + \frac{\sigma}{4} \begin{cases} U_i^-(k) + U_i^+(k) & i \in L \\ 0 & i \in R \\ U_i^-(k) + \hat{C} & i = d \\ \hat{C} + U_i^+(k) & i = u \end{cases}$$

Thereby, its next expected position is

$$\mathbb{E}\{p_i(k+1)\} = (1-\delta)p_i(k) + \delta(1-\gamma)\left(p_i(k) + \frac{\sigma}{4}\left(U_i^-(k) + U_i^+(k)\right)\right) + \delta\gamma\bar{ar}_i^v(k) \tag{5}$$

Using (5), we can calculate $\mathbb{E}\{\bar{p}(k+1)\}$, the next expected centroid of the agents' constellation for any time step k, as follows:

$$\frac{1}{n}\sum_i \mathbb{E}\{p_i(k+1)\} = \bar{p}(k) + \frac{\delta\sigma}{4n}\gamma\left(\sum_{u \leq i < d} U_i^-(k) + \sum_{u < i \leq d} U_i^+(k) + 2\hat{C}\right)$$

Notice that we assume that the agents located at the convex-hull corners are marked by successive indices in a counter-clockwise increasing order, and therefore $u < d$. Using the fact that a pair of agents i and $i+1$ located at consequent corners of the convex-hull are the extremal right and left neighbours of each other, we here too have that $U_i^+(k) = -U_{i+1}^-(k)$. Hence,

$$\mathbb{E}\{\bar{p}(k+1)\} = \bar{p}(k) + \frac{\delta\sigma}{4n}\gamma\left(\sum_{u \leq i < d} U_i^-(k) - \sum_{u \leq i < d} U_i^-(k) + 2\hat{C}\right) = \bar{p}(k) + \frac{\delta\gamma\sigma}{2n}\hat{C} \tag{6}$$

This proves that a constellation of agents with a complete visibility graph moves with an expected velocity of $\frac{\delta\gamma\sigma}{2n}$ in the direction of $\hat{C}$. We verified this result in multiple simulations, the results being displayed in Fig. 4.

Let $\Delta\bar{p}(k)$, be the constellation centroid displacement at time step k, i.e. $\Delta\bar{p}(k) = \bar{p}(k+1) - \bar{p}(k)$. Let S_k and X_k be the projections of the distributions of $\bar{p}(k) - k\frac{\delta\gamma\sigma}{2n}\hat{C}$ and $\Delta\bar{p}(k) - \frac{\delta\gamma\sigma}{2n}\hat{C}$ on a unit vector with an arbitrary direction U. Then, by (6), we have that $\mathbb{E}\{\bar{X}_{k+1}|X_k\} = 0$.

Clearly, the increments X_k are uniformly bounded by $n\sigma$. We next prove that, the sum of their variances tends to infinity with probability 1. Denote the distribution of the step of an agent i at time step k by $\Delta p_i(k) = p_i(k+1) - p_i(k)$, and let $Var(A)$ be the variance of a random variable A. Then, due to the fact that any pair of random variables $\Delta p_i(k)$ and $\Delta p_j(k)$ are conditionally independent for $i \neq j$, we have that

$$Var(X_k|P(k)) = Var(\frac{1}{n}U^\mathsf{T}\sum_i \Delta p_i(k) \geq \frac{1}{n^2}Var(U^\mathsf{T}\Delta p_s(k)|P(k))$$

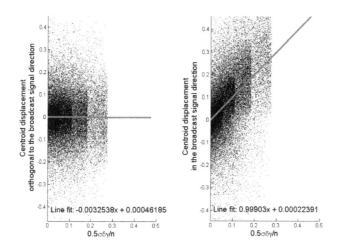

Fig. 4. The displacement of a swarm centroid vs $\frac{\delta\gamma\sigma}{2n}$. Results of 6 batches of 10000 simulations. Each batch ran with fixed number of agents $[2, 3, 5, 10, 15, 30]$, and each simulation ran with homogeneous distributed random values of $0 < \gamma < 1$ and $0 < \delta < 1$. The results are compatible with the theoretic expected velocities $\delta\gamma/(2n)$. Note that the 6 band like patterns are the result of the fixed number of agents in each simulation batch.

where $P(k) = \{p_1(k), p_2(k), ..., p_1(k)\}$, and s is the agent located at the sharpest corner occupied by an agent from the set $\{L, u, d\}$.

The minimal value $Var(U^{\mathsf{T}}\Delta p_s(k))$ can assume is for a unit vector U orthogonal to $\hat{\psi}_s(k)$, i.e. $U^{\mathsf{T}}\hat{\psi}_s(k) = 0$. Calculating it yields:

$$Var(U^{\mathsf{T}}\Delta p_s(k)) \geq \delta^2 \left(\frac{\sigma}{2}\right)^2 \frac{1 - \cos^4\left(\frac{\pi - \psi_s(k)}{2}\right)}{\frac{\pi - \psi_s(k)}{2} - \frac{1}{2}\sin(\pi - \psi_s(k))}$$

Recall that u and d are the agents with the maximal and minimal projections on $\hat{C}^{\perp}$, and L is the set of agents on the convex-hull from the left side of the line crossing through the positions of agents u and d. Denote the virtual agents of u and d by v_u and v_d. Then considering the convex-hull of positions of the set $\{L, u, d, v_u, v_d\}$, we have that the sum of its inner angles is $\pi(m+2-2)$, assuming the cardinality of this set is $m + 2$. Furthermore, the sum of inner-angles of the convex hull corners without those associate with v_u and v_d is $\pi(m + 2 - 2) - \pi$, and their average value is $\pi(m-1)/m = \pi(1-1/m)$. Let n be the number of real agents in the system, then we have that $\pi(1-1/m) \leq \pi(1-1/n) = \varphi_*$. Therefore, $\psi_s(k)$, the angle of the sharpest corner of that convex-hull, without referring the corners associated with the virtual agents, is, clearly, upper bounded by φ_*. Hence, we have that

$$Var(X_k) \geq Var(U^{\mathsf{T}}\Delta p_s(k)) \geq \delta^2 \left(\frac{\sigma}{2}\right)^2 \frac{1 - \cos^4\left(\frac{\pi - \varphi_*}{2}\right)}{\frac{\pi - \varphi_*}{2} - \frac{1}{2}\sin(\pi - \varphi_*)} \triangleq Var^*$$

and as a consequence $\sum_{k=1}^{\infty} Var\{X_k\} \to \infty$. Then, from Theorem 35.11 in [2], we have that $\frac{S_{\tilde{k}}}{\sqrt{\nu}} \xrightarrow{p} N(0,1)$. where $\tilde{k} = \min\{t : \sum_{k=1}^{t} Var(X_k) > \nu\}$ define as the stopping time as ν goes to infinity.

Recall that all the variables X_k are uniformly bounded by $n\sigma$. Hence, their variances are bounded by $(n\sigma)^2$, and we may assume that the mean value of the increments X_ks' variances converges to a finite constant value η^2, i.e. $\frac{1}{k}\sum_k Var(X_k) \to \eta^2$. Then, we have that

$$\frac{S_\nu}{\sqrt{\nu}\eta} \xrightarrow{p} N(0,1)$$

i.e. the distribution of S_n converges in probability to the distribution of a random-walk with steps of the size η, so that the projection of the random vector $\bar{p}(k) - \frac{\delta\gamma\sigma}{2n}$ on an arbitrary (constant) direction U converges to a normal distribution with variance $k\eta^2$.

We ran multiple simulations, and used the results to estimate η. Interestingly, the average random step size, η, is best fit to the following model:

$$\eta \propto \frac{1}{n(1+\gamma)}$$

Note that in [3], we have seen that without the control η is inversely dependent on the numbers of agents. In the guided case the centroid of the swarm behaves is as if each agent adds its own virtual agent to the constellation. This results and the analysis can be found in the technical report [4].

5   Discussion

We showed that a flock/swarm of identical, anonymous and oblivious agents having limited visibility and bearing only sensing with an initial constellation having connected visibility graph, can be directed by exogenous control to move in desired directions. This may be achieved via a simple broadcast control mechanism. Furthermore, we showed that the random constellation's centroid moves in the desired direction with a speed determined by the probability that agents "hear" the broadcast control. In fact, centroid of the flock performs a motion that is similar in probability to a biased random walk, biased in the direction given by the guidance vector. It would be interesting to also consider other gathering processes and augment them by an exogenous control/guidance law of the type considered herein. Also it would be very useful to use such ideas for the design of controlled autonomous flexible but cohesive swarms to be deployed in a variety of surveillance and patrol tasks.

References

1. Bellaiche, L.I., Bruckstein, A.: Continuous time gathering of agents with limited visibility and bearing-only sensing. Swarm Intell. **11**(3–4), 271–293 (2017)
2. Billingsley, P.: Probability and Measure. Wiley Series in Probability and Mathematical Statistics. Wiley, New York (1995)
3. Bruckstein, A., Manor, R.: Discrete time gathering of agents with bearing only and limited visibility range sensors. CIS Technical report, TASP (2017). http://www.cs.technion.ac.il/users/wwwb/cgi-bin/tr-get.cgi/2017/CIS/CIS-2017-01.pdf
4. Bruckstein, A., Manor, R.: Guidance of swarms with agents having bearing only and limited visibility sensors, CIS Technical report, TASP (2017, Submitted to IEEE Intelligent systems). http://www.cs.technion.ac.il/users/wwwb/cgi-bin/tr-get.cgi/2017/CIS/CIS-2017-03.pdf
5. Çelikkanat, H., Şahin, E.: Steering self-organized robot flocks through externally guided individuals. Neural Comput. Appl. **19**(6), 849–865 (2010)
6. Cucker, F., Huepe, C.: Flocking with informed agents. Math. Action **1**(1), 1–25 (2008)
7. Ferrante, E., Turgut, A.E., Mathews, N., Birattari, M., Dorigo, M.: Flocking in stationary and non-stationary environments: a novel communication strategy for heading alignment. In: Schaefer, R., Cotta, C., Kołodziej, J., Rudolph, G. (eds.) PPSN 2010. LNCS, vol. 6239, pp. 331–340. Springer, Heidelberg (2010). https://doi.org/10.1007/978-3-642-15871-1_34
8. Gauci, M., Chen, J., Dodd, T.J., Groß, R.: Evolving aggregation behaviors in multi-robot systems with binary sensors. In: Ani Hsieh, M., Chirikjian, G. (eds.) Distributed Autonomous Robotic Systems. STAR, vol. 104, pp. 355–367. Springer, Heidelberg (2014). https://doi.org/10.1007/978-3-642-55146-8_25
9. Gauci, M., Chen, J., Li, W., Dodd, T.J., Gross, R.: Self-organized aggregation without computation. Int. J. Robot. Res. **33**(8), 1145–1161 (2014)
10. Gazi, V., Passino, K.M.: Stability analysis of social foraging swarms. IEEE Trans. Syst. Man Cybern. Part B: Cybern. **34**(1), 539–557 (2004)
11. Gordon, N., Wagner, I.A., Bruckstein, A.M.: A randomized gathering algorithm for multiple robots with limited sensing capabilities. In: Proceedings of MARS 2005 workshop at ICINCO Barcelona (2005)
12. Johnson, M., Brown, D.: Evolving and controlling perimeter, rendezvous, and foraging behaviors in a computation-free robot swarm. In: Proceedings of the 9th EAI International Conference on Bio-inspired Information and Communications Technologies (formerly BIONETICS), pp. 311–314. ICST (Institute for Computer Sciences, Social-Informatics and Telecommunications Engineering) (2016)
13. Manor, R., Bruckstein, A.M.: Chase your farthest neighbour. In: Groß, R., Kolling, A., Berman, S., Frazzoli, E., Martinoli, A., Matsuno, F., Gauci, M. (eds.) Distributed Autonomous Robotic Systems. SPAR, vol. 6, pp. 103–116. Springer, Cham (2018). https://doi.org/10.1007/978-3-319-73008-0_8
14. Olfati-Saber, R.: Flocking for multi-agent dynamic systems: algorithms and theory. IEEE Trans. Autom. Control. **51**(3), 401–420 (2006)
15. Pimenta, L.C., Michael, N., Mesquita, R.C., Pereira, G.A., Kumar, V.: Control of swarms based on hydrodynamic models. In: IEEE International Conference on Robotics and Automation, ICRA 2008, pp. 1948–1953. IEEE (2008)
16. Pimenta, L.C., et al.: Swarm coordination based on smoothed particle hydrodynamics technique. IEEE Trans. Robot. **29**(2), 383–399 (2013)

17. Rahmani, A., Ji, M., Mesbahi, M., Egerstedt, M.: Controllability of multi-agent systems from a graph-theoretic perspective. SIAM J. Control. Optim. **48**(1), 162–186 (2009)
18. Segall, I., Bruckstein, A.: On stochastic broadcast control of swarms. In: Dorigo, M., Birattari, M., Li, X., López-Ibáñez, M., Ohkura, K., Pinciroli, C., Stützle, T. (eds.) ANTS 2016. LNCS, vol. 9882, pp. 257–264. Springer, Cham (2016). https://doi.org/10.1007/978-3-319-44427-7_23
19. Yu, C.H., Werfel, J., Nagpal, R.: Collective decision-making in multi-agent systems by implicit leadership. In: Proceedings of the 9th International Conference on Autonomous Agents and Multiagent Systems-Volume 3, pp. 1189–1196. International Foundation for Autonomous Agents and Multiagent Systems (2010)

Hybrid Control of Swarms
for Resource Selection

Marco Trabattoni[1]([⊠]) [iD], Gabriele Valentini[2] [iD], and Marco Dorigo[1] [iD]

[1] IRIDIA, Université Libre de Bruxelles, Brussels, Belgium
{mtrabatt,mdorigo}@ulb.ac.be
[2] School of Earth and Space Exploration, Arizona State University, Tempe, AZ, USA
gvalentini@asu.edu

Abstract. The design and control of swarm robotics systems generally relies on either a fully self-organizing approach or a completely centralized one. Self-organization is leveraged to obtain systems that are scalable, flexible and fault-tolerant at the cost of reduced controllability and performance. Centralized systems, instead, are easier to design and generally perform better than self-organizing ones but come with the risks associated with a single point of failure. We investigate a hybrid approach to the control of robot swarms in which a part of the swarm acts as a control entity, estimating global information, to influence the remaining robots in the swarm and increase performance. We investigate this concept by implementing a consensus achievement system tasked with choosing the best of two resource locations. We show (i) how estimating and leveraging global information impacts the decision-making process and (ii) how the proposed hybrid approach improves performance over a fully self-organizing approach.

1 Introduction

Swarm robotics is a promising approach to the design and control of systems composed of large numbers of embodied agents [9]. Robot swarms have shown potential for solving tasks which are deemed too dangerous or too demanding for humans, such as search and rescue, de-mining, underwater surveillance or environment patrolling. Inspired by nature [3,5], robot swarms are generally designed and controlled through the principles of self-organization with the aim to obtain systems that are flexible, fault-tolerant and scalable [4,9]. Typically, robot swarms do not have a leader, do not use global information, and are highly redundant thanks to a large number of constituent robots. Robots in a swarm rely on local sensing and communication to solve the tasks they are given. Having a large number of robots acting in an unsupervised manner, however, often results in a system that is hard to control and/or to predict and whose performance can vary greatly over a same task.

Centralized control, on the opposite, relies on a control entity with access to global information and with the authority to correct the behavior of the system to reach the desired goal. In general, centralized systems are easier to manage and

© Springer Nature Switzerland AG 2018
M. Dorigo et al. (Eds.): ANTS 2018, LNCS 11172, pp. 57–70, 2018.
https://doi.org/10.1007/978-3-030-00533-7_5

predict than self-organizing ones and often achieve better performances. Centralized approaches to the control of large groups of robots rely on a central entity, for example to provide the robots with directives regarding the task to execute, the motions required, or information about the position of objects of interest in the environment [15,17,40]. While centralized control provides us with more manageability over the system, as well as a more stable and trusted performance, the presence of a centralized entity in charge of controlling the functioning of the whole system reduces parallelism and scalability and introduces a single point of failure.

In this paper we investigate a different approach to the control of a robot swarm that we refer to as Swarm Hybrid Control System (SHCS). SHCS combines localized elements of centralized control with self-organizing behaviors performed by the remaining elements of the system with the aim to obtain the best of both design approaches. In our approach, the control authority is not an entity external to the swarm; rather, it consists of a group of robots of the swarm which cooperate in a self-organizing way to provide services akin to those of a central authority. In this way, we are able to exploit the advantages associated to a central authority without introducing a single point of failure into the system. The control entity is thus a formation of robots, created through a self-organizing process, that exchanges information locally to obtain an estimate of the global state of the system and that uses this information to influence the future behavior of the swarm. We investigate this idea by implementing a SHCS for a problem of consensus achievement.

Consensus achievement is a common problem that robot swarms are required to solve in many different application scenarios (e.g., to choose which area to explore in a de-mining scenario or which target requires the most attention in a search and rescue situation). Also known as the best-of-n problem [38], it requires the swarm to choose the best option over a set of n available possibilities which (generally) differ in their quality and cost. The problem of consensus achievement for a robot swarm has been studied in many different application scenarios and modeled with a variety of mathematical tools (e.g., ODE [19], chemical master equations [39]). Additionally, various decision-making strategies have been proposed to address this problem, most of which take inspiration from nature [30]. We consider a binary resource-selection scenario, in which the swarm is foraging between a central location (the *nest*) and two locations (*sources*) containing resources that have the same quality but different costs in terms of time necessary to collect/extract them. That is, the cost of a resource location corresponds to the time required by a robot to collect resources from that specific location. For example, robots might be collecting minerals buried underground and the cost may represent how deep the robot needs to dig to reach the minerals. The scenario we have chosen is a binary consensus achievement problem with indirect modulation of robots opinions resulting from the different cost associated to each resource location [35], in which robots alternate between foraging from their preferred source and disseminating their preference in the nest. Before returning to forage, robots pool the opinions of their neighbors and apply a decision rule (either the majority rule or the voter model) to decide whether or not to change their current preference.[1]

A well-mixed state of robots' opinions is generally assumed to be one of the condition necessary to address distributed decision-making problems [24]. Well-mixed systems are systems in which each robot in the swarm has the same probability to interact with any other robot in the swarm. The necessity for the robots to be well-mixed when disseminating is due to their limited interaction range which limits the information they can perceive about the opinions of other robots. Poor mixing of robots' opinions may result in the fragmentation of the system in parties with contrasting opinions and prevent the achievement of consensus. While robots, when disseminating their opinion, are usually programmed to move randomly in the environment for an amount of time sufficiently long to properly mix inside the swarm [39], random motion does not guarantee that the resulting system will be well-mixed. Moreover, increasing the amount of time that the robots spend disseminating (and thus mixing) their opinions increases the overall duration of the decision-making process as well. In our implementation, the SHCS collects information about the opinions in the swarm through local interactions, and merges them in order to obtain an estimate of the global state of the system in the form of a database of robot opinions. By giving the rest of the swarm access to this information, the SHCS tries to approximate the information that robots would have access to in a well-mixed system. We show the potential of this idea by comparing the SHCS with a fully self-organizing approach over the same task.

The remaining of this paper is organized as follows. In Sect. 2, we discuss related work. In Sect. 3, we describe the chosen decision-making scenario and the controllers of the robots for both the self-organizing approach and the SHCS one. In Sect. 4, we present the results of our experiments performed in simulation. In Sect. 5, we discuss the effect of the SHCS based on our experimental results. Finally, in Sect. 6 we draw our conclusions and discuss our future directions of research.

2 Related Work

2.1 Control of Robot Swarms

Brambilla et al. [4] reviewed the literature of swarm robotics focusing on self-organizing approaches and proposed a taxonomy summarizing different design and analysis methodologies adopted in the field. Most of these design methods are bottom-up approaches in which the controller of each single robot is iteratively refined in order to obtain a desired behavior of the swarm as a whole. Recently, different design methods have been proposed to automatically derive the robot controllers for a given task. Trianni et al. [34] use a generational evolutionary algorithm to evolve robot controllers for a clustering behavior. Francesca et al. [11] proposed AutoMoDe, an approach to automatically generate modular

---

[1] In the paper we use the terms 'robot opinion' and 'robot preference' interchangeably.

control software in the form of probabilistic finite state machines, starting from a set of predefined atomic behaviors and conditional state transitions through an optimization process. Bottom-up approaches have been used to program a number of different robot swarm behaviors: pattern formation behaviors, aimed at distributing the swarm in space according to desired properties [21–23,31,32]; navigation behaviors, aimed at coordinating the movement of the swarm in the environment [10]; and collective decision-making behaviors, in which the swarm has to take a decision about how to distribute its components (i.e., the robots) among different tasks [6] or which option to unanimously choose [30].

Centralized methods for the control of multi robot systems have also been proposed, in particular for navigation problems, such as deployment of robots in cooperative surveillance [33], target tracking [14], path planning [1,28], or formation control [8]. The purpose of central control can vary between different tasks, but generally it includes calculating the motion plans for the single robots, allowing the robots to localize themselves by sensing and providing global information, or simply providing updated mission goals [41]. Some approaches can be found in which a distributed swarm behavior also relies on an external control entity to initiate or correct its functioning, such as in the work of Berman et al. [2], where a central unit broadcasts updated transition parameters for task allocation.

One notable exception to the above-mentioned approaches where the control is either fully self-organizing or centralized, is the recent work by Mathews et al. [18]. In this work, robots in a swarm are able to physically merge into a single entity, named a 'mergeable nervous system robot' (MNS-robot for short), comprising one single brain robot which acts as central controller for the robot aggregate. While both our work and the one of Mathews et al. share the idea of a centralized form of control internal to the swarm, the MNS aims at obtaining swarms able to morphologically adapt to the task of interest, while our focus is on designing a swarm able to monitor and influence its own behavior so as to increase its performance.

2.2   Consensus Achievement

Consensus achievement is one of the two branches of collective decision-making, the other being task allocation [4], and refers to the problem of having a robot swarm select a single option among different alternatives to maximizes the benefits of the swarm [35]. Many scenarios have been proposed by the community, mostly inspired by biological systems such as ants choosing the shortest path connecting a pair of locations [7], or honeybees collectively selecting the best site for relocation of the swarm [25]. Montes de Oca et al. [19] proposed a collective decision-making strategy based on the majority rule and the concept of latent voters (i.e. after updating their opinions, agents do not take part in the decision making process for a stochastic amount of time) first described by Lambiotte et al. [16]. We utilize a similar concept in our scenario: after updating their opinion, agents enter a latent phase during which first they forage from the source indicated by their opinion and then disseminate their opinion to other robots. Valentini et al. [38] reviewed the best-of-n problem for robot swarms in all of its variations, proposing two taxonomies to classify the literature, one based on the relation between cost and quality of each option, and one based on the design approaches. Despite the variety of methods proposed for consensus achievement problems in robot swarms [12,13,29], to our knowledge, our work is the first one that proposes to use an emerging control entity to estimate and leverage global information to influence the collective decision-making process.

3 Methods

3.1 Experimental Setup

We consider a binary resource selection problem for a robot swarm performing a foraging task. We define an environment consisting of an arena of size 200×100 cm^2 divided into three areas: a nest (80×100 cm^2) positioned in the center of the arena, and two resource locations (60×100 cm^2 each) on each side of the nest. These locations, called source A and source B, have different costs σ_A and σ_B, with $\sigma_A < \sigma_B$ in our experiments. The cost of a resource location reflects the time required to collect resources from that source, representing features such as how deep a robot would have to dig for minerals, or how far the source is from the central nest. Two light sources are positioned on one side of the arena in order to provide the robots with a light gradient and to enable them to navigate the environment. The robots are initially placed in the nest and have an initial opinion for a preferred source when the experiment starts. Initially, robots in the swarm are equally split among the two options. Robots perform the foraging task by collecting resources from their currently preferred source, and then returning to the nest. Robots in the nest can change their opinion based on the opinions of neighboring robots by applying a decision mechanism. The goal of the swarm is to achieve consensus on the best source (which is always source A in our experiments).

We implemented this scenario using the ARGoS3 simulator [27] and the ARGoS3 Kilobot plug-in [26]. Figure 1a shows a view of the environment and of the swarm of Kilobots implemented inside the simulation, where source A and source B are represented, respectively, by the blue and red areas. The Kilobot [31] is a low cost and small size (3.3 cm diameter) autonomous robot. It is able to communicate with other Kilobots at a distance of up to 10 cm via infrared communication, to sense ambient light, and to move by means of 2 vibrating motors and 3 rigid legs. By means of an ARGoS loop function, we provide the Kilobots with the ability to detect whether or not they are in close proximity of a wall, in which area of the environment they currently are (i.e., nest, source A, source B), and, in case they are in one of the two sources, the source quality.

3.2 Self-organizing Behavior

We implement the self-organizing behavior with indirect modulation of the latent phase in the decision-making process [35]. In this phase the robots alternate between dissemination and exploration. During the exploration phase, robots forage from their preferred resource location for a time drawn from a normal distribution with mean $g \cdot \sigma_i$ and standard deviation $g/10$, where $\sigma_A = 1$ and $\sigma_B \geqslant 1$ are the costs, respectively, of source A and source B. In the dissemination phase, robots broadcast their current opinion inside the nest and listen to the opinions of neighboring robots for a time drawn from an exponential distribution with mean q; differently from the exploration time, the dissemination time is not modulated. At the end of the dissemination phase, the robots apply a decision rule on a set of opinions containing the last G opinions received from their neighbors with the aim to decide whether or not to switch their current opinion. After that, robots enter the exploration phase. We implemented two decision rules: the voter model, where a robot changes its opinion to the one of a randomly selected neighbor, and the majority rule, where a robot selects its opinion to be the one of the majority of its neighbors.

During both the dissemination and the exploration phases, robots move randomly, by alternating periods of straight motion with periods of rotating motion. Forward motion lasts for an amount of time drawn from a normal distribution with mean 20 s and standard deviation 5 s while the rotation motion lasts for an amount of time drawn from a normal distribution with mean 3 s and standard deviation 0.5 s. Additionally, when robots move closer than 5 cm to the edges of the arena, they perform wall avoidance by turning on the spot in a random direction and then moving forward. Between dissemination and exploration, robots have to move from the nest to the foraging sites and vice-versa. To do so, robots perform a gradient-following routine, by sensing the light intensity received from the light sources. Robots following the light gradient move forward while keeping track of the minimum and maximum light intensities sampled in intervals of 5 s. If a robot detects that it is not following the light gradient in the desired direction, it turns on the spot (using the same parameters as the random walk rotation) and then moves forward again, until it finds the correct direction of motion. Robots always show their current opinion by switching their on-board LED to the color of their preferred source.

Because of the shorter time required to forage from the source with lower cost, robots foraging from the best source will return to the nest more frequently and have more chances to disseminate their opinion in the nest: this results in a higher chance for their opinion to be observed from other robots as they apply the decision rule which biases the swarm towards consensus for the best option. The swarm is thus able to slowly achieve consensus on the best source as robots repeat the exploration-dissemination-decision rule cycle. In the following, we will refer to robots performing the behavior described in this section as SO robots.

3.3   SHCS Implementation

In our hybrid implementation, we introduce a second behavior in addition to the self-organizing one described in the previous section. Robots of the swarm can be either part of the control entity (SHCS robots) or be SO robots. Moreover,

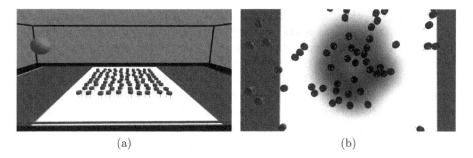

Fig. 1. View of the environment and Kilobot swarm implemented with the ARGoS3 simulator (a) and of the SHCS during a simulated experiment (b). The shaded area shows the communication range of the SHCS considered as a whole. SHCS robots show their LED in green (seed robot) or white (remaining SHCS robots); robots showing blue and red LEDs are SO robots, with the color representing their current opinion. (Color figure online)

robots can switch between these two modalities. At the beginning of the experiment, the swarm allocates its workforce between SHCS robots and SO robots. To do so, the robots select a seed robot around which they start an aggregation process to form the SHCS entity. The seed robot is selected through a self-exclusion process starting with a connected swarm[2] placed in the nest. The connectivity requirement strongly reduces the probability of selecting multiple robot seeds. Each robot spends the first 10 s of the experiment turning on the spot and sampling light values. Then, for the next 10 s, robots broadcast the minimum and maximum light measurement perceived in the swarm, initially set to their own perceived value and later updated according to the received messages. Additionally, robots also broadcast a randomly generated number between 0 and 255. Robots who find themselves outside of a 10% range from the mean value of the light perceived by the swarm (based on information received from neighbors) exclude themselves from the selection process and become SO robots. The purpose of this initial procedure is to obtain a selection of candidate seed robots that is positioned at an intermediate distance from the light source. Then, these candidate seed robots compare their own randomly generated number with those received from their neighbors and, if they receive a lower value, they exclude themselves from the process and become SO robots. After an additional 10 s, all remaining robots in the process become SHCS robots. The aim of this final part of the procedure is to maximize the probability to select a single seed robot. The total procedure to select the seed robot requires abound 30 s.

SHCS robots, initially represented by the sole seed robot selected with the above procedure, maintain a representation of their position h inside the aggregate in a manner similar to that of Nagpal et al. [20], and share this value with their neighbors as part of a *heartbeat* protocol. The seed robot has a posi-

---

[2] A swarm is connected if a path of communicating robots can be found between any two robots in the swarm.

tion $h = 0$. All other SHCS robots in the aggregate set their position h to $h = h'_{min} + 1$ where h'_{min} is the minimum position received from neighboring SHCS robots. In our experiments, we limit the size of the SHCS aggregate by imposing a maximum position $h = 3$, that is, 3 levels of SHCS robots surrounding the seed robot. SO robots that perceive SHCS robots join the SHCS aggregate with probability $p = \frac{0.1}{h+1}$ if $h \in \{0, 1, 2\}$, where h is the position of the SHCS robot broadcasting the message. If the perceived position of the SHCS robot broadcasting the message is $h = 3$, SO robots do not join the aggregate. Once joined the SHCS aggregate, robots estimate their distance from neighboring SHCS robots by measuring the intensity of the infrared signal of received messages. If a SHCS robot with position h is too close (i.e., distance $<40\,\mathrm{mm}$) or too far (i.e., distance $>70\,\mathrm{mm}$) from his neighbors at position $h' = h - 1$, the SHCS robot will try to reposition itself at a favorable distance by moving in a random direction while it does not move otherwise. SHCS robots may lose connectivity from the aggregate during repositioning or due to collisions with other robots. If an SHCS robot loses connectivity for more than 10 s, it becomes an SO robot. This process allows the SHCS aggregate to initially form around the seed robot in a distributed manner and to maintain a stable dimension robust to connectivity failures. Figure 1b shows a top-view of the SHCS aggregate and its communication range during a simulated experiment.

SHCS robots continuously broadcast a *heartbeat* message with the aim (i) to maintain a database of the last 30 source preferences received from SO robots and (ii) to use this database to influence the preference of SO robots. A heartbeat message is composed of the id of the sending SHCS robot, its position h, a robot preference taken from its database, and a decision-making outcome. Whenever an SHCS robot receives a new opinion, either from a heartbeat message or from an SO robot, it adds the received opinion to its database (in a first-in first-out manner) and sets this opinion as the robot preference to share in the heartbeat message. SHCS robots generate a new decision-making outcome each time they send a new heartbeat; to do so, they use either the majority rule or the voter model applied to a set of G preferences randomly selected from their database.

SO robots behave as described in Sect. 2.2 except when receiving a heartbeat message. In this case, if a SO robot is in the dissemination phase, it immediately changes its opinion to match that contained in the decision-making outcome of the heartbeat message, terminates the dissemination phase, and returns to the foraging task. This mechanism improves the efficiency of the swarm as SO robots spend more time foraging and less time disseminating their opinions.

4   Experiments

We perform a series of simulation experiments to compare the hybrid control system (SHCS) approach with the fully self-organizing (SO) approach. In our experiments, we keep the cost of source A constant to $\sigma_A = 1$ and vary the cost of source B in $\{1.11, 1.25, 1.43, 1.67, 2\}$. We use a swarm of 100 robots of which 50 have initial opinion A and 50 have initial opinion B. The mean duration of

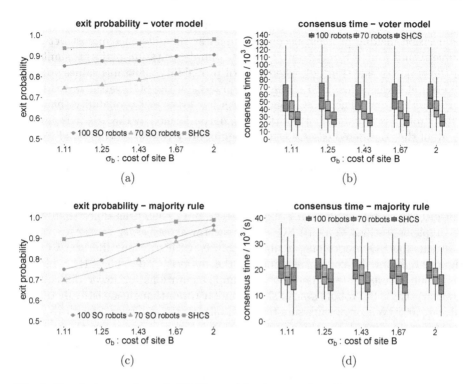

Fig. 2. Simulation results with SHCS, 100 SO robots, and 70 SO robots for varying σ_B: exit probability (a) and consensus time (b) for the voter model; exit probability (c) and consensus time (d) for the majority rule. Results obtained running 1000 simulations for each tested condition.

the dissemination and exploration phases is set, respectively, to $q = 300\,s$ and $g = 600\,s$. We test two decision rules, the majority rule and the voter model, with a group size of $G = 3$ preferences. We perform 1000 simulation runs for each value of σ_B for both the SO approach and the SHCS one. We consider two metrics: exit probability, computed as the proportion of simulations converging to a consensus for source A, and the mean consensus time, computed over all simulations. Since the average SHCS size during experiments was approximately 30 robots, we performed an additional set of experiments implementing the SO approach with a swarm of $100 - 30 = 70$ robots, in order to compare the performance of the SHCS with the SO approach over a similar number of SO robots actively pursuing the foraging task.

Figure 2 shows the exit probability and mean consensus time obtained with the three implementations (SHCS for a swarm of 100 robots, 100 SO robots, 70 SO robots) for the two tested decision rules: voter model and majority rule.

Figure 2a shows the exit probability for the voter model. The SHCS implementation maintains a value above 0.95 for all of the considered values of σ_B while the SO implementations are considerably worse. The accuracy for all the three systems increases as the cost of source B increases. This is because the decision-making problem becomes simpler as the difference in cost between source A and source B increases. The SHCS implementation performs similarly for the majority rule (Fig. 2c), where its exit probability maintains values around 0.9 even at lower values of σ_B; the performances of the SO swarms instead are significantly worse. The 70 SO robot swarm has lower exit probability than the 100 SO robot swarm, and both of them are outperformed by the SHCS approach for all of the considered cases. Overall, the majority rule when compared to the voter model obtains a higher exit probability for the easier cases and a lower exit probability for the more difficult ones, in agreement with what reported in previous works [39].

Figure 2b shows the consensus time for the voter model. The SHCS shows significantly ($p < .001$, Wilcoxon rank-sum test) lower consensus times than both SO implementations. The 70 SO robots swarm shows lower consensus times than the 100 SO robots one, again coherently with previous literature work. Figure 2d shows the consensus time for the majority rule. The SHCS implementation results faster than both SO implementations for lower difficulties; however, the 70 SO robots swarm shows similar (even though statistically different, $p < .001$, Wilcoxon rank-sum test) consensus time at higher difficulties. The consensus time for all the three implementation slowly decreases as the cost σ_B increases, for both decision rules. Overall, the majority rule shows a significantly ($p < .001$, Wilcoxon rank-sum test) lower consensus time than the voter model, resulting in a speed vs accuracy trade-off between the two decision rules [37].

5   Discussion

The results of our experiments show the potential of the SHCS approach which is able to improve the performance of a fully self-organizing robot swarm in a collective decision-making problem. The SHCS approach leverages information regarding the global state of the opinions in the swarm to influence the individual decisions of SO robots. This results in a higher accuracy of the swarm in terms of the probability to choose the best resource location compared to the accuracy of the SO swarm (Figs. 2a and c). Additionally, the SHCS speeds up the decision-making process by allowing robots to terminate the dissemination phase as soon as they get in contact with the SHCS aggregate, since the dissemination of their opinion is performed by the SHCS. The faster convergence to a collective decision shown in Figs. 2b and d derives from a combination of the shorter dissemination phase and the more accurate information provided by the SHCS robots. In future work, we intend to investigate the extent of the contribution of each of the two mechanisms.

One may conjecture that the difference in performance between the SHCS approach and the fully SO approach is due to the fact that the SHCS swarm is actually relying on a smaller swarm size to actively perform the decision-making task. In our experiments, we measured an average of 30 robots composing the SHCS aggregate, leaving 70 SO robots to perform the self-organizing behavior.

However, the results obtained with a swarm of 70 SO robots are significantly different and of lower quality than those obtained with the SHCS approach. These results rule out the above conjecture that the difference in number of SO robots is responsible for the different performances between the SHCS and the SO approach.

It should be noted that in our experiments we use constant probabilities for SO robots to join the SHCS and limit the SHCS to three levels, preventing the SHCS from extending to the entire swarm. However, it would be interesting to extend this approach to include perceived features in the environment, for example by changing the probabilities with which SO robots join the SHCS depending on ambient light values, in order to obtain a more dynamic system.

6   Conclusions and Future Work

In this paper, we proposed a new control strategy for a robot swarm based on a combination of centralized information and self-organized behaviors. We called this control strategy Swarm Hybrid Control System (SHCS) and we investigated this idea with a preliminary implementation of the SHCS approach for a problem of consensus achievement in a binary resource-selection scenario. Our system is characterized by a control entity, having the form of an aggregate of SHCS robots and arising through a self-organizing process, with the purpose to estimate information about the global state of the swarm and to use this information to influence the collective decision-making process. We have shown how, for both the majority rule and the voter model, our system is able to outperform the fully self-organizing approach by achieving a shorter consensus time while providing higher accuracy of the collective decision in terms of exit probability. In the near future, we plan to implement the consensus achievement scenario presented in this paper on a real swarm of 100 Kilobots by leveraging the potential of a $2\,m^2$ Kilogrid system [36].

As future work, we are interested in investigating how the proportion of SHCS robots, the shape of the SHCS or the usage of multiple smaller SHCS each controlling a portion of the swarm can impact the performance of the system, as well as how our control approach can be applied to different scenarios, such as task allocation and pattern formation. We also intend to investigate whether automatic design techniques can be used to generate the controllers for the robots of our hybrid swarm.

Acknowledgements. Gabriele Valentini acknowledges support from the NSF grant No. PHY-1505048. Marco Dorigo acknowledges support from the Belgian F.R.S.-FNRS, of which he is a Research Director. The work presented in this paper was partially supported by the FLAG-ERA project RoboCom++ and by the European Research Council (ERC) under the European Union's Horizon 2020 research and innovation programme (grant agreement number 681872).

References

1. Antonelli, G., Chiaverini, S.: Kinematic control of platoons of autonomous vehicles. IEEE Trans. Robot. **22**(6), 1285–1292 (2006)
2. Berman, S., Halasz, A., Hsieh, M., Kumar, V.: Optimized stochastic policies for task allocation in swarms of robots. IEEE Trans. Robot. **25**(4), 927–937 (2009)
3. Bonabeau, E., Dorigo, M., Theraulaz, G.: Swarm Intelligence: From Natural to Artificial Systems. Oxford University Press, New York (1999)
4. Brambilla, M., Ferrante, E., Birattari, M., Dorigo, M.: Swarm robotics: a review from the swarm engineering perspective. Swarm Intell. **7**(1), 1–41 (2013)
5. Brutschy, A., Scheidler, A., Ferrante, E., Dorigo, M., Birattari, M.: "Can ants inspire robots?" Self-organized decision making in robotic swarms. In: 2012 IEEE/RSJ International Conference on Intelligent Robots and Systems (IROS), pp. 4272–4273. IEEE Press (2012)
6. Brutschy, A., Pini, G., Pinciroli, C., Birattari, M., Dorigo, M.: Self-organized task allocation to sequentially interdependent tasks in swarm robotics. Auton. Agents Multi-Agent Syst. **28**(1), 101–125 (2014)
7. Campo, A., Gutiérrez, Á., Nouyan, S., Pinciroli, C., Longchamp, V., Garnier, S., Dorigo, M.: Artificial pheromone for path selection by a foraging swarm of robots. Biol. Cybern. **103**(5), 339–352 (2010)
8. De La Cruz, C., Carelli, R.: Dynamic modeling and centralized formation control of mobile robots. In: IECON 2006–32nd Annual Conference on IEEE Industrial Electronics, pp. 3880–3885. IEEE (2006)
9. Dorigo, M., Birattari, M., Brambilla, M.: Swarm robotics. Scholarpedia **9**(1), 1463 (2014)
10. Ferrante, E., Turgut, A.E., Huepe, C., Stranieri, A., Pinciroli, C., Dorigo, M.: Self-organized flocking with a mobile robot swarm: a novel motion control method. Adapt. Behav. **20**(6), 460–477 (2012)
11. Francesca, G., Brambilla, M., Brutschy, A., Trianni, V., Birattari, M.: AutoMoDe: a novel approach to the automatic design of control software for robot swarms. Swarm Intell. **8**(2), 89–112 (2014)
12. Francesca, G., Brambilla, M., Trianni, V., Dorigo, M., Birattari, M.: Analysing an evolved robotic behaviour using a biological model of collegial decision making. In: Ziemke, T., Balkenius, C., Hallam, J. (eds.) SAB 2012. LNCS (LNAI), vol. 7426, pp. 381–390. Springer, Heidelberg (2012). https://doi.org/10.1007/978-3-642-33093-3_38
13. Gutiérrez, Á., Campo, A., Monasterio-Huelin, F., Magdalena, L., Dorigo, M.: Collective decision-making based on social odometry. Neural Comput. Appl. **19**(6), 807–823 (2010)
14. Hausman, K., Müller, J., Hariharan, A., Ayanian, N., Sukhatme, G.S.: Cooperative multi-robot control for target tracking with onboard sensing. Int. J. Robot. Res. **34**(13), 1660–1677 (2015)
15. King, J., Pretty, R.K., Gosine, R.G.: Coordinated execution of tasks in a multiagent environment. IEEE Trans. Syst. Man Cybern.-Part A: Syst. Hum. **33**(5), 615–619 (2003)
16. Lambiotte, R., Saramäki, J., Blondel, V.D.: Dynamics of latent voters. Phys. Rev. E **79**, 046107 (2009)
17. Lindsey, Q., Mellinger, D., Kumar, V.: Construction with quadrotor teams. Auton. Robot. **33**(3), 323–336 (2012)

18. Mathews, N., Christensen, A.L., O'Grady, R., Mondada, F., Dorigo, M.: Mergeable nervous systems for robots. Nat. Commun. **8**(1), 439 (2017)
19. Montes de Oca, M.A., Ferrante, E., Scheidler, A., Pinciroli, C., Birattari, M., Dorigo, M.: Majority-rule opinion dynamics with differential latency: a mechanism for self-organized collective decision-making. Swarm Intell. **5**(3–4), 305–327 (2011)
20. Nagpal, R., Shrobe, H., Bachrach, J.: Organizing a global coordinate system from local information on an ad hoc sensor network. In: Zhao, F., Guibas, L. (eds.) IPSN 2003. LNCS, vol. 2634, pp. 333–348. Springer, Heidelberg (2003). https://doi.org/10.1007/3-540-36978-3_22
21. Nouyan, S., Campo, A., Dorigo, M.: Path formation in a robot swarm: self-organized strategies to find your way home. Swarm Intell. **2**(1), 1–23 (2008)
22. Nouyan, S., Dorigo, M.: Chain based path formation in swarms of robots. In: Dorigo, M., Gambardella, L.M., Birattari, M., Martinoli, A., Poli, R., Stützle, T. (eds.) ANTS 2006. LNCS, vol. 4150, pp. 120–131. Springer, Heidelberg (2006). https://doi.org/10.1007/11839088_11
23. Nouyan, S., Groß, R., Bonani, M., Mondada, F., Dorigo, M.: Teamwork in self-organized robot colonies. IEEE Trans. Evol. Comput. **13**(4), 695–711 (2009). https://doi.org/10.1109/TEVC.2008.2011746
24. Nowak, M.A.: Five rules for the evolution of cooperation. Science **314**(5805), 1560–1563 (2006)
25. Parker, C.A.C., Zhang, H.: Cooperative decision-making in decentralized multiple-robot systems: the best-of-N problem. IEEE/ASME Trans. Mechatron. **14**(2), 240–251 (2009)
26. Pinciroli, C., Talamali, M.S., Reina, A., Marshall, J.A.R., Trianni, V.: Simulating Kilobots within ARGoS: models and experimental validation. In: Dorigo, M. (ed.) ANTS 2018. LNCS, vol. 11172, pp. 176–187. Springer, Heidelberg (2018)
27. Pinciroli, C., et al.: ARGoS: a modular, parallel, multi-engine simulator for multi-robot systems. Swarm Intell. **6**(4), 271–295 (2012)
28. Preiss, J.A., Honig, W., Sukhatme, G.S., Ayanian, N.: Crazyswarm: a large nano-quadcopter swarm. In: 2017 IEEE International Conference on Robotics and Automation (ICRA), pp. 3299–3304. IEEE (2017)
29. Reina, A., Dorigo, M., Trianni, V.: Towards a cognitive design pattern for collective decision-making. In: Dorigo, M., et al. (eds.) ANTS 2014. LNCS, vol. 8667, pp. 194–205. Springer, Cham (2014). https://doi.org/10.1007/978-3-319-09952-1_17
30. Reina, A., Valentini, G., Fernández-Oto, C., Dorigo, M., Trianni, V.: A design pattern for decentralised decision making. PLoS One **10**(10), e0140950 (2015)
31. Rubenstein, M., Cornejo, A., Nagpal, R.: Programmable self-assembly in a thousand-robot swarm. Science **345**(6198), 795–799 (2014)
32. Şahin, E., et al.: SWARM-BOT: pattern formation in a swarm of self-assembling mobile robots. In: 2002 IEEE International Conference on Systems, Man and Cybernetics, vol. 4, pp. 1–6. IEEE Press, Piscataway (2002)
33. Saska, M., Vonásek, V., Chudoba, J., Thomas, J., Loianno, G., Kumar, V.: Swarm distribution and deployment for cooperative surveillance by micro-aerial vehicles. J. Intell. Robot. Syst. **84**(1–4), 469–492 (2016)
34. Trianni, V., Groß, R., Labella, T.H., Şahin, E., Dorigo, M.: Evolving aggregation behaviors in a swarm of robots. In: Banzhaf, W., Ziegler, J., Christaller, T., Dittrich, P., Kim, J.T. (eds.) ECAL 2003. LNCS (LNAI), vol. 2801, pp. 865–874. Springer, Heidelberg (2003). https://doi.org/10.1007/978-3-540-39432-7_93

35. Valentini, G.: Achieving Consensus in Robot Swarms: Design and Analysis of Strategies for the Best-of-N Problem. Springer International Publishing, Cham (2017). https://doi.org/10.1007/978-3-319-53609-5
36. Valentini, G., et al.: Kilogrid: a novel experimental environment for the kilobot robot. Swarm Intell. **12**(3), 245–266 (2018)
37. Valentini, G., Brambilla, D., Hamann, H., Dorigo, M.: Collective perception of environmental features in a robot swarm. In: Dorigo, M., et al. (eds.) ANTS 2016. LNCS, vol. 9882, pp. 65–76. Springer, Cham (2016). https://doi.org/10.1007/978-3-319-44427-7_6
38. Valentini, G., Ferrante, E., Dorigo, M.: The best-of-n problem in robot swarms: formalization, state of the art, and novel perspectives. Front. Robot. AI **4**, 9 (2017)
39. Valentini, G., Ferrante, E., Hamann, H., Dorigo, M.: Collective decision with 100 Kilobots: speed versus accuracy in binary discrimination problems. Auton. Agents Multi-Agent Syst. **30**(3), 553–580 (2016)
40. Weigel, T., Gutmann, J.S., Dietl, M., Kleiner, A., Nebel, B.: CS Freiburg: coordinating robots for successful soccer playing. IEEE Trans. Robot. Autom. **18**(5), 685–699 (2002)
41. Winfield, A.F., Holland, O.: The application of wireless local area network technology to the control of mobile robots. Microprocess. Microsyst. **23**(10), 597–607 (2000)

Local Communication Protocols for Learning Complex Swarm Behaviors with Deep Reinforcement Learning

Maximilian Hüttenrauch[1(✉)], Adrian Šošić[2], and Gerhard Neumann[1]

[1] School of Computer Science, University of Lincoln, Lincoln, UK
{mhuettenrauch,gneumann}@lincoln.ac.uk
[2] Department of Electrical Engineering, Technische Universität Darmstadt, Darmstadt, Germany
adrian.sosic@spg.tu-darmstadt.de

Abstract. Swarm systems constitute a challenging problem for reinforcement learning (RL) as the algorithm needs to learn decentralized control policies that can cope with limited local sensing and communication abilities of the agents. While it is often difficult to directly define the behavior of the agents, simple communication protocols can be defined more easily using prior knowledge about the given task. In this paper, we propose a number of simple communication protocols that can be exploited by deep reinforcement learning to find decentralized control policies in a multi-robot swarm environment. The protocols are based on histograms that encode the local neighborhood relations of the agents and can also transmit task-specific information, such as the shortest distance and direction to a desired target. In our framework, we use an adaptation of Trust Region Policy Optimization to learn complex collaborative tasks, such as formation building and building a communication link. We evaluate our findings in a simulated 2D-physics environment, and compare the implications of different communication protocols.

1 Introduction

Nature provides many examples where the performance of a collective of limited beings exceeds the capabilities of one individual. Ants transport prey of the size no single ant could carry, termites build nests of up to nine meters in height, and bees are able to regulate the temperature of a hive. Common to all these phenomena is the fact that each individual has only basic and local sensing of its environment and limited communication capabilities to its neighbors.

Inspired by these biological processes, swarm robotics [1,4,5] tries to emulate such complex behavior with a collective of rather simple entities. Typically, these robots have limited movement and communication capabilities and can sense only a local neighborhood of their environment, such as distances and bearings to neighbored agents. Moreover, these agents have limited memory systems, such that the agents can only access a short horizon of their perception. As a

© Springer Nature Switzerland AG 2018
M. Dorigo et al. (Eds.): ANTS 2018, LNCS 11172, pp. 71–83, 2018.
https://doi.org/10.1007/978-3-030-00533-7_6

consequence, the design of control policies that are capable of solving complex cooperative tasks becomes a non-trivial problem.

In this paper, we want to learn swarm behavior using deep reinforcement learning [10,17,21–23] based on the locally sensed information of the agents such that the desired behavior can be defined by a reward function instead of hand-tuning controllers of the agents. Swarm systems constitute a challenging problem for reinforcement learning as the algorithm needs to learn decentralized control policies that can cope with limited local sensing and communication abilities of the agents.

Most collective tasks require some form of active cooperation between the agents. For efficient cooperation, the agents need to implement basic communication protocols such that they can transmit their local sensory information to neighbored agents. Using prior knowledge about the given task, simple communication protocols can be defined much more easily than directly defining the behavior. In this paper, we propose and evaluate several communication protocols that can be exploited by deep reinforcement learning to find decentralized control policies in a multi robot swarm environment.

Our communication protocols are based on local histograms that encode the neighborhood relation of an agent to other agents and can also transmit task-specific information such as the shortest distance and direction to a desired target. The histograms can deal with the varying number of neighbors that can be sensed by a single agent depending on its current neighborhood configuration. These protocols are used to generate high dimensional observations for the individual agents that is in turn exploited by deep reinforcement learning to efficiently learn complex swarm behavior. In particular, we choose an adaptation of Trust Region Policy Optimization [21] to learn decentralized policies.

In summary, our method addresses the emerging challenges of decentralized swarm control in the following way:

1. **Homogeneity:** explicit sharing of policy parameters between the agents
2. **Partial Observability:** efficient processing of action-observation histories through windowing and parameter sharing
3. **Communication:** usage of histogram-based communication protocols over simple features

To demonstrate our approach, we formulate two cooperative learning tasks in a simulated swarm environment. The environment is inspired by the Colias robot [2], a modular platform with two wheel motor-driven movement and various sensing systems.

Paper Outline. In Sect. 2, we review the concepts of Trust Region Policy Optimization and describe our problem domain. In Sect. 3, we show in detail how we tackle the challenges of modeling observations and the policy in the partially observable swarm context, and how to adapt Trust Region Policy Optimization to our setup. In Sect. 4, we present the model and parameters of our agents and introduce two tasks on which we evaluate our proposed observation models and policies.

2   Background

In this section, we provide a short summary of Trust Region Policy Optimization and formalize our learning problem domain.

2.1   Trust Region Policy Optimization

Trust Region Policy Optimization (TRPO) is an algorithm to optimize control policies in single-agent reinforcement learning problems [21]. These problems are formulated as Markov decision processes (MDP) which are compactly written as a tuple $\langle S, A, P, R, \gamma \rangle$. In an MDP, an agent chooses an action $a \in A$ via some policy $\pi(a \mid s)$, based on its current state $s \in S$, and progresses to state $s' \in S$ according to a transition function $P(s' \mid s, a)$. After each step, the agent is assigned a reward $r = R(s, a)$, provided by a reward function R which judges the quality of its decision. The goal of the agent is to find a policy which maximizes the expected cumulative reward $\mathbb{E}[\sum_{k=t}^{\infty} \gamma^{k-t} R(s_k, a_k)]$, discounted by factor γ, achieved over a certain period of time.

In TRPO, the policy is parametrized by a parameter vector θ containing weights and biases of a neural network. In the following, we denote this parameterized policy as π_θ. The reinforcement learning objective is expressed as finding a new policy that maximizes the expected advantage function of the current policy, i.e., $J^{\mathrm{TRPO}} = \mathbb{E}\left[\frac{\pi_\theta}{\pi_{\theta_{\mathrm{old}}}} \hat{A}(s, a)\right]$, where $\hat{A}$ is an estimate of the advantage function of the current policy π_{old} which is defined as $\hat{A}(s, a) = Q^{\pi_{\mathrm{old}}}(s, a) - V_{\mathrm{old}}^{\pi}(s)$. Herein, state-action value function $Q^{\pi_{\mathrm{old}}}(s, a)$ is typically estimated by a single trajectory rollout while for the value function $V^{\pi_{\mathrm{old}}}(s)$ rather simple baselines are used that are fitted to the monte-carlo returns. The objective is to be maximized subject to a fixed constraint on the Kullback-Leibler (KL) divergence of the policy before and after the parameter update, which ensures the updates to the new policy's parameters θ are bounded, in order to avoid divergence of the learning process. The overall optimization problem is summarized as

$$\underset{\theta}{\mathrm{maximize}} \quad \mathbb{E}\left[\frac{\pi_\theta}{\pi_{\theta_{\mathrm{old}}}} \hat{A}(s, a)\right]$$
$$\mathrm{subject\ to} \quad \mathbb{E}[D_{\mathrm{KL}}(\pi_{\theta_{\mathrm{old}}} || \pi_\theta)] \leq \delta.$$

The problem is approximately solved using the conjugate gradient optimizer after linearizing the objective and quadratizing the constraint.

2.2   Problem Domain

Building upon the theory of single-agent reinforcement learning, we can now formulate the problem domain for our swarm environments. Because of their limited sensory input, each agent can only obtain a local observation o from the vicinity of its environment. We formulate the swarm system as a swarm MDP (see [24] for a similar definition) which can be seen as a special case of a decentralized partially observed Markov decision process (Dec-POMDP) [20]. An agent in the swarm MDP is defined as a tuple $\mathbb{A} = \langle \mathcal{S}, \mathcal{O}, \mathcal{A}, O \rangle$, where, $\mathcal{S}$ is a set of local states, $\mathcal{O}$ is the space of local observations, and $\mathcal{A}$ is a set of local actions for each agent. The observation model $O(o|s, i)$ defines the observation probabilities for agent i given the global state s. Note that the system is invariant to the order of the agents, i.e., given the same local state of two agents, the observation probabilities will be the same. The swarm MDP is then defined as $\langle N, \mathcal{E}, \mathbb{A}, P, R \rangle$, where N is the number of agents, $\mathcal{E}$ is the global environment state consisting of all local states $\mathcal{S}^N$ of the agents and possibly of additional states of the environment, and $P : \mathcal{S}^N \times \mathcal{S}^N \times \mathcal{A}^N \rightarrow [0, \infty)$ is the transition density function. Each agent maintains a truncated history $h_t^i = (a_{t-\eta}^i, o_{t-\eta+1}^i, \ldots, a_{t-1}^i, o_t^i)$ of the current and past observations $o^i \in \mathcal{O}$ and actions $a^i \in \mathcal{A}$ of length η. All swarm agents are assumed to be identical and therefore use the same distributed policy π (now defined as $\pi(a \mid h)$) which yields a sample for the action of each agent given its current history of actions and observations. The reward function R of the swarm MDP depends on the *global state* and, optionally, all actions of the swarm agents, i.e., $R : \mathcal{S}^N \times \mathcal{A}^N \rightarrow \mathbb{R}$. Instead of considering only one single agent, we consider multiple agents of the same type, which interact in the same environment. The global system state is in this case comprised of the local states of all agents and additional attributes of the environment. The global task of the agents is encoded in the reward function $R(s, a)$, where we from now on write a to denote the joint action vector of the whole swarm.

2.3   Related Work

A common approach to program swarm robotic systems is by extracting rules from the observed behavior of their natural counterparts. Kube et al. [13], for example, investigate the cooperative prey retrieval of ants to infer rules on how a swarm of robots can fulfill the task of cooperative box-pushing. Similar work can be found e.g. in [12,16,19]. However, extracting these rules can be tedious and the complexity of the tasks that we can solve via explicit programming is limited. More examples of rule based behavior are found in [5] where a group of swarming robots transports an object to a goal. Further comparable work can be found in [6] for aggregation, [18] for flocking, or [9] for foraging.

In deep RL, currently, there are only few approaches tackling the multi-agent problem. One of these approaches can be found in [15], where the authors use a variation of the deep deterministic policy gradient algorithm [14] to learn a centralized Q-function for each policy, which, as a downside, leads to a linear increase in dimensionality of the joint observation and action spaces therefore scales poorly. Another algorithm, tackling the credit assignment problem, can be found in [8]. Here, a baseline of other agents' behavior is subtracted from a centralized critic to reason about the quality of a single agent's behavior. However, this approach is only possible in scenarios with discrete action spaces since it requires marginalization over the agents' action space. Finally, a different line of work concerning the learning of communication models between agents can be found in [7].

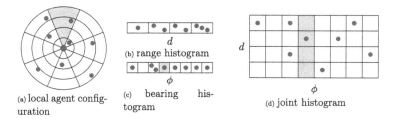

Fig. 1. This Figure shows an illustration of the histogram-based observation model. Figure 1a shows an agent in the center of a circle whose neighborhood relations are to be captured by the histogram representation. The shaded green area is highlighted as a reference for Figs. 1c and d. Figure 1b hereby shows the one dimensional histogram of agents over the neighborhood range d into four bins, whereas Fig. 1c shows the histogram over the bearing angles ϕ into eight bins. Figure 1d finally shows the two dimensional joint histogram over range and bearing.

3 Multi-agent Learning with Local Communication Protocols

In this section, we introduce different communication protocols based on neighborhood histograms that can be used in combination to solve complex swarm behaviors. Our algorithm relies on deep neural network policies of special architecture that can exploit the structure of the high-dimensional observation histories. We present this network model and subsequently discuss small adaptations we had to make to the TRPO algorithm in order to apply it to this cooperative multi-agent setting.

3.1 Communication Protocols

Our communication protocols are based on histograms that can either encode neighborhood relations or distance relations to different points of interest.

Neighborhood Histograms

The individual agents can observe distance and bearing to neighbored agents if they communicate with this agent. We assume that the agents are constantly sending a signal, such that neighbored agents can localize the sources. The arising neighborhood configuration is an important source of information and can be used as observations of the individual agents. One of the arising difficulties in this case is to handle changing number of neighbors which would result in a variable length of the observation vector. Most policy representations, such as neural networks, expect a fixed input dimension.

One possible solution to this problem is to allocate a fixed number of neighbor relations for each agent. If an agent experiences fewer neighborhood relations, standard values could be used such as a very high distance and 0 bearing. However, such an approach comes with several drawbacks. First of all, the size of the resulting representation scales linearly with the number of agents in the system and so does the number of parameters to be learned. Second, the execution of the learned policy will be limited to scenarios with the exact same number of agents as present during training. Third, a fixed allocation of the neighbor relation inevitably destroys the homogeneity of the swarm, since the agents are no longer treated interchangeably. In particular, using a fixed allocation rule requires that the agents must be able to discriminate between their neighbors, which might not even be possible in the first place.

To solve these problems, we propose to use *histograms over observed neighborhood relations*, e.g., distances and bearing angles. Such a representation inherently respects the agent homogeneity and naturally comes with a fixed dimensionality. Hence, it is the canonical choice for the swarm setting. For our experiments, we consider two different types of representations: (1) concatenated one-dimensional histograms of distance and bearing and (2) multidimensional histograms. Both types are illustrated in Fig. 1. The one-dimensional representation has the advantage of scalability, as it grows linearly with the number of features. The downside is that potential dependencies between the features are completely ignored.

Shortest Path Partitions

In many applications, it is important to transmit the location of a point of interest to neighbored agents that can currently not observe this point due to their limited sensing ability.

We assume that an agent can observe bearing and distance to a point of interest if it is within its communication radius. The agent then transmits the observed distance to other agents. Agents that can not see the point of interest might in this case observe a message from another agent containing the distance to the point of interest. The distance of the sending agent is added to the received distance to obtain the distance to the point of interest if we would use the sending agent as a via point. Each agent might now compute several of such distances and transmits the minimum distance it has computed to indicate the length of the shortest path it has seen.

The location of neighbored agents including their distance of the shortest path information is important knowledge for the policy, e.g. for navigating to the point of interest. Hence, we adapt the histogram representation. Each partition now contains the minimum received shortest path distance of an agent that is located in this position.

3.2   Weight Sharing for Policy Networks

The policy maps sequences of past actions and observations to a new action. We use histories of a fixed length as input to our policy and a feed-forward deep neural network as architecture. To cope with such high input dimensionality, we propose a weight sharing approach. Each action-observation pair in an agent's history is first processed independently with a network using the same weights.

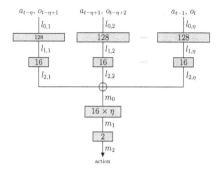

Fig. 2. This diagram shows a model of our proposed policy with three hidden layers. The numbers inside the boxes denote the dimensionalities of the hidden layers. The plus sign denotes concatenation of vectors.

After this initial reduction in dimensionality, the hidden states are concatenated in a subsequent layer and finally mapped to an output. The homogeneity of agents is achieved by using the same set of parameters for all policies. A diagram of the architecture is shown in Fig. 2.

3.3 Adaptations to TRPO

In order to apply TRPO to our multi-agent setup, some small changes to the original algorithm have to be made, similar to the formulation of [11]. First, since we assume homogeneous agents, we can have one set of parameters of the policy shared by all agents. Since the agents cannot rely on the global state, the advantage function is redefined as $A(h, a)$. In order to estimate this function, each agent is assigned the same global reward r in each time step and all transitions are treated as if they were executed by a single agent.

4   Experimental Setup

In this section, we briefly discuss the used model and state representation of a single agent. Subsequently, we describe our two experimental setups and the policy architecture used for the experiments.

4.1   Agent Model

The local state of a single agent is modeled by its 2D position and orientation, i.e., $s^i = [x^i, y^i, \phi^i] \in \mathcal{S} = \{[x, y, \phi] \in \mathbb{R}^3 : 0 \le x \le x_{\max}, \ 0 \le y \le y_{\max}, \ 0 \le \phi \le 2\pi\}$. The robot can only control the speed of its wheels. Therefore, we apply a force to the left and right side of the agent, similarly to the wheels of the real robot. Our model of a single agent is inspired by the Colias robot (a detailed description of the robot specifications can be found in [2]), but the underlying principles can be straightforwardly applied to other swarm settings with limited observations. Generally, our observation model is comprised of the sensor readings of the short and long range IR sensors (later denoted as 'sensor' in the evaluations). Furthermore, we augment this observation representation with the communication protocols developed in the following section. Our simulation is using a 2D physics engine (Box2D), allowing for correct physical interaction of the bodies of the agents.

4.2   Tasks

The focus of our experiments is on tasks where agents need to collaborate to achieve a common goal. For this purpose, we designed the following two scenarios:

Task 1: Building a Graph

In the first task, the goal of the agents is to find and maintain a certain distance to each other. This kind of behavior is required, for example, in surveillance tasks, where a group of autonomous agents needs to maximize the coverage of a target area while maintaining their communication links. We formulate the task as a graph problem, where the agents (i.e. the nodes) try to maximize the number of active edges in the graph. Herein, an edge is considered active whenever the distance between the corresponding agent lies in certain range. The setting is visualized in Fig. 3a. In our experiment, we provide a positive reward for each edge in a range between 10 cm and 16 cm, and further give negative feedback for distances smaller than 7 cm. Accordingly, the reward function is

$$R(s, a) = \sum_{i=1}^{M} \sum_{m>i}^{M} \mathbf{1}_{[0.1m,\ 0.16m]}(d_m^i) - 5 \sum_{i=1}^{M} \sum_{m>i}^{M} \mathbf{1}_{[0m,\ 0.07m]}(d_m^i), \qquad (1)$$

where $d_m^i = \sqrt{(x_i - x_m)^2 + (y_i - y_m)^2}$ denotes the Euclidean distance between the centers of agent i and agent m and

$$\mathbf{1}_{[a,b]}(x) = \begin{cases} 1 & \text{if } x \in [a, b], \\ 0 & \text{else} \end{cases}$$

is an indicator function. Note that we omit the dependence of d_m^i on the system state s to keep the notion simple.

Task 2: Establishing a Communication Link

The second task adds another layer of difficulty. While maintaining a network, the agents have to locate and connect two randomly placed points in the state space. A link is only established successfully if there are communicating agents connecting the two points. Figure 3b shows an example with an active link spanned by three agents between the two points. The task resembles the problem of establishing a connection between two nodes in a wireless ad hoc network [3, 25]. In our experiments, the distance of the two points is chosen to be larger than 75 cm, requiring at least three agents to bridge the gap in between. The reward is determined by the length of the shortest distance between the two points d_{opt} (i.e. a straight line) and the length of the shortest active link d_{sp} spanned by the agents,

$$R(s, a) = \begin{cases} \frac{d_{opt}}{d_{sp}} & \text{if link is established} \\ 0 & \text{otherwise.} \end{cases}$$

In this task, we use the shortest path partitions as communication protocol. Each agent communicates the shortest path it knows to both points of interests, resulting in two 2-D partitions that are used as observation input for a single time step.

4.3 Policy Architecture

We decided for a policy model with three hidden layers. The first two layers process the observation-action pairs (a_{k-1}, o_k) of each timestep in a history individually and map it into hidden layers of size 128 and 16. The output of the second layer is then concatenated to form the input of the third hidden layer which eventually maps to the two actions for the left and right motor.

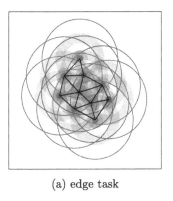 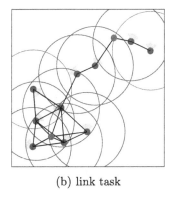

(a) edge task   (b) link task

Fig. 3. Illustration of the two cooperative tasks used in this paper. The green dots represent the agents, where the green ring segments located next to the agents indicate the short range IR front sensors. The outer green circles illustrate the maximum range in which distances/bearings to other agents can be observed, depending on the used observation model. **(a) Edge task:** The red rings show the penalty zones where the agents are punished, the outer green rings indicate the zones where legal edges are formed. **(b) Link task:** The red dots correspond to the two points that need to be connected by the agents. (Color figure online)

5   Results

We evaluate each task in a standardized environment of size $1m \times 1m$ where we initialize ten agents randomly in the scene. Of special interest is how the amount of information provided to the agents affects the overall system performance. Herein, we have to keep in mind the general information-complexity trade-off, i.e., high-dimensional local observations generally provide more information about the global system state but, at the same time, result in a more complex learning task. Recall that the information content is mostly influenced by two factors: (1) the length of the history, and (2) the composition of the observation.

5.1   Edge Task

First, we evaluate how the history length η affects the system performance. Figure 4a shows an evaluation for $\eta = \{2, 4, 8\}$ and a weight sharing policy using a two-dimensional histogram over distances and bearings. Interestingly, we observe that longer observation histories do not show an increase in the performance. Either the increase in information could not counter the effect of increased learning complexity, or a history length of $\eta = 2$ is already sufficient to solve the task. We use these findings and set the history length to $\eta = 2$ for the remainder of the experiments.

Next, we analyze the impact of the observation model. Figure 4b shows the results of the learning process for different observation modalities. The first observation is that, irrespective of the used mode, the agents are able to establish a certain number of edges. Naturally, a complete information of distances and bearing yields the best performance. However, the independent histogram representation yields comparable results to the two dimensional histogram. Again, this is due to the aforementioned complexity trade-off where a higher amount of information makes the learning process more difficult.

5.2   Link Task

We evaluate the link task with raw sensor measurements, count based histograms over distance and bearing, and the more advanced shortest path histograms over distance and bearing. Based on the findings of the edge task we keep the history length at $\eta = 2$. Figure 4c shows the results of the learning process where each observation model was again tested and averaged over 8 trials. Since at least three agents are necessary to establish a link between the two points, the models without shortest path information struggle to reliably establish the connection. Their only chance is to spread as wide as possible and, thus, cover the area between both points. Again, it is interesting to see that independent histograms over counts seem to be favorable over the 2D histogram. However, both versions are surpassed by the 2D histogram over shortest paths which yields information about the current state of the whole network of agents, currently connected to each of the points.

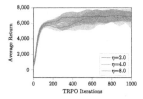 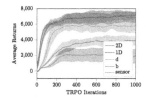 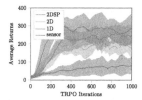

(a) Comparison of different history lengths η. (2D histogram)

(b) Edge task: Comparison of different observation models. ($\eta = 2$)

(c) Link task: Comparison of different observation models. ($\eta = 2$)

Fig. 4. Learning curves for (a), (b) the edge task and (c) the link task. The curves show the mean values of the average undiscounted return of an episode (i.e. the sum of rewards of one episode, averaged over the number of episodes for one learning iteration) over the learning process plus/minus one standard deviation, computed from eight learning trials. Intuitively, the return in the edge task corresponds to the number of edges formed during an episode of length 500 steps. In the link task, it is a measure for the quality of the link. **Legend:** 2DSP: two dimensional histogram over shortest paths, 2D: two-dimensional histogram over distances and bearings, 1D: two independent histograms over distances and bearing, d: distance only histogram, b: bearing only histogram, sensor: no histogram.

6   Conclusions and Future Work

In this paper, we demonstrated that histograms over simple local features can be an effective way for processing information in robot swarms. The central aspect of this new model is its ability to handle arbitrary system sizes without discriminating between agents, which makes it perfectly suitable to the swarm setting where all agents are identical and the number of agents in the neighborhood varies with time. We use these protocols and an adaptation of TRPO for the swarm setup to learn cooperative decentralized control policies for a number of challenging cooperative task. The evaluation of our approach showed that this histogram-based model leads the agents to reliably fulfill the tasks.

Interesting future directions include, for example, the learning of an explicit communication protocol. Furthermore, we expect that assigning credit to agents taking useful actions should speedup our learning algorithm.

Acknowledgments. The research leading to these results has received funding from EPSRC under grant agreement EP/R02572X/1 (National Center for Nuclear Robotics). Calculations for this research were conducted on the Lichtenberg high performance computer of the TU Darmstadt.

References

1. Alonso-Mora, J., Montijano, E., Schwager, M., Rus, D.: Distributed multi-robot formation control among obstacles: a geometric and optimization approach with consensus. In: Proceedings of the IEEE International Conference on Robotics and Automation, pp. 5356–5363 (2016)

2. Arvin, F., Murray, J., Zhang, C., Yue, S.: Colias: an autonomous micro robot for swarm robotic applications. Int. J. Adv. Robot. Syst. **11**(7), 113 (2014)
3. Basu, P., Redi, J.: Movement control algorithms for realization of fault-tolerant ad hoc robot networks. IEEE Netw. **18**(4), 36–44 (2004)
4. Bayındır, L.: A review of swarm robotics tasks. Neurocomputing **172**(C), 292–321 (2016)
5. Chen, J., Gauci, M., Groß, R.: A strategy for transporting tall objects with a swarm of miniature mobile robots. In: Proceedings of the IEEE International Conference on Robotics and Automation, pp. 863–869 (2013)
6. Correll, N., Martinoli, A.: Modeling and designing self-organized aggregation in a swarm of miniature robots. Int. J. Robot. Res. **30**(5), 615–626 (2011)
7. Foerster, J., Assael, Y.M., de Freitas, N., Whiteson, S.: Learning to communicate with deep multi-agent reinforcement learning. Adv. Neural Inf. Process. Syst. **29**, 2137–2145 (2016)
8. Foerster, J., Farquhar, G., Afouras, T., Nardelli, N., Whiteson, S.: Counterfactual multi-agent policy gradients. arXiv:1705.08926 (2017)
9. Goldberg, D., Mataric, M.J.: Robust behavior-based control for distributed multi-robot collection tasks (2000)
10. Gu, S., Lillicrap, T., Ghahramani, Z., Turner, R.E., Levine, S.: Q-prop: sample-efficient policy gradient with an off-policy critic. In: Proceedings of the 5th International Conference on Learning Representations (2017)
11. Gupta, J.K., Egorov, M., Kochenderfer, M.: Cooperative multi-agent control using deep reinforcement learning. In: Sukthankar, G., Rodriguez-Aguilar, J.A. (eds.) AAMAS 2017. LNCS (LNAI), vol. 10642, pp. 66–83. Springer, Cham (2017). https://doi.org/10.1007/978-3-319-71682-4_5
12. Hoff, N.R., Sagoff, A., Wood, R.J., Nagpal, R.: Two foraging algorithms for robot swarms using only local communication. In: Proceedings of the IEEE International Conference on Robotics and Biomimetics, pp. 123–130 (2010)
13. Kube, C., Bonabeau, E.: Cooperative transport by ants and robots. Robot. Auton. Syst. **30**(1), 85–101 (2000)
14. Lillicrap, T.P., et al.: Continuous control with deep reinforcement learning. arXiv:1509.02971 (2015)
15. Lowe, R., Wu, Y., Tamar, A., Harb, J., Abbeel, P., Mordatch, I.: Multi-agent actor-critic for mixed cooperative-competitive environments. arXiv:1706.02275 (2017)
16. Martinoli, A., Easton, K., Agassounon, W.: Modeling swarm robotic systems: a case study in collaborative distributed manipulation. Int. J. Robot. Res. **23**(4–5), 415–436 (2004)
17. Mnih, V., et al.: Human-level control through deep reinforcement learning. Nature **518**(7540), 529–533 (2015)
18. Moeslinger, C., Schmickl, T., Crailsheim, K.: Emergent flocking with low-end swarm robots. In: Dorigo, M., et al. (eds.) ANTS 2010. LNCS, vol. 6234, pp. 424–431. Springer, Heidelberg (2010). https://doi.org/10.1007/978-3-642-15461-4_40
19. Nouyan, S., Gross, R., Bonani, M., Mondada, F., Dorigo, M.: Teamwork in self-organized robot colonies. IEEE Trans. Evol. Comput. **13**(4), 695–711 (2009)
20. Oliehoek, F.A.: Decentralized POMDPs. In: Wiering, M., van Otterlo, M. (eds.) Reinforcement Learning. Adaptation, Learning, and Optimization, vol. 12, pp. 471–503. Springer, Heidelberg (2012). https://doi.org/10.1007/978-3-642-27645-3_15
21. Schulman, J., Levine, S., Moritz, P., Jordan, M., Abbeel, P.: Trust region policy optimization. In: Proceedings of the 32nd International Conference on Machine Learning, pp. 1889–1897 (2015)

22. Schulman, J., Wolski, F., Dhariwal, P., Radford, A., Klimov, O.: Proximal policy optimization algorithms. arXiv:1707.06347 (2017)
23. Teh, Y.W., et al.: Distral: robust multitask reinforcement learning. arXiv:1707.04175 (2017)
24. Šošić, A., KhudaBukhsh, W.R., Zoubir, A.M., Koeppl, H.: Inverse reinforcement learning in swarm systems. In: Proceedings of the 16th Conference on Autonomous Agents and MultiAgent Systems, pp. 1413–1421 (2017)
25. Witkowski, U., et al.: Ad-hoc network communication infrastructure for multi-robot systems in disaster scenarios. In: Proceedings of the IARP/EURON Workshop on Robotics for Risky Interventions and Environmental Surveillance (2008)

Morphogenesis as a Collective Decision of Agents Competing for Limited Resource: A Plants Approach

Payam Zahadat[(✉)], Daniel Nicolas Hofstadler, and Thomas Schmickl

Artificial Life Lab, University of Graz, Graz, Austria
payam.zahadat@uni-graz.at

Abstract. Competition for limited resource is a common concept in many artificial and natural collective systems. In plants, the common resources – water, minerals and the products of photosynthesis – are a subject of competition for individual branches striving for growth. The competition is realized via a dynamic vascular system resulting in the dynamic morphology of the plant that is adapting to its environment. In this paper, a distributed morphogenesis algorithm inspired by the competition for limited resources in plants is described and is validated in directing the growth of a physical structure made out of braided modules. The effects of different parameters of the algorithm on the growth behavior of the structure are discussed analytically and similar effects are demonstrated in the physical system.

1 Introduction

Nature is full of patterns and forms. A huge diversity of natural patterns emerges from self-organization of several components interacting with each other and with their environment. Many patterns are regular repetitions of semi-identical units of forms, e.g. regular patterns on the outer skin of animals, or nonlinear non-equilibrium chemical oscillators, i.e., the Belousov-Zhabotinsky reaction [4,12]. Such patterns can be described by self-organizing "Turing processes" [23,28]. More complex patterns are usually multi-level hierarchies of forms. A mechanism of developing such complex structures in nature is morphogenesis— a generative process starting the system from single units and developing it into a complex organism as a result of interactions between several components of the system and the environment, driven by the laws of physics and chemistry and directed by encoded information in the genome [12]. The wide diversity of patterns in both natural and artificial developmental systems and their inherent adaptivity to environmental conditions are investigated by many researchers [4,12,33]. Various models of developing systems have been introduced and used for artificial systems. One example are L-systems [22] which are abstract generative encodings devised to describe development of multicellular organisms, particularly plants. Variations of the model are used in developing structures of artificial organisms [16,26]. Other examples of morphogenesis are

© Springer Nature Switzerland AG 2018
M. Dorigo et al. (Eds.): ANTS 2018, LNCS 11172, pp. 84–96, 2018.
https://doi.org/10.1007/978-3-030-00533-7_7

models that are inspired by cells, i.e. cell types and division, gene regulatory networks, and diffusion [10], or cellular automata with different types of cells [19].

A related area of research dealing with the development of complexity from local interactions is the field of multi-agent systems. Multi-agent systems span from swarm intelligence [2] that is widely inspired from social insects, to swarm robotics [11,14,34] and to distributed approaches in microeconomics and market-based methods [6,7,20]. A common subject of interest in all systems is the distribution of resources including distribution of labor [2,17,29,30]. Individual agents in a swarm consume or contribute to common resources available for the swarm while pursuing their personal motivations. Having to share resources imposes dependencies between agents and thus the mechanisms of resource distribution can steer the behavior of the swarm in various ways. Such mechanisms are widely investigated in microeconomics and market-based control [6,20]. The mechanisms of division of labor and task allocation in swarm intelligence and swarm robotics share similar challenges, e.g. how to distribute the agents as the limited resource to handle sets of given tasks [3,18,30].

Here we use a morphogenesis algorithm called Vascular Morphogenesis Controller (VMC) [32] which is inspired by distribution of common resources between branches of a plant by means of vascular dynamics. The algorithm acts based on competition of individual agents and via local interactions. The negative feedback mechanisms due to scarcity of resources for the branches and positive feedback loops reinforcing vessels that transfer resources to the favorable paths govern the dynamics of the growing system. The result is a dynamic system of vessels that allows exploration of the environment and leads to stronger pathways of common resource between the root and the tips located in more favorable regions of the environment. The concept has strong similarities with other swarm systems of self-organized path formation, e.g. the formation of pheromone trails connecting the nest of ants to patches of foods [8] which has inspired optimization algorithms [9] and is implemented in many robotic swarms (e.g., [5,24,27]).

This work is in the context of the project *flora robotica* [13] that explores the symbiosis between plants and artificial structures for developing adaptive bio-hybrid architectural artifacts. The VMC is used as an embodied distributed algorithm reflecting environmental features in directing the growth of the artificial structures. The growth process is realized here manually by adding new modules to the structure based on the collective decision of the distributed controller. However the process is reversible meaning removal of modules is also possible. As a method of additive construction of artificial structures, the old technique of braiding is used. The braids consisting of reciprocally interwoven filaments posses attributes of flexibility of topology and are well-suited for incorporation of wires and distributed electronics.

In the following, a general formulation of the VMC algorithm is introduced and the effects of parameters in the morphogenesis behavior of the structures are described following a formal approach. The parameter effects are then demonstrated in a set of experiments with physical structures built out of braided modules hosting sensors and VMC controllers.

2   The Model: Vascular Morphogenesis Controller

Vascular Morphogenesis Controller (VMC) is inspired by the mechanisms of growth and branching in plants. Individual branches in a plant act as agents of a swarm competing with each other for shared resources. Each branch explores its local environment and according to the modality of the local resources in the environment (e.g., light) it produces amounts of a hormone, called auxin [21]. The hormone flows root-wards and adjusts the quality of the vessels along its way. The vascular system of a plant is responsible for distributing essential resources (e.g., water and minerals) from the roots to all the branches. According to the *canalization* hypothesis [1,25], a well-positioned branch (wrt. environmental resources, i.e., light) produces high amounts of auxin which leads to better quality of vessels and therefore more share of the common resources and ultimately more growth for the branch. The larger share of the resource for well-positioned branches means lower shares being distributed among the others. The growth of a well-positioned branch can locate it in even better regions of the environment and gives it more new branches which leads to a positive feedback loop of auxin production and growth. The collective decision making process enables the plant to find the favorable regions of the environment and to benefit the growth in those regions.

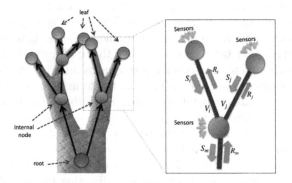

Fig. 1. An example structure guided by VMC. A value, called Successin, is produced at the leaves based on the sensor values and encoded parameters in the genome and flows root-wards through the internal nodes. The flow is modified at those nodes based on sensor values and parameters. The Successin flow adjusts the thickness of vessels which in turn are responsible for distributing the common resource from the root.

The VMC abstracts the above mentioned dynamics by introducing the growth process of acyclic directed graphs. Figure 1 shows a schematic representation of VMC. The figure shows the flow of a value, we call Successin in analogy to auxin in a plant, produced at the leaves of the graph and propagating towards the root. The flow of Successin (S) regulates the thickness of vessels (weights of the edges of the graph). A common Resource (R) starts at the root and is distributed between the children of each node proportional to their vessel thickness (V). Growth happens at the leaves by adding new nodes.

Similar to production of auxin in growing tips of plants, Successin is produced at the leaves (of the VMC graph) based on the local sensory inputs and constant parameters:

$$S_{leaf} := \text{PRODUCTION}(\text{params}, \text{sensors}) \tag{1}$$

Successin flows towards the root. At an internal node i, the flow of Successin is influenced by the inputs from the local sensors and constant parameters via a transfer function in the range of $[0, 1]$:

$$S_{\text{non-leaf}} := \text{TRANSFER}(\text{params}, \text{sensors}) \sum_{b \in \text{children}} S_b. \tag{2}$$

The weight of each connection (i, j) (thickness of the vessel) is adjusted based on Successin passing the connection (vessel) and the parameters determining the competition rate between the siblings:

$$V_{i,j} := V_{i,j} + \alpha(S_j^{\beta_i} - V_{i,j}), \quad with \quad \beta_i = \text{COMPETITION}(\text{params}, \text{sensors}), \tag{3}$$

where $V_{i,j}$ is the connection between node i and its child node j. S_j is the Successin of node j flowing towards i.

The above mentioned functions are implemented in this work as follows. The production rate of the Successin at the leaves is defined as

$$\text{PRODUCTION}(\text{params}, \text{sensors}) = f(\omega_{\text{const}} + \sum_{s \in \text{sensors}} \omega_s I_s), \tag{4}$$

where $f(x) = \max(0, x)$, ω_{const} is the constant production rate of Successin at a leaf and ω_s is the sensor dependent production rate which is the coefficient determining the dependency of Successin production on the sensor input I_s.

The transfer rate of Successin passing a node is defined as

$$\text{TRANSFER}(\text{params}, \text{sensor}) = g(\rho_{\text{const}} + \sum_{s \in \text{sensors}} \rho_s I_s), \tag{5}$$

where ρ_{const} is a constant transfer rate, ρ_s is the sensor-dependent transfer rate for sensor s, and $g(x) = \max(0, \min(1, x))$.

The competition rate is defined as:

$$\text{COMPETITION}(\text{params}, \text{sensor}) = \beta_{\text{const}} + \sum_{s \in \text{sensors}} \beta_s I_s, \tag{6}$$

where β_{const} is the constant competition rate and β_s is the sensor-dependent competition rate for sensor s.

Each of the parameters above can be set to zero depending on the particular applications.

Resource Distribution Over the Structure. Common resource starts at the root and is distributed throughout the structure according to vessel thickness (weight of connections). A part of the resource, R_i, reaching node i can be consumed at that node and the remaining is divided among its children proportional to the thickness of their vessels. A given child j with vessel thickness $V_{i,j}$ receives

$$R_j := (R_i - c)\frac{V_{i,j}}{\sum_{b\in\text{children}} V_{i,b}}, \tag{7}$$

where c is the constant consumption rate of the resource at a node and 'children' is the set of children of node i. c can be set to zero in order to use the resource only at the leaves (for growth). The common resource initiated at the root can be a constant value or a function of the environment and/or Successin that reaches the root from anywhere within the graph. In the current implementation, the R_{root} is fixed to a constant value.

Addition of Nodes. When the graph grows at a leaf, a number of new leaves appear as the children of the old leaf. The decision about the occurrence of growth on a particular leaf follows a growth strategy based on the share of the common resource reaching the leaf. Different strategies can be used to make the growth decision. For example, one strategy is to use a threshold th_{add} on the value of resource at the leaves to determine whether or not they should grow. In this case, the consumption rate of the nodes (c) in relation to the amount of resource at the root (R_{root}) puts a constraint on the overall graph size. Another example strategy is to consider the resource at the leaves as the probability of growth. In the current implementation, the leaf with the maximum resource value is the candidate node to grow next.

Deletion of Nodes. Leaves can be removed from the VMC graph following a deletion strategy based on the resource reaching the nodes. A threshold th_{del} can be used to decide on the deletion of a node's children. For example, a leaf i can be removed if $R_i < th_{\text{del}}$. Another example strategy is to remove all the children of a node i if they are all leaves and the amount of the resource at the node i is below the threshold. In the implementation used in this work, there is no deletion of nodes.

3   A Closer Look on the Effects of Parameters

The parameters described in the previous section and their meanings are summarized in Table 1. Here we use a formal approach to look into the effect of some of these parameters.

Intrinsic Tendency Towards Shorter Paths. A simplified 1-dimensional VMC structure is defined in Fig. 2. The root in this setup has two children and all the other nodes have a single child at most. The number of nodes between the leaves and the root on the left and the right side are n and m respectively. The sensor-dependent transfer and competition rates are set to zero ($\rho_s = \beta_s = 0$).

In a structure as in Fig. 2, the amount of Successin reaching the root from the left and right branches converge to $S_{\text{mainL}} = S_{\text{L}} \cdot \rho_c{}^n$ and $S_{\text{mainR}} = S_{\text{R}} \cdot \rho_c{}^m$

Table 1. List of parameters

Parameter	Description
α	Adaptation rate of vessels
β_c	Competition rate of sibling vessels, constant rate
β_s	Competition rate of sibling vessels, sensor-dependent
ρ_c	Transfer rate of Successin at the internal nodes, constant rate
ρ_s	Transfer rate of Successin at the internal nodes, sensor-dependent
ω_c	Production rate of Successin at the leaves, constant rate
ω_s	Production rate of Successin at the leaves, sensor-dependent
c	Consumption rate of resource in every node
R_{root}	Constant resource value at the root

Fig. 2. An example 1-dimensional VMC graph

correspondingly. If the resource value at the root is $R_{\text{root}} = R$, and with the competition rate β_c, the vessel thicknesses for the branches of the root converge to $V_{\text{mainL}} = S_{\text{mainL}}^{\beta_c}$ and $V_{\text{mainR}} = S_{\text{mainR}}^{\beta_c}$ with a speed of α as the adaptation rate. The amount of the resource reaching each leaf converges to

$$R_{\text{L}} = R \frac{(S_{\text{L}}\rho_c{}^n)^{\beta_c}}{(S_{\text{L}}\rho_c{}^n)^{\beta_c} + (S_{\text{R}}\rho_c{}^m)^{\beta_c}} - n \cdot c, \qquad R_{\text{R}} = R \frac{(S_{\text{R}}\rho_c{}^m)^{\beta_c}}{(S_{\text{L}}\rho_c{}^n)^{\beta_c} + (S_{\text{R}}\rho_c{}^m)^{\beta_c}} - m \cdot c,$$

$$(8)$$

In the case of $S_{\text{L}} = S_{\text{R}}$, the equations are simplified to

$$R_{\text{L}} = R \frac{\rho_c{}^{n\beta_c}}{\rho_c{}^{n\beta_c} + \rho_c{}^{m\beta_c}} - n \cdot c, \qquad R_{\text{R}} = R \frac{\rho_c{}^{m\beta_c}}{\rho_c{}^{n\beta_c} + \rho_c{}^{m\beta_c}} - m \cdot c, \qquad (9)$$

and therefore, the leaf with the shorter path to the root gets more of the resource. The preference for shorter paths is previously demonstrated in a case study of a maze scenario with a simulated VMC-controlled organism [31].

Regulation of Growth in Particular Branches by Using the Sensor-Dependent Transfer Rates. In the previous example, the transfer rate, ρ, was assumed to be identical in all nodes. However, the transfer rate can be also dependent on sensors (see Eq. 5). For instance, one can imagine a scenario with using light sensors influencing the production rate of Successin at the leaves, and accelerometers (providing the tilting angle of branches) or stress sensors (associated to physical joints) for influencing the transfer rate at the internal nodes. As an example, in the structure of Fig. 2, with $S_{\text{L}} = S_{\text{R}}$ and $m = n$ (see Eq. 9), a high stress or

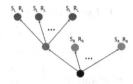

Fig. 3. An example VMC graph with n children for the root and its leftmost child. (Color figure online)

bending that influences an internal node at the left branch may decrease the ρ for that node and leads to $S_{\text{mainL}} < S_{\text{mainR}}$ and consequently $R_{\text{L}} < R_{\text{R}}$, which results in a preference for growth at the right branch.

Combined Effect of the Number of Nodes, Competition Rate and Transfer Rate. Figure 3 shows an example VMC graph with n children for each non-leaf node. Let's assume that all the leaves of the left branch (represented in blue color) have the same sensor values and thus the same Successin production S_{L}, and all the other leaves (represented in orange color) also have the same Successin production, S_{R}. The ratio between the resources reaching a leaf at the left branch and one of the other leaves, depends on the ratio between their Successin values, as well as the competition and transfer rates and the value of n, and is computed as follows:

$$R_{\text{L}} = \frac{R_{\text{root}}}{V_{\text{sum}}} n^{\beta-1}(S_{\text{L}}\rho)^\beta, \quad R_{\text{R}} = \frac{R_{\text{root}}}{V_{\text{sum}}} S_{\text{R}}^\beta, \implies \frac{R_{\text{L}}}{R_{\text{R}}} = n^{\beta-1}\rho^\beta \left(\frac{S_{\text{L}}}{S_{\text{R}}}\right)^\beta \quad (10)$$

where $V_{\text{sum}} = (nS_{\text{L}}\rho)^\beta + (n-1)S_{\text{R}}^\beta$ is the sum of all the vessels at the root node, R_{L} is the resource reaching a leaf of the left branch, and R_{R} is the resource reaching one of the other leaves.

In an environment with $S_{\text{L}} = S_{\text{R}}$, $\frac{R_{\text{L}}}{R_{\text{R}}} = n^{\beta-1}\rho^\beta$. This shows a potential tendency for growing children in branches that already hold larger number of nodes with large values of β and a potential tendency towards growing at the shorter branches with small values of ρ. The condition for the preference of the large branches is $n^{\frac{1-\beta}{\beta}} < \rho$. Considering that $\rho \leq 1$, the above condition never holds for $\beta \leq 1$.

4 Experiments with Physical Structures

Here we present a set of experiments representing the growth behavior of structures with various parameterizations. Most of the experiments are designed to demonstrate the parameter effects discussed in the previous section. The VMC is embodied in a set of controller nodes mounted on Y-shaped braided modules. A controller board is attached to the main part of the module, and two sensor boards, containing 4 light sensors and an accelerometer, are each attached to one of the branches. Each branch of a module can have a child module connected to it (see Fig. 4). The controller board maintains the communications with the children via the sensor boards and with its parent module. The detailed implementation of the modules are described in [15]. Each controller board contains a main VMC node. If a branch of a module has no child, the controller additionally keeps a leaf node associated to that branch. Otherwise, it adopts the main node of the child module as a child node of itself locating in a different module. This way, the VMC graph is formed and distributed over the structure. Growth of the structure is carried out by manually attaching a new braided module to the branch that contains the VMC leaf with the maximum resource value. Other selection strategies could be used here (see *addition of nodes* in Sect. 2), e.g. a threshold on the resource value at a leaf can determine whether or not the leaf should grow. Unless stated otherwise, in all the experiments here, the parameter settings are as follows: $\alpha = 0.9$, $\beta_c = 2$, $\rho_c = 0.5$, $\rho_{tilt} = 0.5$, $\omega_{light} = 1$, $R_{root} = 1$. All the other parameters are set to zero. The values from all the 4 light sensors are averaged and scaled to $[0, 1]$ to make the input variable I_{light} used in production of Successin at the leaf nodes. The value of the accelerometers indicating the tilting of the branches are also scaled to $[0, 1]$ to make the input variable I_{tilt} influencing the transfer rate at the internal nodes. Due to technical reasons regarding the communication protocol between the modules, the value of the Successin at all the leaves are rescaled with a factor of 0.167. In all the experiments $I_{tilt} \simeq 0.99$ unless stated otherwise.

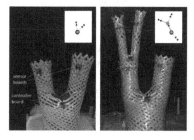

Fig. 4. An example braided module (left), and two connected modules (right), with their underlying VMC graphs (inset images). The circles with thick outline indicate the root nodes. The small circles are the leaf nodes associated with the branches of the modules with no child modules connected to them.

Growing a Structure with Different Competition Rates. The effects of the competition rate β is investigated in this experiment with $\beta_c \in \{1, 2\}$ and with a light source at the top-left of the structure. Figure 5 shows the growth of the braided structure with $\beta_c = 2$. At each step of the growth, a new module is added to the leaf branch with maximum resource among all the leaves. Figure 6 shows the growth of the structure with $\beta_c = 1$. Since β_c cannot have any influence on the behavior of the first single module, we started the experiment with a second one already connected (step A in Fig. 5). As can be seen in the figures, the structure with larger competition rate grows strongly towards the brighter region of the

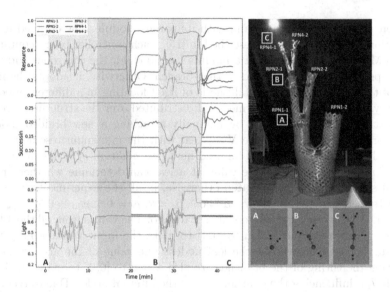

Fig. 5. The variables over the course of the growth (left) with $\beta = 2$, the final structure (right), and the VMC graphs of each growth step (bottom-right). The A-C labels in the plots mark the steps right before the start of manual growth. In the photo of the final structure, the labels indicate the position of the growth at each step. The shaded parts of the plots indicate the periods when the growth was physically realized.

environment while the other structure tends to grow all the branches with slight preference for the brighter region (see a video[1]).

Combined Effect of Transfer Rate and Competition Rate. The combined effect of transfer rate and competition rate are investigated here. The experiments are performed in room light (no directional light is used). The final structure from Fig. 5 is used with $\rho_c \in \{0.25, 0.5\}$ and $\beta_c \in \{1, 2\}$. Considering that $\rho_{tilt} = 0.5$ and $I_{tilt} \simeq 0.99$, then $\rho = \text{TRANSFER} \in \{0.74, 0.99\}$. Table 2 shows the resource and the light value of each leaf, with the maximum resource value of each setup in bold and the maximum light values in italic fonts. The experiment shows a tendency towards shorter paths with smaller transfer rate and the tendency for further growth at the already grown branches with larger competition rate which is in line with the discussion in the previous section.

Regulating Growth in Particular Branches by using a Sensor-Dependent Transfer Rate. In this experiment the effect of sensor-dependent transfer rate is shown. The final structure of Fig. 5 is used in room light. After the first few minutes of the experiment with the intact structure, the leftmost branch is bent such that the I_{tilt} decreases considerably. Figure 7 shows the variable values over the course of the experiment. It shows that bending a branch leads to small values of

---

[1] https://youtu.be/-niKFhrXocI.

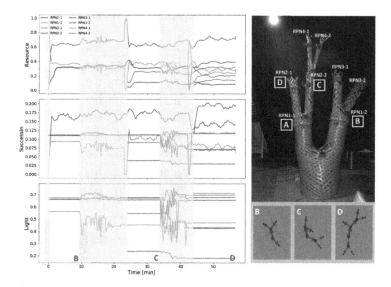

Fig. 6. The variables over the course of the growth (left) with $\beta = 1$, the final structure (right), and the VMC graphs of each growth step (bottom-right). The A-D labels in the plots mark the steps right before the start of manual growth. In the photo of the final structure, the labels indicate the position of the growth at each step. The shaded parts of the plots indicate the periods when the growth was physically realized.

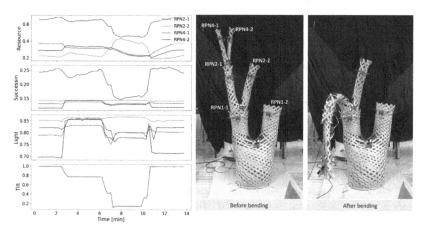

Fig. 7. Different variables in the course of the experiment with sensor-dependent transfer rate. The structure is first intact, then a branch is bent and then released again.

I_{tilt}, decreases the transfer rate in the associated internal node and results in a lower share of resource for that branch which may eventually restrict its growth.

The Effect of Adaptation Rate. In this experiment, the effect of different adaptation rates on the speed of dynamics of the system is investigated. A directional

Table 2. Combined effect of competition and transfer rates.

ρ	β_c	State var.	1-2	2-2	4-1	4-2
0.99	2.0	Resource	0.112	0.227	**0.364**	0.287
		Light	0.806	*0.849*	0.787	0.698
0.99	1.0	Resource	0.252	**0.268**	0.247	0.219
		Light	0.809	*0.857*	0.793	0.702
0.74	1.0	Resource	**0.358**	0.279	0.189	0.168
		Light	0.806	*0.854*	0.791	0.699
0.74	2.0	Resource	0.231	**0.289**	0.263	0.205
		Light	0.804	*0.853*	0.791	0.698

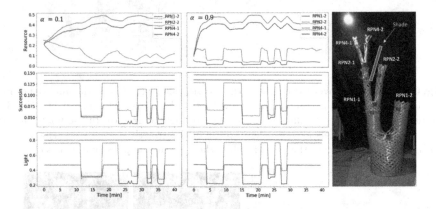

Fig. 8. The effects of small and large adaptation rates

light source is located at the topleft of the structure. A cardboard is used to cast shades on the right branches in different time intervals in order to investigate the response time of the system to the shade/no-shade conditions. Two different adaptation rates $\alpha \in \{0.1, 0.9\}$ are tested. Figure 8 demonstrates the variable values during the course of the experiments. It indicates slower changes in the resource values, reflecting slower change in the vessels, for the smaller α. Since the vessels act as a spatial memory for the system, the slow dynamics of the vessels can be beneficial in filtering out environmental noise.

5   Conclusions and Future Work

Morphogenesis of artificial structures is investigated here by using VMC, a recently introduced plant-inspired controller. The collective decision of the controller is based on the environmental and structural features and intrinsic properties of the controller determined by its parameters. The general formulation of the algorithm is described here and the effects of some parameters are analytically discussed. The algorithm is validated by implementation in a physical braided structure. The parameter effects demonstrated by the physical structure follow the results of the formal analysis. In the future, other behaviors of the controlled system, e.g., the tendency towards asymmetry or dynamics of the structure (deletion and addition of nodes over time) will be investigated. Although the VMC has been so far only used in tree-like structures, nothing prevents the implementation on other acyclic directional graphs with several incoming connections to the nodes and several roots.

Acknowledgments. This work was supported by EU-H2020 project 'florarobotica', no. 640959.

References

1. Bennett, T., Hines, G., Leyser, O.: Canalization: what the flux? Trends Genet. **30**(2), 41–48 (2014)
2. Bonabeau, E., Dorigo, M., Theraulaz, G.: Swarm Intelligence: From Natural to Artificial Systems. Oxford University Press, Oxford (1999)
3. Bonabeau, E., Sobkowski, A., Theraulaz, G., Deneubourg, J.L.: Adaptive task allocation inspired by a model of division of labor in social insects. In: Biocomputing and Emergent Computation: Proceedings of BCEC97, pp. 36–45. World Scientific Press (1997)
4. Camazine, S., et al.: Self-organizing Biological Systems. Princeton University Press, Princeton (2001)
5. Campo, A., et al.: Artificial pheromone for path selection by a foraging swarm of robots. Biol. Cybern. **103**(5), 339–352 (2010)
6. Clearwater, S.H. (ed.): Market-Based Control: A Paradigm for Distributed Resource Allocation. World Scientific Publishing Co., Inc., River Edge (1996)
7. Deconinck, G., Craemer, K.D., Claessens, B.: Combining market-based control with distribution grid constraints when coordinating electric vehicle charging. Engineering **1**(4), 453–465 (2015)
8. Detrain, C., Deneubourg, J.L.: Self-organized structures in a superorganism: do ants like molecules? Phys. Life Rev. **3**(3), 162–187 (2006)
9. Dorigo, M., Maniezzo, V., Colorni, A.: Ant system: optimization by a colony of cooperating agents. Trans. Syst. Man Cyber. Part B **26**(1), 29–41 (1996)
10. Doursat, R., Sánchez, C., Dordea, R., Fourquet, D., Kowaliw, T.: Embryomorphic engineering: emergent innovation through evolutionary development. In: Doursat, R., Sayama, H., Michel, O. (eds.) Morphogenetic Engineering. Understanding Complex Systems, pp. 275–311. Springer, Heidelberg (2012). https://doi.org/10.1007/978-3-642-33902-8_11
11. Ferrante, E., Turgut, A.E., Duenez-Guzman, E., Dorigo, M., Wenseleers, T.: Evolution of self-organized task specialization in robot swarms. PLOS Comput. Biol. **11**(8), 1–21 (2015)
12. Goodwin, B.: How the Leopard Changed its Spots: The Evolution of Complexity. Princeton University Press, Princeton (2001)
13. Hamann, H., et al.: Flora robotica - an architectural system combining living natural plants and distributed robots. arXiv preprint arXiv:1709.04291 (2017)
14. Hamann, H.: Swarm Robotics: A Formal Approach. Springer, Berlin (2018). https://doi.org/10.1007/978-3-319-74528-2

15. Hofstadler, D.N., et al.: Artificial plants - vascular morphogenesis controller-guided growth of braided structures. arXiv preprint arXiv:1804.06343 (2018)

16. Hornby, G.S., Pollack, J.B.: Body-brain co-evolution using l-systems as a generative encoding. In: Proceedings of the Genetic and Evolutionary Computation Conference (GECCO-2001), pp. 868–875. Morgan Kaufmann, San Francisco, July–November 2001

17. Huberman, B.A., Hogg, T.: Distributed computation as an economic system. J. Econ. Perspect. **9**(1), 141–152 (1995)

18. Karsai, I., Schmickl, T.: Regulation of task partitioning by a "common stomach": a model of nest construction in social wasps. Behav. Ecol. **22**, 819–830 (2011)

19. Kowaliw, T., Banzhaf, W.: Mechanisms for complex systems engineering through artificial development. In: Doursat, R., Sayama, H., Michel, O. (eds.) Morphogenetic Engineering. Understanding Complex Systems. Springer, Heidelberg (2012). https://doi.org/10.1007/978-3-642-33902-8_13

20. Kurose, J.F., Simha, R.: A microeconomic approach to optimal resource allocation in distributed computer systems. IEEE Trans. Comput. **38**(5), 705–717 (1989)

21. Leyser, O.: Auxin, self-organisation, and the colonial nature of plants. Curr. Biol. **21**(9), R331–R337 (2011)

22. Lindenmayer, A.: Developmental algorithms for multicellular organisms: a survey of L-systems. J. Theor. Biol. **54**(1), 3–22 (1975)

23. Murray, J.D.: On the mechanochemical theory of biological pattern formation with application to vasculogenesis. Comptes Rendus Biol. **326**(2), 239–252 (2003)

24. Payton, D., Daily, M., Estowski, R., Howard, M., Lee, C.: Pheromone robotics. Auton. Robot. **11**(3), 319–324 (2001)

25. Sachs, T.: The control of the patterned differentiation of vascular tissues. Adv. Bot. Res. **9**, 151–262 (1981)

26. Sims, K.: Evolving 3D morphology and behavior by competition. In: Brooks, R., Maes, P. (eds.) Artificial Life IV, pp. 28–39. MIT Press (1994)

27. Sperati, V., Trianni, V., Nolfi, S.: Self-organised path formation in a swarm of robots. Swarm Intell. **5**(2), 97–119 (2011)

28. Turing, A.M.: The chemical basis of morphogenesis. Philos. Trans. R. Soc. London. Ser. B Biol. Sci. **B237**(641), 37–72 (1952)

29. Waldspurger, C.A., Hogg, T., Huberman, B.A., Kephart, J.O., Stornetta, S.: Spawn: a distributed computational economy. IEEE Trans. Softw. Eng. **18**(2), 103–117 (1992)

30. Zahadat, P., Hahshold, S., Thenius, R., Crailsheim, K., Schmickl, T.: From honeybees to robots and back: division of labor based on partitioning social inhibition. Bioinspiration Biomim. **10**(6), 066005 (2015)

31. Zahadat, P., Hofstadler, D.N., Schmickl, T.: Development of morphology based on resource distribution: finding the shortest path in a maze by vascular morphogenesis controller. In: 14th European Conference on Artificial Life (ECAL-2017), vol. 14, pp. 428–429 (2017)

32. Zahadat, P., Hofstadler, D.N., Schmickl, T.: Vascular morphogenesis controller: a generative model for developing morphology of artificial structures. In: Proceedings of the Genetic and Evolutionary Computation Conference, GECCO 2017, pp. 163–170. ACM, New York (2017)

33. Zahadat, P., Schmickl, T.: Generation of diversity in a reaction-diffusion-based controller. Artif. Life **20**(3), 319–342 (2014)

34. Zahadat, P., Schmickl, T.: Division of labor in a swarm of autonomous underwater robots by improved partitioning social inhibition. Adapt. Behav. **24**(2), 87–101 (2016)

Negative Updating Combined with Opinion Pooling in the Best-of-n Problem in Swarm Robotics

Chanelle Lee[1,2,3]([✉]), Jonathan Lawry[2], and Alan Winfield[1,3]

[1] Bristol Robotics Laboratory, Bristol, UK
c.l.lee@brl.ac.uk
[2] Department of Engineering Mathematics, University of Bristol, Bristol, UK
[3] University of the West of England, Bristol, UK

Abstract. There is a need for effective collective decision making in decentralised multi-agent and robotic systems. This paper introduces a novel approach to the *best-of-n* decision problem with large n. It utilises negative feedback obtained from direct pairwise comparison of options and evidence preserving opinion pooling. We present agent-based simulation experiments that explore the effects of pool size and the number of options on the speed of consensus. Robotic simulation experiments are then used to investigate the potential of the approach as a method for solving the *best-of-n* decision problem in swarm robotic applications. Overall, the results suggest that the proposed approach is highly scalable with regards to n.

1 Introduction and Background

There is a widely acknowledged and growing need for effective collective decision-making in decentralised multi-agent and robotic systems [1,13]. Of particular interest is the class of *best-of-n* decision problems [9], where a system needs to achieve consensus on the most desirable option drawn from a number of n distinct possibilities. For example, the choice could be between different nesting sites, foraging locations [15] or which action to perform next [12]. Each option, i, has an associated option quality, ρ_i, which is used by the members of the system to guide the collective decision in favour of the best option. There are three key challenges to this problem. Firstly, the system must reach consensus on a single option based on only local communications. Secondly it needs to ensure that convergence is to the best possible option. Finally, the third challenge is achieving the first two within an application appropriate time frame.

Study of collective decision-making in artificial systems is often heavily influenced by solutions found in nature, such as those of social insects like bees or ants [9,11]. Scheidler et al. point out in [10] that these solutions tend to be based on positive feedback, i.e. good options are reinforced more than bad ones. For example, in [9] the rate at which an agent recruits others to an option is proportional to the option quality. The greater the quality of an option, the more frequently

© Springer Nature Switzerland AG 2018
M. Dorigo et al. (Eds.): ANTS 2018, LNCS 11172, pp. 97–108, 2018.
https://doi.org/10.1007/978-3-030-00533-7_8

the agent will advocate for it, thus making it more likely that other agents in the population will be recruited to that option. For many applications, positive feedback with raw values is very successful; however, its effectiveness may be limited in cases where there is little difference in the range of option qualities and as such the best option has insufficient advantage over the others. Valentini et al. notes that there is a lack of research extending into the $n > 2$ cases, leading to the suspicion that this becomes a potentially damaging limitation as n increases and the option quality space becomes more saturated. Furthermore, in the few such examples in the literature, [2,4,8,10], no case larger than $n = 7$ is discussed.

With this in mind, this paper presents a novel approach to the *best-of-n* problem based on negative information obtained from pairwise comparisons. Rather than updating agent opinions using the raw values of the option qualities to inform a positive feedback mechanism, agents instead compare pairs of options and update their opinions based on which is the worst. By means of this direct comparison, agents determine which option is not the best overall and thus acquire negative feedback with which to update their opinions. We show that by combining such negative updating with opinion pooling the system will converge significantly faster than exhaustive comparative search, wherein each agent samples all option qualities and compares them all. This is achieved by using the opinion pooling operators discussed in [6].

This paper is organized as follows: The next section outlines a Bayesian evidential updating method based on negative feedback. We discuss a particular opinion pooling operator in Sect. 3 and explore the effect of combining evidential updating with opinion pooling on system level consensus and convergence. In the fourth section, we present agent-based simulation results on the speed and reliability of consensus and convergence for the cases of $n = 10, 20, 50$ and 100 with varying pooling sizes. In section five, we present robot simulation experiments with a fixed population size and spontaneous pooling and explore the results as the number of options n is increased. Finally, in section six we give some conclusions and further work.

2   Evidential Updating with Comparisons

We now introduce a mechanism for evidential updating focused on utilising negative feedback from direct option comparisons. The model uses an opinion-based approach as introduced in [14]; extended to the general case of $n > 2$, where agent opinions will be represented as probability vectors across the set of exclusive and exhaustive hypotheses $H = \{\mathcal{H}_i : i = 1, \ldots, n\}$ where $\mathcal{H}_i$ denotes the claim *option i is the best*. As such, an agent A_r, represents their opinion as a probability vector $\mathbf{x}_r$ where $P_{A_r}(\mathcal{H}_i) = x_{ri}$ for $i = 1, \ldots, n$ with $\sum_{i=1}^{n} x_{ri} = 1$, i.e. x_{ri} is the probability with which agent A_r believes $\mathcal{H}_i$ to be true.

An agent samples two options, i and j, and receives qualities ρ_i and ρ_j. Further suppose, without loss of generality, that $\rho_j > \rho_i$. In this case, the agent does not have enough information to know whether j is the best option, but does learn that i cannot be the best possible option, i.e. it receives the evidence $E_i = \{\mathcal{H}_i\}^c$ the complement of $\{\mathcal{H}_i\}$ with respect to H. The agent can now update their prior belief $\mathbf{x}$ to obtain the posterior $\mathbf{x}|E_i$ using Bayes' theorem as follows:

Definition 1 (Evidential Updating). *Assume we have a set of exclusive and exhaustive hypotheses* $\{\mathcal{H}_i : i = 1, \ldots, n\}$. *Then for* $\mathbf{x} \in [0,1]^n$, $E_i = \{\mathcal{H}_i\}^c$ *and* $\alpha \in [0, \frac{1}{2}]$, *we have,*

$$
\mathbf{x}|E_i = \begin{cases} \dfrac{(1-\alpha)x_j}{\alpha x_i + (1-\alpha)(1-x_i)}, & j \neq i, \\[2mm] \dfrac{\alpha x_j}{\alpha x_i + (1-\alpha)(1-x_i)}, & j = i. \end{cases}
$$

Here α quantifies the agent's belief in the reliability of the evidence source. For $\alpha = 0$, the evidence source is completely reliable evidence source and only provide E_i if i was the worse of the two options. Alternatively, for $\alpha = 0.5$ the evidence source is completely unreliable and so is as likely to provide E_i if i is the best or the worst option.

Unfortunately, even in the best conditions a system using evidential updating on comparisons alone will need all agents to make at least $n - 1$ pairwise comparisons before reaching consensus. Now if we are considering perfect conditions with no noise in the sensed quality values, this is significantly worse than exhaustive comparative search with agents visiting two sites at a time. In the next section we introduce an approach to opinion pooling which allows evidence to be efficiently propagated across the swarm and significantly enhances the effectiveness of negative updating.

3   Combining with Opinion Pooling

In this section, we describe the benefits of combining evidential updating and opinion pooling as we suggested in [6]. We speculate that the use of opinion pooling to propagate evidence between agents in the system will significantly reduce the number of comparisons agents need to make before reaching consensus.

For this study, we limit ourselves to evidence preserving propagation and so use the Product Operator [3,5]. Below we present an extended version for the case of multiple hypotheses.

Definition 2 (The Multi-Option Product Operator (MProdOP)). *Assume we have a set of exclusive and exhaustive hypotheses* $\{\mathcal{H}_i : i = 1, \ldots, n\}$. *The Product Operator for k agents is the function* $c : [0,1]^k \to [0,1]$, *such that for agents* $A_1, \ldots, A_k$ *with opinions* $P_{A_r}(\mathcal{H}_i) = x_{ri}$ *for* $r = 1, \ldots, k$,

$$
c(\mathbf{x}_1, \ldots, \mathbf{x}_k) = \frac{\prod_{r=1}^{k} \mathbf{x}_i}{\left\| \prod_{r=1}^{k} \mathbf{x}_i \right\|_1},
$$

where $\mathbf{x}_r = [x_{r1}, \ldots, x_{rn}]$ *for all* $r = 1, \ldots, k$, $\prod$ *is the Hadamard product and* $\|-\|_1$ *is the L_1 norm.*

Given a pool of k agents with prior beliefs $\mathbf{x}_r$ for $r = 1, \ldots, k$, we suppose that each samples two distinct options and consequently receives evidence E_{i_r}. They then each update their opinion to $\mathbf{x}_r | E_{i_r}$ and aggregate to form the pooled opinion $c(\mathbf{x}_1 | E_{i_1}, \ldots, \mathbf{x}_k | E_{i_k})$. Since MProdOP is evidence preserving this is equivalent to $c(\mathbf{x}_1, \ldots, \mathbf{x}_k) | E_{i_1} \ldots | E_{i_k}$.

If we consider the case where all the agents are initialised with uniform opinions where $\mathbf{x}_r = [\frac{1}{n}, \ldots, \frac{1}{n}]$ for $r = 1, \ldots, k$. Furthermore, suppose that the evidence E_i is received by m_i of the agents. Now, without loss of generality, we can also assume that $\rho_1 > \ldots, > \rho_n$, so we have $m_1 = 0$ and $\sum_{i=2}^{n} m_i = k$. This leads to:

$$c(\mathbf{x}_1 | E_{i_1}, \ldots, \mathbf{x}_k | E_{i_k})_i = \frac{\alpha^{m_i}(1-\alpha)^{k-m_i}}{\sum_{i=1}^{n} \alpha^{m_i}(1-\alpha)^{k-m_i}}. \tag{1}$$

We can also calculate the probability that an agent receives the evidence E_i. To receive evidence E_i, an agent must sample two options, one of which is i and the other is some $j > i$. There are $2(i-1)$ of these pairings out of a total $n(n-1)$ distinct pairs. Hence, provided distinct pairs of options are selected at random, then $P(E_i) = \frac{2(i-1)}{n(n-1)}$. Thus, the probability that there are m_i occurrences of E_i amongst the k agents for $i = 1, \ldots, n$ is,

$$\frac{k!}{\prod_{i=1}^{n} m_i!} \prod_{i=1}^{n} (\frac{2(i-1)}{n(n-1)})^{m_i}. \tag{2}$$

Combining both of these it follows that,

$$\mathbb{E}(c(\mathbf{x}_1 | E_{i_1}, \ldots, \mathbf{x}_k | E_{i_k})_i)$$

$$= \sum_{\mathbf{m}:m_1=0, \sum_i m_i=k} \frac{k!}{\prod_{i=1}^{n} m_i!} \prod_{i=1}^{n} (\frac{2(i-1)}{n(n-1)})^{m_i} \frac{\alpha^{m_i}(1-\alpha)^{k-m_i}}{\sum_{i=1}^{n} \alpha^{m_i}(1-\alpha)^{k-m_i}}, \tag{3}$$

giving the expected value of the pooled opinion in option i after a single pooling of the whole population.

4   Agent-Based Simulations Experiments

For initial simulation experiments, we present a simple event based multi-agent model exploring the consensus attainment properties of the decision making algorithm proposed in Sects. 2 and 3. Specifically, we are interested in the performance of our proposed algorithm versus an exhaustive comparison of all options. We hypothesise that there will be an optimal pool size k^* for each n value, below which our algorithm will take longer on average, and above which it will be faster. The simulation has no physical representation of the environment and as such the options have no associated cost. However, the spatial distribution of the population is represented and at every time step the agents are shuffled to emulate random movement, this approach being consistent with the well-stirred assumption as described in [9].

We assume that a population of N agents begin with no prior knowledge of the option qualities and all opinions are initialised uniformly with probabilities $P(\mathcal{H}_1), \ldots, P(\mathcal{H}_n) = 1/n$ to ensure no initial bias. At every iteration, each agent makes a weighted random choice of two options, i and j say, based on their current probability distribution. The agent compares the qualities of this pair, then uses the updating method as described in Definition 1 to update on the evidence $E = \{\mathcal{H}_i\}^c$ where $\rho_i < \rho_j$. We set $\alpha = 0$ to indicate that agents have total trust in the evidence, a not unreasonable assumption as there is currently no noise. We assign the qualities $\rho_i = \frac{(n-1)-(i-1)}{n-1} \in [0, 1]$ to the options $i \in \{1, \ldots, n\}$, assuming that option 1 is the best with maximal quality, i.e. $\rho_1 > \rho_i \; \forall i \in \{1, \ldots, n\}$.

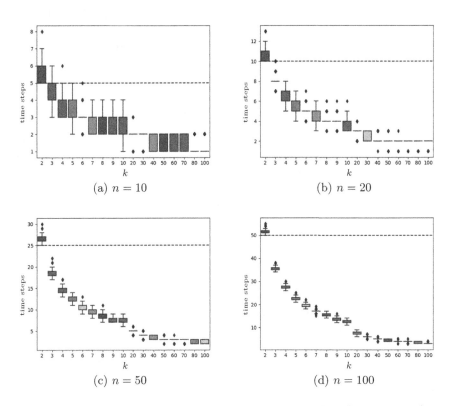

Fig. 1. Box and whisker plots showing the time to consensus plotted against population size with (a) $n = 10$, (b) $n = 20$, (c) $n = 50$ and (d) $n = 100$. The dashed lines at $y = \frac{n}{2}$ show the number of time steps $\hat{t}$ needed for the agents to sample every option.

In addition, all agents in the population pool their opinions using the MProdOp operator from Definition 2 with every agent adopting the resulting pooled opinion. Thus pooling size k is fixed and equal to the population size N. For an embodied system, this set-up could be envisaged as a population of robots visiting potential new nest sites, receiving some sensory data indicating that site's quality and updating on this comparison before returning to the original nest to pool opinions. It is thus reminiscent of many similar experiments in the *best-of-n* literature [13]. For each set of parameter values, 1,000 independent runs are carried out with each lasting for 100 iterations. We judge that consensus has been attained in a run once all agents in the population have $P(\mathcal{H}_1) = 1$, i.e. $x_{r1} = 1 \forall r$. Each agent is assumed to be able to sample a pair of options every time step and the number of time steps needed to reach consensus is recorded for each run. This ensures that the two conditions of effective collective decision making are met; the population has converged to a consensus on a single option and that option is the best one possible. If the run fails for either of these conditions, its consensus time is recorded as 100 iterations for ease of comparison. Results are averaged over all 1,000 runs, giving a consensus time for each set of parameter values.

Figures 1(a)–(d) show the number of time steps needed for the system to reach consensus for varying populations size and when the number of site is $n = 10$, $n = 20$, $n = 50$ and $n = 100$ respectively. This is compared with the number of time steps that the system would need if each agent were to sample the quality of every option two at a time, i.e. $\hat{t} = \frac{n}{2}$ time steps. As expected the number of time steps decreases as the pooling size k increases, this effect plateaus once the best performance of a single time step is reached. We show that our hypothesis was correct, that for each different n there would be some optimum k_n^* where performance is better than or equal to $\hat{t}_n$; with $k_{10}^* = 5$ and $k_{20}^* = k_{50}^* = k_{100}^* = 3$. We can see that this would be the case, as if we substitute our requirement that $\alpha = 0$ into Eq. 1 gives us:

$$
c(\mathbf{x}_1|E_{i_1}, \ldots, \mathbf{x}_k|E_{i_k})_i = \begin{cases} 0, & m_i \geq 1, \\ \dfrac{1}{|\{m_i : m_i = 0 \text{ for } i = 2, \ldots, n\}| + 1}, & m_i = 0. \end{cases} \tag{4}
$$

Thus only one agent in the pool needs to have received evidence E_i for all agents to completely disregard the option, i.e. $x_{ri} = 0$ and this becomes more likely with larger pooling sizes. This effect can be seen in Fig. 2.

The values for k^* are surprisingly small with $k = 3$ performing optimally for almost all n tested. For example, with $n = 100$ for $k = 3$ consensus was achieved on average within 35 time steps, an improvement of 15 time steps when compared to the 50 that would be needed for each agent to visit every site. Furthermore, considering the $n = 100$ results again, when $k = 5$ the system achieves consensus within on average 23 time steps, this is less than half $\hat{t}_{100}$ with only a $1 : 20$ ratio between pooling agents and the number of options. These results suggest that the evidence propagation is very effective and that our method is highly scalable to large n. Unexpectedly, k_{10}^* was greater than the optimal pooling sizes for larger n. This could be due to the accumulative effect of evidence preservation within the system. Every time an agent pools its opinion, it receives all the evidence that every other agent in the pool has. For example, if every agents $A_1, \ldots, A_{k-1}$ all update on evidence E_i and agent A_k updates on some different evidence E_j then

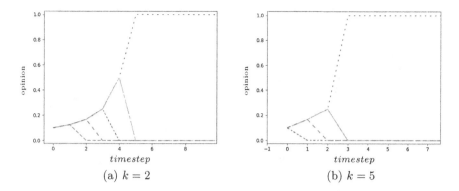

Fig. 2. Plots showing a single agent's opinion of different options changing over time when (a) $k = 2$ and (b) $k = 3$.

we would have $c(\mathbf{x}_1|E_i, \ldots, \mathbf{x}_{k-1}|E_i, \mathbf{x}_k|E_j) = c(\mathbf{x}_1, \ldots, \mathbf{x}_k)|E_i|E_j$. A secondary explanation for this is that there is a greater chance for diversity with larger n and so less instances of 'redundant pooling', i.e. when all agents have the same evidence and hence receive no gain from pooling.

5   Robot-Based Simulation Experiments

In this section, we present robot-based simulation experiments where we have a fixed population size, spontaneous pooling sizes k and varying n in order to test the feasibility of our approach in a swarm robotics scenario. We use e-puck robots [7] since they are small, mobile and equipped with a range of sensors making them well-suited to small scale swarm experiments. Experiments are conducted in the V-Rep[1] simulation environment which models many of the required physical characteristics of the e-pucks, such as motion, communication and sensory feedback. Figure 3 shows the experimental arena consisting of n sites equally spaced around a 1.5 m disc with a central 'nest' site. Each site is coloured a different shade of red or blue indicating site quality. Site i is given quality $\rho_i = \frac{(n-1)-(i-1)}{n-1} \in [0, 1]$. Sites are coloured a proportion of blue equal to $1-\rho_i$ to help visually distinguish between sites. Noise has not been added into the simulation and physical communication limitations have not been considered.

As in Sect. 4, the swarm is initialised with uniform probabilities for all sites and each robot makes a weighted random choice of two sites to visit. The robots are given the locations of all sites and use a simple path planning algorithm to travel between sites and the nest. At each site the robots use their colour camera to return a value indicating the amount of red visible, i.e. the site quality. They then compare qualities and update on this negative feedback using Eq. 1, before returning to the nest site for aggregation. Once a robot has reached the nest site, it will listen for a message from the transceiver located there to confirm

---
[1] http://www.coppeliarobotics.com/

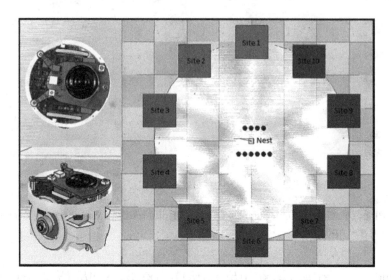

Fig. 3. Top-down and side close-ups of the e-puck model and the arena set-up for $n = 10$.

it has reached the nest. This ensures that all robots who have returned for aggregation are within communication range of each other. The robot broadcasts its opinion, while also listening for any neighbouring robots broadcasting their own opinions. As the system has no centralised controller, pooling between the robots is spontaneous and so the pooling size k could range from zero to the whole population, depending on which robots happen to be at the nest site. This differs from the agent-based simulation experiments where k was fixed, and allows us to investigate the effect pooling size variance has on the system. To reduce the communication requirements between robots we employ neighbourhosod-based pooling, wherein each robot has their own pool of opinions on which only they update. The robot then uses its updated belief to make a weighted random choice about the next two sites to visit. This process is repeated until $x_{ri} = 1$ for some i, at which point they move to their chosen site and stop. Each parameter set of ten robots and $n \in [5, 8, 10]$ was run ten times, with the pooling size of each aggregation and the number of time steps needed for each agent to reach convergence being recorded.

For all runs the swarm was successful with respect to the first two key challenges of the *best-of-n* decision problem; all robots reached consensus on a single option and that option was the best possible one. Figure 4 gives the time frame for reaching consensus, showing for each run the number of time steps the swarm needed before reaching a final decision. As in Sect. 4, we use the number of time steps that a robot would need to visit all sites, two at a time, as a benchmark for performance. Many of the runs across all values of n achieved consensus within $\hat{t}$ time steps, with the best performance for $n = 10$, where the swarm reached consensus within $\hat{t}$ time steps for all runs. Moreover, in 60% of the runs, the

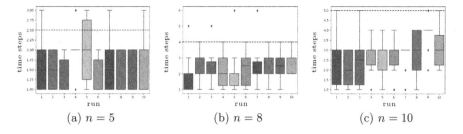

(a) $n = 5$ (b) $n = 8$ (c) $n = 10$

Fig. 4. Box and whisker plots showing the spread of the number of time steps to consensus for individual robots in a population of size $N = 10$ for ten different runs with number of sites (a) $n = 5$, (b) $n = 8$ and (c) $n = 10$. The dashed lines at $\frac{n}{2}$ show the number of time steps $\hat{t}$ necessary for a robot to sample every option.

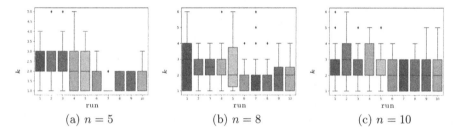

(a) $n = 5$ (b) $n = 8$ (c) $n = 10$

Fig. 5. Box and whisker plots showing range of pooling sizes k for robots in a population of size $N = 10$ for ten different runs with (a) $n = 5$, (b) $n = 8$ and (c) $n = 10$.

swarm achieved consensus in less than $\hat{t}$ and, in particular, in Run 2 consensus is achieved in just three time steps. This is consistent with our findings in Sect. 4 that increasing n can have a positive effect on pooling due to the decreased likelihood of redundant pooling. These results have also been achieved with relatively low pooling sizes with Fig. 5 showing that no run achieved an average pool size greater than three. This suggests that alterations in the control architecture leading to a higher average pooling size, such as increasing the time the swarm spends sharing opinions with neighbouring robots during aggregation, could lead to even faster consensus times.

One of the worst results is Run 5 for $n = 8$ where the swarm took six time steps to reach consensus, two more than $\hat{t}$. A closer look at this result shows that this was caused by a single outlying agent, with the rest of the swarm achieving consensus within three time steps. Consideration of the average pooling sizes in this run, as seen in Fig. 5(b), reveals that while there were some very large pools with six robots, the average was much lower at only two robots. From this we conjecture that some robots who were part of the larger pools converged very quickly, and thus essentially removing themselves and the evidence they had gained from the system too early for the other robots to benefit. This suggests that while larger k will give faster consensus in general, care has to be taken with the potential variance of k values so as not to isolate robots from the system. A way of alleviating this effect in future work could be to have a period of opinion broadcasting after a robot has reached their decision for the benefit of other robots. This could also have the additional benefit of reducing the range of time steps needed for the swarm to reach consensus, which in the run above was as high as six and as low as one.

6   Conclusion

In this paper, we have introduced a novel approach to solving the *best-of-n* decision problem that uses evidence updating from negative feedback combined with opinion pooling. We present an evidential updating method that utilises negative feedback obtained from direct pairwise comparisons of options. We then introduced the multi-option pooling operator MProdOp, with the expectation that its evidence preserving property would efficiently propagate evidence throughout a swarm. In simulation experiments, we explore the effect of pooling size on the time to consensus and the scalability of our approach to increasing values of n. Finally, we investigated our approach in a typical swarm robotics scenario in simulation to test its applicability.

The simulation experiments presented in Sect. 4 suggest that our approach is highly scalable with regards to n. Indeed, successful and effective consensus was reached even with $n = 50$ and $n = 100$ options. We also found that although performance improves with larger k, the system can achieve consensus faster than exhaustive comparison even with very small k. For example, with $n = 100$ for $k = 3$ consensus was achieved on average within 35 time steps, a considerable improvement on the 50 time steps that would be needed for each agent to visit every site.

Overall, the robot simulation experiments indicate that our approach has potential as a method for solving the *best-of-n* decision problem in swarm robotics applications. We have presented a simplified scenario where the swarm needed to pick the reddest of n sites with $n = 5, 8$, and 10. The first two key challenges facing the *best-of-n* problem were met in all runs. Additionally, for a majority of runs the swarm was able to achieve consensus faster than exhaustive comparative search. This demonstrates a level of robustness to pooling size, with possible improvements if average pooling sizes could be increased. Furthermore, as not all agents had to visit all sites to achieve consensus, this approach could potentially work in a scenario where each agent is unable to visit all sites in the environment. We also saw improvements with an increase in n suggesting that our approach will scale well with large n in the swarm robotics environment.

An observed limitation of our proposed method is that it currently only works in environments for which both n and the location of sites is known. For many possible applications, such as a search a rescue site checking task, this is of minor significance as all sites would be known; however, it does restrict the adaptability of the approach in uncertain or changing environments. Further work would look to address this by introducing the ability to increase n upon the discovery of new sites.

Parker and Zhang argue in [9] that agents should not be performing such direct comparisons of options as it can leave the system exposed to potential stagnation from evaluation errors and hence in future work we will investigate how our algorithm performs in the presence of noise, e.g. in sensed quality values. We hypothesise that by introducing distrust, both by setting $\alpha > 0$ and using a diluting pooling operator [6], our system could be robust to such noise. Furthermore, we intend to explore the robustness of the system in a dynamic environment where the best option may change, much as in [11]. In addition, we plan to replicate our experiments on a physical robotic platform and investigate what happens with much larger swarms, for example when $N > 500$.

Acknowledgements. This research was partially funded by an EPRSC PhD studentship as part of the Centre for Doctoral Training in Future Autonomous and Robotic Systems (grant number EP/L015293/1). The authors would like to thank Michael Crosscombe for many useful discussions and valuable comments. All underlying data is included in full within the paper.

References

1. Brambilla, M., Ferrante, E., Birattari, M., Dorigo, M.: Swarm robotics: a review from the swarm engineering perspective. Swarm Intell. **7**(1), 1–41 (2013)
2. Crosscombe, M., Lawry, J.: Exploiting vagueness for multi-agent consensus. In: Bai, Q., Ren, F., Fujita, K., Zhang, M., Ito, T. (eds.) Multi-agent and Complex Systems. SCI, vol. 670, pp. 67–78. Springer, Singapore (2017). https://doi.org/10.1007/978-981-10-2564-8_5
3. Easwaran, K., Fenton-Glynn, L., Hitchcock, C., Velasco, J.D.: Updating on the credences of others: disagreement, agreement, and synergy. Philos. Imprint **16**(11), 1–39 (2016)
4. Garnier, S., Combe, M., Jost, C., Theraulaz, G.: Do ants need to estimate the geometrical properties of trail bifurcations to find an efficient route? A swarm robotics test bed. PLoS Comput. Biol. **9**(3), e1002903 (2013)
5. Genest, C., Zidek, J.V.: Combining probability distributions: a critique and an annotated bibliography. Stat. Sci. **1**(1), 114–135 (1986)
6. Lee, C., Lawry, J., Winfield, A.: Combining opinion pooling and evidential updating for multi-agent consensus. International Joint Conference on Artificial Intelligence, pp. 347–353 (2018)
7. Mondada, F., et al.: The E-puck, a robot designed for education in engineering. In: Proceedings of the 9th Conference on Autonomous Robot Systems and Competitions, vol. 1, pp. 59–65. IPCB: Instituto Politécnico de Castelo Branco (2009)
8. Parker, C.A., Zhang, H.: Active versus passive expression of preference in the control of multiple-robot decision-making. In: IEEE/RSJ International Conference on Intelligent Robots and Systems 2005, (IROS 2005), pp. 3706–3711. IEEE (2005)
9. Parker, C.A., Zhang, H.: Cooperative decision-making in decentralized multiple-robot systems: the best-of-n problem. IEEE/ASME Trans. Mechatron. **14**(2), 240–251 (2009)
10. Scheidler, A., Brutschy, A., Ferrante, E., Dorigo, M.: The k-unanimity rule for self-organized decision-making in swarms of robots. IEEE Trans. Cybern. **46**(5), 1175–1188 (2016)

11. Schmickl, T., et al.: Get in touch: cooperative decision making based on robot-to-robot collisions. Auton. Agents Multi-Agent Syst. **18**(1), 133–155 (2009)

12. Seth, A.K., Bryson, J.J.: Natural action selection, modeling. In: Pashler, H. (ed.) Encyclopedia of the Mind, pp. 557–559. Sage (2013)

13. Valentini, G., Ferrante, E., Dorigo, M.: The best-of-n problem in robot swarms: formalization, state of the art, and novel perspectives. Front. Robot. AI **4**, 9 (2017)

14. Wessnitzer, J., Melhuish, C.: Collective decision-making and behaviour transitions in distributed ad hoc wireless networks of mobile robots: target-hunting. In: Banzhaf, W., Ziegler, J., Christaller, T., Dittrich, P., Kim, J.T. (eds.) ECAL 2003. LNCS (LNAI), vol. 2801, pp. 893–902. Springer, Heidelberg (2003). https://doi.org/10.1007/978-3-540-39432-7_96

15. Winfield, A.F.: Foraging robots. In: Meyers, R. (ed.) Encyclopedia of Complexity and Systems Science. Springer, New York (2009). https://doi.org/10.1007/978-0-387-30440-3

On Mimicking the Effects of the Reality Gap with Simulation-Only Experiments

Antoine Ligot🆔 and Mauro Birattari(✉)🆔

IRIDIA, Université Libre de Bruxelles, Brussels, Belgium
{aligot,mbiro}@ulb.ac.be

Abstract. One issue in the automatic design of control software for robot swarms is the so-called reality gap—the difference between reality and the simulation models used in the automatic design process. It is commonly understood that the reality gap manifests itself as a drop in performance when control software developed in simulation is used to control physical robots. Yet, often disregarded is the relative nature of this performance drop: the reality gap does not affect equally all instances of control software. Indeed, one might observe a rank inversion: control software A might perform better than control software B in simulation, but perform worse on robots. The possibility of rank inversion undermines any performance comparison made in simulation. It would thus seem the only way to assess control software is in robot experiments, which are costly and time consuming. We argue it is unnecessary to assume reality is more complex than simulation models for the effects of the reality gap to occur. Indeed, we show that performance drop and rank inversion can occur if one automatically designs control software in simulation using a model and then assesses it in simulation on another model—what we call a pseudo-reality. Our results suggest that an appropriately conceived pseudo-reality could be used to test automatically-generated control software for performance drop and rank inversion, without performing robot experiments.

1 Introduction

The reality gap is one of the main issues in the automatic design of robot swarms [17]. A robot swarm is a highly redundant, self-organized, and decentralized system [1,13,39]. Designing the individual rules that lead to the desired collective behavior is difficult. Methods to guide the designers exist for some specific collective behaviors and under some hypotheses [2,8,23,38]. However, a generally applicable methodology is still missing.

Automatic design methods [7,17] eliminate the burden of manually decomposing the desired global behavior into the appropriate microscopic behaviors of the individuals. By maximizing a mission-dependent performance measure, an

---

All experiments were performed by AL. The paper was drafted by AL and revised by the two authors. The research was directed by MB.

M. Dorigo et al. (Eds.): ANTS 2018, LNCS 11172, pp. 109–122, 2018.
https://doi.org/10.1007/978-3-030-00533-7_9

optimization algorithm searches for an appropriate instance of control software to be installed on each individual robot. Generally, the optimization process relies on simulation. Methods have been proposed that (could possibly or have been demonstrated to) operate directly on robot hardware [9,12,22,28,30,41,43]. Although these methods are promising to adapt/fine-tune behaviors to the environment, they do not appear to be an alternative to simulation-based design due to safety concerns and to the limited solution space they can explore [17]. When the design is performed in simulation, a resulting instance of control software is likely to be fine-tuned to the specific simulation model [15], which should not be expected to perfectly reproduce the real world. Due to the differences between simulation and reality, which are commonly referred to as the *reality gap* [10,27], a performance drop typically occurs when an instance of control software designed in simulation is assessed on physical robots.

An issue that is often overlooked is that the occurrence of performance drops due to the reality gap is a relative problem: each instance of control software might be affected to a different extent. The relative nature of performance drops might result in what we shall call a *rank inversion*: control software A outperforms control software B in simulation, but B outperforms A when assessed on the physical robots. Rank inversions can be observed when comparing instances of control software produced by different design methods [19], or by the same one at different steps along the optimization process [4]. Indeed, Birattari et al. [4] observed a phenomenon that they called *overdesign*: past an optimal number of steps of the optimization process, the performance obtained in reality diverges from the one obtained in simulation.

In the literature, performance drops due to the reality gap are commonly explained by saying that reality is more complex than simulations—or equivalently, that simulations are too simplistic [29,35].

In this work, we argue that it is not necessary to assume that reality is more complex than simulation for the effect of the reality gap to occur. More precisely, we contend that performance drops that lead to rank inversion can be observed even if the model under which control software is designed is not a simplistic version of the context/conditions under which it is eventually assessed. We support our contention with a set of simulation-only experiments in which we create an artificial reality gap.

Creating an artificial, simulation-only reality gap is not a novel contribution we make here for the first time. Koos et al. [29] already created a simulation-only reality gap between a simple simulator—used to design control software—and an accurate one—used for assessing it. The choice of creating a reality gap between a simple and a more complex simulator clearly reflects the common understanding discussed above, which is precisely what we challenge here. We maintain that it is not necessary to assume that control software is assessed under context/conditions that are more complex than those experienced in the design for the effects of the reality gap to manifest.

The artificial reality gap we create is based on two robot models: M_A and M_B. We design control software in simulation on model M_A and then we assess it, always in simulation, but relying on model M_B. We shall call a *pseudo-reality* any secondary model that we use for assessing control software—and that therefore plays the role of reality. Model M_A has been proposed by Francesca et al. [19] who used it to design control software that was eventually assessed on robots. We introduce here model M_B, which we conceived so that, when used as pseudo-reality to assess control software designed on M_A, it produces performance drops and rank inversions that are qualitatively similar to those observed by Francesca et al. [19].

A priori, it could be argued that M_A and M_B are equally complex as they share the same nature—see Sect. 3. Nonetheless, to completely exclude the possibility that the observed effects are the results of an undesired higher complexity of M_B, we consider both the case in which we use M_A for the design and M_B for the assessment, and the case in which we invert the roles of the two models. As we show in Sect. 4, qualitatively similar drops and inversions appear in both cases. This substantiates our contention, and indicates that the effects of the reality gap can manifest even when the design model is not a simplistic version of the one used in the assessment, possibly due to the fact that control software *overfits* the former.

Besides shedding further light on the nature of the reality gap, this study suggests that creating an artificial, simulation-only version of it could have useful practical implications. For example, it would dispense researchers from costly and time consuming robot experiments that, at the moment, are necessary to tell whether a design method is more prone than another one to performance drops, whether a rank inversion should be expected, or whether to stop an optimization process to prevent the *overdesign* phenomenon to occur.

2   Related Work

Several approaches have been proposed to cross the reality gap effectively—that is, to limit the performance drop of control software. However, none of these approaches have been studied in details, no extensive comparison has been made, and the reality gap remains a major issue in the automatic design of robot control software [17,40]. Approaches to cross the reality gap have mainly been proposed in the context of evolutionary robotics for single robots. Nonetheless, they are typically general enough to be relevant to any design method based on off-line simulation, both for single- and multi-robot systems.

Behind these approaches, we see two main lines of reasoning. On the one hand, some researchers have aimed at reducing the differences between simulation and reality as much as possible [6,27,29,33,44]. They were driven by the assumption that a smooth transition from simulation to reality would occur if simulation reproduced relevant real-world dynamics accurately. On the other hand, other researchers have striven to make control software robust to differences [18,19,25,26,42]. They were driven by the assumption that differences between simulation and reality are eventually unavoidable. Each of these lines of reasoning were developed with a focus either on simulation models [6,25,27,33,44] or on the design method [18,19,29,42]. In the first case,

Table 1. Taxonomy of the most significant approaches proposed in the literature to cross the reality gap. We group the approaches according to the main line of reasoning followed in their development.

Focus on	Reducing differences between simulation and reality	Enhancing robustness of control software
Simulation models	Miglino et al. [33] Jakobi et al. [27] Bongard and Lipson [6] Zagal and Ruiz-Del-Solar [44]	Jakobi [25, 26]
Design methods	Koos et al. [29]	Floreano et al. [14, 16, 42] Francesca et al. [18, 19]

researchers focused on making simulation models more realistic or more general so as to render the design process more robust. In the second case, researchers focused on conceiving methods that either exploit regions of the search space that are accurately reproduced by the simulator or that are intrinsically more robust than traditional methods. See Table 1 for a taxonomy.

Reducing Differences Between Simulation and Reality—Focus on Simulation Models. Miglino et al. [33] were the first to propose guidelines for reducing differences between simulation models and reality. They suggested to (i) use samples from the robot's sensors and actuators; (ii) add conservative noise to models; and (iii) continue the design process in reality, should an unacceptable performance drop be observed. Similarly, Jakobi et al. [27] insisted on the importance of adding appropriate levels of noise to models. Since then, using real data in simulation and fine-tuning noise models have become common practice [40]. Bongard and Lipson [6] proposed a method based on the co-evolution of control software and simulator. While optimizing the control software, the method improves the simulation models using sensor readings gathered in robot experiments. Zagal and Ruiz-Del-Solar [44] developed a method in which differences between performance observed in simulation and in reality are used to tune the parameters of the simulation.

Reducing Differences Between Simulation and Reality—Focus on Design Methods. Koos et al. [29] proposed a multi-objective method that aims at constraining the design process to instances of control software whose behavior is accurately simulated. The method relies on a model to estimate the differences between performance in simulation and reality. The model is updated based on physical-robot evaluations of instances of control software generated by the design process. To assess the proposed method, the authors performed experiments with two different robotic platforms. They also performed experiments in a fully simulated setting in which the role of the physical-robot evaluations was played by highly-realistic simulations. In other terms, the authors artificially created a simulation-only reality gap problem between a simple and a more accurate simulator.

Enhancing Robustness of Control software—Focus on Simulation models. Jakobi [25,26] was the first to explicitly aim at producing control software that is robust to differences between simulation and reality. The method he proposed is based on two devices: (i) model only the robot-robot and robot-environment interactions that are meaningful to obtain the desired behavior, and (ii) apply random variations on all aspects of the simulation.

Enhancing Robustness of Control Software—Focus on Design Methods. Floreano et al. [16,42] applied an on-line adaptation mechanism to the parameters of a neuro-controller. The behavior developed was observed to transfer smoothly from simulation to reality [14]. Francesca et al. [18,19] observed that the reality gap resembles the generalization problem of supervised learning. They conjectured that evolutionary robotics is seriously affected by the reality gap due to an excessive representational power of neural networks. As a result, it overfits the conditions experienced during the design process. Guided by their conjecture, the authors developed design methods with restricted representational power: Vanilla [19] and Chocolate [18]. Their experiments have shown that the control software produced by these methods crosses the reality gap more satisfactorily than a traditional evolutionary robotics method they called EvoStick [18,19].

3   Materials and Methods

In this section, we describe the robots, the automatic design methods, the simulation models and the protocol used in the experiments presented hereafter.

Robots (Simulated). We simulate an extended version of the e-puck robot [20,34] using the ARGoS3 simulator [36] (version 3.0.0-beta45). For the purpose of this study, we consider a subset of the sensors and actuators the robot is equipped with. The control software has access to variables that abstract sensors and actuators. These variables are updated every 100 ms. The reference model RM1.1 [24] of Table 2 formally defines the sensors and actuators and the corresponding variables.

The accessible sensors comprise eight infrared proximity sensors for detecting obstacles ($prox_i$) and for measuring ambient light ($light_i$), three ground sensors for sampling the grayscale color of the ground situated under the robot ($ground_i$), and a range-and-bearing board used for local communication between robots [21]. Upon reception of a message via the range-and-bearing board, an e-puck can estimate the relative distance and angle of the emitting robot. At each time step, the relative distance and angle of all perceived neighbors are lumped into a vector (V_d) representing a virtual attraction force towards the neighbors. In addition to this direction vector V_d, the control software has also access to the number of perceived neighbors (n).

The control software also controls actuators: the motors of the wheels. The e-pucks are driven by a two-wheeled differential steering system. The control software dictates the displacement of the robot via two velocity variables (v_l and v_r).

Table 2. Reference model RM1.1 [24]. Sensors and actuators of the extended version of the e-puck robot simulated in the experiments.

Sensor/actuator	Variables
Proximity	$prox_i \in [0, 1]$, with $i \in \{0, ..., 7\}$
Light	$light_i \in [0, 1]$, with $i \in \{0, ..., 7\}$
Ground	$ground_i \in \{white, gray, black\}$, with $i \in \{0, ..., 2\}$
Range-and-bearing	$n \in \{0, ..., 19\}$ and $V_d \in ([0, 0.7]\, m, [0, 2\pi])$
Wheels	$v_l, v_r \in [-0.12, 0.12]\, m/s$

Design Methods. In this section, we briefly describe the three automatic design methods considered in the experiments: EvoStick [19], Vanilla [19], and Chocolate [18]. We refer the readers to the original papers for their detailed description. The implementations are publicly available [31].

EvoStick is an implementation of the classical evolutionary robotics setup. An evolutionary algorithm optimizes the parameters of a fully connected, feedforward, neural network. The neural network comprises 24 input and 2 output nodes that are directly connected. The inputs and outputs are defined on the basis of the reference model RM1.1 (see Table 2). More precisely, the inputs are allocated as follows: 8 for the readings of the proximity sensors, 8 for the readings of the light sensors, 3 for the readings of the ground sensors, 1 for the number of neighbors, and 4 for the scalar projections of the vector V_d on four unit vectors distributed around the robot. The outputs of the neural network are the speed of the left and right wheels of the e-puck.

Vanilla produces control software in the form of probabilistic finite state machines by assembling preexisting modules. A module is either a *behavior* or a *transition*. A behavior is an action that can be performed by the robot, while a transition is a condition on the environment perceived by the robot. All modules operate on the variables presented in the reference model RM1.1 of Table 2, and some of the modules have parameters that adjust their functioning. In a probabilistic finite state machine, the transitions (i.e., the edges) regulate the succession of behaviors (i.e., states) that alternatively control the robot by determining the values of the output variables.

Similarly to Vanilla, Chocolate is a modular automatic design method. The methods differ by the optimization algorithm they use: Vanilla uses F-race [3, 5] and Chocolate uses Iterated F-race [32]. In order to conceive probabilistic finite state machines, Vanilla and Chocolate have at their disposal the same set of preexisting modules: six behaviors and six transitions. In addition to the topology of the probabilistic finite state machine, Vanilla and Chocolate also tune the parameters of the modules. The design space explored by the two methods is restricted to all possible probabilistic finite state machines composed of up to four states (i.e., behaviors) and up to four edges (i.e., transitions) departing from each state. Chocolate has been shown to outperform Vanilla [18].

Models. We use the two e-puck models, namely M_A and M_B, described in Table 3. Model M_A is the same model used during the design process of the experiments ran by Francesca et al. [19]. We generated model M_B by modifying actuator and sensor noise of model M_A. We did so via trial-and-error so that, when model M_B is used as a pseudo-reality to assess the performance of control software automatically generated on the basis of model M_A, we obtain a rank inversion that qualitatively resembles the one observed by Francesca et al. [19].

Table 3. The two e-puck models. The values for the proximity, light and ground sensors are the range of the uniform white noise added to the readings of the sensors. The value for the range-and-bearing sensor is the probability of failing to receive a message sent by a robot within communication range. The value for the wheels actuator is the standard deviation of Gaussian white noise added to the speed of the left and right wheels.

Sensor/actuator	M_A	M_B
Proximity	$[-0.05, 0.05]$	$[-0.05, 0.05]$
Light	$[-0.05, 0.05]$	$[-0.90, 0.90]$
Ground	$[-0.05, 0.05]$	$[-0.05, 0.05]$
Range-and-bearing	0.85	0.90
Wheels	0.05	0.15

Protocol. We consider two missions: AGGREGATION and FORAGING. For each mission, we define an objective function to be maximized. The same objective function is used for both designing control software and assessing its performance. We run experiments in which the control software is designed by the three design methods described above: `EvoStick`, `Vanilla`, and `Chocolate`. We consider a homogeneous swarm composed of $N = 20$ robots operating in a dodecagonal arena for a time period of 250 s. The arena is delimited by walls and its surface area is $4.91 \, \text{m}^2$.

For each mission, we consider two stages: S_{AB} and S_{BA}. In stage S_{AB}, each automatic design method produces control software via simulations based on model M_A; the control software is then assessed with simulations based on model M_B. To study the generalization capability of the control software produced, the performance evaluated on model M_B is compared to the one evaluated on model M_A. In stage S_{BA}, the roles of the two models are inverted: control software is produced on M_B and then assessed on M_A. Also in this case, the performance on M_A is compared to the one on M_B to study the generalization capability of the control software. In other terms, in stage S_{AB} the pseudo-reality is model M_B; whereas in stage S_{BA}, it is model M_A.

Each design method is run with a design budget of 200 000 simulations. For each mission and each stage S_{xy}—where by x and y we indicate A and B, or viceversa—each design method is run 20 times on model M_x and produces therefore a total of 20 instances of control software. For the assessment, each of

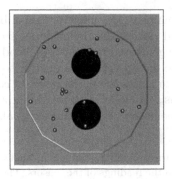 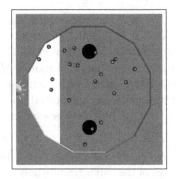

Fig. 1. Simulated environments: AGGREGATION (*left*) and FORAGING (*right*).

these instances is evaluated 20 times on model M_x, and 20 times on model M_y to study their generalization capability.

We present the results by means of notched box-and-whiskers boxplots. The notches indicate the 95% confidence interval around the median. If the notches of two boxes do not overlap, the difference between their medians is significant [11]. Moreover, we aggregate the results of the two stages to estimate the performance drop experienced by each design method. For each method, we report a 95% confidence interval on the difference between the performance observed on models M_x and M_y.[1] We also highlight a lower bound D on the difference between the performance drop of EvoStick and Vanilla—confidence 95%. We focus on EvoStick and Vanilla as Francesca et al. [19] assessed their performances for the same mission in robot experiments.

4   Experiments

In this section, we provide details on the two missions considered and we report the results of our experiments. Figure 1 shows the simulated environments in which the swarm operates. The missions have already been studied in [19]. We report in the following only the information that is strictly needed to understand the results. We refer the reader to the original paper for the details.

4.1   Aggregation

In this experiment, the swarm must aggregate on one of two black areas, named a or b. These black areas have a radius of $0.35\,\mathrm{m}$. The performance of the swarm is measured via the following objective function:

$$F_A = \max(N_a, N_b)/N,$$

---

[1] Confidence intervals are computed based on the statistic of the paired Wilcoxon signed rank test. The normal approximation is adopted as the sample size is larger than 50. The implementation used is the one of R's *stats* package [37].

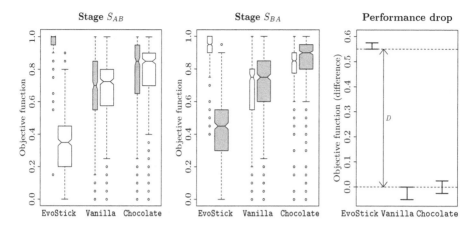

Fig. 2. AGGREGATION. *Left* and *center*: narrow boxes represent the performance assessed on the model used during the design step; wide boxes represent the performance assessed in pseudo-reality. Gray boxes represent performance assessed on model M_A; white boxes represent performance assessed on model M_B. *Right*: the segments represent a 95% confidence interval on the performance drop experienced by each method—aggregated across the two stages. D is a bound on the difference between the performance drop of EvoStick and Vanilla.

where $N = 20$ is the total number of robots composing the swarm; and N_a and N_b are the number of robots that at the end of the experimental run are located on a and b, respectively. The objective function is maximized when, at the end of a run, all robots are either on a or on b.

The results of this experiment show a rank inversion—see Fig. 2 (*left* and *center*). In each stage S_{xy}, EvoStick performs significantly better than both Vanilla and Chocolate when the performance of the control software they produced is assessed on model M_x. On the other hand, EvoStick performs significantly worse than both Vanilla and Chocolate when the performance is assessed on model M_y.

Indeed, the performance of the control software designed by EvoStick drops noticeably when assessed in pseudo-reality: the drop is at least 0.55 (confidence 95%). In the case of Vanilla and Chocolate, the drop is significantly smaller: at most 0.00 and 0.02 respectively (confidence 95%). See Fig. 2 (*right*).

In both stages, the rank inversion between EvoStick and Vanilla is similar to the one observed by Francesca et al. [19] on the same mission. This corroborates further the conjecture of Francesca et al. [19] according to which EvoStick is affected by the reality gap more seriously than Vanilla and Chocolate because of its higher representational power. At least in this experiment, the artificial reality gap we created with the models M_A and M_B was able to qualitatively predict performance drop and rank inversion for EvoStick and Vanilla.

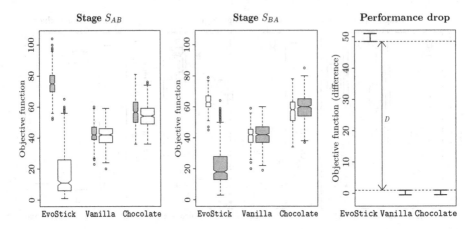

Fig. 3. FORAGING. See caption of Fig. 2 for the explanation of width and color of boxes.

4.2 Foraging

In this experiment, the swarm must perform an idealized form of foraging. We consider that an individual robot has retrieved an object when it enters the nest after having visited a foraging source. Two sources are available, and are represented by black circular areas of radius 0.15 m. The nest is represented by a white area situated at a distance of 0.45 m from the two black areas. A light source is placed behind the nest to help the robots locate it.

The performance of the swarm is measured by the number of objects retrieved during the whole experimental run. It is computed via the following objective function:

$$F_F = N_o,$$

where N_o is the total number of objects retrieved.

Also in this experiment, we observe a rank inversion—see Fig. 3 (*left* and *center*). EvoStick performs significantly better than Vanilla and Chocolate when the performance of the control software produced is assessed on model M_x, but significantly worse when the performance is assessed on model M_y.

When assessed in pseudo-reality, the performance of the control software designed by EvoStick drops by at least 48 objects (confidence 95%), whereas in the case of Vanilla and Chocolate, the drop is at most of 1 object (confidence 95%). See Fig. 3 (*right*).

Also on this mission, the rank inversion between EvoStick and Vanilla is similar to the one observed by Francesca et al. [19], which corroborates further their conjecture. Also here, the artificial reality gap yields good qualitative predictions.

5   Conclusions

With this paper, we shed further light on the reality gap. Specifically, we investigated how and under what conditions the effects of the reality gap manifest. We contend that, for the effects of the reality gap to manifest, it is unnecessary to assume that the control software is assessed under context/conditions that are more complex than those experienced in the design.

To substantiate our contention, we conceived a set of simulation-only experiments in which we created an artificial reality gap based on two robot models M_A and M_B. We used M_A for the design and M_B for the assessment; we then inverted the role of the two models. In both cases, we observed performance drop and rank inversion: a design method (EvoStick) performed significantly better than the others (Vanilla and Chocolate) when the control software they produced was assessed on the same model used in the design, but significantly worse on the other one. Having observed performance drop and rank inversion both when (i) designing on M_A and assessing on M_B, and when (ii) designing on M_B and assessing on M_A, we can exclude that the effects of the reality gap emerge only due to the fact that the design is performed on a simplistic model that fails to reproduce the complexity of the environment in which the final assessment is performed.

Furthermore, our results indicate that simulation-only experiments could be used to tell whether and to what extent automatic design methods are prone to performance drop and rank inversion. This might have useful practical implications. Indeed, we foresee that an artificial, simulation-only reality gap could be used to *validate* automatically-generated control software and to predict its real-world performance. We have in mind here a development process that mimics the classical machine learning procedure based on training, validation, and test set. We imagine a development process in which control software is generated using a model, validated on another model to predict its ability to cross the reality gap, and eventually tested in the real world.

Future work will be dedicated to study whether an artificial reality gap can reliably predict real-world performance. Moreover, future work should be dedicated to defining reliable and meaningful ways to generate a pair of models that can properly serve as an artificial reality gap. In this work, we considered two models that differ in the noise level. Other differences between the models could be considered, which could be more appropriate. Finally, future research should be dedicated to quantifying the difference between two models. A quantity measuring the difference between two models could be used to characterize the severity of the artificial reality gap associated with them.

Acknowledgements. The project has received funding from the European Research Council (ERC) under the European Union's Horizon 2020 research and innovation programme (grant agreement No 681872). Mauro Birattari acknowledges support from the Belgian *Fonds de la Recherche Scientifique* – FNRS.

References

1. Beni, G.: From swarm intelligence to swarm robotics. In: Şahin, E., Spears, W.M. (eds.) SR 2004. LNCS, vol. 3342, pp. 1–9. Springer, Heidelberg (2005). https://doi.org/10.1007/978-3-540-30552-1_1
2. Berman, S., Kumar, V., Nagpal, R.: Design of control policies for spatially inhomogeneous robot swarms with application to commercial pollination. In: Zexiang, L. (ed.) IEEE International Conference Robotics and Automation, ICRA, pp. 378–385. IEEE Press, Piscataway NJ (2011)
3. Birattari, M.: Tuning Metaheuristics: A Machine Learning Perspective. Springer, Berlin Heidelberg, Germany (2009). https://doi.org/10.1007/978-3-642-00483-4
4. Birattari, M., Delhaisse, B., Francesca, G., Kerdoncuff, Y.: Observing the effects of overdesign in the automatic design of control software for robot swarms. In: Dorigo, M., Birattari, M., Li, X., López-Ibáñez, M., Ohkura, K., Pinciroli, C., Stützle, T. (eds.) ANTS 2016. LNCS, vol. 9882, pp. 149–160. Springer, Cham (2016). https://doi.org/10.1007/978-3-319-44427-7_13
5. Birattari, M., Stützle, T., Paquete, L., Varrentrapp, K.: A racing algorithm for configuring metaheuristics. In: Langdon, W., et al. (eds.) Proceedings of the Genetic and Evolutionary Computation Conference, GECCO, pp. 11–18. Morgan Kaufmann, San Francisco (2002)
6. Bongard, J., Lipson, H.: Once more unto the breach: co-evolving a robot and its simulator. In: Pollack, J., et al. (eds.) Artificial Life IX: Proceedings of the Conference on the Simulation and Synthesis of Living Systems, pp. 57–62 (2004)
7. Bozhinoski, D., Birattari, M.: Designing control software for robot swarms: software engineering for the development of automatic design methods. In: ACM/IEEE 1st International Workshop on Robotics Software Engineering, RoSE, pp. 33–35. ACM, New York (2018)
8. Brambilla, M., Brutschy, A., Dorigo, M., Birattari, M.: Property-driven design for swarm robotics: a design method based on prescriptive modeling and model checking. ACM Trans. Auton. Adapt. Syst. 9(4), 17.1–17.28 (2015)
9. Bredeche, N., Montanier, J.M., Liu, W., Winfield, A.F.: Environment-driven distributed evolutionary adaptation in a population of autonomous robotic agents. Math. Comput. Model. Dyn. Syst. 18(1), 101–129 (2012)
10. Brooks, R.: Artificial life and real robots. In: Varela, F.J., Bourgine, P. (eds.) Towards a Practice of Autonomous Systems. In: Proceedings of the First European Conference on Artificial Life, pp. 3–10. MIT Press, Cambridge (1992)
11. Chambers, J., Cleveland, W., Kleiner, B., Tukey, P.: Graphical Methods For Data Analysis. Wadsworth, Belmont (1983)
12. Di Mario, E., Martinoli, A.: Distributed particle swarm optimization for limited-time adaptation with real robots. Robotica 32(02), 193–208 (2014)
13. Dorigo, M., Birattari, M., Brambilla, M.: Swarm robotics. Scholarpedia 9(1), 1463 (2014)
14. Floreano, D., Urzelai, J.: Evolution of plastic control networks. Auton. Robot. 11(3), 311–317 (2001)
15. Floreano, D., Husbands, P., Nolfi, S.: Evolutionary robotics. In: Siciliano, B., Khatib, O. (eds.) Springer Handbook of Robotics. Springer, Heidelberg (2008). https://doi.org/10.1007/978-3-540-30301-5_62
16. Floreano, D., Mondada, F.: Evolution of plastic neurocontrollers for situated agents. In: Maes, P., et al. (eds.) From animals to animats 4: Proceedings of the International Conference on Simulation of Adaptive Behavior, ETH Zurich (1996)

17. Francesca, G., Birattari, M.: Automatic design of robot swarms: achievements and challenges. Front. Robot. AI **3**(29), 1–9 (2016)
18. Francesca, G., et al.: AutoMoDe-Chocolate: automatic design of control software for robot swarms. Swarm Intell. **9**(2/3), 125–152 (2015)
19. Francesca, G., Brambilla, M., Brutschy, A., Trianni, V., Birattari, M.: AutoMoDe: a novel approach to the automatic design of control software for robot swarms. Swarm Intell. **8**(2), 89–112 (2014)
20. Garattoni, L., Francesca, G., Brutschy, A., Pinciroli, C., Birattari, M.: Software infrastructure for e-puck (and TAM). Technical report TR/IRIDIA/2015-004, IRIDIA, Université libre de Bruxelles, Belgium (2015)
21. Gutiérrez, Á., Campo, A., Dorigo, M., Donate, J., Monasterio-Huelin, F., Magdalena, L.: Open E-puck range and bearing miniaturized board for local communication in swarm robotics. In: Kosuge, K. (ed.) IEEE Int. Conf. Robot. Autom. ICRA, pp. 3111–3116. IEEE Press, Piscataway NJ (2009)
22. Haasdijk, E., Bredeche, N., Eiben, A.: Combining environment-driven adaptation and task-driven optimisation in evolutionary robotics. PloS One **9**(6), e98466 (2014)
23. Hamann, H., Wörn, H.: A framework of space-time continuous models for algorithm design in swarm robotics. Swarm Intell. **2**(2–4), 209–239 (2008)
24. Hasselmann, K., Ligot, A., Francesca, G., Birattari, M.: Reference models for AutoMoDe. Technical report TR/IRIDIA/2018-002, IRIDIA, Université libre de Bruxelles, Belgium (2018)
25. Jakobi, N.: Evolutionary robotics and the radical envelope-of-noise hypothesis. Adapt. Behav. **6**(2), 325–368 (1997)
26. Jakobi, N.: Minimal simulations for evolutionary robotics. Ph.D. thesis, University of Sussex, Falmer, UK (1998)
27. Jakobi, N., Husbands, P., Harvey, I.: Noise and the reality gap: the use of simulation in evolutionary robotics. In: Morán, F., Moreno, A., Merelo, J.J., Chacón, P. (eds.) ECAL 1995. LNCS, vol. 929, pp. 704–720. Springer, Heidelberg (1995). https://doi.org/10.1007/3-540-59496-5_337
28. König, L., Mostaghim, S.: Decentralized evolution of robotic behavior using finite state machines. Int. J. Intell. Comput. Cybern. **2**(4), 695–723 (2009)
29. Koos, S., Mouret, J.B., Doncieux, S.: The transferability approach: crossing the reality gap in evolutionary robotics. IEEE Trans. Evol. Comput. **17**(1), 122–145 (2013)
30. Lee, J.B., Arkin, R.C.: Adaptive multi-robot behavior via learning momentum. In: George Lee, C.S. (ed.) IEEE/RSJ International Conference on Intelligent Robots - IROS, pp. 2029–2036. IEEE Press, Piscataway (2003)
31. Ligot, A., Hasselmann, K., Delhaisse, B., Garattoni, L., Francesca, G., Birattari, M.: AutoMoDe, NEAT, and EvoStick: implementations for the E-puck robot in ARGoS3. Technical report TR/IRIDIA/2017-002, IRIDIA, Université libre de Bruxelles, Belgium (2017)
32. López-Ibáñez, M., Dubois-Lacoste, J., Pérez Cáceres, L., Birattari, M., Stützle, T.: The irace package: iterated racing for automatic algorithm configuration. Oper. Res. Perspect. **3**, 43–58 (2016)
33. Miglino, O., Lund, H., Nolfi, S.: Evolving mobile robots in simulated and real environments. Artif. Life **2**(4), 417–434 (1995)
34. Mondada, F., et al.: The E-puck, a robot designed for education in engineering. In: Gonçalves, P., Torres, P., Alves, C. (eds.) Proceedings of the 9th Conference on Autonomous Robot Systems and Competitions, pp. 59–65. Instituto Politécnico de Castelo Branco (2009)

35. Nolfi, S., Floreano, D., Miglino, G., Mondada, F.: How to evolve autonomous robots: different approaches in evolutionary robotics. In: Brooks, R.A., Maes, P. (eds.) Artificial Life IV: Proceedings of the Workshop on the Synthesis and Simulation of Living Systems, pp. 190–197. MIT Press, Cambridge (1994)
36. Pinciroli, C., et al.: ARGoS: a modular, parallel, multi-engine simulator for multi-robot systems. Swarm Intell. **6**(4), 271–295 (2012)
37. R Development Core Team: R: A Language and Environment for Statistical Computing. R Foundation for Statistical Computing, Vienna, Austria (2008). http://www.R-project.org
38. Reina, A., Valentini, G., Fernández-Oto, C., Dorigo, M., Trianni, V.: A design pattern for decentralised decision making. PLoS One **10**(10), e0140950 (2015)
39. Şahin, E.: Swarm robotics: from sources of inspiration to domains of application. In: Şahin, E., Spears, W.M. (eds.) SR 2004. LNCS, vol. 3342, pp. 10–20. Springer, Heidelberg (2005). https://doi.org/10.1007/978-3-540-30552-1_2
40. Silva, F., Duarte, M., Correia, L., Oliveira, S., Christensen, A.: Open issues in evolutionary robotics. Evol. Comput. **24**(2), 205–236 (2016)
41. Silva, F., Urbano, P., Correia, L., Christensen, A.L.: odNEAT: an algorithm for decentralised online evolution of robotic controllers. Evol. Comput. **23**(3), 421–449 (2015)
42. Urzelai, J., Floreano, D.: Evolutionary robotics: coping with environmental change. In: Whitney, L.D., et al. (eds.) Proceedings of Conference on the Genetic and Evolutionary Computation Conference, GECCO, pp. 941–948. Morgan Kaufmann, San Francisco (2000)
43. Watson, R., Ficici, S., Pollack, J.: Embodied evolution: distributing an evolutionary algorithm in a population of robots. Robot. Auton. Syst. **39**(1), 1–18 (2002)
44. Zagal, J.C., Ruiz-Del-Solar, J.: Combining simulation and reality in evolutionary robotics. J. Intell. Robot. Syst. **50**(1), 19–39 (2007)

Optimization of Swarm Behavior Assisted by an Automatic Local Proof for a Pattern Formation Task

Mario Coppola$^{(\boxtimes)}$ and Guido C. H. E. de Croon

Faculty of Aerospace Engineering,
Delft University of Technology, Delft, The Netherlands
{m.coppola,g.c.h.e.decroon}@tudelft.nl

Abstract. In this work, we optimize the behavior of swarm agents in a pattern formation task. We start with a local behavior, expressed as a local state-action map, that has been formally proven to lead the swarm to always eventually form the desired pattern. We seek to optimize this for performance while keeping the formal proof. First, the state-action map is pruned to remove unnecessary state-action pairs, reducing the solution space. Then, the probabilities of executing the remaining actions are tuned with a genetic algorithm. The final controllers allow the swarm to form the patterns up to orders of magnitude faster than with the original behavior. The optimization is found to suffer from scalability issues. These may be tackled in future work by automatically minimizing the size of the local state-action map with a further direct focus on performance.

1 Introduction

Collaboration between autonomous agents, while already a difficult task in itself, becomes increasingly challenging when dealing with swarm of robots with limited on-board sensing and computing capacity. In recent work, detailed in [1], we introduced a method to extract local behaviors with which very limited agents could arrange into a desired shape. The agents were: homogeneous (identical and without hierarchy), anonymous (did not have identities), reactive (memoryless), could not communicate, did not have global position information, did not (explicitly) know the goal of the swarm, and operated asynchronously in an unbounded space. The only knowledge available to the agents in the decision making was: (1) a common heading direction (i.e., North), (2) the relative location of their neighbors within a maximum range (e.g., similar to the robotic system in [2]). Despite such limited agents, it was possible to define the local agent behavior such that a desired pattern would always emerge, with a formal proof that this would be reached from any initial configuration.

Simulations in [1] further showed that the swarms indeed always eventually reached the desired formation by assuming random (but feasible) actions on the part of the agents. However, as the agents moved randomly and asynchronously,

© Springer Nature Switzerland AG 2018
M. Dorigo et al. (Eds.): ANTS 2018, LNCS 11172, pp. 123–134, 2018.
https://doi.org/10.1007/978-3-030-00533-7_10

even simple patterns with a few agents were found to take hundreds of actions before completion, and this number appeared to grow exponentially with the size of the swarm and, in turn, with the complexity of the pattern. This becomes an issue if the algorithm is to be used on real robots with limited battery life. In this work, we thus explore how the behavior of the agents can be optimized so that they do not just "eventually" form the pattern, but do so efficiently. In doing so, we also explore a novel use of evolutionary algorithms in the context of swarm intelligence, as we perform an optimization procedure while maintaining the conditions for the formal proof that the goal will always eventually be achieved.

This paper is organized as follows. In Sect. 2 we review relevant literature and introduce the context of this research. Then, Sect. 3 summarizes the framework used to enable the swarm to form a desired pattern. The optimization methodology is detailed in Sect. 4, followed by an assessment in Sect. 5. In Sect. 6, we summarize the findings and discuss future work.

2   Related Work and Research Context

Evolutionary algorithms can search through vast solution spaces and discover solutions to complex problems, and are thus a popular approach to dealing with the intricacies of swarm robotics and extracting valid local behaviors [6,14]. They have been used for numerous architectures, including: neural networks [3,11], state machines [7], behavior trees [12], and grammar rules [5]. When applied to swarms, the following issues typically arise:

1. As the number of agents grows, the complexity of the solution and the size of the potential solution space grow [10,16].
2. The evolutionary algorithm is likely to drift into undesired local optima. This may happen due to deceptive fitness functions or bootstrap issues [9,17].
3. As the size of the swarm grows, the iteration time needed to find a solution grows. This can be due to, for instance:
   (a) The computational requirements needed to evaluate the fitness of a controller are higher because of the need to simulate a larger swarm.
   (b) Depending on the task, it might take longer for the desired behavior to emerge, requiring a longer simulation time upon each evaluation trial.
   (c) Each controller may have to be simulated multiple times in order to accurately assess its expected average fitness [18].

In state of the art, the problems above have mostly been tackled in two ways. First, there are methods that try to deal with the broad solution space. For example, Gomes et al. [8] used novelty search to encourage a broader exploration of the solution space. The second way is to use global knowledge and insights to aid the evolutionary process. For example, Duarte et al. [3] partitioned complex swarm behavior into simpler sub-behaviors. Hüttenrauch et al. [10], with a focus on deep reinforcement learning, used global information to guide the learning process towards a solution. Alternatively, Trianni et al. [18] and Ericksen et al.

[4] explored whether evolved behaviors for smaller swarms could generalize to larger swarms.

Evolutionary approaches are thus typically used to establish the behavior needed to achieve the global goal, but it is not known whether the final behavior generalizes to all initial conditions. In this work we present the first steps to an alternate approach towards achieving an optimum swarm behavior: optimizing a behavior that is formally proven to always eventually lead to the emergent global goal. In this approach, the proof that the goal will always eventually be achieved remains preserved throughout. The focus of the optimization procedure is not on figuring out *how* to solve the problem, but on how to do it more efficiently while ensuring that the resulting behavior still guarantees that any initial condition will always eventually lead to the goal. Using the framework from [1], and limited agents as introduced therein, we attempt to optimize the local behavior of the agents in finite pattern formation tasks of increasing complexity. We begin from a local state-action map given to the agents. This state-action map can be verified to always eventually lead to the goal, but is not optimized for performance, as any agent in a given state can select its action randomly from several options, with equal probability. We then tune this state-action map with the goal of simplifying the behavior and minimizing the number of actions needed, on average, to achieve the final pattern when starting from an arbitrary initial configuration. More specifically, we do the following:

1. Restrict the possible actions that an agent can take when in a given state, subject to the constraint that it must still be provable that the global goal will emerge. This minimizes the size of the local state-action map of the agents, and in turn the size of the possible solution space.
2. We take the minimized state-action map and we apply an evolutionary algorithm to optimize the probability of executing each action.

The desired final outcome is a probabilistic local state-action map that enables the agents to arrange into the desired pattern most efficiently when starting from a random initial configuration.

3   Framework and Approach to Pattern Formation

This section summarizes the pattern formation methodology, which can be found in more detail in [1]. For the sake of brevity, in this work we will assume that the swarm operates in a grid world and in discrete time. However, as demonstrated in [1], the behavior can also be used in continuous time and space with asynchronous agents.

Consider N robots that exist in an unbounded discrete grid world and operate in discrete time. In the case studied in this paper, each robot $\mathcal{R}_i$ can sense the location of its neighbors in the 8 grid points that surround it, as depicted in Fig. 1a. This is the local state s_i of the agent (which is all the information that it has). The local state space $\mathcal{S}$ consists of all combinations of neighbors that it could sense, such that $|\mathcal{S}| = 2^8$. At time step $k = 0$, we assume the swarm begins

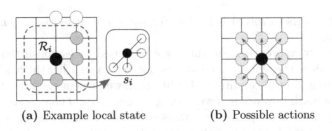

(a) Example local state   (b) Possible actions

Fig. 1. Depictions of local state and the actions that an agent can take

in a connected topology forming an arbitrary pattern P_0. At each time step, one random robot in the swarm takes an action, whereby it can move to any of the 8 grid points surrounding it, as depicted in Fig. 1b. This is the action space of the agents, denoted $\mathcal{A}$. Moreover, if a robot takes an action, then it will not take an action at the next time step (unless no other robot can take an action).

The goal of the swarm is to rearrange from its initial arbitrary pattern P_0 into a desired pattern P_{des}. This is achieved using the following principle. The local states that the agents are in when P_{des} is formed are extracted, this forms a set of local desired states $\mathcal{S}_{des} \in \mathcal{S}$, as depicted by the examples in Fig. 2. If robot $\mathcal{R}_i$ finds itself in any state $s_i \in \mathcal{S}_{des}$ then it is instructed to not move, because, from its perspective, the goal has been achieved. In [1], it is shown that, given a P_{des} and the corresponding $\mathcal{S}_{des}$, it can be automatically verified whether the local desired states will uniquely form P_{des}, or whether they can also can give rise to spurious global patterns. In the following, we assume that set of local desired states has passed this verification. Therefore, until P_{des} is formed, at least one agent will be in a state $s \notin \mathcal{S}_{des}$ and will seek to amend the situation. The swarm will then keep reshuffling until P_{des} forms.

When an agent $\mathcal{R}_i$ is in a state $s_i \notin \mathcal{S}_{des}$, it can execute an action. From the state space and action space, we can extract a state-action map $\mathcal{Q} = (\mathcal{S} \backslash \mathcal{S}_{des}) \times \mathcal{A}$. However, not all actions should be allowed. The actions that: (a) cause collisions and (b) cause local separation of the swarm are eliminated from $\mathcal{Q}$, because they are not safe. From this, we extract a *safe* state-action map $\mathcal{Q}_{safe}$, where $\mathcal{Q}_{safe} \subseteq \mathcal{Q}$. From this process, there will also emerge some local states that cannot take any safe actions. An agent in such a state will not be able to move or else it will either collide with other agents or possibly cause separation of the swarm. We refer to such states as *blocked* states. The set of blocked states is denoted $\mathcal{S}_{blocked}$. By contrast, there are states where an agent will be capable of moving away from its neighborhood without issues. We call these states *simplicial*. The set of simplicial states is denoted $\mathcal{S}_{simplicial}$. Figure 3 shows examples of blocked states and simplicial states.

Now consider a graph $G_{\mathcal{S}} = (V, E)$. Let the nodes of $G_{\mathcal{S}}$ be all states that the agents can be in, such that $V = \mathcal{S}$. The edges of $G_{\mathcal{S}}$ are all local transitions between states. These are all the state transitions that an agent can locally experience as a result of the changing environment when it, or any other agent in the swarm, moves. More specifically, $G_{\mathcal{S}}$ is the union of three subgraphs:

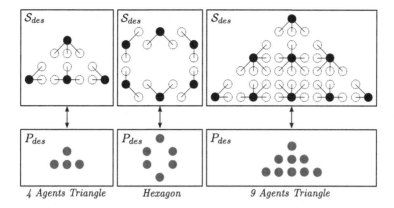

Fig. 2. Set of desired states $\mathcal{S}_{des}$ for the exemplary patterns treated in this paper, featuring patterns of increasing complexity and size (from left to right)

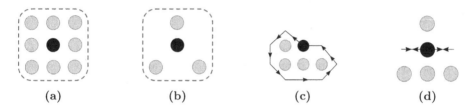

Fig. 3. Examples of: (a) a state $s \in \mathcal{S}_{blocked}$, due to it being surrounded; (b) a state $s \in \mathcal{S}_{blocked}$, because any motion will cause the swarm to locally disconnect; (c) a state $s \in \mathcal{S}_{active} \cap \mathcal{S}_{simplicial}$, because it can travel around all its neighbors; (d) a state $s \in \mathcal{S}_{active}$ but $s \notin \mathcal{S}_{simplicial}$, because it can move but it cannot travel around all its neighbors or else it might disconnect the swarm

$G_{\mathcal{S}}^1$ indicates all state transitions that an agent could go through by an action of its own, based on $\mathcal{Q}_{safe}$. $G_{\mathcal{S}}^2$ indicates all state transitions that an agent could go through by an action of its neighbors (which could also move out of view). $G_{\mathcal{S}}^3$ indicates all state transitions that an agent could go through if another agent, previously out of view, were to move into view and become a new neighbor. Furthermore, let $G_{\mathcal{S}}^{2r}$ be a subgraph of $G_{\mathcal{S}}^2$. $G_{\mathcal{S}}^{2r}$ only indicates the state transitions in $G_{\mathcal{S}}^2$ where a neighbor moves about the central agent, but *not* out of view. By analyzing certain properties of these graphs, it can be verified that the pattern P_{des} will eventually form starting from any initial pattern P_0. Specifically, the following conditions need to be met:

1. $G_{\mathcal{S}}^1 \cup G_{\mathcal{S}}^2$ shows that each state in $\mathcal{S}$ features a path to each state in $\mathcal{S}_{des}$.
2. For all states $s \in \mathcal{S}_{blocked} \cap \mathcal{S}_{\neg des}$, none of the cliques[1] of each state can be formed uniquely by agents that are in a state $s \in \mathcal{S}_{des} \cap \mathcal{S}_{simplicial}$.

---

[1] A *clique* is a connected set of an agent's neighbors. Without the central agent, the agents in each clique would remain connected with each other, but the different cliques would not be connected.

3. $G_{\mathcal{S}}^{2r}$ shows that all static states with two neighbors can directly transition to an active state.
4. $G_{\mathcal{S}}^{1}$ shows that any agent in state $s \in \mathcal{S}_{active} \cap \mathcal{S}_{simplicial}$ could move around all its local neighbors (as exemplified in Fig. 3c).
5. $G_{\mathcal{S}}^{3}$ shows that any agent in any state $s \in \mathcal{S}_{des} \cup \mathcal{S}_{blocked}$ will always, by the arrival of a new neighbor in an open position, transition into an active agent (with the exception of any agent that is, or becomes, surrounded).

The motivations behind these conditions can be found in [1]. They are not repeated here due to page restrictions. However, they essentially ensure that all agents will keep moving around with sufficient freedom for the swarm to reshuffle without deadlocks or endless loops until the pattern is achieved. These conditions are local in nature; they focus on the local perception in an agent's limited sensing range and the actions that the agent could take as a result. The advantage of this is that checking whether the conditions are met is independent of the size of the swarm, avoiding the combinatorial explosion that would otherwise ensue. This makes it possible to verify them within a heuristic optimization process. This proof, combined with the fact that it can deal with very limited agents (anonymous, homogeneous, memoryless, with limited range sensing, and without needing any communication, global knowledge, or seed agents) moving in space, sets the work in [1] apart from other works such as [13,15,19].

4   Optimization Methodology

Following the framework in Sect. 3, we can know whether a given pattern P_{des} will eventually form if the agents act based on its corresponding $\mathcal{Q}_{safe}$. However, this may take a significant amount of actions, due to the fact that any active agent could move at any time step and select a random action from its options in $\mathcal{Q}_{safe}$. The objective of this article is to minimize the number of actions that the agents will take, on average, to form P_{des} when starting from an arbitrary pattern P_0. In Sects. 4.1 and 4.2 we take two preliminary steps to automatically, at the local level, prune $\mathcal{Q}_{safe}$ from unnecessary actions. This will lead us to a new set $\mathcal{Q}_{reduced} \subseteq \mathcal{Q}_{safe}$ which is minimally sufficient to achieve the global goal. This reduces randomness in the system and restricts the solution space. Then, in Sect. 4.3, we use an evolutionary algorithm to tune the probability of taking each action in $\mathcal{Q}_{reduced}$, leading to a final controller. Throughout all steps, measures will be taken to ensure that the conditions of the proof (as detailed at the end of Sect. 3) remain respected. We apply this procedure to the patterns from Fig. 2.

4.1   Step 1: A-Priori Local Reduction of Active States

$\mathcal{S}$ can be sub-divided in two sets: $\mathcal{S}_{active}$, in which agents take an action based on $\mathcal{Q}_{safe}$, and $\mathcal{S}_{des} \cup \mathcal{S}_{blocked}$, in which the agents do not take actions. For simplicity, the latter is grouped under the umbrella set $\mathcal{S}_{static} = \mathcal{S}_{des} \cup \mathcal{S}_{blocked}$.

In this step, we aim to move states from $\mathcal{S}_{active}$ to $\mathcal{S}_{static}$. This will reduce the number of agents in the swarm that are likely to move, decreasing the size of $\mathcal{Q}_{safe}$.

As explained in Sect. 3, an important axiom needed to guarantee that P_{des} will form is that, for a swarm of N agents, N instances of the local states in $\mathcal{S}_{static}$, with repetition, must uniquely rearrange into P_{des}. If this is not the case, another pattern could emerge where all agents are in a state $\mathcal{S}_{static}$ and do not move. Here, because we are already at the optimization stage, we consider the case where the original $\mathcal{S}_{static}$ already guarantees that P_{des} is unique. From this starting point, we present a method to augment $\mathcal{S}_{static}$ based only on a local analysis, while keeping P_{des} as the unique static pattern.

Consider a state $s \in \mathcal{S}_{active}$. For s, we locally check whether it could be fully surrounded by agents with a state within $\mathcal{S}_{static}$. If this is not possible, because s is such that at least one of its neighbors would be in an active state, then we add s to $\mathcal{S}_{static}$. This is because we know that, if this state s were static, there would always be one active agent somewhere next to it anyway, so P_{des} still remains the only unique pattern that can be formed by static states. Then, when this active neighbor moves, the local state of the agent will also change and it will also no longer be static. We run this process iteratively for all states until no more states from $\mathcal{S}_{active}$ can be moved to $\mathcal{S}_{static}$. As an exception, due to the importance of active simplicial states remaining active to guarantee that there is motion in the swarm, these are not included in the process.

Using this approach, it is possible to significantly increase the size of $\mathcal{S}_{static}$, and in turn reduce the size of $\mathcal{Q}_{safe}$. Additionally, one can also add to $\mathcal{S}_{static}$ all states that expect more neighbors than are present in the swarm. For instance, for a swarm of 4 robots, all states with 4 or more neighbors may be discarded, because they cannot happen in the first place and we need not consider them. Table 1 shows the results of Step 1 for the patterns in Fig. 2.

4.2   Step 2: Local Elimination of Unnecessary Actions

In this step, individual state-action pairs that are not necessary towards achieving the final pattern, in accordance with the proof, are discarded. The objective is to minimize $|\mathcal{Q}_{safe}|$ while keeping the conditions listed at the end of Sect. 3. The minimization was performed with a Genetic Algorithm (GA) in order to avoid

Table 1. Results of Step 1 on the size of $\mathcal{S}_{static}$ (which increases) and $\mathcal{Q}_{safe}$ (which decreases)

		4 agents triangle	*Hexagon*	*9 agents triangle*
Before	$|\mathcal{S}_{static}|$	28	30	33
	$|\mathcal{Q}_{safe}|$	543	550	531
After	$|\mathcal{S}_{static}|$	188	128	87
	$|\mathcal{Q}_{safe}|$	172	381	439

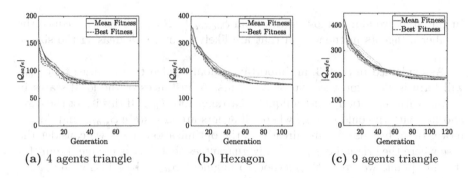

Fig. 4. Results of the evolutionary reductions of $\mathcal{Q}_{safe}$ from Step 2

local minima. The fitness function to be minimized was $f = |\mathcal{Q}_{safe}|$, subject to the following constraints:

1. $\mathcal{S}_{static}$ must not change. This is because, following Step 1, we know that all remaining states must be active, else a spurious pattern might form.
2. The conditions at the end of Sect. 3 must be respected.

The population of the GA was formed by 100 binary genomes. Each gene in a genome represented a state-action pair in $\mathcal{Q}_{safe}$, with a 1 indicating that the state-action pair is kept and a 0 indicating that it is eliminated. All genomes in the initial population were such that the constraints were respected. Then, the new generation consisted of: elite members (30%), new offspring (40%), and mutated members (30%). Offspring genomes were the result of an AND operation between two parent genomes. This automatically meant that any offspring would be at least as fit as its parents, because the AND operator natively either reduced or kept the quantity of activated bits. Offspring were only kept if they complied with the constraints, else the parents were forced to look for new mates to perform the AND operation with. This also made for a convenient stopping criterion, which is when all children are equally as fit as the parents or when all parents are unable to find any mate that will result in a valid offspring. On each generation round, mutation was applied to a random portion of the population, for which the NOT operator was randomly applied to 10% of each selected member's genome (thus changing 1s to 0s and viceversa). Similarly as to the offspring, a mutation was kept only if it returned a genome for which the constraints were met, else it was discarded and a new mutation was attempted. This way there was a guarantee that the population always consisted of valid genomes.

We executed 5 evolutionary runs for each pattern from Fig. 2, with similar results. The results of the runs are shown in Fig. 4. Thanks to the local nature of the proof, the evolution time was not dependent on the number of agents in the swarm.

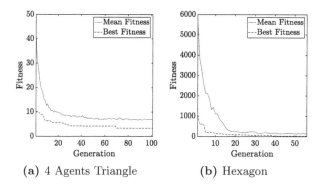

(a) 4 Agents Triangle   **(b)** Hexagon

Fig. 5. Optimization results of Step 3 from the best evolutionary runs

4.3   Step 3: Behavior Optimization

Steps 1 and 2 lead to reduced state-action maps $\mathcal{Q}_{reduced}$ that are minimally sufficient to guarantee that the patterns will be achieved. In Step 3, we tune the probability of executing each action in $\mathcal{Q}_{reduced}$. This is done with a more classical evolutionary robotics approach to swarm robotics: the swarm is simulated and evaluated based on its statistical performance, and this information is used in the fitness function of a GA.

The fitness function to be minimized is the expected number of actions needed to achieve the goal. This was evaluated by the mean over 10 trials. The GA used a population of 100 scalar genomes. Each gene in a genome held a value $0 < p \leq 1$, indicating the probability of taking the corresponding action from $\mathcal{Q}_{reduced}$. By means of the inequality, it is not possible to bring the probability of a state-action pair down to 0 and deactivate it (keeping the proof intact). Each new generation was produced by elite members (30%), offspring (40%), and mutated members (30%), as in Step 2. Offspring resulted from mixing two parents' genomes via a uniform crossover strategy, where each gene of an offspring's genome is randomly selected from the genes of either parent with equal probability. Mutation was applied to random genomes, for which 10% of their genes were replaced by random values from a uniform distribution. The members of the initial population were produced randomly from uniform distributions.

Using this scheme, we optimized the behavior for the 4 agents triangle and the hexagon, running 5 evolutionary runs each. The best evolutionary runs are shown in Fig. 5. For the triangle, 3 out of 5 runs converged. For the hexagon, 2 out of 5 runs converged, one considerably lower than the other. We associate the convergence issues to bootstrap and noise issues during evaluation, which grow with the size of the swarm. In light of this, we were unable to establish an optimal solution for the triangle with 9 agents. This is due to two problems: (1) the controllers in early generations took a very long time to evaluate, which made executing the GA troublesome, (2) the fitness metric was subject to considerable variance, leading to inaccurate controller evaluations. These problems and their implications are discussed further in Sect. 5.

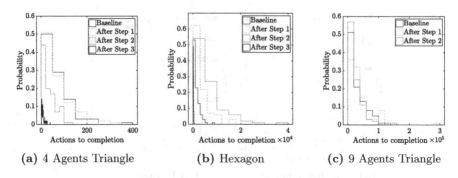

(a) 4 Agents Triangle   **(b)** Hexagon   **(c)** 9 Agents Triangle

Fig. 6. Normalized histograms of the performance of the system through all steps of the optimization

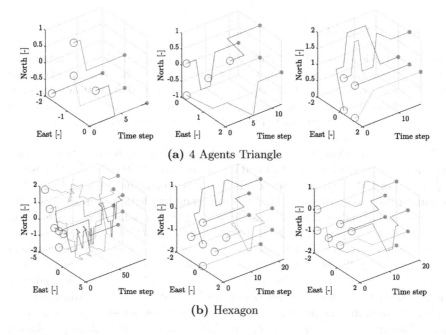

Fig. 7. Exemplary simulations showing pattern formation of the triangle with 4 agents and the hexagon after optimization

5   Results and Discussion

We tested the performance of the original baseline controller against the performance of the controllers after Step 1, Step 2, and Step 3. Each controller was tested 100 times. The normalized distributions for the number of actions to completion are shown in Fig. 6. Exemplary simulations of swarms as they create a pattern using the evolved behaviors from Step 3 are shown in Fig. 7. In all tests, the desired pattern was eventually achieved. Reducing the size of Q_{safe}

in Steps 1 and 2 simplified the state-action map and did not have an impact on whether the goal could be achieved. This result serves as empirical evidence for the proofs in [1], which were apt guards to ensure that the swarm could form the pattern. Then, when the minimized state-action map $Q_{reduced}$ was optimized, we were able to significantly improve the performance of the system for the triangle with 4 agents and the hexagon, achieving a fast average performance while also respecting the proof.

There remain issues to be investigated. The first issue is that Step 1 and 2 only modified Q_{safe} with the goal to minimize its size and simplify the agent's behavior. As seen in Fig. 6, this in itself does not necessarily aid performance. In future work, there should be efforts to understand how to reduce Q_{safe} while also improving performance. The second issue is scalability, as encountered in Step 3. Most notably, this prevented us from completing Step 3 for the triangle with 9 agents. It is possible that these problems can be mitigated by improving Steps 1 and 2 to minimize $|Q_{safe}|$ while also assessing performance. Another option could be to stop simulations before completion and use a fitness measure that favors global patterns closer to P_{des} over other less similar patterns. However, it might also be possible that the scalability issue is intrinsic to the system. As the size of the swarm grows, then the relative information that each agent has of the whole swarm decreases, and it become increasingly difficult for an agent to predict whether an action is the best for the good of the whole swarm. It would be interesting to explore this limitation in future work.

6   Conclusions and Future Work

The approach presented in this paper is a first step towards optimizing swarm behavior of severely limited agents by aid of an automatic proof, where a local proof allows for the fast verification of certain properties, and can thus be included within the optimization process. The focus was not on how to achieve the goal, but on how to achieve it more efficiently. This led to efficient controllers where the number of actions needed to achieve the patterns were significantly lower than the original controllers, making them more suitable for use in the real world. In the meanwhile, the controllers remained such that eventual success by the swarm is guaranteed.

The approach encountered problems with scalability in the final step. This could be tackled by using the automatic minimization steps, prior to the final optimization, to reduce the solution space in a way that is more favorable for performance. However, there remains the issue that, as the size of the swarm grows, each agent becomes less empowered to take an optimal action, given that it has relatively less information on the state of the swarm. For this reason, it would also be valuable to explore how scalability improves when the agents have more information of their surroundings (e.g., they can sense further away), or some limitations are lifted (e.g. memory).

References

1. Coppola, M., Guo, J., Gill, E.K., de Croon, G.C.H.E.: Provable emergent pattern formation by a swarm of anonymous, homogeneous, non-communicating, reactive robots with limited relative sensing and no global knowledge or positioning. ArXiv Preprint arXiv:1804.06827 (2018). (Submitted to Swarm Intelligence, Springer)
2. Coppola, M., McGuire, K.N., Scheper, K.Y.W., de Croon, G.C.H.E.: On-board communication-based relative localization for collision avoidance in micro air vehicle teams. Auton. Robots (2018)
3. Duarte, M., et al.: Evolution of collective behaviors for a real swarm of aquatic surface robots. PloS One **11**(3), e0151834 (2016)
4. Ericksen, J., Moses, M., Forrest, S.: Automatically evolving a general controller for robot swarms. In: 2017 IEEE Symposium Series on Computational Intelligence, SSCI, pp. 1–8 (2017)
5. Ferrante, E., Duéñez Guzmán, E., Turgut, A.E., Wenseleers, T.: GESwarm: grammatical evolution for the automatic synthesis of collective behaviors in swarm robotics. In: Proceedings of the 15th Annual Conference on Genetic and Evolutionary Computation, GECCO 2013, pp. 17–24. ACM, New York (2013)
6. Francesca, G., Birattari, M.: Automatic design of robot swarms: achievements and challenges. Front. Robot. AI **3**, 29 (2016)
7. Francesca, G., et al.: AutoMoDe-Chocolate: automatic design of control software for robot swarms. Swarm Intell. **9**(2), 125–152 (2015)
8. Gomes, J., Urbano, P., Christensen, A.L.: Introducing novelty search in evolutionary swarm robotics. In: Dorigo, M., et al. (eds.) ANTS 2012. LNCS, vol. 7461, pp. 85–96. Springer, Heidelberg (2012). https://doi.org/10.1007/978-3-642-32650-9_8
9. Gomes, J., Urbano, P., Christensen, A.L.: Evolution of swarm robotics systems with novelty search. Swarm Intell. **7**(2), 115–144 (2013)
10. Hüttenrauch, M., Šošić, A., Neumann, G.: Guided deep reinforcement learning for swarm systems. ArXiv Preprint arXiv:1709.06011 (2017)
11. Izzo, D., Simões, L.F., de Croon, G.C.H.E.: An evolutionary robotics approach for the distributed control of satellite formations. Evol. Intell. **7**(2), 107–118 (2014)
12. Jones, S., Studley, M., Hauert, S., Winfield, A.: Evolving behaviour trees for swarm robotics. In: Groß, R., et al. (eds.) Distributed Autonomous Robotic Systems. SPAR, vol. 6, pp. 487–501. Springer, Cham (2018). https://doi.org/10.1007/978-3-319-73008-0_34
13. Klavins, E.: Programmable self-assembly. IEEE Control Syst. **27**(4), 43–56 (2007)
14. Nolfi, S.: Power and the limits of reactive agents. Neurocomputing **42**(1–4), 119–145 (2002)
15. Rubenstein, M., Cornejo, A., Nagpal, R.: Programmable self-assembly in a thousand-robot swarm. Science **345**(6198), 795–799 (2014)
16. Saska, M., Vonásek, V., Chudoba, J., Thomas, J., Loianno, G., Kumar, V.: Swarm distribution and deployment for cooperative surveillance by micro-aerial vehicles. J. Intell. Robot. Syst. **84**(1), 469–492 (2016)
17. Silva, F., Duarte, M., Correia, L., Oliveira, S.M., Christensen, A.L.: Open issues in evolutionary robotics. Evol. Comput. **24**(2), 205–236 (2016)
18. Trianni, V., Nolfi, S., Dorigo, M.: Cooperative hole avoidance in a swarm-bot. Robot. Auton. Syst. **54**(2), 97–103 (2006)
19. Yamins, D., Nagpal, R.: Automated global-to-local programming in 1-D spatial multi-agent systems. In: Proceedings of the 7th International Joint Conference on Autonomous Agents and Multiagent Systems, AAMAS 2008, vol. 2, pp. 615–622. International Foundation for Autonomous Agents and Multiagent Systems, Richland (2008)

Quality-Sensitive Foraging by a Robot Swarm Through Virtual Pheromone Trails

Anna Font Llenas[1,2], Mohamed S. Talamali[1] , Xu Xu[2,3] ,
James A. R. Marshall[1] , and Andreagiovanni Reina[1($\boxtimes$)]

[1] Department of Computer Science, University of Sheffield, Sheffield, UK
{mstalamali1,james.marshall,a.reina}@sheffield.ac.uk
[2] Department of Engineering and Mathematics, Sheffield Hallam University,
Sheffield, UK
[3] MERI, Sheffield Hallam University, Sheffield, UK

Abstract. Large swarms of simple autonomous robots can be employed to find objects clustered at random locations, and transport them to a central depot. This solution offers system parallelisation through concurrent environment exploration and object collection by several robots, but it also introduces the challenge of robot coordination. Inspired by ants' foraging behaviour, we successfully tackle robot swarm coordination through indirect stigmergic communication in the form of virtual pheromone trails. We design and implement a robot swarm composed of up to 100 Kilobots using the recent technology Augmented Reality for Kilobots (ARK). Using pheromone trails, our memoryless robots rediscover object sources that have been located previously. The emerging collective dynamics show a throughput inversely proportional to the source distance. We assume environments with multiple sources, each providing objects of different qualities, and we investigate how the robot swarm balances the quality-distance trade-off by using quality-sensitive pheromone trails. To our knowledge this work represents the largest robotic experiment in stigmergic foraging, and is the first complete demonstration of ARK, showcasing the set of unique functionalities it provides.

1 Introduction

The task of collecting objects clustered at random locations and transporting them to a central depot can benefit from a decentralised solution. In contrast to a single large vehicle dedicated to load/unload all the objects, an interesting solution consists in having a large number of simple autonomous vehicles, or robots, each carrying a single object and coordinating with each other. Advantages of this solution are parallel exploration of the environment and possibility to distribute the resources among various source locations. Controlling the robot behaviour via a decentralised algorithm adds the advantage of scalability, by which the system throughput can be calibrated with increase/removal of robots without need for a system redesign. We study a decentralised solution

M. Dorigo et al. (Eds.): ANTS 2018, LNCS 11172, pp. 135–149, 2018.
https://doi.org/10.1007/978-3-030-00533-7_11

that employs a swarm of simple robots that coordinate via stigmergic communication to collect items clustered in the environment in various source areas and to transport them to a central depot. Exploiting nearer sources would increase the system throughput, however objects may also have different qualities. We study how our system can balance the trade-off between quality and distance.

The task of finding and collecting objects is known in the multi-robot literature as *foraging* due to the resemblance to the activity that some animals perform when hunting food. In foraging terms, the source areas are food sources and the depot area is the animal's nest. We exploit this analogy with biology to also design our solution. Inspired by foraging behaviour of ant colonies, we design a robot swarm that relies on a stigmergic communication medium similar to the pheromone trails used by ants [12,26,56,60]. Initially, scout ants randomly search the environment, when one finds food she returns to the nest carrying a food item and leaving a pheromone trail on her path. Through this process, ants create pheromone trails between their nest and food sources located by scout ants. The pheromone trails are used by other members of the colony to avoid further random exploration and to exploit the sources that have been already found. Similarly, we design a solution where robots start to randomly search the environment and, later, they converge to exploiting the object sources by relying on stigmergic communication. It has been observed that ants modulate pheromone deposition as a function of the food source quality [22,26,60] (or of the nest-site quality during nest hunting [28]). In a similar fashion, our robots deposit pheromone proportionally to the estimated objects' quality. In this study, we focus on the strategies to coordinate the robot motion leaving in abstract terms the object load/unload issues.

Previous work investigated the use of pheromone as a form of indirect communication between robots (see a review in Sect. 2). Our study includes analysis of the quality-distance trade-off which is an aspect that has not been previously explored in multi-robot foraging studies (see the problem description in Sect. 3). We implement a foraging swarm composed of simple robots (Sect. 3.1), the Kilobots [50], that operate in a virtual environment where they can deposit/sense virtual pheromone (Sects. 3.2 and 3.3). In Sect. 4, we evaluate the system performance through simulations and we employ the *Augmented Reality for Kilobots* (ARK) [47] system to showcase the functioning with a set of demos with swarms up to 100 Kilobots. We finally discuss the relevance of the work for engineering and biology in Sect. 5.

2   Related Work

Several studies employed a form of stigmergy similar to the ants' pheromone trails to coordinate the robots' movement. A pivotal point of these studies concerns the way in which pheromone is implemented, that is how the environment stores/updates the pheromone and how the robots deposit/sense pheromone in the environment. We identify and discuss three main categories which we name: beacon robots, smart-environment based, and on-board pheromone.

The first robotic systems that used pheromone communication to coordinate the group motion allocated a set of robots as static **beacon robots** [5,10,19,25,36,39,59]. The role of the beacons was to store and communicate pheromone levels to robots that moved in their surroundings. The advantage of this solution is that it can be implemented by simple robots in unknown unstructured environments. The drawback is that part of the robots are not actively contributing to the main task (e.g. foraging and collecting items) but need to stop and act as beacons. This strategy may limit the functioning in vast environments. Mobile robot beacons overcome the sacrifice of robots and allow robots to be concurrently beacons and active foragers [11,53]. However, the correct functioning relies on the tuning of the swarm size and communication range as a function of the environment size.

Several studies, similarly to ours, implemented pheromone communication through a *smart* **environment** which was capable to store virtual pheromone information and to provide this information in real-time to the robots [1,17,18, 21,54,58]. Within this category, several studies implemented virtual pheromone through the use of RFID tags which were deployed in the environment and stored pheromone information [3,23,24,29,32,33]. Our study relies on a different form of smart environment: Kilobots perceive and deposit virtual pheromone via ARK which has similarities with implementations of [1,17,18,54]. Similarly to ARK, robots were real-time tracked with an overhead camera, although, differently from ARK, their robots used the light sensors to read the virtual pheromone that was projected as light on the floor. In [58], pheromone foraging was implemented on a Kilobot swarm using a different augmented reality system, the Kilogrid.

Researchers designed various solutions to equip robots with **on-board sensors and actuators** customised to mark the environment and thus shifted the pheromone mechanism from the smart-environment to the robots. In an early work [55], the robot used a marker pen to draw lines on the floor to improve its performance in the area coverage task. This technology had the drawback of not allowing evaporation or diffusion of pheromone. Differently, in [45], the robots could emit and read gas which was used to guide other robots towards a source area. A limitation of this work was the high volatility of the gas. In [34], the E-Puck robots were equipped with phosphorescent glowing paint to temporarily mark the environment. Robots had to operate in a dark environment and follow light to move between two areas. Finally, in [15,16], robots used alcohol to mark the environment and improve the collective performance in the foraging task.

Most work discussed in this Section, as ours does, aims to implement a robotic system for the foraging task where robots are asked to move between two (or more) locations (mimicking the activity of objects collection). In our study, we include the aspect of objects' quality that has not been taken in consideration earlier and we analyse how the system can balance the quality-distance trade-off. Previous work included robot swarms up to a maximum of 50 robots [58], in this study we scale to 100 robots.

3   Problem Description

A robot swarm is asked to collect objects from n source areas deployed in a 2D environment and transport them to a central depot area. In this study, we ignore the details relative to object picking, deposition, and storage; instead, we focus on the coordinated activity of the robots to move between sources and depot areas. Each source area A_i ($i \in \{1, 2, \ldots, n\}$) is an infinite source of one type of object characterised by a quality $v_i \in [0, 10]$. The objective is to maximise the throughput of objects weighted by their quality.

3.1   Robots

The swarm is completely decentralised and composed of S simple autonomous robots that have minimal knowledge about the environment and limited sensory and memory capabilities. We assume that the robots do not know and cannot keep memory of the number, location, and quality of the source areas. The robots do not communicate with each other or cannot perceive other robots and obstacles in the environment. The robots coordinate and collaborate with each other only through stigmergic communication, i.e. by leaving temporary traces in the environment that can be read by other robots. The robots are equipped with the following sensors and actuators: (i) *differential drive motors* to move in the 2D environment, (ii) *area sensor* to detect source and depot areas when the robot is within the area, (iii) *object quality sensor* to estimate the quality of the collected object, (iv) *depot direction sensor* to know the relative orientation towards the depot area, (v) *pheromone gland* to leave in the environment temporary traces (i.e. pheromone), and (vi) *pheromone antennae* to perceive the presence of pheromone in the robot's immediate surroundings.

3.2   The Kilobots and ARK

We implemented the robot swarm using Kilobots [50] which are inexpensive simple robots designed to perform large-scale swarm robotic studies. The Kilobot modulates the frequency of its two vibration motors to move on a flat surface. The motors have been automatically calibrated via ARK [47] to move at an average speed of $\sim$1 cm/s and rotate in place at $\sim$40 °/s. The robots have a limited set of sensors and actuators therefore we relied on the ARK system to enhance the Kilobot's capabilities. The ARK system allows the user to equip the Kilobot with a customised set of virtual sensors and virtual actuators to sense and modify simulated virtual environments shared by all robots in realtime. While the ARK system has global information on the environment and the process, its function is limited to enhance the Kilobots' abilities, and enrich the experiments for they can be used within; ARK lets the Kilobots operate autonomously in a decentralised fashion without any central control.

Via ARK, we equipped the Kilobots with the required virtual sensors and actuators. The Kilobot periodically receives a message with the relative direction to the depot (coded in 4 bits). When the Kilobot is within an area (either source or depot), ARK informs the robot of the type of area (2 bits) and, if within source, of the object's quality (4 bits). The Kilobots, through their LED, signal to ARK when they want to deposit a drop of pheromone in their current position. Finally, ARK signals the Kilobot if it has pheromone within detection range (4 bits). The detection range of the virtual pheromone antennae is depicted in Fig. 1(a); the robot can perceive a binary value (presence/absence of pheromone) in four areas 45° wide in front of itself at a maximum distance of ∼3.5 cm.

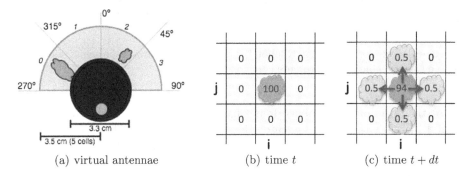

(a) virtual antennae   (b) time t (c) time $t + dt$

Fig. 1. (a) The robot can perceive the presence of pheromone in its immediate surrounding through virtual pheromone antennae implemented via ARK [47]. Kilobots sense a binary value (presence/absence of pheromone) in each area. In the depicted example, the cyan shapes show traces of virtual pheromone and the robot's reading is [1, 0, 1, 0]. (b)–(c) Example of pheromone dynamics as from Eq.(1); at time t, the cell $c(i, j)$ has a pheromone value of 100, and at time $t + dt$ (with $dt = 0.5$ s) the pheromone evaporated at rate $\epsilon = 0.08$ and diffused at rate $\gamma = 0.01$ to the four neighbouring cells.

The virtual environment is updated in realtime by the ARK system that increases pheromone level when a robot deposits a pheromone drop $\phi = 100$ and computes evaporation and diffusion of pheromone over time. The pheromone is stored in a matrix that discretises the 2D environment in 6.7 mm cells (i.e. 150 cells per metre). At each time-step (of length $dt = 0.5$ s), ARK updates each matrix cell $c(i, j)$ (with generic indices (i, j)) as follows:

$$c(i, j) = c(i, j)[1 - (\epsilon + 4\gamma)dt] + \gamma[c(i, j \pm 1) + c(i \pm 1, j)]dt, \qquad (1)$$

where parameter $\epsilon = 0.08$ is the evaporation rate and $\gamma = 0.01$ the diffusion rate, and $c(i, j) \geq 0$. Figure 1(b)–(c) show an example of the pheromone dynamics where at time t a drop $\phi = 100$ is deposited at cell $c(i, j)$. Equation (1) is a simplification of the exponential decay observed in ant's pheromone [8,17].

3.3 Robot Behaviour

The proposed solution has been designed taking inspiration from the foraging behaviour of ants that use pheromone trails to mark the environment. This form of stigmergic communication allows the colony to limit unnecessary independent exploration and to coordinate among peers to collectively exploit the found food resources. The individual behaviour of the Kilobot is implemented as the finite state machine (FSM) of Fig. 2.

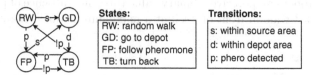

Fig. 2. FSM of the individual Kilobot behaviour. The arrows represent transitions between states which are represented as circles.

At the beginning, the robots do not have information about the source location(s) therefore they start searching the environment. Given the Kilobot's limited capabilities, an easy and efficient method to search an unknown environment is through an isotropic random walk [9] which we implemented with alternate straight motion for 7.5 s and uniformly random rotation in $[-\pi, \pi]$. Once a source area A_i has been found, the robot (virtually) collects one object and carries it towards the depot area. On its way towards the depot, the robot deposits drops of pheromone with probability P_i proportional to the object quality v_i, i.e. $P_i = v_i/v_{max}$. The Kilobot updates its decision to deposit pheromone every ~2 s (which it signals to ARK via its LED), therefore a medium-quality object will lead the robot to lay down intermittent pheromone trails. Once the Kilobot returned to the depot, it unloads the object, turns 180°, and resumes exploration because it cannot store in its memory the source area location. However, through pheromone trails the Kilobot exploits a form of collective memory which is stored in the environment in the form of temporary stigmergic information. In fact, once a Kilobot perceives pheromone in any of the four antennae areas (of Fig. 1(a)), it follows the trail by moving in the direction of the triggered antennae area. If more than one antennae area detect pheromone (as in the example depicted in Fig. 1(a)), the robot selects the area in the most opposite direction from the depot. This selection relies on the assumption that robots only deposit pheromone in their straight path from a source area to the depot and that they have access to the depot vector.

As in every study, we make the experiment code available online; download it at https://github.com/DiODeProject/PheromoneKilobot.

4   Experiments and Results

We measured the system performance through accurate physics-based simulation of the Kilobot swarm. We ran our simulations via *ARGoS* [41] which is a simulator tailored to swarm robotics needs that allows high speed and accurate simulation of the physics dynamics. ARGoS allows the simulation of the Kilobot robots and the ARK system through its dedicated plugin [40]. Using ARGoS is particularly advantageous because it allows the experimenter to use the same identical code in simulation and on the robots.

4.1 Simulation Scenarios

We investigated the performance of a Kilobot swarm of size $S \in \{50, 100, 200\}$ in a 2.5 m × 2.5 m environment with a central circular depot area and $n \in \{1, 2, 4\}$ circular source areas with a radius of 10 cm. The source areas' positions and object's qualities were varied to study the quality-distance trade-off. We varied the distance from the depot of the source areas $d_i \in [0.5, 1.5]$ m and object's qualities $v_i \in [0, 10]$, with $i \in n$. The robots were initially deployed with (uniformly) random position and orientation within a square 70 cm × 70 cm region centred on the depot area. The experiments length was 20 simulated minutes. We report the mean number of objects retrieved from each source and the mean number of robots on each path (computed as the number of robots at a maximum distance of 20 cm from the straight line between depot and source).

4.2 Results for Varying Distance and Quality

We investigated the effects of distance and quality through a scenario with $n = 2$ source areas with diametrically opposed positions. To investigate the effects of distance, we positioned the source A_1 at distance $d_1 = 1$ m from the depot and we varied the distance $d_2 \in [0.5, 1.5]$ m of source A_2. Both sources have objects with maximum quality $v_1 = v_2 = 10$. On the contrary, to investigate the effects of quality, we set the source A_1 with objects of quality $v_1 = 5$, and we varied the object's quality $v_2 \in [0, 10]$ of source A_2. Both sources were placed at distance $d_1 = d_2 = 1$ m from the depot. Figure 3(a), (c) show the number of items collected from each source after 20 min by swarms composed of $S = \{50, 100, 200\}$ simulated Kilobots for distance and quality experiments, respectively. The closer, or better-quality, source area always has a larger number of collected object. As expected, source A_1 has approximately constant throughput while the throughput of source A_2 decreases with distance d_2 in Fig. 3(a), and increases with quality v_2 in Fig. 3(c). Figure 3(b) shows the allocation of robots among the two paths. Large swarms (e.g. $S = 200$) have a redundancy of robots and by moving away the source area, more robots are allocated to it. On the other hand, smaller swarms (e.g. $S = 50$) reduce the robots on that path as it gets further than 1 m in length. Similarly, Fig. 3(d) shows similar dynamics for the quality experiments. Swarms of $S = 50$ robots reallocate robots to the highest quality, whereas larger swarms of $S = 200$ robots saturate the source paths for qualities $v_2 > 3$. Still the collection is directly proportional to the object's quality (Fig. 3(c)) because (especially at the beginning of the experiment) the pheromone trails are more continuous and easier to follow for areas with better quality objects.

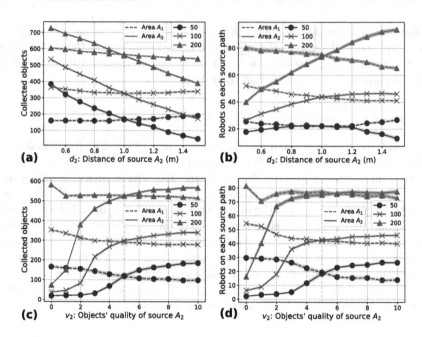

Fig. 3. Results from simulation experiments with $S = \{50, 100, 200\}$ Kilobots and two source areas with (a)–(b) equal quality, varying distance, and (c)–(d) equal distance, varying quality. In (a), (c) we report the number of collected objects after 20 min; in both cases, the closest, or better quality, source area has a larger number of collected items. In (b), (d) we report the number of robots on each source path after 20 min. Small swarms allocate resources differently than larger swarms. Lines are mean of 100 simulations and the lighter colour fill is the 95% confidence interval. (Color figure online)

4.3 Effects of the Swarm Size S

Figure 4 shows the number of collected items for varying swarm size $S \in [10, 250]$ in scenarios with $n = 2$ or $n = 4$ sources with equal objects' qualities $v_i = 10$ and equal distances $d_i = [1]$ m, with $i \in n$. On one hand, increasing the swarm size S results in an increasing absolute throughput of objects (blue lines on left y-axis). On the other hand, adding more robots increases the swarm density and causes more collisions. This physical interference among robots reduces the individual robot efficiency (green lines on right y-axis). In fact, Kilobots do not have any collision sensor and, in dense environments, they may lose time pushing each other without moving. A similar trade-off between benefits and costs of adding individuals has been already observed in collective behaviour studies [6,14,31].

4.4 Quality-Distance Trade-Off

As shown in Sect. 4.2, our system favours nearer over further source areas, and better over worse object qualities. Here, we explore how the system compromises between far, better-quality sources versus nearer, lower-quality sources.

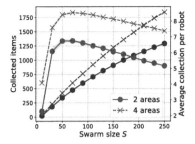

Fig. 4. Effect of the swarm size S on the number of collected objects. Absolute throughput increases with S but physical interference caused by collisions reduces individual efficiency for increasing S. Lines are mean of 100 simulations and the lighter colour fill is the 95% confidence interval for $v_i = 10, d_i = 1$ m, $n \in \{2,4\}, i \in n$. (Color figure online)

We investigate two-sources scenarios with A_1 fixed and varying A_2. Figure 5(a)–(b) have $d_1 = 1$m, $v_1 = 10$, $v_2 = 5$, and varying $d_2 \in [0.5, 1.5]$m. We can appreciate that the swarm collects more objects from the lower quality area A_2 than from the better quality A_1 only when the difference in distance is $d_1 - d_2 > 0.2$ m. Figure 5(b) shows the number of robots on each path; small swarms (i.e. $S = 50$) always allocate more resources (robots) to the better quality option, instead larger swarms (i.e. $S = 200$) have such a redundancy of resources that robots can fill both paths without need to selectively chose the best. In a similar analysis we fixed the best quality A_1 at $d_1 = 1.5$ m, $v_1 = 10$ and varied the quality $v_2 \in [1, 10]$ of the closest area A_2 in $d_2 = 0.75$ m. Figure 5(c) shows that large swarms are minimally influenced by quality variation, whereas smaller swarms select the further and best quality when the closest area has objects of poor quality $v_2 < 4$.

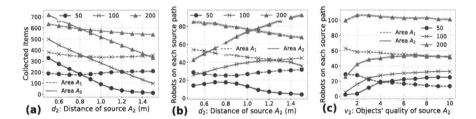

Fig. 5. Trade-off between closer, lower-quality areas and further, better-quality areas. (a)–(b) Scenario with $v_1 = 10, d_1 = 1$ m, $v_2 = 5$ and $d_2 \in [0.5, 1.5]$ m. (c) Scenario with $v_1 = 10, d_1 = 1.5$ m, $d_2 = 0.75$ m and $v_2 \in [1, 10]$. When robots are overabundant the trade-off is ignored, whereas smaller swarms prioritise higher quality resources. Lines are mean of 100 simulations and the lighter colour fill is the 95% confidence interval. (Color figure online)

4.5   Kilobot Swarm Demonstrations

The real Kilobot demonstrations are run in scenarios almost identical to the one described in Sect. 4.1 except for the environment size which is $2\,\text{m} \times 2\,\text{m}$, and the experiment length which is 30 min or longer. We run four demos $D1$, $D2$, $D3$, and $D4$. Demos $D1$ and $D2$ investigate how 50 Kilobots respond to different qualities and distances, respectively. Demos $D3$ and $D4$ show how the system scales with increasing number of sources (i.e. $n = 4$) and robots ($S = 100$). Figure 6(a) show an image of $D3$ where in the closeup the ARK screen visualise the camera stream and the virtual environment information. Figure 6(b) shows a screenshot of $D4$. The video complete videos are available at http://diode.group.shef.ac.uk/FontLlenas2018.html.

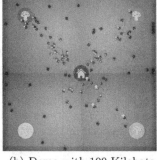

(a) Demo with 50 Kilobots   (b) Demo with 100 Kilobots

Fig. 6. (a) Image from demo $D3$ with $n = 4$ source areas. In the closeup, the computer screen shows the ARK's virtual environment, on the background the Kilobots move between virtual sources following virtual pheromone trails (the virtual environment has been superimposed to the image). (b) Screenshot from demo $D4$. Full videos are available at http://diode.group.shef.ac.uk/FontLlenas2018.html.

Demo $D1$ shows 50 Kilobots foraging from two sources placed at the same distance $d_1 = d_2 = 0.6\,\text{m}$ with different qualities $v_1 = 10$ and $v_2 = 5$. In contrast, demo $D2$ shows 50 Kilobots foraging from two sources with equal quality $v_1 = v_2 = 10$ but placed at different distances $d_1 = 0.6\,\text{m}$ and $d_2 = 1\,\text{m}$. In both cases, the swarm response is similar to the one observed in simulation. A noticeable difference consists in lower number of retrieved object and robots on paths in comparison with the results of Fig. 3. This difference is due to the large actuation noise of the Kilobots that was not included in the noise-free simulations. Additionally, the ARGoS Kilobot plugin [40] used for this study was not yet finely tuned on the real speed/friction of the robot and resulted in largely faster robots. Instead, real Kilobots spent considerable time to resolve collisions between robots moving in opposite directions. Larger commuting time should be balanced by more stable pheromone trails which could be achieved by letting the robot autonomously increase the amount of pheromone deposited in a decentralised fashion.

Demos $D3$ and $D4$ showcase the system with $n = 4$ sources and up to 100 Kilobots. The four sources have a set of qualities and distances that allows the viewer to appreciate the quality-distance trade-off investigated in this study. The considered qualities are $v_1 = 10$, $v_2 = 8$, $v_3 = 5$, $v_4 = 3$ for both demos, while distance are $d_1 = d_3 = 0.6\,\mathrm{m}$, $d_2 = 0.8\,\mathrm{m}$, $d_4 = 0.5\,\mathrm{m}$ for $D3$, and $d_1 = d_2 = d_3 = d_4 = 1\,\mathrm{m}$ for $D4$.

5   Discussion and Conclusion

Foraging is a general task that consists of the two main activities of searching the environment to locate objects and of transporting the objects to a central depot area. This task is widely studied in robotics because it entails activities relevant for several robot applications [61]. Employing multi-robot systems to solve the foraging problem has the clear advantage of offering system parallelisation through concurrent object collection by several robots. At the same time, it introduces the challenge of robot coordination. We successfully tackle the robot swarm coordination through stigmergic communication, although we acknowledge that this is not the only solution as several previous studies explored various alternatives, e.g. [2,13,20,42,44,48,51,61]. In this study, we assume that robots cannot directly communicate or sense each other; they are totally unaware of the other swarm members, still they cooperate with each other through indirect communication. Pheromone trails allowed memoryless robots to create a form of collective memory; robots stored information in the environment that they used to repeatedly find previously discovered sources. The simplicity of the individual robot behaviour allowed a direct transfer of the noise-free simulation code to the Kilobot experiments and minimised the impact of the reality gap [27].

We successfully implemented the system on swarms of 50 and 100 Kilobots supported by the ARK [47] system which allowed robots to operate in a virtual environment. This work exploits the full potential of the ARK infrastructure and showcases ARK's unique functionalities. Even though the robot experiments included virtual components, the physical implementation of the system has been useful to display the solution robustness and validate the simulation results. While we acknowledge that hybrid experiments (in between reality and simulation) do not correspond to real-world applications, we still believe they represent useful test-beds to validate and demonstrate theories within research labs.

This study included the source quality as a factor influencing the foraging behaviour, this may relate to the priority to fetch each type of object. Further work should better investigate how to control the balance between quality and distance. This investigation could relate to *optimal foraging* theory [30,46] which considers the net energy intake as the energy gain discounted by the foraging cost. This type of 'economical' analysis of the foraging behaviour allows determination of the best theoretical foraging strategy as a function of various components, such as food-distance, prey-payload, and food-quality. Optimal foraging theory has been applied to predict a large variety of foraging behaviour including the central place foraging [37,38,52] investigated here. We acknowledge that previous work has employed optimal foraging theories to engineer multi-robot systems [4,43,57] and we believe that this research line should be continued.

Our results are in-line with previous investigations and show that system performance is dependent to the strength of the positive feedback, the swarm size, and discoverability of sources. Robots are in control of only the first factor and it might be useful to identify if the robot could prioritise quality or distance by modulating the positive feedback strength (i.e. pheromone drop size and deposition frequency as a function of source's quality and discoverability). Additionally, in our study, the robots do not perceive differences in pheromone concentrations, in contrast, ants have a nonlinear response to pheromone that can result in a collective decisions in favor of one food source over another [35]. We hypothesise that the swarm could achieve a similar selective allocation of all resources to the best available source by exploiting a negative feedback in the form of repellent pheromone [7,49].

Acknowledgments. This work was funded by the ERC under the EU-H2020 research and innovation programme (grant agreement 647704). The authors thank Michael Port, Alex Cope, and Carlo Pinciroli for their crucial help and support in tackling the hardware and software challenges of this project.

References

1. Arvin, F., Yue, S., Xiong, C.: Colias-ϕ: an autonomous micro robot for artificial pheromone communication. Int. J. Mech. Eng. Robot. Res. **4**(4), 349–353 (2015)
2. Berman, S., Kumar, V., Nagpal, R.: Design of control policies for spatially inhomogeneous robot swarms with application to commercial pollination. In: Proceedings of the IEEE/RSJ International Conference on Robotics and Automation, ICRA 2011, pp. 378–385. IEEE Press (2011)
3. Bosien, A., Turau, V., Zambonelli, F.: Approaches to fast sequential inventory and path following in RFID-enriched environments. Int. J. Radio Freq. Identif. Technol. Appl. **4**(1), 28 (2012)
4. Campo, A., Dorigo, M.: Efficient multi-foraging in swarm robotics. In: Almeida e Costa, F., Rocha, L.M., Costa, E., Harvey, I., Coutinho, A. (eds.) ECAL 2007. LNCS, vol. 4648, pp. 696–705. Springer, Heidelberg (2007). https://doi.org/10.1007/978-3-540-74913-4_70
5. Campo, A., et al.: Artificial pheromone for path selection by a foraging swarm of robots. Biol. Cybern. **103**(5), 339–352 (2010)
6. Couzin, I.D., Krause, J., James, R., Ruxton, G.D., Franks, N.R.: Collective memory and spatial sorting in animal groups. J. Theoret. Biol. **218**(1), 1–11 (2002)
7. Detrain, C., Deneubourg, J.L.: Self-organized structures in a superorganism: do ants "behave" like molecules? Phys. Life Rev. **3**(3), 162–187 (2006)
8. Detrain, C., Deneubourg, J.L.: Collective decision-making and foraging patterns in ants and honeybees. Adv. Insect Physiol. **35**(08), 123–173 (2008)
9. Dimidov, C., Oriolo, G., Trianni, V.: Random walks in swarm robotics: an experiment with Kilobots. In: Dorigo, M., et al. (eds.) ANTS 2016. LNCS, vol. 9882, pp. 185–196. Springer, Cham (2016). https://doi.org/10.1007/978-3-319-44427-7_16

10. Ducatelle, F., Di Caro, G.A., Pinciroli, C., Gambardella, L.M.: Self-organized cooperation between robotic swarms. Swarm Intell. **5**(2), 73–96 (2011)

11. Ducatelle, F., Di Caro, G.A., Pinciroli, C., Mondada, F., Gambardella, L.M.: Communication assisted navigation in robotic swarms: self-organization and cooperation. In: Proceedings of the IEEE/RSJ International Conference on Intelligent Robots and Systems, IROS 2011, pp. 4981–4988. IEEE Press (2011)

12. Dussutour, A., Nicolis, S.C., Shephard, G., Beekman, M., Sumpter, D.J.T.: The role of multiple pheromones in food recruitment by ants. J. Exp. Biol. **212**(15), 2337–2348 (2009)

13. Ferrante, E., Turgut, A.E., Duéñez-Guzmán, E., Dorigo, M., Wenseleers, T.: Evolution of self-organized task specialization in robot swarms. PLoS Comput. Biol. **11**(8), 1–21 (2015)

14. Flanagan, T.P., Letendre, K., Burnside, W.R., Fricke, G.M., Moses, M.E.: Quantifying the effect of colony size and food distribution on harvester ant foraging. PLoS One **7**(7), e39427 (2012)

15. Fujisawa, R., Dobata, S., Kubota, D., Imamura, H., Matsuno, F.: Dependency by concentration of pheromone trail for multiple robots. In: Dorigo, M., Birattari, M., Blum, C., Clerc, M., Stützle, T., Winfield, A.F.T. (eds.) ANTS 2008. LNCS, vol. 5217, pp. 283–290. Springer, Heidelberg (2008). https://doi.org/10.1007/978-3-540-87527-7_28

16. Fujisawa, R., Dobata, S., Sugawara, K., Matsuno, F.: Designing pheromone communication in swarm robotics: group foraging behavior mediated by chemical substance. Swarm Intell. **8**(3), 227–246 (2014)

17. Garnier, S., Combe, M., Jost, C., Theraulaz, G.: Do ants need to estimate the geometrical properties of trail bifurcations to find an efficient route? A swarm robotics test bed. PLoS Comput. Biol. **9**(3), e1002903 (2013)

18. Garnier, S., Tâche, F., Combe, M., Grimal, A., Theraulaz, G.: Alice in pheromone land: an experimental setup for the study of ant-like robots. In: Proceedings of the 2007 IEEE Swarm Intelligence Symposium, SIS 2007, pp. 37–44. IEEE Press (2007)

19. Goss, S., Deneubourg, J.L., Bourgine, P., Varela, E.: Harvesting by a group of robots. In: 1st European Conference on Artificial Life, pp. 195–204. MIT Press, Cambridge (1992)

20. Hamann, H., Wörn, H.: An analytical and spatial model of foraging in a swarm of robots. In: Şahin, E., Spears, W.M., Winfield, A.F.T. (eds.) SR 2006. LNCS, vol. 4433, pp. 43–55. Springer, Heidelberg (2007). https://doi.org/10.1007/978-3-540-71541-2_4

21. Hecker, J.P., Letendre, K., Stolleis, K., Washington, D., Moses, M.E.: *Formica ex Machina*: ant swarm foraging from physical to virtual and back again. In: Dorigo, M., et al. (eds.) ANTS 2012. LNCS, vol. 7461, pp. 252–259. Springer, Heidelberg (2012). https://doi.org/10.1007/978-3-642-32650-9_25

22. Heredia, A., Detrain, C.: Influence of seed size and seed nature on recruitment in the polymorphic harvester ant Messor barbarus. Behav. Process. **70**(3), 289–300 (2005)

23. Herianto, Kurabayashi, D.: Realization of an artificial pheromone system in random data carriers using RFID tags for autonomous navigation. In: Proceedings of the IEEE/RSJ International Conference on Robotics and Automation, ICRA 2009, pp. 2288–2293. IEEE Press (2009)

24. Herianto, Sakakibara, T., Kurabayashi, D.: Artificial pheromone system using RFID for navigation of autonomous robots. J. Bion. Eng. **4**(4), 245–253 (2007)

25. Hoff, N., Wood, R., Nagpal, R.: Distributed colony-level algorithm switching for robot swarm foraging. In: Martinoli, A. (ed.) Distributed Autonomous Robotic Systems. STAR, vol. 83, pp. 417–430. Springer, Heidelberg (2013)
26. Hölldobler, B., Wilson, E.O.: The Ants. Harvard University Press, Cambridge (1990)
27. Jakobi, N., Husbands, P., Harvey, I.: Noise and the reality gap: the use of simulation in evolutionary robotics. In: Morán, F., Moreno, A., Merelo, J.J., Chacón, P. (eds.) ECAL 1995. LNCS, vol. 929, pp. 704–720. Springer, Heidelberg (1995). https://doi.org/10.1007/3-540-59496-5_337
28. Jeanson, R., Deneubourg, J.L., Grimal, A., Theraulaz, G.: Modulation of individual behavior and collective decision-making during aggregation site selection by the ant Messor barbarus. Behav. Ecol. Sociobiol. **55**(4), 388–394 (2004)
29. Khaliq, A.A., Di Rocco, M., Saffiotti, A.: Stigmergic algorithms for multiple minimalistic robots on an RFID floor. Swarm Intell. **8**(3), 199–225 (2014)
30. Macarthur, R.H., Pianka, E.R.: On optimal use of a patchy environment. Am. Nat. **100**(916), 603–609 (1966)
31. Mailleux, A.C., Deneubourg, J.L., Detrain, C.: Regulation of ants' foraging to resource productivity. Proc. Roy. Soc. Lond. B: Biol. Sci. **270**(1524), 1609–1616 (2003)
32. Mamei, M., Zambonelli, F.: Physical deployment of digital pheromones through RFID technology. In: Proceedings of the 2005 IEEE Swarm Intelligence Symposium, SIS 2005, pp. 281–288. IEEE Press (2005)
33. Mamei, M., Zambonelli, F.: Pervasive pheromone-based interaction with RFID tags. ACM Trans. Auton. Adapt. Syst. **2**(2), 4 (2007)
34. Mayet, R., Roberz, J., Schmickl, T., Crailsheim, K.: Antbots: a feasible visual emulation of pheromone trails for swarm robots. In: Dorigo, M., et al. (eds.) ANTS 2010. LNCS, vol. 6234, pp. 84–94. Springer, Heidelberg (2010). https://doi.org/10.1007/978-3-642-15461-4_8
35. Nicolis, S.C., Deneubourg, J.L.: Emerging patterns and food recruitment in ants: an analytical study. J. Theoret. Biol. **198**(4), 575–592 (1999)
36. Nouyan, S., Groß, R., Bonani, M., Mondada, F., Dorigo, M.: Teamwork in self-organized robot colonies. IEEE Trans. Evol. Comput. **13**(4), 695–711 (2009)
37. Olsson, O., Brown, J.S., Helf, K.L.: A guide to central place effects in foraging. Theoret. Popul. Biol. **74**(1), 22–33 (2008)
38. Orians, G.H., Pearson, N.E.: On the theory of central place foraging. Anal. Ecol. Syst. 154–177 (1979)
39. Payton, D.W., Daily, M., Estowski, R., Howard, M., Lee, C.: Pheromone robotics. Auton. Robots **11**(3), 319–324 (2001)
40. Pinciroli, C., Talamali, M.S., Reina, A., Marshall, J.A.R., Trianni, V.: Simulating Kilobots within ARGoS: models and experimental validation. In: Dorigo, M., et al. (eds.) ANTS 2018. Lecture Notes in Computer Science, vol. 11172, pp. 176–187. Springer, Heidelberg (2018)
41. Pinciroli, C., et al.: ARGoS: a modular, parallel, multi-engine simulator for multi-robot systems. Swarm Intell. **6**(4), 271–295 (2012)
42. Pini, G., Brutschy, A., Scheidler, A., Dorigo, M., Birattari, M.: Task partitioning in a robot swarm: object retrieval as a sequence of subtasks with direct object transfer. Artif. Life **20**(3), 291–317 (2014)
43. Pitonakova, L., Crowder, R., Bullock, S.: Information flow principles for plasticity in foraging robot swarms. Swarm Intell. **10**(1), 33–63 (2016)
44. Pitonakova, L., Crowder, R., Bullock, S.: The Information-Cost-Reward framework for understanding robot swarm foraging. Swarm Intell. **12**(1), 71–96 (2018)

45. Purnamadjaja, A.H., Russell, R.A.: Guiding robots' behaviors using pheromone communication. Auton. Robots **23**(2), 113–130 (2007)
46. Pyke, G.H.: Optimal foraging theory: a critical review. Annu. Rev. Ecol. Evol. Syst. **15**, 523–75 (1984)
47. Reina, A., Cope, A.J., Nikolaidis, E., Marshall, J.A.R., Sabo, C.: ARK: augmented reality for Kilobots. IEEE Robot. Autom. Lett. **2**(3), 1755–1761 (2017)
48. Reina, A., Miletitch, R., Dorigo, M., Trianni, V.: A quantitative micro-macro link for collective decisions: the shortest path discovery/selection example. Swarm Intell. **9**(2–3), 75–102 (2015)
49. Robinson, E.J., Ratnieks, F.L., Holcombe, M.: An agent-based model to investigate the roles of attractive and repellent pheromones in ant decision making during foraging. J. Theoret. Biol. **255**(2), 250–258 (2008)
50. Rubenstein, M., Ahler, C., Hoff, N., Cabrera, A., Nagpal, R.: Kilobot: a low cost robot with scalable operations designed for collective behaviors. Robot. Auton. Syst. **62**(7), 966–975 (2014)
51. Scheidler, A., Brutschy, A., Ferrante, E., Dorigo, M.: The k-unanimity rule for self-organized decision-making in swarms of robots. IEEE Trans. Cybern. **46**(5), 1175–1188 (2016)
52. Seeley, T.D.: Honey bee foragers as sensory units of their colonies. Behav. Ecol. Sociobiol. **34**(1), 51–62 (1994)
53. Sperati, V., Trianni, V., Nolfi, S.: Self-organised path formation in a swarm of robots. Swarm Intell. **5**(2), 97–119 (2011)
54. Sugawara, K., Kazama, T., Watanabe, T.: Foraging behavior of interacting robots with virtual pheromone. In: Proceedings of the IEEE/RSJ International Conference on Intelligent Robots and Systems, IROS 2004, vol. 3, pp. 3074–3079. IEEE Press (2004)
55. Svennebring, J., Koenig, S.: Building terrain-covering ant robots: a feasibility study. Auton. Robots **16**(3), 313–332 (2004)
56. Theraulaz, G., Bonabeau, E.: A brief history of stigmergy. Artif. Life **5**(2), 97–116 (1999)
57. Ulam, P., Balch, T.: Using optimal foraging models to evaluate learned robotic foraging behavior. Adapt. Behav. **12**(3–4), 213–222 (2004)
58. Valentini, G., et al.: Kilogrid: a novel experimental environment for the kilobot robot. Swarm Intell. 1–22 (2018)
59. Werger, B.B., Matarić, M.J.: Robotic "food" chains: externalization of state and program for minimal-agent foraging. In: From Animals to Animats 4. Proceedings of the 4th International Conference on Simulation of Adaptive Behavior, SAB 1996, pp. 625–634. MIT Press, Cambridge (1996)
60. Wilson, E.O.: Chemical communication among workers of the fire ant Solenopsis saevissima (Fr. Smith): the organization of mass-foraging. Anim. Behav. **10**(1–2), 134–147 (1962)
61. Winfield, A.F.T.: Foraging robots. In: Encyclopedia of Complexity and System Science, pp. 3682–3700. Springer, Heidelberg (2009). https://doi.org/10.1007/978-0-387-30440-3

Search in a Maze-Like Environment with Ant Algorithms: Complexity, Size and Energy Study

Zainab Husain[1(✉)], Dymitr Ruta[2], Fabrice Saffre[2], Yousof Al-Hammadi[1], and Abdel F. Isakovic[3]

[1] ECE Department, Khalifa University - KUST, Abu Dhabi, UAE
[2] EBTIC, Khalifa University - KUST, Abu Dhabi, UAE
[3] Physics Department, Khalifa University - KUST, Abu Dhabi, UAE
{zainab.husain,dymitr.ruta,fabrice.saffre,yousof.alhammad}@ku.ac.ae,
iregx137@gmail.com

Abstract. We demonstrate the applicability of inverted Ant Algorithms (iAA) for target search in a complex unknown indoor environment with obstructed topology, simulated by a maze. The colony of autonomous ants lay repellent pheromones according to the novel local interaction policy designed to speed up exploration of the unknown maze instead of reinforcing presence in already visited areas. The role of a target-collocated beacon emitting a rescue signal within the maze is evaluated in terms of its utility to guide the search. Different models of iAA were developed, with beacon initialization (iAA-B), and with increased sensing ranges (iAA-R with a 2-step far-sightedness) to quantify the most effective one. Initial results with mazes of various sizes and complexity demonstrate our models are capable of localizing the target faster and more efficiently than other open searches reported in the literature, including those that utilized both AA and local path planning. The presented models can be implemented with self-organizing wireless sensor networks carried by autonomous drones or vehicles and can offer life-saving services of localizing victims of natural disasters or during major infrastructure failures.

1 Introduction

With Mobile Ad Hoc Networks becoming the core of the IT infrastructure [14, 26], static communication topologies become extended, or even replaced by the mobile nodes of ad hoc networks where they can be deployed as per the exact requirement, thereby reducing cost and energy expenditure in providing services [3,12]. Such setups become particularly attractive in situations and tasks where human lives might be at peril like for civilian defense, disaster relief and/or rescue operations [2].

Electronic supplementary material The online version of this chapter (https://doi.org/10.1007/978-3-030-00533-7_12) contains supplementary material, which is available to authorized users.

© Springer Nature Switzerland AG 2018
M. Dorigo et al. (Eds.): ANTS 2018, LNCS 11172, pp. 150–162, 2018.
https://doi.org/10.1007/978-3-030-00533-7_12

From an optimization point of view, the problem of searching for a victim in a randomized environment is modeled as a maze solving problem where search time can be optimized by optimizing the path taken to reach the target where the maze walls can represent the randomly occurring obstacles. Maze solving, in a known environment, using global path planning has been a widely attacked research problem over the years, with several main issues having been addressed [4,19,22], but maze solving in an unknown environment is a more complex issue where local path planning is intuitively a better strategy [13].

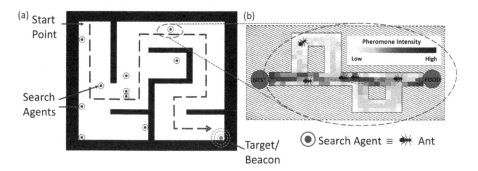

Fig. 1. (a) A Search and Rescue (SAR) operation modeled as a maze solving problem. The SAR can be implemented as an adaption of AA based decision making performed at each step in the maze. (b) illustrates a natural Ant Colony phenomenon

Following swarm inspired work in [1] and [7], we develop a self adaptive evolutionary path planning strategy for applications in critical, unknown environments like Search and Rescue (SAR) operations. This paper will be answering the following questions in developing a local path planning strategy:

1. The use of AA in reducing the number of steps needed to solve a maze
2. The influence of a beacon signal to guide the search
3. How does the path planning strategy scale with the size and complexity of the search space.

The coming sections discuss the already existing research in the field, the research methodology and simulation setup followed by a discussion of the results and a conclusion.

2   Overview of Current Approaches

In our research problem, a good path planning algorithm is one that would allow exploration, while still maintaining a level of coherence in terms of overall direction of motion.

Path planning strategies are of broadly two types: Global Path Planning, and Local Path Planning. Global path planning requires complete information of the area of interest and is performed before the time of injection into the maze. Several of these global planning techniques, like wall-follower [15], Dijkstra [23], flood-fill algorithm [22], among others have been proven to solve a maze with optimal results. On the other hand, local path planning, performed in real time, as the agents continue moving through the unknown environment and mapping it on-the-go [18], is a more complex problem that has been opened several discussions to dynamic optimization. Several researchers have attacked this problem with different techniques like the Point Bug Algorithm [6], Rapidly Exploring Random Trees [11], Artificial Potential Field Method [5], real time learning with Neural Networks [19], but research in the field continues to make the solution robust to obstacles, and quicker in both computation and convergence.

There has been a growing interest in deploying swarms of robots to solve real-world problems, by exploiting their collective decision making [9, 20, 21]. SAR in a maze closely resembles a colony of ants exploring an unknown territory in search of food. Like a natural ant colony search, SAR too is constrained by time and energy to ensure discovery of food or target before complete exhaustion.

Table 1. Comparison of research in the field of deterministic and bio-inspired path planning strategies

Ref.	Year	Uses local path planning?	Uses any AA?	Algorithms used
[9]	2014	✓	✓	Repellent Pheromone for coverage
[20]	2005	✓	✓	Combination of multiple pheromones
[1]	2010	✓	✓	Fuzzy Logic with Counter ACO
[7]	2014	✓	✓	Improved ACO Heuristic Function
[4]	2013	X	✓	Simple ACO Hybrid ACO with Random Search RL based search
[10]	2015	X	✓	Simple ACO
[16]	2016	X	✓	Improved ACO

Table 1 summarizes prominent research in the past decade with a focus on the use of AA or just pheromone based models. Ants, in a colony, work together by exchanging information via pheromones deposited in the shared space to speed up their search for food. This concept of using pheromones to optimize coverage in the Area of Interest (AoI) has been used by [9], with a detailed study on the influence of ant density and evaporation rate on the search time. [20] and [18] uses a combination of attractive and repulsive pheromones to also optimize coverage in the AoI for the purpose of surveillance operations. However, in our SAR, guided by a beacon, the objective is to quicken target location, without necessarily maximizing the AoI coverage. We believe, using an AA based decision making, with a heuristic pheromone model, could outperform the other pheromone guided models introduced.

In this research we shall target optimizing number of steps taken to locate the target, regardless of coverage, while keeping in mind the limitations of working in arbitrary indoor conditions.

3   This Work

We assume that the victim passively or actively guides the Search and Rescue (SAR) operation by the strength of the beacon signal automatically broadcasted, from his location, in all directions, as depicted in Fig. 1(a). Using this signal as the only guidance, the SAR agents autonomously carry out the search and locate the victim in a 2D randomized obstacle course modeled using a maze. The search starts with the injection of all agents from the start point at the top left of the maze in Fig. 1(a), and terminates on the co-location of at least one agent with the target.

The following subsections go through the simulation setup, indoor signal propagation and decision-making models employed in this research.

3.1   Simulation Environment Setup

To simulate a search in an indoor environment interspersed obstacles to model the randomness encountered after a natural disaster in urban areas, a maze was used to model the environment because of its complexity and heterogeneity in signal propagation media. Parameters of the maze, like its dimensions, and space to wall ratio can be modified to test the performance of the system under different conditions. Figure 2 shows the sample mazes that were generated for testing purposes. These mazes differ in both size and complexity, but have the same space to wall ratio of 5:1. For comparing efficiency as a function of size, the medium sized maze was used in simulations in its original 39×39, and re-scaled 78×78 (scaled by 2) and 156×156 (scaled by 4) versions.

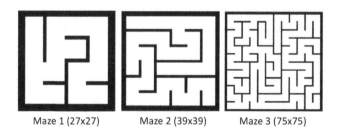

Maze 1 (27x27)   Maze 2 (39x39)   Maze 3 (75x75)

Fig. 2. Sample mazes used for simulations

To simplify the computation and implementation of the system, the mazes were discretized as explained in [25]. The grid unit size is the same as the agent size, thereby making the step size also equal to a grid unit.

To model indoor propagation, the ITU model, proposed by the International Telecommunication Union, described in [17], was modified through the addition of the $(w \times c)$ term to account for attenuation caused by the internal walls/obstacles to the following form

$$PL[dB] = 20log_{10}(f) + 28log_{10}(d) - 28 + (w \times c) \qquad (1)$$

where $w = 4.4349$ is the wall attenuation factor (for brick wall) [24], c is the wall count encountered, $f = 2400$ MHz is the frequency channel of communication for standard Wi-Fi, and d is the distance, in meters, from the beacon source [17]. Figure 3 shows the signal propagation with a corner located beacon.

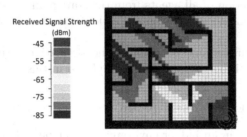

Fig. 3. Color-coded illustration of the beacon signal power received across the maze. The signal attenuates with distance and across the obstacles (walls) shown in black color. (Color figure online)

3.2   Ant Decision Making Models

One of the main contributions of this project is to improve the decision-making model, which would help the agents find the optimal path in the unknown environment in the shortest time possible. To study the impact of AA on the autonomous ants' led search in the maze, a simple AA based algorithm along with a standard random search were implemented for comparison. The AA based algorithm applies the AA probability density function (PDF) at each step to determine the next best step. As pictorially depicted in Fig. 4, the movement introduced in the ants is a 4-directional movement with an added option to stay in the current location, which means they have a choice of either moving to one of the 4 immediate neighboring cells, or continue hovering over the same position at each step. This choice between the 5 (4 nearest neighbors + 1 current position) possible cells is where the AA probability function is introduced. The general probability equation for Classic AA was modified to fit this decision making process. The modified equation is as follows

$$P_i(t+1) = \frac{(c + n_i(t))^\alpha}{\sum_{i=1}^{5}(c + n_i(t))^\alpha}$$

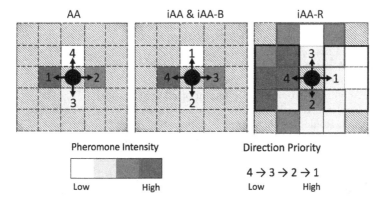

Fig. 4. All AA based decision making models prioritize their 4 possible directions of movement based on pheromone concentration associated with the 3 directions. AA prioritizes moving to a cell with a higher pheromone level, while all iAA based models prefer moving to a cell with lower pheromone concentration. AA, iAA and iAA-B check pheromone levels in only the immediate neighboring cells, while iAA-R checks for average pheromone in neighboring regions

where $P_i(t+1)$ is the probability of moving in direction i, $n_i(t)$ is the amount of pheromone in block i. Also, c is the degree of attraction to an unexplored path, and α is the bias to using a pheromone concentrated path. Here, one would consider adopting the values $\alpha = 2$ and $c = 20$ as per empirical evidence [8].

With simple AA, the ant is influenced to make a move towards a more pheromone concentrated cell, and therefore, ants were noticed to be clustering together. To counter this effect, and to quicken exploration, an inverted AA (iAA) model was also developed, where the pheromone was designed to be repulsive, encouraging ants to venture into unexplored territory. With promising initial results, 2 new versions of iAA, namely inverted-AA with beacon initialization (iAA-B) and inverted-AA with an increased sensing range (iAA-R) were also developed.

AA, iAA and iAA-B are all designed with ants having a sensing range of 1 cell. Therefore, decisions are made based on pheromone levels in the immediate 4 neighboring cells only. The general working of the AA algorithms is illustrated as a pseudocode in Algorithm 1. In iAA-R, on the other hand, ants have an increased sensing range, allowing them to base their decision on 3×2 neighboring regions and their corresponding average pheromone levels. This increased sensing range incorporates a slight far-sightedness in the ants' decision helping it move to a region with lower pheromone density which would improve its area coverage in 2 steps time.

All 4 decision making models were simulated on MATLAB on the 3 sample mazes. The search initiates with the agents being injected, all at once, from the top left of the maze and terminates when at least one of the agents collocates with the target located at the end of the maze (bottom right), as was illustrated

```
initialization;
possible moves = [stay, right, left, forward, backward];
while target not found do
    for each ant do
        Generate list of all possible next states;
        Acquire pheromone information of all next states;
        Roulette Wheel ← generate probabilities of moving to each next state;
        Spin Roulette Wheel to pick next state;
        Update current position and pheromone levels;
    end
end
```

Algorithm 1: Basic AA Algorithm used for all decision making models, where each model differs in probability generation

in Fig. 1(a). All simulations are performed on a 2.6 GHz Intel Core i7 processor running a 2016a 64-bit version of MATLAB.

4   Results and Discussion

All AA based decision-making models have an inherent element of randomness thanks to the AA probability equation. Therefore, all 4 models were simulated 30 times each to average out the effect of the randomness. The performances of the 4 models are compared, along with a purely random movement solution, in terms of number of iterations needed to locate the target, which is the equivalent of solving the maze. We believe the number of steps is a better measure of performance and it can be compared with other works on different platforms.

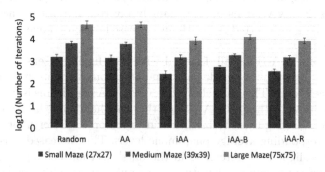

Fig. 5. Comparing the performance of the 4 AA based models and a random movement solution in solving the 3 mazes, of different sizes and complexities, with 100 ants each.

Figure 5 compares the performance of the 5 models when simulated with 100 ants each on the 3 mazes, with 30 repetitions each. The pure AA based model did not introduce much of an improvement, compared to a purely random solution. This lack of improvement can be attributed to the clustering effect of pheromone in pure AA that is likely limiting the exploration of the maze. iAA is the best performing algorithm among the 5, with a closely following performance of iAA-R.

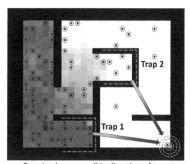

Growing beacon pull in direction of target

Fig. 6. Inverted AA with beacon initialization for pheromone intensities can lead agents into the traps, marked as green-dashed U-shaped walls in the figure, as they tend to blindly follow a positive pheromone gradient and are unable to go around these obstacles. (Color figure online)

Contrary to intuitive preconception, iAA-B did not positively add to the performance of the iAA algorithm with its beacon initialization that was supposed to better guide the ants to the target. This is due to a trapping effect noticed in the simulation, as illustrated in Fig. 6, where ants get trapped in nooks of the maze while being pulled towards the target.

We suggest a useful primitive measure of the complexity of the maze to be a number of 90-degrees turns an ant may make while performing a search, as can be surmised from Fig. 2. In this context, a study of the number of iterations, with a simulation of the best contender iAA, as a function of the maze complexity is shown in Fig. 7(a). Two approaches were used: (a) the fixed number of ants, and (b) the variable number of ants. For the latter case, we scaled the number of ants with the size of the maze, which leads to qualitatively different behavior, and points towards a practical recipe of how to organize the iAA based search on mazes of increasing size. A well scaled group of ants with a rough measure of the search space would be energy-conserving in the search, without much deterioration in performance, as observed especially in the simulations with the small and medium sized mazes. A similar set of simulations and comparisons were carried out for iAA-B and iAA-R algorithms as well, which can be accessed in the supplementary material.

The study of the number of iterations as a function of the maze size is shown in Fig. 7(b) for three different ant colony populations. As one might expect, the increase in the size of the colony requires fewer iterations to reach the target,

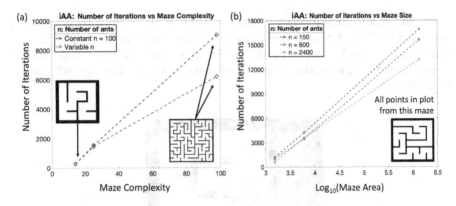

Fig. 7. (a) Number of iterations needed to solve the maze as a function of the maze complexity, with complexity defined as the number of 90° turns an ant can make in the maze. The simulations were carried out on the mazes of complexities 14, 25 and 98, respectively. (b) Number of iterations needed to solve the maze as a function of the size of the maze of the fixed complexity, that of 25. The size of this maze was re-scaled from $39 \times 39 \,\mathrm{m}^2$, to $78 \times 78 \,\mathrm{m}^2$, and $156 \times 156 \,\mathrm{m}^2$. Insets show respective maze layouts.

but, we note that the 28% improvement (a decrease in the number of iterations) comes at the cost of increase the colony size by the factor of 16. This result is now an input into our ongoing effort to optimize the energy of the search operation.

5   Energy Expenditure Analysis

Energy expenditure is an important aspect to consider in the simulation of the search to factor in real life battery limitations on mobile robotic agents. The agents' battery powers both its mechanical and communication functions, which means a faster depletion and makes the issue crucial in the success of the operation. The direction of motion plays an important influence in the energy expended in flight control.

Consequently, a set of experiments were set up to simulate AA and iAA algorithms with variable group sizes ranging between 100 and 600, with 100 unit increments. the results are shown in Fig. 8.

Continuing in the same direction of motion is always cheaper than introducing a displacement in direction. As our agents are limited to a 4-directional motion (in addition to an option to hover), the only direction changes possible are a 90°, to either side, or a 180° turn, meaning a backward motion. Considering the different amounts of energy needed to slow down (control speed), and change direction in each motion possible, a movement cost of 0.5, 1, 1.5, and 2 were assigned to hovering, forward, 90° turning and backward movements respectively.

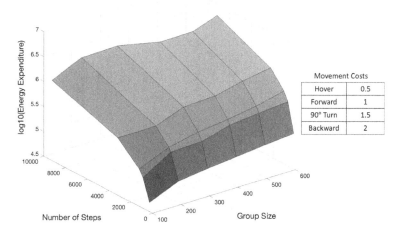

Fig. 8. A 3D depiction of variation in performance and energy expenditure with variable group sizes in 5 different mazes with complexities 14, 25, 36, 74 and 98

6   Search for Appropriate AA Parameters on the Maze

For the AA Eq. 1, empirically, colonies of Argentine ants are observed to adhere to the model with values of $\alpha = 2$ and $c = 20$. However, with initial trials, AA and iAA didn't respond well with high values of c. This could be due to a forced randomness that overpowers the influence of pheromones in our particular simulation set up. Therefore, to simplify initial simulations, we started by using $\alpha = 1$ and $c = 1$. However, we find it necessary to address the selection of these 2 parameters for an optimized working of the system. As the probability equation is a function of both α and c (in addition to pheromone intensity), we optimized either parameter by keeping the other constant for a set of trials.

The influence of α in AA is erratic and non-conclusive, which is in correlation with our observations earlier in Sect. 4, where a stronger pheromone attraction (larger α) resulted in lesser exploration and thus poor performance. On the other hand, the influence of α on the performance of iAA is much more encouraging. The performance continues to improve (as the number of steps decrease) with a growing α with the lowest recorded at $\alpha = 3$. A 7th degree polynomial was fitted to the generated values to estimate the trend of the graph for $\alpha > 3$ which showed a saturation, without much change in the number of steps with further increase in α. Therefore, the study was concluded with the selection of $\alpha = 3$ as the best choice for an iAA simulation, and an observation of poor influence of α on AA.

The influence of c is quite erratic on the performance in iAA, as shown in Fig. 9(b), owing to the heuristic nature of the algorithm. Over averaged results, the performance graph shows a local minima at around $c = 1.3$ and $c = 2.7$.

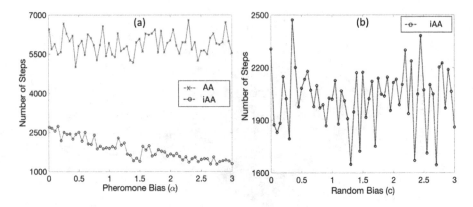

Fig. 9. AA and iAA performances, in terms of number of steps, as a function α. Results are an average of 20 repeated simulations.

However, due to the highly erratic trend, it's difficult to pinpoint a single optimal value of c, to finalize the empirical study. For now, the results shown are for $c = 1$.

The test for the influence of variable c on AA performance was not successful as the search agents stalled around obstacles, with no further progress for all values of $1 < c < 1.5$. Since simulations with only 3 mazes have been attempted so far, there's ongoing work to ascertain the nature of the dependence more precisely.

7   Conclusions and Future Work

Real time search on the maze in an unknown bounded environment with a local path planning is studied with several versions of AA, such as inverted AA (iAA), iAA-R, and iAA-B. It is shown that a traditional AA is much closer to random search in performance and measurably weaker than the iAA. In addition to elucidating which version of AA is most applicable, the report analyzes the influence of the following parameters: (a) the complexity of the maze, (b) the number of ants, and (c) the size of the maze (with the number of ants and the maze size both changing by approximately an order of magnitude). Energy expenditure for such search is studied as a function of the ants group size and the number of steps, demonstrating steep changes in the energy as the function of the group size. Additionally, empirical studies were conducted demonstrating that traditional values for AA parameters may need to be modified for the maze search in this particular task.

Acknowledgement. We gratefully acknowledge the support from UAE ICT Fund through the grant "Biologically Inspired Self-organizing Network Services" and Prof. Sami Muhaidat (KUST) for advices with the models of indoor signal propagation.

References

1. Ahuja, M.: Fuzzy counter ant algorithm for maze problem. Master's thesis. University of Cincinnati (2010)
2. Aljehani, M., Inoue, M.: Communication and autonomous control of multi-UAV system in disaster response tasks. In: Jezic, G., Kusek, M., Chen-Burger, Y.-H.J., Howlett, R.J., Jain, L.C. (eds.) KES-AMSTA 2017. SIST, vol. 74, pp. 123–132. Springer, Cham (2018). https://doi.org/10.1007/978-3-319-59394-4_12
3. Andryeyev, O., Mitschele-Thiel, A.: Increasing the cellular network capacity using self-organized aerial base stations. In: Proceedings of the 3rd Workshop on Micro Aerial Vehicle Networks, Systems, and Applications, pp. 37–42. ACM (2017)
4. Aurangzeb, M., Lewis, F.L., Huber, M.: Efficient, swarm-based path finding in unknown graphs using Reinforcement Learning. In: 2013 10th IEEE International Conference on Control and Automation, ICCA, pp. 870–877. IEEE (2013)
5. Bounini, F., Gingras, D., Pollart, H., Gruyer, D.: Modified Artificial Potential Field method for online path planning applications. In: 2017 IEEE Intelligent Vehicles Symposium, IV, pp. 180–185. IEEE (2017)
6. Buniyamin, N., Ngah, W., Sariff, N., Mohamad, Z.: A simple local path planning algorithm for autonomous mobile robots. Int. J. Syst. Appl. Eng. Dev. 5(2), 151–159 (2011)
7. Cao, J.: Robot global path planning based on an Improved Ant Colony Algorithm. J. Comput. Commun. 4(02), 11 (2016)
8. Engelbrecht, A.P.: Fundamentals of Computational Swarm Intelligence. Wiley, Hoboken (2006)
9. Fossum, F., Montanier, J.M., Haddow, P.C.: Repellent pheromones for effective swarm robot search in unknown environments. In: 2014 IEEE Symposium on Swarm Intelligence, SIS, pp. 1–8. IEEE (2014)
10. Krentz, T., Greenhagen, C., Roggow, A., Desmond, D., Khorbotly, S.: A modified Ant Colony Optimization algorithm for implementation on multi-core robots. In: 2015 Swarm/Human Blended Intelligence Workshop, SHBI, pp. 1–6. IEEE (2015)
11. Lavalle, S.M.: Rapidly-exploring random trees: A new tool for path planning. TR 98–11. Computer Science Deparment, Iowa State University, October 1998
12. Li, Y., Cai, L.: UAV-assisted dynamic coverage in a heterogeneous cellular system. IEEE Netw. 31(4), 56–61 (2017)
13. Mac, T.T., Copot, C., Tran, D.T., De Keyser, R.: Heuristic approaches in robot path planning: a survey. Robot. Auton. Syst. 86, 13–28 (2016)
14. Mainetti, L., Patrono, L., Vilei, A.: Evolution of wireless sensor networks towards the Internet of Things: a survey. In: 2011 19th International Conference on Software, Telecommunications and Computer Networks, SoftCOM, pp. 1–6. IEEE (2011)
15. Mishra, S., Bande, P.: Maze solving algorithms for micro mouse. In: IEEE International Conference on Signal Image Technology and Internet Based Systems, SITIS 2008, pp. 86–93. IEEE (2008)
16. Wang, Z.W.: Robot path planning for mobile robot based on Improved Ant Colony Algorithm. Appl. Mech. Mater. 385–386
17. Rappaport, T.S., et al.: Wireless Communications: Principles and Practice, vol. 2. Prentice Hall PTR, Upper Saddle River (1996)
18. Ravankar, A., Ravankar, A.A., Kobayashi, Y., Emaru, T.: On a bio-inspired hybrid pheromone signalling for efficient map exploration of multiple mobile service robots. Artif. Life Robot. 21(2), 221–231 (2016)

19. Rivera, G.: Path planning for general mazes. Master's thesis. Missouri University of Science and Technology (2012)
20. Sauter, J.A., Matthews, R., Parunak, H.V.D., Brueckner, S.A.: Performance of digital pheromones for swarming vehicle control. In: Proceedings of the Fourth International Joint Conference on Autonomous Agents and Multiagent Systems, pp. 903–910. ACM (2005)
21. Shiltagh, N.A., Jalal, L.D.: Optimal path planning for intelligent mobile robot navigation using modified Particle Swarm Optimization. Int. J. Eng. Adv. Technol. **2**(4), 260–267 (2013)
22. Tjiharjadi, S., Setiawan, E.: Design and implementation of a path finding robot using Flood Fill algorithm. Int. J. Mech. Eng. Robot. Res. **5**(3), 180–185 (2016)
23. Wang, H., Yu, Y., Yuan, Q.: Application of Dijkstra algorithm in robot path-planning. In: 2011 Second International Conference on Mechanic Automation and Control Engineering, MACE, pp. 1067–1069. IEEE (2011)
24. Wilson, R.: Propagation Losses Through Common Building Materials 2.4 GHz vs 5 GHz. Magis Networks Inc., San Diego (2002)
25. Yi, G., Feng-ting, Q., Fu-jia, S., Wei-ming, H., Peng-ju, Z.: Research on path planning for mobile robot based on ACO. In: 2017 29th Chinese Control and Decision Conference, CCDC, pp. 6738–6743. IEEE (2017)
26. Zanella, A., Bui, N., Castellani, A., Vangelista, L., Zorzi, M.: Internet of Things for smart cities. IEEE Internet Things J. **1**(1), 22–32 (2014)

Self-adaptive Quantum Particle Swarm Optimization for Dynamic Environments

Gary Pampará[1]([✉])[iD] and Andries P. Engelbrecht[2][iD]

[1] Department of Computer Science, University of Pretoria, Pretoria, South Africa
gpampara@gmail.com
[2] Institute for Big Data and Data Science, University of Pretoria,
Pretoria, South Africa
engel@cs.up.ac.za

Abstract. The quantum-inspired particle swarm optimization (QPSO) algorithm has been developed to find and track an optimum for dynamic optimization problems. Though QPSO has been shown to be effective, despite its simplicity, it does introduce an additional control parameter: the radius of the quantum cloud. The performance of QPSO is sensitive to the value assigned to this problem dependent parameter, which basically limits the area of the search space wherein new, better optima can be detected. This paper proposes a strategy to dynamically adapt the quantum radius, with changes in the environment. A comparison of the adaptive radius QPSO with the static radius QPSO showed that the adaptive approach achieves desirable results, without prior tuning of the quantum radius.

1 Introduction

Optimization algorithms have very different performance characteristics based on the underlying problem search space. Static environments have been widely studied and several algorithms have been shown to be applicable to a wide variety of static optimization problems. Notable examples include genetic algorithms [14], differential evolution [22] and particle swarm optimization (PSO) [16]. In the case where the underlying problem search space changes over time, the behavior characteristics of these algorithms are not always still applicable. A dynamic optimization problem describes such a problem search space, where the optimal value of the search space not only changes over time, but can drastically change in value and/or location. When applying an algorithm designed for a static environment to a dynamic environment, the inefficiencies of the algorithms become apparent with the algorithms unable to adapt to the changing search space. Common reasons for the inefficiency include a loss of diversity, outdated memory of previous best positions, and the inability of the algorithm to detect that the search space has actually changed [21].

Quantum particle swarm optimization (QPSO) [3] was proposed as an inherently dynamic variant of the PSO, capable of handling underlying environment

© Springer Nature Switzerland AG 2018
M. Dorigo et al. (Eds.): ANTS 2018, LNCS 11172, pp. 163–175, 2018.
https://doi.org/10.1007/978-3-030-00533-7_13

changes by categorizing a set of particles within the algorithm as "quantum" particles. These quantum particles follow a different position update strategy from normal particles: Quantum particles move probabilistically within a predefined quantum cloud around the global best particle position. The size of the quantum cloud determines the area within which a quantum particle is allowed to move. Unfortunately, the solution tracking ability of the QPSO is sensitive to the size of the radius. If the radius is too small, the area the quantum particles can explore for changes in the optimum, will be restricted; preventing the detection of an optima change outside of the cloud radius. Conversely, if the cloud radius is too large, a larger portion of the search space will be covered by the cloud, resulting in unnecessary exploration for small change severities, and resulting more in a random search.

This paper proposes a strategy to adapt the cloud radius value, starting with a large radius and reducing it to a small value. As soon as the environment changes, the radius is increased to a large value. This strategy helps to increase the exploration by the quantum particles whenever a change in the environment occurs, while moving towards exploitation the longer the environment remains unchanged.

The remainder of the paper is organized as follows: Sect. 2 discusses the necessary background information for the algorithm alteration proposed in Sect. 3. Following in Sect. 4, is a discussion on the experimental approach with results provided in Sect. 5. Lastly, concluding remarks are presented in Sect. 6.

2   Background

This section provides background information for the remainder of the paper. Section 2.1 discusses dynamic environments, with the moving peaks benchmark (MPB) discussed in Sects. 2.2, 2.3 and 2.4 respectively discuss the PSO and QPSO algorithms, with previous radius management strategies discussed in Sect. 2.5.

2.1   Dynamic Environments

A problem space wherein optima move over time is regarded as a dynamic environment. The changes experienced by such environments can vary from subtle movements to extreme changes where optima seemingly move at random. Much work has been done to attempt to classify these problem space changes, based on the frequency and severity of the change observed, the type of movement the change undertook, and the trajectory that a change followed. Eberhart and Shi [10] and Hu and Eberhart [15] describe changes observed for an optimum as (a) Type I, where the measurable value of an optimum remains the same, but its position in the search space changes, (b) Type II, where the value of the optimum changes but the position in the search space remains constant, or (c) Type III, where both the position and the optimum value change. Angeline [1] categorizes the movement of an optimum as either linear, circular, or random.

Duhain and Engelbrecht [9] combine the previous classifications with spatial and temporal severity to create 27 unique environments. These environments are broadly classified into:

1. **Quasi-static** environments which have both low spatial and temporal changes.
2. **Progressive** environments which have small spatial, but frequent changes. The changes result in a search space where optima move gradually over time.
3. **Abrupt** environments which have infrequent and large spatial changes. The problem space almost seems to remain constant for a period of time before the next, large change is observed.
4. **Chaotic** environments experience large spatial adjustments that occur at a frequent interval.

2.2 Moving Peaks Benchmark Function

The MPB [15] has been developed as a generator for dynamic environments, based on a set of input parameters. The generator can create problems spaces that adhere to a variety of classification goals, and is quite popular within literature [5,6,17,19]. The generated problem search spaces are characterized by the number of peaks, within a given domain, with each peak maintaining a width, height, and location within the search space. The dynamic nature of the problem is achieved by varying the parameters of the MPB generator over time, resulting in a changed environment. The movement of the peaks is also determined by the state of the previous environment, where trajectory information of the peaks themselves is maintained.

2.3 Particle Swarm Optimization

The PSO algorithm, introduced by Kennedy and Eberhart [16], describes an optimization process which is modeled on the flocking behavior of birds. PSO is a population-based, stochastic algorithm which maintains a collection of entities, known as "particles". Each particle exhibits a simple behavior which determines its movement: (1) move towards the best position in the immediate neighborhood of particles, and (2) move towards a particle's own best, previously observed, position.

Particles move in an iterative manner throughout the problem search space, determining the magnitude of the movement through the application of a "velocity" vector (representing a step size and direction) and to the currently maintained position vector. The velocity vector implements the previously mentioned particle behavior, and is calculated as:

$$v_{ij}(t+1) = \omega v_{ij}(t) + c_1 r_{1j}(y_{ij}(t) - x_{ij}(t)) + c_2 r_{2j}(\hat{y}_{ij}(t) - x_{ij}(t)) \qquad (1)$$

where $v_{ij}(t)$ is the velocity of particle i in dimension j, with the current particle position given by $x_{ij}(t)$, at time step t; ω denotes the inertia coefficient, determining the extent to which a particle will continue to search in the same direction, with c_1 and c_2 respectively specifying the influence of the social and cognitive components to the resultant velocity vector. The stochastic vectors, $\mathbf{r}_1$ and $\mathbf{r}_2$, are multiplied with the cognitive and social velocity components.

The movement of particle i through the search space, to the next position, is then determined by

$$\mathbf{x}_i(t+1) = \mathbf{x}_i(t) + \mathbf{v}_i(t+1) \tag{2}$$

The classical PSO algorithm displays several disadvantages when applied to dynamic environments, particularly related to the personal best position. There is no way to know, ahead of time, whether an environment change would (potentially) make the particle's best position obsolete or invalid [2]. Additionally, as the swarm converges on a solution, the dispersion of particles within the swarm (diversity) [25] decreases. This loss of diversity results in small step sizes, preventing particle movement to other areas of the search space.

2.4   Quantum Particle Swarm Optimization

The QPSO [3] was introduced to address the problems of the PSO within dynamic environments. The QPSO employs a percentage of particles that use a position update that is different to Eq. (2). These particles, referred to as *quantum particles*, move in a manner similar to electrons orbiting the nucleus of an atom. The movement of the quantum particles is done by sampling a probability distribution, centered at the global best position of the current swarm of particles. The movement is then

$$x_{ij}(t+1) \sim d(\hat{y}_{ij}(t), r_{cloud}) \tag{3}$$

where d is a probability distribution and r_{cloud} is a constant determining the size of the quantum "cloud" around the global best position. Any particle that is not updated using Eq. (3), is termed a neutral particle and follows the update equations of the classical PSO algorithm.

Although this approach does allow for a percentage of particles to behave in a manner that addresses concerns around diversity loss, an additional parameter, r_{cloud}, is introduced. Because r_{cloud} restricts the search area of quantum particles around the global best position, the parameter requires problem dependent tuning. After an environment change, a larger radius is desirable, as more exploration can occur within the quantum cloud, whereas a smaller radius would aid with solution refinement when the environment is unchanging.

2.5   Previously Suggested Radius Management Strategies

In order to prevent the problem dependent tuning of r_{cloud}, several strategies have been proposed. Blackwell *et al.* [4] propose the use of different distributions to influence the movement of quantum particles within the quantum cloud.

Although this approach does not hint at adjusting the radius of the quantum cloud, the quantum particles may move to positions beyond the defined radius, based on the distribution being sampled. For example, sampling a standard Gaussian distribution would restrict 95% of the observed values to be within two standard deviations from the mean, and would on occasion allow for larger values.

Harrison *et al.* [12] examine the effect of different distributions on the performance of the QPSO, concluding that the choice of distribution depends on the type of dynamic environment and that smaller values for r_{cloud} lead to better performances for QPSO. These conclusions also stated that the uniform distribution is, generally, a poor choice. Another study by Harrison *et al.* [13] attempt to remove the parameter r_{cloud} and the probability distribution completely through the use of a parent centric crossover (PCX) [7] operator. The resulting algorithm completely removes the atom metaphor, replacing it with a crossover operator instead. As a result, the algorithm can not truly be classified as a QPSO variant. The algorithm does, however, address the problem of diversity loss because the PCX operator was originally designed to introduce diversity into the swarm of particles, but introduces two additional parameters: the deviations of two Gaussian distributions.

3   Self-adaptive Quantum Particle Swarm Optimization

In order to allow the QPSO to manage a dynamically sized quantum cloud, adjusting automatically based on the current swarm and the current environment, some of the problems associated with the PSO in dynamic environments (outdated memory, diversity loss, etc.) need to be addressed. The resulting algorithm is largely unchanged from the standard QPSO definition, with a few configuration changes applied:

- When a change in the environment occurs, the memory of particles (personal best positions) may have become invalid or obsolete. These invalid values need to be corrected, otherwise the particle may be attracted to an undesirable area of the search space. One such mechanism, is to reset the personal best position of the particle to the current position, and to re-evaluate the particle. Quantum particles do not depend on a personal best position and are re-initialized within the problem domain.
- The personal best position of a particle is updated, if and only if the current position is within the boundaries of the search space. This constraint on the update process prevents solutions that may be seemingly more optimal, but are located outside the defined problem search space to become personal best positions. Allowing infeasible personal best positions will result in particles being attracted to infeasible solutions, and/or fruitless search in infeasible space [11].
- The quantum cloud radius, r_{cloud}, is calculated by taking the maximum between the diversity of neutral and quantum particles. The diversity calculation for neutral particles only considers the particle personal best position,

as these positions will be within the problem search space. Quantum particles are only considered if the current position of the particle is within the problem bounds. Diversity is calculated as:

$$D(S(t)) = \frac{1}{n_s} \Sigma_{i=1}^{n_s} \sqrt{\Sigma_{j=1}^{n_x}(x_{ij}(t) - \bar{x}_j(t))^2} \tag{4}$$

where n_s is the number of the neutral, or quantum particles considered in the diversity calculation; $\bar{x}_j(t)$ is the average j-th dimension of the entire swarm, calculated as

$$\bar{x}_j(t) = \frac{\Sigma_{i=1}^{n_s} x_{ij}(t)}{n_s} \tag{5}$$

The resulting diversity value is then used as the cloud radius value. The cloud radius value is fed into a random distribution as the deviation, from which quantum particle positions are sampled. Scaling the calculated diversity by a constant to determine the cloud radius was not considered as it would introduce an additional constant, which would require problem dependent tuning.

– The neighborhood topology should facilitate information exchange, but at a slower rate than a fully connected topology. Slower information propagation through the swarm using topologies like the local best or Von Neumann, allows for larger parts of the search space to be covered, whilst still allowing for convergence. Fully connected topologies result in faster convergence which results in smaller diversity and cloud radius values.

– Particle neighborhoods should consist of both types of particles. Because quantum particles move within the quantum cloud at each iteration, it is advantageous to share information about the changed search space with the neutral particles in the local particle neighborhood.

4   Experimental Approach

The main objective of this paper is to demonstrate that the performance of the QPSO on a set of dynamic optimization problems is either better than, or unchanged, when the algorithm dynamically adapts the value of r_{cloud} compared to keeping the value at a predefined static value. This section describes the different considerations in order to prepare the experiments to evaluate the QPSO and the self-adaptive QPSO, with experiment design in Sect. 4.1, performance measures in Sect. 4.2, and Sect. 4.3 discussing the statistical process.

4.1   Experimental Design

Due to the complex nature of algorithms that operate in dynamic environments, it is beneficial to elaborate on the software approach used. All experiments used the software library *CIlib* [20] which allows for type-safe, repeatable experimentation, with perfectly reproducible results. The initial seed, from which 30 distinct seeds are then generated for the different independent algorithm runs, is listed

Table 1. Environment parameters

Parameter	Static	Progressive	Abrupt	Chaotic
Peak height	[30, 70]	[30, 70]	[30, 70]	[30, 70]
Peak width	[1, 12]	[1, 12]	[1, 12]	[1, 12]
Height change severity	1	1	10	10
Width change severity	0.05	0.05	0.05	0.05
Change severity	1	1	10	10
Random movement % (λ)	0	0	0	0
Change freq. (iterations)	200	1	200	5

Table 2. QPSO algorithm parameters

Parameter	Value
Particles	40
Proportion quantum particles	50%
ω	0.729844
c_1, c_2	1.496180
Topology	l-best (size 3)
Iteration strategy	Synchronous
PRNG Seed	123456789L
Static radius values	$r_{cloud} \in [5, 10, 50]$
Quantum cloud distribution	Gaussian

in Table 2. Importantly, different pseudo-random number generators are used for algorithms and dynamic environment updates. PSO parameter choices are based on the convergence properties described by Van den Bergh [24], with the l-best topology providing slower information propagation throughout the neutral particles. Sampling a Gaussian distribution centered at the global best position allows for quantum particle movement with a central tendency at the global best position, whilst still allowing unconstrained movement that may exceed the boundaries of the quantum cloud.

To compare the performance of the static QPSO with the self-adaptive QPSO, benchmark environments were defined that match the classifications of Duhain [8] and Duhain and Engelbrecht [9], using the MPB as the search space problem. Each problem space contained 10 peaks and used the parameters defined in Table 1. Each environment was also classified as a Type III environment [15]. Each algorithm configuration was executed 30 times for 1000 iterations. The domain of the problem search space was defined to be [0, 100], with 5 dimensions. The static QPSO variants are identified by the size of the associated radius: QPSO-5, QPSO-10, and QPSO-50 for cloud radius of 5, 10 and 50 respectively. Other parameters, common to the QPSO algorithms, are listed in Table 2.

4.2 Dynamic Environment Performance Measures

Performance measures are required to quantify the performance of an algorithm within a dynamic environment. Performance measures of static environments do not behave in a manner that allows for a valid performance quantification for dynamic environments. Duhain [8] recommend that better choices for performance measurement within a dynamic environment include the *accuracy* of the solutions over time, the *stability* (solution quality after an environment change), and algorithm *exploitative capacity*, which is the quality of the best solution between environment changes. Because the MPB defines a maximum peak value, the error produced by an algorithm, by comparing a solution to the search space optimum, is calculable. A set of "good" [8] measures are:

- The **collective mean error** (CME) [18] records the mean error of the best solution over the entire experiment. The measure is defined as:

$$\text{CME} = \frac{1}{n_t} \Sigma_{t=1}^{n_t} err_{t,m} \qquad (6)$$

  where n_t is the number of iterations within an experiment, and $err_{t,m}$ is the difference between the optimum in environment m and the best solution, at time-step t. The CME is a good overall measure [18] that quantifies the overall performance of an algorithm within a dynamic environment.
- The **average best error before change** (ABEBC) is a measure that records the difference between the optimum value and the quality of the best particle, or error (provided that the target value is known). Knowledge of when an environment change occurs is a prerequisite for using the ABEBC measure, and the measure provides insight about the *exploitative capability* [9] of an algorithm on a given problem. Formally, the measure is defined by:

$$\text{ABEBC} = \frac{1}{n_c} \Sigma_{c=0}^{n_c} (err_{c,r-1}) \qquad (7)$$

  where r is the number of iterations between two environmental changes and $err_{c,r-1}$ is the difference between the best fitness and the optimal fitness at iteration t after the last change c; n_c is the total number of environment changes.
- The **average best error after change** (ABEAC) [23] is a measure that determines the *stability* [5] of an algorithm within a dynamic environment. The measure is similar to the ABEBC, except that the error in fitness compared to the global optimum, determined directly after an environment change. As such, the measure favors algorithms that tend to prefer areas of the search space that do not change all that much. The measure is defined by

$$\text{ABEAC} = \frac{1}{n_c} \Sigma_{c=0}^{n_c} err_{c,0} \qquad (8)$$

where $err_{c,0}$ is the error at the iteration directly after an environment change.

4.3   Statistical Process

The performance of the static QPSO and the self-adaptive QPSO was evaluated using the measures defined in Sect. 4.2. For each combination of error measurement, a Mann-Whitney-U rank sum test with Holm correction was used to determine if a significant difference ($\alpha = 0.05$) existed between the algorithm performances. A value of 1 is allocated to an algorithm if the results are superior to the other, and the inferior algorithm is assigned a score of -1. These scores then determine the win/loss ratio of the algorithms.

5   Results

This section contains the analysis of the obtained results for the four algorithms (QPSO-5, QPSO-10, QPSO-50, and self-adaptive QPSO). An analysis of the CME, ABEBC and ABEAC measures follow in Sects. 5.1, 5.2 and 5.3 respectively. Section 5.4 analyzes the diversity and radius size of the self-adaptive QPSO.

5.1   Analysis of Collective Mean Error

Table 3 provides algorithm rankings with respect to the CME. For the CME measurement, the rankings indicate that the self-adaptive QPSO performed the best across the different environment types. QPSO-50 was the second best performing algorithm, followed by QPSO-10 and QPSO-5. As shown in Fig. 1(a), a similar trend to the ranking data can be observed when comparing algorithm performances. For the abrupt and progressive environments all algorithms achieved similar performances, but the win/loss ratio favors the self-adaptive QPSO within these environments.

After an environment change, the self-adaptive QPSO has an increase in diversity, as illustrated in Fig. 1(d). The increase in diversity results in a larger area for quantum particles to explore, and as the swarm starts to converge on an optimum, the radius value decreases. With a decreasing radius, quantum particles begin exploitation of the search space around the optimum. The error values in Fig. 1(a) also show that the QPSO is sensitive to the frequency of environment change: the lower error values were achieved for the quasi-static environments where the frequency of change is low. Compared to the static QPSO, it should be noted that the self-adaptive QPSO did not perform worse.

5.2   Analysis of Average Best Error Before Change

Figure 1(b) illustrates that all four algorithms manages to achieve values of less than 20 for the ABEBC within the quasi-static environments. For the other environments, the same trend of the CME measurement is evident, with none of the algorithms particularly providing a clearly better solution, and a similar spread of error values. Because the ABEBC demonstrates the exploratory capacity of an algorithm, it is clear that none of the algorithms were able to effectively locate a new solution before the environment changed. The self-adaptive QPSO achieved a comparable performance, compared to the static QPSO.

5.3   Analysis of Average Best Error After Change

After an environment change, the QPSO-10 and self-adaptive QPSO managed to achieve median values that are lower than that of the other QPSO algorithms for the quasi-static environments. Unfortunately, for the other environment types, no one algorithm displayed a clear improvement, and all algorithms (including the self-adaptive QPSO) achieved equally poor results. These performances are illustrated in Fig. 1(c).

5.4   Analysis of the Dynamic Radius Size and Diversity

For the self-adaptive QPSO, the average radius size is illustrated in Fig. 1(d) for each environment type over 1000 iterations. From the graph it is clear that the diversity (which is the cloud radius value), did change over the course of algorithm execution. For environments with high temporal severity (chaotic and progressive environments), the cloud radius size fluctuated at a large value which is roughly half of the problem domain. Due to the frequency of the environment

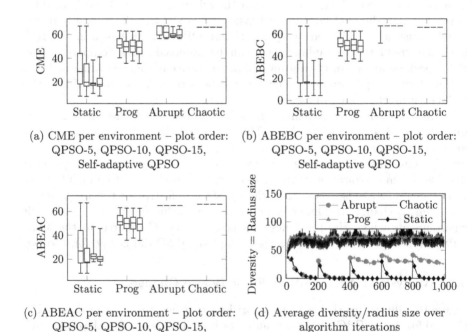

(a) CME per environment – plot order: QPSO-5, QPSO-10, QPSO-15, Self-adaptive QPSO

(b) ABEBC per environment – plot order: QPSO-5, QPSO-10, QPSO-15, Self-adaptive QPSO

(c) ABEAC per environment – plot order: QPSO-5, QPSO-10, QPSO-15, Self-adaptive QPSO

(d) Average diversity/radius size over algorithm iterations

Fig. 1. Measurements over environment types

Table 3. Algorithm performance ranking

Problem	Measure	QPSO-5	QPSO-10	QPSO-50	Self-adaptive QPSO
Quasi-static	CME (win/loss)	$(0/-3)$	$(1/-2)$	$(2/-1)$	$(3/0)$
	ABEBC (win/loss)	$(0/-3)$	$(1/-2)$	$(3/0)$	$(2/-1)$
	ABEAC (win/loss)	$(0/-3)$	$(3/0)$	$(1/-2)$	$(2/-1)$
	Win+loss	-9	-1	3	5
	Rank	1	2	3	4
Progressive	CME (win/loss)	$(0/-2)$	$(2/-1)$	$(0/-1)$	$(2/0)$
	ABEBC (win/loss)	$(0/-2)$	$(2/-1)$	$(0/-1)$	$(2/0)$
	ABEAC (win/loss)	$(0/-2)$	$(2/-1)$	$(0/-1)$	$(2/0)$
	Win+loss	-6	3	-3	6
	Rank	1	3	2	4
Abrupt	CME (win/loss)	$(0/-3)$	$(1/-2)$	$(2/0)$	$(2/0)$
	ABEBC (win/loss)	$(0/-3)$	$(1/-2)$	$(2/-1)$	$(3/0)$
	ABEAC (win/loss)	$(0/-3)$	$(1/-2)$	$(2/-1)$	$(3/0)$
	Win+loss	-9	-3	4	8
	Rank	1	2	3	4
Chaotic	CME (win/loss)	$(0/-3)$	$(1/-2)$	$(2/-1)$	$(3/0)$
	ABEBC (win/loss)	$(0/-3)$	$(1/-2)$	$(2/-1)$	$(3/0)$
	ABEAC (win/loss)	$(0/-3)$	$(1/-2)$	$(2/-1)$	$(3/0)$
	Win+loss	-9	-3	3	9
	Rank	1	2	3	4
Win/loss total		-33	1	7	28

changes, there is not enough time between the environment changes for particles to share enough information in order to attract the swarm to a specific region within the search space. Therefore, the re-initialization process maintains a large diversity.

The quasi-static environment plot shows that the radius reduced to a small value and increased as the environment changed (every 200 iterations), albeit a small change. The size of the cloud radius for the abruptly changing environment reduced similarly to the quasi-static environment, but at 400 iterations, increased to a value under half of the problem domain size and remained there for the remainder of the algorithm execution. It is not clear why this behavior is observed. As expected, the size of the cloud radius remained large for the progressive and chaotic environments where the frequency of environment change is high.

6   Conclusion

This paper investigated if the QPSO could be able to dynamically adapt and maintain the value of the quantum cloud radius, without requiring that the value be defined ahead of time, and without tuning the value for a given optimization problem. A new strategy was suggested, whereby the cloud radius value is based on the diversity of the particle swarm. By allowing the cloud radius value to dynamically adapt during the execution of the algorithm, it was shown that the self-adaptive strategy ranked well against three static quantum cloud radius QPSO algorithms. Even though the results indicated that the self-adaptive strategy did not provide significantly improved results when compared to the static radius QPSO algorithms, the results did indicate that the self-adaptive cloud radius does, generally, perform well and should be preferred. In future work, the influence of different distributions on the performance of the self-adaptive QPSO, and refinements to the adaptive cloud radius strategy, will be explored.

References

1. Angeline, P.J.: Tracking extrema in dynamic environments. In: Angeline, P.J., Reynolds, R.G., McDonnell, J.R., Eberhart, R. (eds.) EP 1997. LNCS, vol. 1213, pp. 335–345. Springer, Heidelberg (1997). https://doi.org/10.1007/BFb0014823
2. Blackwell, T.: Particle swarm optimization in dynamic environments. In: Yang, S., Ong, Y.S., Jin, Y. (eds.) Evolutionary Computation in Dynamic and Uncertain Environments, vol. 51, pp. 29–49. Springer, Heidelberg (2007). https://doi.org/10.1007/978-3-540-49774-5_2
3. Blackwell, T., Branke, J.: Multi-swarm optimization in dynamic environments. In: Raidl, G. (ed.) EvoWorkshops 2004. LNCS, vol. 3005, pp. 489–500. Springer, Heidelberg (2004). https://doi.org/10.1007/978-3-540-24653-4_50
4. Blackwell, T., Branke, J., Li, X.: Particle swarms for dynamic optimization problems. In: Blum, C., Merkle, D. (eds.) Swarm Intelligence, pp. 193–217. Springer, Heidelberg (2008). https://doi.org/10.1007/978-3-540-74089-6_6
5. Branke, J.: Memory enhanced evolutionary algorithms for changing optimization problems. In: Proceedings of the 1999 Congress on Evolutionary Computation-CEC99 (Cat. No. 99TH8406), vol. 3, p. 1882 (1999). https://doi.org/10.1109/CEC.1999.785502
6. Branke, J.: The moving peaks benchmark (1999). http://www.aifb.uni-karlsruhe.de/~jbr/MovPeaks/movpeaks
7. Deb, K., Joshi, D., Anand, A.: Real-coded evolutionary algorithms with parent-centric recombination. In: Proceedings of the 2002 Congress on Evolutionary Computation, CEC 2002, vol. 1, pp. 61–66, May 2002. https://doi.org/10.1109/CEC.2002.1006210
8. Duhain, J.G.: Particle swarm optimisation in dynamically changing environments-an empirical study. Master's thesis, University of Pretoria (2011)
9. Duhain, J.G., Engelbrecht, A.P.: Towards a more complete classification system for dynamically changing environments. In: 2012 IEEE Congress on Evolutionary Computation (CEC), pp. 1–8. IEEE (2012)
10. Eberhart, R.C., Shi, Y.: Tracking and optimizing dynamic systems with particle swarms. In: Proceedings of the 2001 Congress on Evolutionary Computation (IEEE Cat. No.01TH8546), vol. 1, pp. 94–100 (2001). https://doi.org/10.1109/CEC.2001.934376
11. Engelbrecht, A.: Roaming behavior of unconstrained particles. In: Proceedings - 1st BRICS Countries Congress on Computational Intelligence, BRICS-CCI 2013, pp. 104–111 (09 2013)

12. Harrison, K., Ombuki-Berman, B.M., Engelbrecht, A.P.: The effect of probability distributions on the performance of quantum particle swarm optimization for solving dynamic optimization problems. In: 2015 IEEE Symposium Series on Computational Intelligence, pp. 242–250, Decembrer 2015. https://doi.org/10.1109/SSCI.2015.44

13. Harrison, K.R., Ombuki-Berman, B.M., Engelbrecht, A.P.: A radius-free quantum particle swarm optimization technique for dynamic optimization problems. In: 2016 IEEE Congress on Evolutionary Computation (CEC), pp. 578–585, July 2016. https://doi.org/10.1109/CEC.2016.7743845

14. Holland, J.: Adaptation in Natural and Artificial Systems. University of Michigan Press, Ann Arbor (1975)

15. Hu, X., Eberhart, R.: Tracking dynamic systems with PSO: where's the cheese. In: Proceedings of the workshop on particle swarm optimization, pp. 80–83 (2001)

16. Kennedy, J., Eberhart, R.C.: Particle swarm optimization. In: Proceedings of the IEEE International Joint Conference on Neural Networks, vol. IV, pp. 1942–1948. IEEE (1995)

17. Li, C., et al.: Benchmark generator for CEC 2009 competition on dynamic optimization. University of Leicester, UK, Technocal report (2008)

18. Morrison, R.W.: Performance measurement in dynamic environments. In: GECCO workshop on evolutionary algorithms for dynamic optimization problems, pp. 5–8. Citeseer (2003)

19. Moser, I., Chiong, R.: Dynamic function optimization: the moving peaks benchmark. In: Alba, E., Nakib, A., Siarry, P. (eds.) Metaheuristics for Dynamic Optimization, vol. 433, pp. 35–59. Springer, Heidelberg (2013). https://doi.org/10.1007/978-3-642-30665-5_3

20. Pampará, G., Nepomuceno, F., Leonard, B.: Cilib v2.0.1, October 2014. https://doi.org/10.5281/zenodo.12371

21. van der Stockt, S., Engelbrecht, A.P.: Analysis of hyper-heuristic performance in different dynamic environments. In: 2014 IEEE Symposium on Computational Intelligence in Dynamic and Uncertain Environments (CIDUE), pp. 1–8. IEEE (2014)

22. Storn, R., Price, K.: Differential evolution - a simple and efficient heuristic for global optimization over continuous spaces. J. Glob. Optim. $11(4)$, 341–359 (1997). https://doi.org/10.1023/A:1008202821328

23. Trojanowski, K., Michalewicz, Z.: Searching for optima in non-stationary environments. In: Proceedings of the 1999 Congress on Evolutionary Computation, CEC 1999, vol. 3, pp. 1843–1850. IEEE (1999)

24. Van Den Bergh, F.: An analysis of particle swarm optimizers. Ph.D. thesis, Pretoria, South Africa, South Africa (2002). aAI0804353

25. Van Den Bergh, F., Engelbrecht, A.P.: A study of particle swarm optimization particle trajectories. Inf. Sci. $176(8)$, 937–971 (2006)

Simulating Kilobots Within ARGoS: Models and Experimental Validation

Carlo Pinciroli[1]([✉])[iD], Mohamed S. Talamali[2][iD], Andreagiovanni Reina[2][iD], James A. R. Marshall[2][iD], and Vito Trianni[3][iD]

[1] Robotics Engineering, Worcester Polytechnic Institute, Worcester, MA, USA
cpinciroli@wpi.edu
[2] Department of Computer Science, University of Sheffield, Sheffield, UK
{mstalamali1,a.reina,james.marshall}@sheffield.ac.uk
[3] ISTC, National Research Council, Rome, Italy
vito.trianni@istc.cnr.it

Abstract. The Kilobot is a popular platform for swarm robotics research due to its low cost and ease of manufacturing. Despite this, the effort to bootstrap the design of new behaviours and the time necessary to develop and debug new behaviours is considerable. To make this process less burdensome, high-performing and flexible simulation tools are important. In this paper, we present a plugin for the ARGoS simulator designed to simplify and accelerate experimentation with Kilobots. First, the plugin supports cross-compiling against the real robot platform, removing the need to translate algorithms across different languages. Second, it is highly configurable to match the real robot behaviour. Third, it is fast and allows running simulations with several hundreds of Kilobots in a fraction of real time. We present the design choices that drove our work and report on experiments with physical robots performed to validate simulated behaviours.

1 Introduction

Simulators are key tools for swarm robotics research. Many studies were performed mainly (when not exclusively) in simulation [3,7]. Swarm simulations tend to be "minimalistic", in that they include only few relevant features of the robots. Often, robots are modeled as abstract particle-like agents. For instance, a simulator often used for swarm robotics research is MASON [14], specifically developed for multi-agent systems research and not tailored to represent any specific robotic platform. This kind of simulations are useful to prove the validity of decentralised coordination algorithms, but fall short when physical interactions need to be taken into account, e.g., to simulate pulling and pushing forces among robots [23]. Simulations in this case need to move beyond simple kinematics and include the full dynamics of modern rigid-body simulation engines [16,17,19].

When a reference robotic hardware is available, simulations need to account for all its components, including sensors, actuators and communication devices. These can make simulations very costly in terms of computational requirements,

M. Dorigo et al. (Eds.): ANTS 2018, LNCS 11172, pp. 176–187, 2018.
https://doi.org/10.1007/978-3-030-00533-7_14

placing a tradeoff between the accuracy of the simulation and its speed [16]. However, even with accurate simulations, the "reality gap" cannot be completely filled [11]. This is especially apparent when using automatic design techniques that might exploit the idiosyncrasies of the simulated experimental setup [7,8]. To support swarm robotics experimentation and minimise errors when moving to real-world experiments, simulations not only need to implement techniques that reduce the reality gap (e.g., sensor sampling [15] and introduction and configurability of noise [10]), but should also provide cross-compiling solutions to directly reuse the control software developed for simulation also with the real robot hardware. In this way, any issue related to translating the algorithm from the one to the other platform can be removed.

Among the simulators developed for swarm robotics research, ARGoS [17] offers a number of desirable features. ARGoS has a modular design conceived to allow the user to select which aspects of the simulation should be assigned more computational resources (thus increasing accuracy) and which aspects can be simulated coarsely (thus improving scalability). For example, ARGoS supports the use of different types of physics engine—from simple kinematics to fully 3D dynamics. In addition, ARGoS is designed for parallel computation, an important feature to fully exploit the computational power of modern multi-core machines. Several robotic platforms, both custom and off-the-shelf, are currently available in ARGoS either as part of the core package or as extensions.

One robotic platform currently having momentum in swarm robotics is the Kilobot [20,22]. The Kilobot design is driven by the need for low cost (to allow for production of a thousand robots), small size (to fit the spaces of a typical research lab), robustness (to reduce faults), and ease of use [20]. Meeting these design goals meant sacrificing important features for swarm robotics research, such as accurate environmental sensing and remote access to the robot state. To compensate for missing features, devices such as the *Kilogrid* [24] and the *ARK* virtualisation environment [18] have been proposed, greatly expanding the realm of experimental activities attainable with Kilobots.

Despite the success of the Kilobot as an experimental platform, a fast and accurate simulation environment is still important to enable fast design cycles and educational activities based on the Kilobot. In the recent past, several simulators have appeared that include the Kilobot platform. Among these, it is worth mentioning V-REP [19], KBSim [9], and Kilombo [12]. V-REP is a generic, modular framework originally designed for complex 3D simulations of robotic arms and mobile robots. V-REP simulations tend to be very accurate, at the cost of long run-times when hundreds of robots are involved in the simulation. KBSim and Kilombo are designed to provide fast simulations for large numbers of robots. Kilombo, in particular, can perform simulations involving thousands of Kilobots hundreds of times faster than real time. Both KBSim and Kilombo, however, achieve scalability by drastically simplifying the motion and communication models of the robots, and proper validation of these models is currently unavailable. Another important aspect is that all these platform impose limitations when one is to transfer code developed in simulation onto real platforms.

V-REP and KBSim do not target the Kilobot API, and thus require a complete rewrite of the code. This is likely to introduce bugs and it makes convincing model validation hard to achieve. Kilombo, on the other hand, achieves direct interoperability with physical Kilobots by modifying the original Kilobot API and imposing limitations on how the code can be written. Specifically, users are not allowed to use global variables and must resort to conditional compilation techniques to transfer the code successfully. Modifying the Kilobot API also means that future improvements and bug fixes must be ported to the Kilombo version, if compatibility is to be preserved.

In this paper, we present a plugin for the ARGoS simulator [17] that offers accurate models, scalability sufficient to run one-thousand-robot experiments in real-time, and full compatibility with the original Kilobot API. We present the Kilobot and the reference behaviour in Sect. 2, along with experiments conducted to determine the real-world behaviour and extract features to be implemented in simulation. In Sect. 3, we validate the simulations against the real-world behaviour in representative experimental conditions. We demonstrate that ARGoS closely predicts the Kilobot behaviour, and offers sufficient efficiency to run large-scale simulations in a fraction of real-time. Section 4 concludes the paper.

2   Kilobot: Reference Behaviour and Simulation Models

The Kilobot is a small autonomous robot with a circular shape (diameter: 3.3 cm) standing on three rigid legs, which moves thanks to two vibrational motors and a slip-stick locomotion principle [20]. The robot is provided with infrared transceivers for communication and range detection of neighbours, as well as with an ambient light sensor and a coloured LED for displaying the robot state. In the following, we detail the simulation design for the various components.

2.1   Body Model

The Kilobot is simulated as a small cylinder with the same radius as the real robot, resting on three thin cylinders representing the legs. When the simulation is performed with ARGoS's 2D physics engine, the robot body is the circular projection of the main board on the plane. The Kilobot locomotion is provided by the two vibrational motors, which implement a classical differential-drive model. The forward and rotational speeds are determined by the difference between the velocity v_ℓ and v_r, resulting respectively from the left and right motor activation.

Although the motors can be finely controlled, the normal practice with Kilobots is to use only 3 motion modes: straight motion, clockwise and counterclockwise rotation. We calibrated the parameters of the ARGoS model so as to result in a forward speed of 1 cm/s and a rotation speed of 45 °/s, corresponding to the nominal speed of real Kilobots [20].

2.2   Noise and Inter-individual Variations

Real Kilobots have strong inter-individual variations, which are due to the simple design and the slip-stick locomotion system. This system is strongly dependent on small variations in the position of the motors and the bending of the legs. To obtain acceptable locomotion, calibration of individual robots is necessary to find good values for the motor parameters that provide straight motion and rotations at the desired speed. Most researchers rely on manual calibration, but this process is cumbersome and dependent on the surface on which the Kilobot maneuver. The ARK platform [18] enables the parallel calibration of tens of robots. While ARK shortens calibration time, it cannot guarantee error-free precision due to the intrinsically high noise of Kilobot motion. As a result, through either manual or automatic calibration, Kilobots hardly move straight or rotate at the nominal speed, and present strong inter-individual variations.

To capture this, the Kilobot model has a noise component for the motion of a single robot. Given the nominal left (ℓ) and right (r) speeds v_i (with $i \in \{\ell, r\}$) of the differential drive model, the actual speed $\hat{v}_i$ of each wheel is computed as $\hat{v}_i = f_i(v_i + b_i)$, where f_i is a per-step actuation noise and b_i is a per-robot bias added to the nominal speed v_i. Both f_i and b_i are Gaussian-distributed random parameters, with mean and standard deviation defined in the experiment configuration file. For each robot, the bias b_i, with $i \in \{\ell, r\}$, are drawn from the specified Gaussian distributions at the beginning of the experiment. Instead, the actuation noise f_i is drawn at each time-step.

The ARGoS default values are set from measurements performed on a sample of 120 real Kilobots. The nominal (noise-free) speed is set to a forward speed of $v_i \simeq 1\,\mathrm{cm/s}$ and a rotation speed of $\sim 45\,°/\mathrm{s}$ (i.e. $v_\ell \simeq 2\,\mathrm{cm/s}$, $v_r = 0$ for right rotation and viceversa for left rotation). To determine the distribution of the noise observed in reality, we conducted experiments on 120 different Kilobots (in batches of 6) that have been previously calibrated (60 manually and 60 automatically—we could not notice any remarkable difference). Robots were asked to move straight for 1 min. Through ARK, we recorded the trajectory of each robot, we derived the robot displacement every 10 s, and through a differential drive model we computed the left and right speeds $\hat{v}_i$ (with $i \in \{\ell, r\}$). Through $\hat{v}_i$, we could compute the bias $b_i^t = \hat{v}_i - v_i$ (ignoring white noise, i.e. $f_i = 1$) for each 10 s motion trajectory (thus $t \in \{1, ..., 6\}$ for our 60 s experiments). Finally, we computed the average bias $b_i = \sum b_i^t / 6$ for each robot and we report the distributions of biases (for both left/right velocities $i \in \{\ell, r\}$) of the 120 tested Kilobots in Fig. 1(a). From this distribution, we computed the mean $\mu_b = 0.015\,\mathrm{mm/s}$ and the standard deviation $\sigma_b = 1.86\,\mathrm{mm/s}$ which we use as default values in ARGoS. Figure 1(b) shows a comparison between the mean square displacement (MSD) of the 120 Kilobots and the simulated robots. Noise-free simulations show the robots moving (as expected) at the nominal speed of $\sim 1\,\mathrm{cm/s}$, whereas the default noise values show that the simulated robots have motion dynamics remarkably similar to reality.

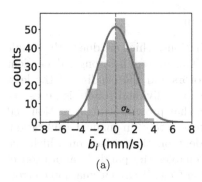
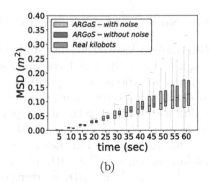

(a)   (b)

Fig. 1. (a) Distribution of the bias in straight motion estimated from measurements over 120 different Kilobots which have been previously calibrated (60 manually, 60 automatically). The bars considers 12 bins in the range $[-6, 6]$ mm. The solid red line shows the approximated Gaussian distribution $\mathcal{N}_b(\mu_b, \sigma_b)$ (with mean μ_b and standard deviation σ_b) used in the default configuration of ARGoS. (b) Comparison between simulation (600 robots) and reality (120 robots) in term of MSD when robots are asked to go straight for 1 min. The default noise values of ARGoS give an accurate match between reality and simulation.

2.3   Robot-Robot Communication

Communication among the Kilobots is implemented exploiting the infrared transceiver positioned under the robot body, which sends a modulated infrared signal that bounces on the ground and can be perceived by neighbours within a distance of about 15 cm. The communication protocol implemented in the most recent firmware (see https://www.kilobotics.com) is Carrier Sense Multiple Access with Collision Detection (CSMA-CD) with exponential backoff, meaning that upon detection of the occupied channel, message sending is delayed within an exponentially increasing range of time slots. To avoid interferences, the maximum transmission frequency is set to 2 Hz. Nevertheless, Kilobots may find the channel busy when transmitting, and collisions can still occur. To evaluate the impact of concurrent communication, we performed an experiment with 25 Kilobots packed in a 5×5 square formation—their bodies touching each other—and attempting to transmit messages at maximum rate. Messages contained just the ID of the sender, so that robots getting a message could update the number of messages received from each other robot. At the same time, robots stored also the number of messages successfully transmitted. We performed 10 independent runs, and found that the transmission probability—i.e., the ratio between successfully transmitted messages between the packed robots and solitary robots—was close to maximum (0.992 ± 0.002) and independent from the position of the Kilobot in the 5×5 formation. On the reception side, we observed that the probability of receiving a messages depended strongly on the position of the receiving robot (see Fig. 2). Generally speaking, robots in the periphery where affected less by interferences than robots in the center (see the left panel in Fig. 2 showing the reception probability of each robot with respect to all other

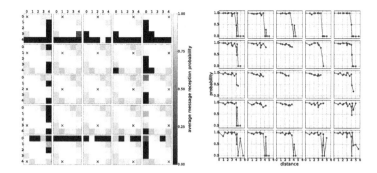

Fig. 2. Communication interference measured on 25 Kilobots packed in a 5×5 grid and concurrently sending messages to each other. Left: average message reception probability for each robot in the grid from any other robot. For each panel, the position of the receiving robot on the grid is indicated by a ×. Right: average message reception probability with respect to distance, in body lengths, for every Kilobot in the grid.

robots). Additionally, the decrease of the probability of reception is more pronounced for center robots, indicating a stronger effect from interferences (see the right panel in Fig. 2). Indeed, collisions are more probable in the center, where robots may receive at the same time messages sent by robots at the periphery that do not sense each other.

According to these results, we have implemented the Kilobot communication limiting the maximum transmission frequency to 2 Hz, including a configurable small error on the transmission side, and modelling message collision on the receiver side with a tuneable probability when two robots happen to be concurrently transmitting.

2.4   Light Sensor

Real Kilobots are provided with a photodiode sensor to detect the ambient light. In the commercial version[1], this sensor is placed looking upward on the robot body. Simulating the light sensors is highly dependent on the type and position of the ambient light. ARGoS natively offers light sensor models and can simulate light sources with a tuneable intensity. The sensor readings decrease quadratically with the distance from the lights. In the Kilobot plugin, we rescaled the readings in the range typical of the real robot.

2.5   ARK Simulation

The Augmented Reality for Kilobots system (ARK) [18] overcomes the limitations of Kilobots by providing a flexible set of virtual environments, with which

---

[1] Kilobots are open-hardware and in Europe are produced and sold by K-Team Corporation (see https://www.k-team.com).

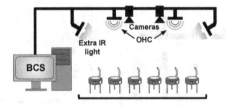

Fig. 3. Graphical representation of the ARK architecture. See more details in [18].

Kilobots can interact through a variety of user-configurable sensors and actuators. ARK comprises a base station interfaced with an array of cameras, and an array of overhead controllers (OHCs) for transmitting IR messages to Kilobots (see the system architecture in Fig. 3). Communication with the Kilobots during an experiment is obtained via broadcast of addressed messages, with each broadcast packet containing three *ARK messages* (3-bytes long) for different Kilobots. Kilobot addresses (10 bits) are uniquely assigned by ARK during a pre-experimental phase. For sensor readings, the other 14 bits of data can be assigned to multiple virtual sensors as desired by the user. Location-specific information from the virtual environment can be determined for a Kilobot by its physical position at the time the message is to be sent, which is determined through robust tracking of each ID-assigned Kilobot over the duration of an experiment. Kilobots communicate virtual actuator commands via signalling through RGB LEDs, which are received via the base station's camera array and translated into operations on the virtual environment. Each augmented reality experiment can be composed of more than one virtual environments. Each environment has user-defined structure and spatio-temporal dynamics, as exemplified in [6]. As well as enabling richer experimental paradigms, ARK's features also lend it to automatic motor calibration and other house-keeping features [18].

ARK is integrated with ARGoS through the ARK Loop Function (ALF), the simulated counterpart of the ARK's base control station. The ALF is executed every ARGoS time-step and is in charge of simulating the virtual environments and of sending IR ARK messages to the simulated Kilobots. To facilitate the transfer from simulation to reality, the ALF uses the same method names and structure of its real counterpart. Similarly to the ARK's base control station, the ALF has real-time access to the state of the simulated Kilobots, i.e., their position, orientation, and LED colour. This information can be used by the user to code the functioning of the virtual actuators and sensors. The virtual actuators update the virtual environments, and virtual sensor readings are computed using the Kilobot's state, then transmitted to the robot. The ALF automatically codes the 3-byte ARK messages within standard 9-byte Kilobot messages in the same way ARK does. Therefore the Kilobot control software needs to decode the ARK messages in ARGoS in the same way it does in reality. This implementation choice is particularly helpful because it allows for the use of identical code in simulation and reality. ALF gives the user the possibility to limit the communication to a maximum frequency of 60 ARK messages per second (to

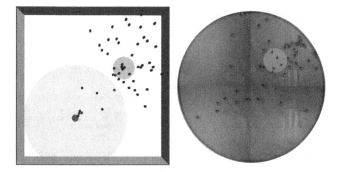

Fig. 4. Two screenshots of the same experiment in simulation (left) and reality (right). We (re)implemented the Demo C from [18] in which 50 Kilobots sense and modify two virtual environments. The full video is available at https://youtu.be/kioZR99hnU4.

match the real ARK's frequency) or to simulate an unlimited ARK message frequency.

To showcase the ALF functioning, we reproduced a simulated version of one experiment based on ARK, the Demo C of [18]. Figure 4 shows two screenshots of the experiment in simulation (left) and reality (right) featuring 50 Kilobots that operate in two virtual environments (flower field and nest).

3 Experimental Validation

We run a set of experiments to assess the reliability of ARGoS in simulating Kilobot swarms and to compare the ARGoS performance with the existing Kilobot simulator Kilombo [12]. In Sect. 3.1, we tested how reliably ARGoS and Kilombo simulate physical interactions between Kilobots through a random diffusion experiment. In Sect. 3.2, we compared the simulation speed of the two simulators. Finally, in Sect. 3.3, we show that ARGoS successfully simulates physical interactions between the robots and physical objects (e.g. a box) and that force factors are taken into consideration in the simulation.

3.1 Random Diffusion Experiment

Kilobots are equipped with minimal sensing capabilities which do not allow robots to implement robust mechanisms of collision avoidance, therefore collisions between robots and objects in the environment are frequent. In this experiment, we assess how realistically the collisions between Kilobots are simulated in ARGoS and Kilombo. To perform this study, we designed an experiment that maximises the number of collisions in a repeatable setup. The 50 Kilobots have an initial compact distribution as illustrated in Fig. 5 (left). The robots are placed in concentric circles heading toward the centre of the group (more precisely, the robot are placed on the vertices of four concentric regular polygons,

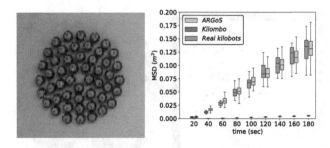

Fig. 5. (left) Initial distribution in four concentric circles with all 50 Kilobots facing towards the centre. (right) Comparison between real 50 Kilobots (19 runs) and 50 simulated Kilobots (100 runs) in ARGoS and Kilombo. We show the average mean square displacement (MSD) in a highly dense environment. ARGoS shows a good agreement with reality, whereas Kilombo does not. Video footage is available at https:// youtu.be/6HYti0ABuxc.

starting from the innermost regular pentagon with radius 35 mm, each polygon has the same centre, twice the number of vertices, and double radius of its internal polygon). In this experiment, the 50 Kilobots perform an isotropic random walk [5] through which they repetitively move forward for ∼10s and turn in a random direction (left or right) for a random time drawn from a uniform distribution $\mathcal{U}(0,4)$s. We performed 19 runs of this experiment with real robots and 100 runs with simulated Kilobots in both ARGoS and Kilombo. For every run, we recorded the trajectory of each Kilobot for a period of 3 min to compute the mean square displacement (MSD) of each robot. We combined the 50 MSD in each experiment and we show in Fig. 5 (right) how the average MSD changes over time. Figure 5 (right) clearly shows that ARGoS correctly simulates physical interactions between robots while Kilombo does not.

3.2   Speed and Scalability

To test the scalability of the Kilobot plugin, we performed experiments based on the `disperse.c` example provided on the Kilobot website. This behaviour allows a group of Kilobots to evenly disperse in the environment. We deemed `disperse.c` a good benchmark because it involves both motion and communication, and the robots are initially deployed in a tight cluster that stress-tests the collision management of ARGoS's 2D physics engine.

Scalability is measured through the *wall clock time*, which is the real time elapsed between the beginning and the end of the simulation of 60 virtual seconds. We performed these experiments on a node of a computing cluster with 48 Intel Xeon Platinum 8168 CPUs at 2.70 GHz. The experiments were performed without graphical visualisation.

We considered several parameters: (i) the number of robots $N \in \{10, 100, 1000\}$; (ii) the number of "worker" threads used by ARGoS $T \in \{0, 4, 16\}$; and

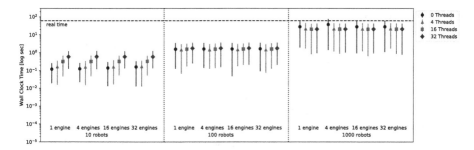

Fig. 6. Scalability experiments. We report the median, max and min wall clock time in a log sec scale for experiments simulating 60 virtual seconds.

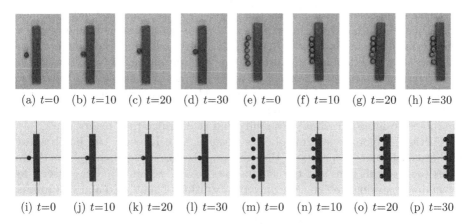

(a) t=0   (b) t=10   (c) t=20   (d) t=30   (e) t=0   (f) t=10   (g) t=20   (h) t=30

(i) t=0   (j) t=10   (k) t=20   (l) t=30   (m) t=0   (n) t=10   (o) t=20   (p) t=30

Fig. 7. Box-pushing experiments. Top: real robots. Bottom: ARGoS simulation. Time in seconds. Video footage at https://youtu.be/fwL9ePWttiU.

(iii) the number of physics engines in which the environment is partitioned $P \in \{1, 4, 16\}$. Every parameter set $\langle N, T, P \rangle$ was tested 30 times.

The results are reported in Fig. 6. The use of multiple threads and physics engines is beneficial when large swarms are simulated—particularly with 1000 robots. Conversely, with small swarms of 10 robots, multi-threading and multiple physics engines decrease performance. With 1000 robots, 1 physics engine and 0 working threads, the runtime mean was 27 s (45% of real time); when 16 physics engines and 16 working threads were used, the simulation was completed in 20 s (33% of real time).

3.3   Accuracy: Box Pushing

Remarkable collective transport experiments have been demonstrated with the Kilobots [1,21]. Our plugin also supports simulations that involve robots that push objects. In our experiments, we found that one robot pushing an 18 g box is not sufficient to move the box. Rather, it is necessary to use at least 5 robots to exert sufficient force to move the box. Figure 7 shows a real-world box pushing experiment and its simulated counterpart.

4   Conclusions

An essential feature of any simulator are usability and realism, and scalability. Kilobots, despite their simple hardware design, present specificities that make their simulation non-trivial. As a matter of fact, most of the simulators available for the Kilobots are rather minimalistic, and prove useful only for proof-of-concept studies without guarantees of respecting real-world behaviour. Our effort in developing a Kilobot model for ARGoS fills the need of a usable and reliable simulation, as demonstrated by the validation experiments we performed. We believe this will be a precious tool for swarm robotics research. To this end, the simulator is released open source for the benefit of the community (available at https://github.com/ilpincy/argos3-kilobot).

Besides the Kilobot simulation, this paper also introduces methodologies for tuning the simulation that can be replicated whenever a close matching between simulation and reality is desired. In particular, inter-individual variation between Kilobots (due to differences in hardware and calibration) needs to be considered as it represent an important problem when moving from simulation to reality. Tuning the simulation as discussed in Sect. 2.2 brings the full complexity of real-world Kilobots into the simulations, allowing for the design of controllers that are robust to inter-individual variability and that bridge the "reality gap".

Further support to swarm robotics research with Kilobots can be obtained by automatisation of practices that are now performed manually. The ARK offers tools to this aim, and the integration of ARK within ARGoS is useful to streamline experimentation. A stronger integration of simulated and real robots can be performed through ARK, letting simulated and real Kilobots run in parallel, therefore opening the way for online learning [13], self-modelling [2] or embodied evolution [4].

Acknowledgments. This work was partially supported by the European Research Council (ERC) under the European Union's Horizon 2020 research and innovation programme under Grant 647704 to James Marshall. Vito Trianni acknowledges support from the project DICE (FP7 Marie Curie Career Integration Grant, ID: 631297). The authors thank Alex Cope for assistance in the preparation of Fig. 3.

References

1. Becker, A., Habibi, G., Werfel, J., Rubenstein, M., McLurkin, J.: Massive uniform manipulation: controlling large populations of simple robots with a common input signal. In: 2013 IEEE/RSJ International Conference on Intelligent Robots and Systems (IROS), pp. 520–527. IEEE (2013)
2. Bongard, J., Zykov, V., Lipson, H.: Resilient machines through continuous self-modeling. Science **314**(5802), 1118–1121 (2006)
3. Brambilla, M., Ferrante, E., Birattari, M., Dorigo, M.: Swarm robotics: a review from the swarm engineering perspective. Swarm Intell. **7**(1), 1–41 (2013)
4. Bredeche, N., Haasdijk, E., Prieto, A.: Embodied evolution in collective robotics: a review. Front. Robot. AI **5**, 12 (2018)

5. Dimidov, C., Oriolo, G., Trianni, V.: Random walks in swarm robotics: an experiment with kilobots. In: Dorigo, M. (ed.) ANTS 2016. LNCS, vol. 9882, pp. 185–196. Springer, Cham (2016). https://doi.org/10.1007/978-3-319-44427-7_16
6. Font Llenas, A., Talamali, M.S., Xu, X., Marshall, J.A.R., Reina, A.: Quality-sensitive foraging by a robot swarm through virtual pheromone trails. In: Dorigo, M., et al. (ed.) Swarm Intelligence (ANTS 2018), LNCS, vol. 11172, pp. X-XY. Springer, Heidelberg (2018). In press
7. Francesca, G., Birattari, M.: Automatic design of robot swarms: achievements and challenges. Front. Robot. AI **3**, 224–9 (2016)
8. Francesca, G., Brambilla, M., Brutschy, A., Trianni, V., Birattari, M.: AutoMoDe: a novel approach to the automatic design of control software for robot swarms. Swarm Intell. **8**(2), 89–112 (2014)
9. Halme, A.: Kilobot app–a kilobot simulator and swarm pattern designer. https://github.com/ajhalme/kbsim (2012). Accessed 20 Apr 2018
10. Jakobi, N.: Evolutionary robotics and the radical envelope-of-noise hypothesis. Adapt. Behav. **6**(2), 325 (1997)
11. Jakobi, N., Husbands, P., Harvey, I.: Noise and the reality gap: the use of simulation in evolutionary robotics. In: Morán, F., Moreno, A., Merelo, J.J., Chacón, P. (eds.) ECAL 1995. LNCS, vol. 929, pp. 704–720. Springer, Heidelberg (1995). https://doi.org/10.1007/3-540-59496-5_337
12. Jansson, F., et al.: Kilombo: a Kilobot simulator to enable effective research in swarm robotics. arXiv.org:1511.04285 (2015)
13. Li, W., Gauci, M., Gross, R.: Turing learning: a metric-free approach to inferring behavior and its application to swarms. Swarm Intell. **10**(3), 211–243 (2016)
14. Luke, S., Cioffi-Revilla, C., Panait, L., Sullivan, K., Balan, G.: MASON: a multi-agent simulation environment. Simulation **81**(7), 517–527 (2005)
15. Miglino, O., Lund, H.H., Nolfi, S.: Evolving mobile robots in simulated and real environments. Artif. Life **2**(4), 417–434 (1995)
16. Mondada, F., et al.: SWARM-BOT: a new distributed robotic concept. Auton. Robots **17**(2), 193–221 (2004)
17. Pinciroli, C., et al.: ARGoS: a modular, parallel, multi-engine simulator for multi-robot systems. Swarm Intell. **6**(4), 271–295 (2012)
18. Reina, A., Cope, A.J., Nikolaidis, E., Marshall, J.A.R., Sabo, C.: ARK: augmented Reality for Kilobots. IEEE Robot. Autom. Lett. **2**(3), 1755–1761 (2017)
19. Rohmer, E., Singh, S.P.N., Freese, M.: V-REP: a versatile and scalable robot simulation framework. In: 2013 IEEE/RSJ International Conference on Intelligent Robots and Systems (IROS), pp. 1321–1326 (2013)
20. Rubenstein, M., Ahler, C., Hoff, N., Cabrera, A., Nagpal, R.: Kilobot: a low cost robot with scalable operations designed for collective behaviors. Robot. Auton. Syst. **62**(7), 966–975 (2014)
21. Rubenstein, M., Cabrera, A., Werfel, J., Habibi, G., McLurkin, J., Nagpal, R.: Collective transport of complex objects by simple robots: theory and experiments. In: Proceedings of the 12th International Conference on Autonomous Agents and Multiagent Systems (AAMAS 2013), pp. 47–54. International Foundation for Autonomous Agents and Multiagent Systems, Richland (2013)
22. Rubenstein, M., Cornejo, A., Nagpal, R.: Programmable self-assembly in a thousand-robot swarm. Science **345**(6198), 795–799 (2014)
23. Trianni, V., Dorigo, M.: Self-organisation and communication in groups of simulated and physical robots. Biol. Cybern. **95**(3), 213–231 (2006)
24. Valentini, G., et al.: Kilogrid: a novel experimental environment for the Kilobot robot. Swarm Intell. **4**(4), 1–22 (2018)

Simulating Multi-robot Construction in ARGoS

Michael Allwright$^{1(\boxtimes)}$ (ID), Navneet Bhalla^2 (ID), Carlo Pinciroli^3 (ID),
and Marco Dorigo^1 (ID)

^{1} IRIDIA, Université Libre de Bruxelles, Brussels, Belgium
{mallwrig,mdorigo}@ulb.ac.be
^{2} Department of Computer Science, University College London, London, UK
n.bhalla@cs.ucl.ac.uk
^{3} Department of Computer Science/Robotics Engineering,
Worcester Polytechnic Institute, Worcester, MA, USA
cpinciroli@wpi.edu

Abstract. Running hardware-based experiments in multi-robot construction is an expensive and time-consuming endeavor. Furthermore, it is difficult to disseminate the results from hardware-based experiments in a way that other researchers can build upon. In this paper, we present a number of plug-ins for a multi-robot simulator that we have developed to enable a high-fidelity simulation of the multi-robot construction systems typically found in laboratory settings. We validate these plug-ins qualitatively by repeating a hardware-based experiment in simulation where a single robot assembles a staircase from blocks [1]. We then show how we can use the plug-ins to scale up the complexity of the construction scenario and demonstrate multi-robot construction in simulation. To enable other researchers to replicate our experiments and to promote collaboration, we have made our plug-ins open source.

1 Introduction

In our research, we study how the coordination mechanisms used by social insects can be applied to a swarm of robots whose task is to collectively assemble structures. This research direction is motivated by the robustness, parallelism, and adaptivity to the environment that social insects exhibit as they construct their nests [4,5]. Construction by social insects is decentralized and is coordinated through stigmergic communication [7], where the probability of an insect performing a construction action is a function of the previously performed construction actions as perceived by that insect [19]. These construction actions may have been performed by the same insect or by other insects.

Using a single robot, we previously studied how stigmergic communication could be used to guide the construction of a staircase [1]. The robot in this study located unused blocks in the environment before attaching them to the staircase according to the arrangement of the blocks already in the structure. The development of this hardware and the running of the experiments, however,

© Springer Nature Switzerland AG 2018
M. Dorigo et al. (Eds.): ANTS 2018, LNCS 11172, pp. 188–200, 2018.
https://doi.org/10.1007/978-3-030-00533-7_15

were both expensive and time-consuming endeavors. Furthermore, since other researchers do not have access to our hardware, it is difficult for them to replicate our experiments and to build upon our results. These challenges make it difficult to answer a number of open research questions regarding, for example, whether it is possible to encode arbitrary structures as sets of assembly rules for swarms of robots and how these encodings can be compared in terms of their robustness and their ability to be executed in parallel by multiple robots.

To address these challenges and to answer these open research questions, we present in this paper a number of plug-ins for ARGoS (Autonomous Robots Go Swarming), an open-source, multi-robot simulator [14]. We show in this paper how we can use these plug-ins to qualitatively reproduce our hardware-based results in simulation and how we can use the plug-ins to scale up to and to study multi-robot construction in simulation. To enable other researchers in the swarm robotics community to build upon our results, we have made these plug-ins open source. In this paper, the presented validation of the plug-ins is strictly qualitative and further quantitative evaluation is required to identify and to mitigate any reality gap found between our hardware and our simulation.

This paper is organized as follows. In the next section, we discuss current practices regarding the use of simulation in the multi-robot construction literature. In Sect. 3, we introduce ARGoS and summarize the plug-ins which we have developed to enable the simulation of multi-robot construction systems. Section 4 presents a case study in which we demonstrate how the plug-ins can be used to simulate multi-robot construction. In the final section, we conclude this paper by discussing the open research questions that we would like to answer in our future work.

2 Background

While previous work in multi-robot construction has used simulation to mitigate some of the challenges of hardware, simulations are often limited to lattice-based two- or three-dimensional worlds where the interaction dynamics is unrealistic as are the perceptual capabilities of the robots [17,18,21,22].

In the work by Jones and Matarić, a high-fidelity simulator was used, but the simulation was tweaked so that the robots simply requested the simulation engine to add construction material directly to the structure, sidestepping several challenges present in a multi-robot construction system such as manipulating the construction materials and avoiding collisions with other robots during construction [8]. In two other cases, results from simulation were presented but not discussed. Werfel et al., for example, included two videos from simulation in their supplementary material but did not discuss these results in either the main article or in the supplementary report [23]. Lindsey et al. also presented results from simulation without detailing their nature [10].

While the aforementioned works significantly contributed to the state of the art in multi-robot construction, the means to replicate the experiments is problematic since other researchers are unable to build upon the presented results.

This lack of a means to replicate experiments and to build upon previous results is a significant barrier to pushing the state of the art in multi-robot construction further forward. One way to remove this barrier is by using an open-source, high-fidelity simulator that can run multi-robot construction experiments. This simulator would need to support the range of sensors, actuators, and mechanisms that are commonly used by multi-robot construction systems in laboratory settings. For example, the simulator would need to support camera systems that can detect tags or light-emitting diodes (LEDs), rangefinders, prismatic and revolute actuators, and magnetism. Furthermore, the simulator would ideally provide good performance when running experiments with large numbers of robots.

3   Contributions to the ARGoS Simulator

ARGoS (Autonomous Robots Go Swarming) is a multi-robot simulator built around the principle of tunable accuracy. In contrast to general-purpose simulators such as USARSim [6], Gazebo [9], and Webots [11] whose performance quickly degrades with increasing numbers of robots, benchmarks for ARGoS have demonstrated the simulation of 10,000 robots running faster than real time [14]. ARGoS is also widely used in the swarm robotics community by more than 21 robotics laboratories in 13 different countries[1].

At the time of writing, however, the latest release of ARGoS (3.0.0-beta49) only comes with plug-ins for two-dimensional dynamics and a limited number of generic sensors and actuators. The implementation of new robots in ARGoS also requires a significant amount of coding and the compilation of new classes that represent a robot's sensors, actuators, physics model, and visualization. However, ARGoS has been purposefully designed to be modular. Most aspects of a simulation can be overridden and developers can add functionality to the simulator in the form of new sensors, actuators, visualizations, physics engines.

We have enhanced three of the plug-ins from the ARGoS core package and have implemented two completely new plug-ins that enable the simulation of three-dimensional dynamics and the rapid prototyping of new robots. The plug-ins are designed to support a high-fidelity simulation of the dynamics, sensors, actuators, and other mechanisms commonly used by the multi-robot construction systems found in robotics laboratories. By making these plug-ins open source and by including them in the ARGoS core package (from 3.0.0-beta50), we hope to increase the pace of the research in multi-robot construction by enabling researchers to implement their own multi-robot construction systems in simulation and by enabling other researchers to replicate the presented results.

In the remainder of this section, we summarize our contributions to ARGoS in the context of multi-robot construction. The plug-ins that we have developed, however, are highly flexible and their potential applications go significantly beyond multi-robot construction. For a more in-depth discussion of the plug-ins including examples on how to use them, we refer the reader to the supplementary technical report for this paper [2].

---

[1] Users of ARGoS – http://www.argos-sim.info/users.php.

3.1   Enhancements to the Entities Plug-in

In the ARGoS simulator, entities represent the components of a robot in the simulation. This representation allows plug-ins such as visualizations, physics engines, and controllers to read and to write the state of these components. We have added four new entities to the entities plug-in. The simulation of magnetism is enabled by introducing the magnet entity. Magnetism can be used in multi-robot construction systems for both manipulating the building materials [15,16] and for the self-alignment of the building materials inside a structure [13,24]. A magnet entity is capable of representing a permanent magnet, an electromagnet, or a semi-permanent electromagnet. We have also implemented a radio entity which represents a simple omnidirectional radio that can broadcast messages to other radios within a given range. This radio entity is suitable for simulating communication between robots and for simulating communication between robots and a building material [22,25]. In contrast to the range-and-bearing entity in ARGoS, the radio entity is not based on infrared light and its messages are not obstructed by other entities in the simulation. A tag and a directional light-emitting diode (LED) entity were also implemented and can be attached to any object in the simulation. The directional LED entity is based on the standard LED entity in ARGoS and includes two additional attributes, namely an orientation and an observable angle. These entities can be detected by camera sensors and can be used by robots to locate building materials.

3.2   Enhancements to the Media Plug-in

Media are used by the simulator to manage entities during a simulation. At a minimum, the implementation of a medium consists of a data structure which is typically queried by a sensor and modified by an actuator. We have defined three new mediums, namely the radio, tag, and directional LED medium. These mediums are required to support the sensors and actuators for a robot, so that it can interact with the radio, tag, and directional LED entities in a simulation.

3.3   Enhancements to the Generic Robot Plug-in

The generic robot plug-in defines generic sensors and actuators that can be used by any robot. We have enhanced this plug-in by implementing a generic camera framework that allows any number of cameras to be attached to a robot. For performance reasons, we have chosen to simulate computer vision algorithms rather than rendering the simulated environment from the perspective of each robot. For example, instead of rendering the tags in a simulation onto a virtual buffer and passing these pixels to a tag detection algorithm, we use a simulated tag detection algorithm that simply queries the tag medium and directly calculates the pixel coordinates of a tag's corners. We have provided three simulated computer vision algorithms for detecting tags, directional LEDs, and the standard LEDs in ARGoS. The output from the simulated tag detection algorithm is consistent with the output from the AprilTag algorithm and enables a robot to estimate the pose of a detected tag [12]. In addition to the generic camera framework, we have also added a generic radio sensor and actuator. In conjunction with the radio entity, this sensor and actuator allow robots to send messages and to receive messages from nearby robots or building material.

3.4 The Three-Dimensional Dynamics Plug-in

To simulate the interaction dynamics that occur during multi-robot construction, we have created a new physics engine plug-in for ARGoS, which is a wrapper around Bullet Physics[2]. This plug-in replaces the deprecated three-dimensional dynamics plug-in from previous versions of ARGoS, which was based on ODE.

The three-dimensional dynamics plug-in provides a number of helper classes for quickly creating new robots. The plug-in also provides a configurable floor, and tunable gravity and magnetism. Magnetic forces and torques are applied to bodies in the simulation where a corresponding magnet entity has been defined. The forces and torques are calculated using a variant of Thomaszewski's algorithm [20], where each magnet is approximated by a single dipole.

3.5 The Prototyping Plug-in

The prototyping plug-in enables the implementation of new robots and building materials in ARGoS without the need to manually code and compile new classes. The plug-in defines a new robot that is entirely described by ARGoS's experiment configuration file.

The plug-in provides new entities for describing links and joints. Links are defined by their mass and geometry, which currently can be either a box, a cylinder, or a sphere. Joints specify how two links are connected to each other. There are four joint types currently supported: fixed, spherical, prismatic, and revolute. For prismatic and revolute joints, it is possible to limit the range of a joint's motion and to assign a sensor and an actuator to the joint. The default sensor can measure either joint position or joint velocity and the default actuator can be configured to use either position control or velocity control. The plug-in also provides a physics model for the three-dimensional dynamics plug-in and a visualization model for the Qt-OpenGL plug-in.

Sensors, actuators, and other robot components such as tags, directional LEDs, standard LEDs, radios, and cameras are also defined in the experiment configuration file and can be attached to any link in the robot.

4 Case Study: Multi-robot Construction

In this section, we demonstrate how our plug-ins for the ARGoS simulator enable us to qualitatively reproduce our hardware-based experiment, in which a single robot constructed a staircase [1]. In addition to reproducing our hardware-based

---

[2] Bullet Physics – http://bulletphysics.org/.

experiment, we show in simulation a more complicated construction scenario involving four robots that collectively assemble a stepped pyramid.

We have used the prototyping plug-in to model the robot and the building material in simulation. The models and sample controllers for the robot and the building material are available on the supplementary material website for this paper [3].

4.1 Summary of the Hardware and the Control Software

The hardware of our autonomous construction system consists of a building material known as stigmergic blocks and an autonomous robot, which assembles the blocks into structures [1]. The blocks are cubes that contain on each face: (i) a tag, (ii) a near-field communication (NFC) interface, and (iii) four multi-color light-emitting diodes (LEDs). The tags are used by the robot to estimate the pose of a block. Spherical magnets are assembled into the corners of the blocks to provide self-alignment and so that the blocks can be picked up by a robot.

The robot consists of two tracks (treads) that form a differential drive, allowing the robot to move around its environment. Using its camera, the robot can locate blocks by detecting their tags and can identify the colors of the LEDs on a block. Semi-permanent electromagnets are attached to an end effector, which enable the robot to pick up a block and to attach it to a structure.

The control software for the robot is implemented using a finite state machine (FSM). In this FSM, a robot starts by locating an unused block (a single block whose LEDs are not illuminated). The robot then searches for, approaches, and inspects structures in its environment (where a structure is defined by one or more blocks whose LEDs are illuminated). If the arrangement and LED colors of the blocks in a structure match a predefined rule, the robot uses its NFC interface to configure the LED colors on the unused block and attaches the unused block to the structure with respect to the matched rule. Since the attachment of this block modifies the arrangement and LED colors of the blocks in the structure, a feedback loop emerges where other predefined rules can now be matched and can continue to coordinate the assembly of a structure. At the time of writing, the robot can only modify the color of the block that is attached to its end effector. However, we are currently enhancing the block's software to allow block-to-block communication that will enable a recently placed block to update the LED colors of adjacent blocks in a structure.

4.2 Modeling the Hardware in Simulation

For the simulation work presented in this paper, both the stigmergic block and the autonomous robot are implemented using the prototyping plug-in. The main body of a block is modeled as a box-shaped link with side lengths of 55 mm. Eight additional sphere-shaped links are defined inside the block to simulate the freely-rotating spherical magnets. Since the magnetism provided by the three-dimensional dynamics plug-in uses a single-dipole approximation, we have tuned

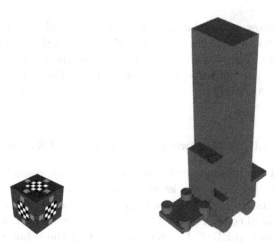

Fig. 1. Visualization of a stigmergic block and an autonomous robot in the ARGoS simulator: these visualizations are generated automatically by the Qt-OpenGL visualization model from the prototyping plug-in

the strength of the magnets based on empirical testing to match the characteristics of the hardware. A tag, radio, and four directional LEDs are added to each face of the block to complete its model in simulation.

The model of the robot consists of 15 links and 15 joints. Four of the links and four of the joints are used to simulate the tracks (treads) of the robot. An additional joint represents the end effector of a robot which is lowered and raised in order to pick up and to assemble unused blocks into a structure. The remaining links are used to simulate the geometry of the robot and the joints between them are fixed. The robot is configured with a camera, rangefinders, radios, semi-permanent electromagnets, and joint sensors and actuators to match the capabilities of the hardware.

These descriptions of the stigmergic block and the autonomous robot are provided to the prototyping plug-in via the experiment configuration file. The prototyping plug-in parses these descriptions and creates models of these objects using the three-dimensional dynamics plug-in and the Qt-OpenGL visualization plug-in (see Fig. 1).

4.3   Reproducing the Hardware Results

In our previous work, we demonstrated the construction of a staircase using a single robot [1]. The staircase consisted of three columns of blocks descending in height with the blocks in each column illuminated with a distinct LED color. The highest column contained three blocks with the LEDs set to green, the middle column contained two blocks with the LEDs set to red, and the last column contained a single block with its LEDs set to violet. To replicate this demonstration in simulation, we create an instance of the finite state machine used by

Fig. 2. Snapshots comparing hardware and simulation results

the hardware inside an ARGoS controller. The controller acts as a wrapper that synchronizes the state machine's data with the sensors and actuators provided by the prototyping and generic robot plug-ins. Snapshots of this demonstration on the hardware and in simulation are shown side-by-side in Fig. 2. We consider

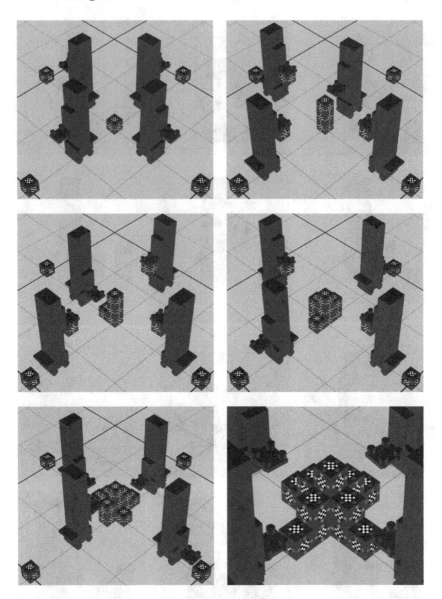

Fig. 3. Snapshots of four robots building a stepped pyramid in simulation

the ability to reproduce the hardware results in simulation as a qualitative validation of the presented plug-ins and of the models of the stigmergic block and the autonomous robot. However, further quantitative testing will be required to identify the extent of any reality gap and to mitigate it. A video of this demonstration is available on the supplementary material website for this paper [3].

4.4   Scaling up to Multi-robot Construction

Following the qualitative validation of our plug-ins and models, we are now able to scale up to a more complex construction scenario where four robots are used to assemble a stepped pyramid in simulation. This demonstration was only possible in simulation due to reliability issues with the robot's drive system.

The stepped pyramid is, in essence, four of the staircases from the previous demonstration which share a common central column. We leverage this symmetry and make only two minor modifications to the control software used by a robot, namely (i) we introduce a random delay state so that the robots do not all approach the structure at the same time and (ii) we introduce a basic collision avoidance mechanism that detects if another robot is attempting to attach a block to the central column of the stepped pyramid. Snapshots of this demonstration running in simulation are shown in Fig. 3. A video of the complete demonstration is available on the supplementary material website [3].

We ran this experiment in ARGoS and gathered data on the construction throughput for five runs. The plot in Fig. 4 shows the construction progress during each of these runs in gray and the average of the five runs in black. From the data, we observe that the construction throughput is initially low due to the central construction site being saturated as all robots attempt to assemble the central column. Following the construction of the central column, however, the construction throughput increases as the robots start building the wings of the pyramid in parallel. The rate of construction then decreases again towards the end of the construction task since the robots no longer have any work to do.

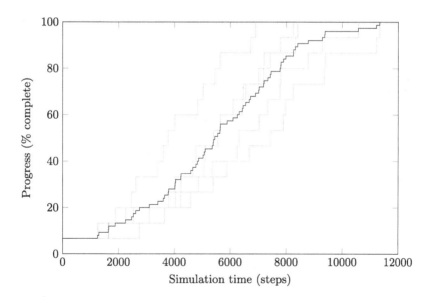

Fig. 4. Construction progress for the stepped pyramid from five runs (shown in gray) and the average progress across all runs (shown in black)

5   Conclusions

In this paper, we have discussed and demonstrated the use of five plug-ins that enable the simulation of our autonomous construction system in ARGoS, a modular, multi-robot simulator. These plug-ins have allowed us to qualitatively reproduce our hardware-based results in simulation and to investigate a more complicated construction scenario.

The plug-ins presented in this paper have been made open source and have been integrated into the ARGoS core package (from 3.0.0-beta50). These plug-ins aim to enable a high-fidelity simulation of the multi-robot construction systems that are commonly found in laboratory settings. We hope that our contribution increases the pace of multi-robot construction research by enabling researchers to disseminate their results and the means to reproduce them more effectively.

In future work, we will use these plug-ins as part of a workflow that aims to answer a number of open research questions regarding the representation of arbitrary structures as assembly rule sets for swarms of robots. We will also investigate and compare different representations of a given structure in terms of its robustness and its ability to be built in parallel by multiple robots.

Acknowledgments. Michael Allwright was supported by the Australian Government through the Endeavour Scholarships and Fellowships Program. Navneet Bhalla was partially supported by a postdoctoral fellowship from the Natural Sciences and Engineering Research Council of Canada. Marco Dorigo acknowledges support from the Belgian F.R.S.-FNRS, of which he is a Research Director. The work presented in this paper was partially supported by the FLAG-ERA project RoboCom++ and by the European Research Council (ERC) under the European Union's Horizon 2020 research and innovation programme (grant agreement number 681872). We would like to thank Haitham El-faham and Weixu Zhu for their help with implementing and testing the magnetism code in the three-dimensional dynamics plug-in.

References

1. Allwright, M., Bhalla, N., Dorigo, M.: Structure and markings as stimuli for autonomous construction. In: Proceedings of the Eighteenth International Conference on Advanced Robotics, pp. 296–302. IEEE (2017). https://doi.org/10.1109/icar.2017.8023623
2. Allwright, M., Bhalla, N., Pinciroli, C., Dorigo, M.: ARGoS plug-ins for experiments in autonomous construction. Technical report TR/IRIDIA/2018-007, IRIDIA, Université Libre de Bruxelles, Brussels, Belgium (2018)
3. Allwright, M., Bhalla, N., Pinciroli, C., Dorigo, M.: Simulating multi-robot construction in ARGoS (supplementary material website) (2018). http://iridia.ulb.ac.be/supp/IridiaSupp2017-004/index.html
4. Bonabeau, E., Dorigo, M., Theraulaz, G.: Swarm Intelligence: From Natural to Artificial Systems. Oxford University Press, Oxford (1999)
5. Camazine, S., Deneubourg, J.L., Franks, N.R., Sneyd, J., Theraulaz, G., Bonabeau, E.: Self-Organization in Biological Systems. Princeton University Press, Princeton (2001)

6. Carpin, S., Lewis, M., Wang, J., Balakirsky, S., Scrapper, C.: USARSim: a robot simulator for research and education. In: 2007 IEEE International Conference on Robotics and Automation, pp. 1400–1405. IEEE (2007). https://doi.org/10.1109/robot.2007.363180

7. Grassé, P.P.: Reconstruction of the nest and coordination between individuals in terms. Bellicositermes Natalensis and Cubitermes sp. the theory of stigmergy: test interpretation of termite constructions. Insectes Soc. **6**(1), 41–80 (1959). https://doi.org/10.1007/bf02223791

8. Jones, C., Matarić, M.J.: Automatic synthesis of communication-based coordinated multi-robot systems. In: 2004 IEEE/RSJ International Conference on Intelligent Robots and Systems, pp. 381–387. IEEE (2004). https://doi.org/10.1109/iros.2004.1389382

9. Koenig, N., Howard, A.: Design and use paradigms for Gazebo, an open-source multi-robot simulator. In: 2004 IEEE/RSJ International Conference on Intelligent Robots and Systems, pp. 2149–2154. IEEE (2004). https://doi.org/10.1109/iros.2004.1389727

10. Lindsey, Q., Mellinger, D., Kumar, V.: Construction with quadrotor teams. Auton. Robots **33**(3), 323–336 (2012). https://doi.org/10.1007/s10514-012-9305-0

11. Michel, O.: Cyberbotics Ltd., Webots^{TM}: professional mobile robot simulation. Int. J. Adv. Robot. Syst. **1**(1), 39–42 (2004). https://doi.org/10.5772/5618

12. Olson, E.: AprilTag: a robust and flexible visual fiducial system. In: 2011 IEEE International Conference on Robotics and Automation, pp. 3400–3407. IEEE (2011). https://doi.org/10.1109/icra.2011.5979561

13. Petersen, K., Nagpal, R., Werfel, J.: TERMES: an autonomous robotic system for three-dimensional collective construction. In: Proceedings of Robotics: Science and Systems, pp. 257–264. RSS Foundation (2011). https://doi.org/10.15607/rss.2011.vii.035

14. Pinciroli, C., et al.: ARGoS: a modular, parallel, multi-engine simulator for multi-robot systems. Swarm Intell. **6**(4), 271–295 (2012). https://doi.org/10.1007/s11721-012-0072-5

15. Soleymani, T., Trianni, V., Bonani, M., Mondada, F., Dorigo, M.: Autonomous construction with compliant building material. In: Menegatti, E., Michael, N., Berns, K., Yamaguchi, H. (eds.) Intelligent Autonomous Systems. AISC, vol. 302, pp. 1371–1388. Springer, Cham (2016). https://doi.org/10.1007/978-3-319-08338-4_99

16. Sugawara, K., Doi, Y.: Collective construction by cooperation of simple robots and intelligent blocks. In: Kubota, N., Kiguchi, K., Liu, H., Obo, T. (eds.) ICIRA 2016. LNCS (LNAI), vol. 9834, pp. 452–461. Springer, Cham (2016). https://doi.org/10.1007/978-3-319-43506-0_40

17. Sugawara, K., Doi, Y.: Collective construction of dynamic equilibrium structure through interaction of simple robots with semi-active blocks. In: Chong, N.-Y., Cho, Y.-J. (eds.) Distributed Autonomous Robotic Systems. STAR, vol. 112, pp. 165–176. Springer, Tokyo (2016). https://doi.org/10.1007/978-4-431-55879-8_12

18. Theraulaz, G., Bonabeau, E.: Coordination in distributed building. Science **269**(5224), 686–688 (1995). https://doi.org/10.1126/science.269.5224.686

19. Theraulaz, G., Bonabeau, E.: A brief history of stigmergy. Artif. Life **5**(2), 97–116 (1999). https://doi.org/10.1162/106454699568700

20. Thomaszewski, B., Gumann, A., Pabst, S., Straßer, W.: Magnets in motion. ACM Trans. Graph. **27**(5), 162:1–162:9 (2008). https://doi.org/10.1145/1409060.1409115

21. Werfel, J., Nagpal, R.: Extended stigmergy in collective construction. IEEE Intell. Syst. **21**(2), 20–28 (2006). https://doi.org/10.1109/mis.2006.25
22. Werfel, J., Nagpal, R.: Three-dimensional construction with mobile robots and modular blocks. Int. J. Robot. Res. **27**(3–4), 463–479 (2008). https://doi.org/10.1177/0278364907084984
23. Werfel, J., Petersen, K., Nagpal, R.: Designing collective behavior in a termite-inspired robot construction team. Science **343**(6172), 754–758 (2014). https://doi.org/10.1126/science.1245842
24. Wismer, S., Hitz, G., Bonani, M., Gribovskiy, A., Magnenat, S.: Autonomous construction of a roofed structure: synthesizing planning and stigmergy on a mobile robot. In: 2012 IEEE/RSJ International Conference on Intelligent Robots and Systems, pp. 5436–5437. IEEE (2012). https://doi.org/10.1109/iros.2012.6386278
25. Worcester, J., Ani Hsieh, M., Lakaemper, R.: Distributed assembly with online workload balancing and visual error detection and correction. Int. J. Robot. Res. **33**(4), 534–546 (2014). https://doi.org/10.1177/0278364913509125

Stability Analysis of the Multi-objective Multi-guided Particle Swarm Optimizer

Christopher W. Cleghorn$^{(\boxtimes)}$ ⓘ, Christiaan Scheepers ⓘ,
and Andries P. Engelbrecht ⓘ

Department of Computer Science, University of Pretoria, Pretoria, South Africa
{ccleghorn,engel}@cs.up.ac.za, cscheepers@acm.org

Abstract. At present particle swarm optimizers (PSO) designed for multi-objective optimization have undergone no form of theoretical stability analysis. This paper derives the sufficient and necessary conditions for order-1 and order-2 stability of the recently proposed multi-guided PSO (MGPSO), which was designed specifically for multi-objective optimization. The paper utilizes a recently published theorem for performing stability analysis on PSO variants, which requires minimal modeling assumptions. It is vital for PSO practitioners to know the actual criteria for particle stability of the given PSO variant being used, as it been shown that particle stability has a considerable impact on PSO's performance. This paper empirically validates its theoretical findings by comparing the derived stability criteria against those of an assumption free MGPSO algorithm. It was found that the derived criteria for order-1 and order-2 stability are an accurate predictor of the unsimplified MGPSO's particle behavior.

1 Introduction

Recently, a particle swarm optimizer (PSO) variant, the multi-guided PSO (MGPSO) was proposed for multi-objective optimization [14,15]. It was found that MGPSO was highly competitive to the current state of the art PSO based multi-objective optimization algorithms, such as speed-constrained multi-objective particle swarm optimization (SMPSO) [10], optimized multi-objective particle swarm optimization (OMOPSO) [13], and the vector evaluated particle swarm optimizer (VEPSO) [11,12]. Furthermore, the MGPSO was also shown to be highly competitive with the current state of the art evolutionary multiple objective optimizers, such as the non-dominated sorting genetic algorithm II (NSGA-II) [8], strength Pareto evolutionary algorithm 2 (SPEA2) [18], pareto envelope-based selection algorithm II (PESA-II) [7], and the multi-objective evolutionary algorithm based on decomposition (MOEA/D) [17].

With the introduction of any new optimization algorithm comes an array of unknown algorithm characteristics to be understood. The characteristic that this paper focuses on is particle stability. It has been empirically shown that order-1 and order-2 particle stability has a considerable impact on performance

© Springer Nature Switzerland AG 2018
M. Dorigo et al. (Eds.): ANTS 2018, LNCS 11172, pp. 201–212, 2018.
https://doi.org/10.1007/978-3-030-00533-7_16

[4]. Furthermore, it was shown that parameter configurations that resulted in particle instability almost always caused PSO to perform worse than random search [4]. The clear relationship between PSO particle stability and the algorithm's performance, implies that knowing the criteria for particle stability is needed for effective use of a PSO variant.

Given that MGPSO has a similar structure to that of the original PSO [9] with the presence of inertia as proposed by Shi [16], existing PSO theory can be readily applied to the stability analysis of MGPSO. Specifically, Cleghorn and Engelbrecht [6] recently proposed a general theorem for deriving stability criteria for a class of PSO variants under minimal modeling assumptions. To the authors' knowledge this is the first paper to perform stability analysis of a multi-objective PSO. The theoretically derived region for particle stability of MGPSO is also empirically validated utilizing the assumption for free methodology for stability region validation, as presented in [1,3], and used in [2,5].

A description of MGPSO is provided in Sect. 2. The theoretical derivation of the order-1 and order-2 stable regions of MGPSO are presented in Sect. 3. The experimental setup and results empirically validating the derived stable regions are presented in Sects. 4 and 5 respectively. A summary of the findings of this paper are presented in Sect. 6.

2   Multi-guided Particle Swarm Optimizer

The MGPSO algorithm was proposed by Scheepers and Engelbrecht [14,15] and is inspired by the vector evaluated particle swarm optimizer (VEPSO) as proposed by Parsopoulos and Vrahatis [11,12]. MGPSO is a multi-swarm multi-objective PSO variant, where each objective is optimized by a sub-swarm. Similar to VEPSO, the Pareto-optimal solutions are stored in an archive. Scheepers and Engelbrecht proposed that a third attractor be added to the velocity update equation, in addition to the usual social and cognitive attractors. The aim of the new attractor is to pull particles towards the Pareto-optimal front (POF). The third attractor is selected from the archive of non-dominated solutions. The archive attractor is selected from the tournament pool as the one with the largest crowding distance [8] to promote convergence to a diverse pareto-front.

The velocity and position update equations of MGPSO are defined as follows:

$$\mathbf{v}_i(t+1) = w\mathbf{v}_i(t) + c_1\mathbf{r}_1 \otimes (\mathbf{y}_i(t) - \mathbf{x}_i(t)) + \lambda_i c_2\mathbf{r}_2 \otimes (\hat{\mathbf{y}}_i(t) - \mathbf{x}_i(t))$$
$$+ (1 - \lambda_i)c_3\mathbf{r}_3 \otimes (\hat{\mathbf{a}}_i(t) - \mathbf{x}_i(t)) \tag{1}$$
$$\mathbf{x}_i(t+1) = \mathbf{x}(t) + \mathbf{v}_i(t+1), \tag{2}$$

where $\mathbf{r}_1$, $\mathbf{r}_2$, $\mathbf{r}_3 \sim U(0,1)^d$, and d is the dimension of the objective functions PSO. The operator $\otimes$ is used to indicate component-wise multiplication of two vectors. The positions $\mathbf{y}_i$ and $\hat{\mathbf{y}}_i$ are respectively the "best" positions that particle i and particle i's neighborhood of particles have visited. In this paper, "best" is defined as the location where a particle has obtained the lowest objective function evaluation. The coefficients c_1, c_2, c_3, and w are the cognitive, social, archive, and inertia weights respectively. λ_i is the exploitation trade-off coefficient for particle i, is initialized as a random constant sampled uniformly from $(0, 1)$ (λ_i does not vary over iterations). The MGPSO algorithm is summarized in Algorithm 1.

Algorithm 1. Multi-guided Particle Swarm Optimization

1: **for** each objective $m = 1, ..., n_m$ **do**
2:     Initialize S_m, of n_{s_m} particles uniformly in a hypercube of dimension n_x
3:     Let f_m be the objective function;
4:     Let $S_m.\mathbf{y}_i$ be the personal best position of particle $S_m.\mathbf{x}_i$;
5:     Let $S_m.\hat{\mathbf{y}}_i$ be the neighborhood best position of particle $S_m.\mathbf{x}_i$;
6:     Initialize $S_m.\mathbf{v}_i(0)$ to $\mathbf{0}$; $S_m.\mathbf{y}_i = S_m.\mathbf{x}_i(0)$; $S_m.\hat{\mathbf{y}}_i = S_m.\mathbf{x}_i(0)$; $S_m.\lambda_i \sim U(0, 1)$;
7: **end for**
8: Let $t = 0$;
9: **repeat**
10:     **for** each objective $m = 1, ..., n_m$ **do**
11:         **for** each particle $i = 1, ..., S_m.n_s$ **do**
12:             **if** $f_m(S_m.\mathbf{x}_i) < f_m(S_m.\mathbf{y}_i)$ **then**
13:                 $S_m.\mathbf{y}_i = S_m.\mathbf{x}_i(t)$;
14:             **end if**
15:             **for** particles $\hat{i}$ with particle i in their neighborhood **do**
16:                 **if** $f_m(S_m.\mathbf{y}_i) < f_m(S_m.\hat{\mathbf{y}}_i)$ **then**
17:                     $S_m.\hat{\mathbf{y}}_i = S_m.\mathbf{y}_i$;
18:                 **end if**
19:             **end for**
20:             Update the archive with the solution $S_m.\mathbf{x}_i$;
21:         **end for**
22:     **end for**
23:     **for** each objective $m = 1, ..., n_m$ **do**
24:         **for** each particle $i = 1, ..., S_m.n_s$ **do**
25:             Select a solution, $S_m.\hat{\mathbf{a}}_i(t)$, from the archive using tournament selection;
26:             $S_m.\mathbf{v}_i(t+1) = wS_m.\mathbf{v}_i(t) + c_1\mathbf{r}_1(S_m.\mathbf{y}_i(t) - S_m.\mathbf{x}_i(t))$
                    $+S_m.\lambda_i c_2\mathbf{r}_2(S_m.\hat{\mathbf{y}}_i(t) - S_m.\mathbf{x}_i(t))$
                    $+(1 - S_m.\lambda_i)c_3\mathbf{r}_3(S_m.\hat{\mathbf{a}}_i(t) - S_m.\mathbf{x}_i(t)))$;
27:             $S_m.\mathbf{x}_i(t+1) = S_m.\mathbf{x}_i(t) + S_m.\mathbf{v}_i(t+1)$;
28:         **end for**
29:     **end for**
30:     $t = t + 1$;
31: **until** stopping condition is true

---

3 Theoretical Derivation

This section presents a theoretical derivation of the order-1 and order-2 stable regions for the MGPSO algorithm.

To derive order-1 and order-2 stable regions for MGPSO, the following general theorem of Cleghorn and Engelbrecht [6] is used:

Theorem 1. *The following properties hold for all PSO variants of the form:*

$$x_k(t+1) = x_k(t)\alpha + x_k(t-1)\beta + \gamma_t \tag{3}$$

where k indicates the vector component, α and β are well defined random variables, and (γ_t) is a sequence of well defined random variables. In the context of this work, a random variable is said to be well defined if it has an expectation and a variance.

1. *Assuming i_t converges, particle positions are order-1 stable for every initial condition if and only if $\rho(\mathbf{A}) < 1$[1], where*

$$\mathbf{A} = \begin{bmatrix} E[\alpha] & E[\beta] \\ 1 & 0 \end{bmatrix} \text{ and } i_t = \begin{bmatrix} E[\gamma_t] \\ 0 \end{bmatrix} \tag{4}$$

2. *The particle positions are order-2 stable if $\rho(\mathbf{B}) < 1$ and (j_t) converges, where*

$$\mathbf{B} = \begin{bmatrix} E[\alpha] & E[\beta] & 0 & 0 & 0 \\ 1 & 0 & 0 & 0 & 0 \\ 0 & 0 & E[\alpha^2] & E[\beta^2] & 2E[\alpha\beta] \\ 0 & 0 & 1 & 0 & 0 \\ 0 & 0 & E[\alpha] & 0 & E[\beta] \end{bmatrix} \text{ and } j_t = \begin{bmatrix} E[\gamma_t] \\ 0 \\ E[\gamma_t^2] \\ 0 \\ 0 \end{bmatrix} \tag{5}$$

under the assumption that the limits of $(E[\gamma_t\alpha])$ and $(E[\gamma_t\beta])$ exist.
3. *Assuming that $x(t)$ is order-1 stable, then the following is a necessary condition for order-2 stability:*

$$1 - E[\alpha] - E[\beta] \neq 0 \tag{6}$$

$$1 - E[\alpha^2] - E[\beta^2] - \left(\frac{2E[\alpha\beta]E[\alpha]}{1 - E[\beta]}\right) > 0 \tag{7}$$

4. *The convergence of $E[\gamma_t]$ is a necessary condition for order-1 stability, and the convergence of both $E[\gamma_t]$ and $E[\gamma_t^2]$ is a necessary condition for order-2 stability.*

The MGPSO's update Eq. (1), can be written in the form of Eq. (3) by setting:

$$\alpha = (1+w) - c_1 r_1 - \lambda c_2 r_2 - (1-\lambda)c_3 r_3, \quad \beta = -w$$
$$\gamma_t = c_1 r_1 y(t) + \lambda c_2 r_2 \hat{y}(t) + (1-\lambda)c_3 r_3 \hat{a}(t)$$

In order to utilize Theorem 1, the following modeling assumption is used:

Definition 1. *Non-stagnant distribution assumption* [6]:
It is assumed that $\hat{y}_i(t)$, $y_i(t)$, and $\hat{a}_i(t)$ are random variables sampled from a time dependent distribution, such that $\hat{y}_i(t)$, $y_i(t)$, and $\hat{a}_i(t)$ have well defined expectations and variances for each t and that $\lim_{t\to\infty} E[\hat{y}_i(t)]$, $\lim_{t\to\infty} E[y_i(t)]$, $\lim_{t\to\infty} E[\hat{a}_i(t)]$, $\lim_{t\to\infty} V[\hat{y}_i(t)]$, $\lim_{t\to\infty} V[y_i(t)]$ and $\lim_{t\to\infty} V[\hat{a}_i(t)]$ exist.

---

[1] $\rho(\mathbf{M})$ denotes the spectral radius of the matrix $\mathbf{M}$.

It is clear from part 4 of Theorem 1 that the non-stagnant distribution assumption is a necessary condition for order-1 and order-2 stability. In order to obtain the criteria for order-1 stability, part 1 of Theorem 1 is used. Specifically, the following expectations are required:

$$E[\alpha] = (1 + w) - \frac{c_1}{2} - \frac{\lambda c_2}{2} - \frac{(1 - \lambda)c_3}{2}, \ E[\beta] = -w,$$

$$E[\gamma_t] = \frac{1}{2}\left(c_1 E[y(t)] + \lambda c_2 E[\hat{y}(t)] + (1 - \lambda)c_3 E[\hat{a}(t)]\right).$$

Given the non-stagnant distribution assumption, it follows by the sum of convergent sequences that $E[\gamma_t]$ converges, and therefore $\mathbf{i}_t$ converges. The criteria for order-1 stability is determined by coefficients that satisfy $\rho(\mathbf{A}) < 1$. After some algebraic manipulation, the following criteria for order-1 stability is obtained:

$$|w| < 1 \text{ and } 0 < c_1 + \lambda c_2 + (1 - \lambda)c_3 < 4(w + 1), \tag{8}$$

or in the case of $c = c_1 = c_2 = c_3$,

$$|w| < 1 \text{ and } 0 < 2c < 4(w + 1). \tag{9}$$

Part 3 of Theorem 1 is used to derive the criteria necessary for order-2 stability. The calculation of additional expected values is needed. In order to calculate $E[\alpha^2]$, α^2 is first calculated as:

$$\begin{aligned}
\alpha^2 &= ((1 + w) - cr_1 - \lambda cr_2 - (1 - \lambda)cr_3)^2 \\
&= (1 + w)^2 - c_1 r_1(1 + w) - \lambda c_2 r_2(1 + w) - (1 + w)(1 - \lambda)c_3 r_3 \\
&\quad - c_1 r_1(1 + w) + c_1^2 r_1^2 + \lambda c_1 c_2 r_1 r_2 + (1 - \lambda)c_1 c_3 r_1 r_3 \\
&\quad - \lambda c_2 r_2(1 + w) + \lambda c_1 c_2 r_1 r_2 + \lambda^2 c_2^2 r_2^2 + \lambda(1 - \lambda)c_2 c_3 r_2 r_3 \\
&\quad - (1 + w)(1 - \lambda)c_3 r_3 + (1 - \lambda)c_1 c_3 r_1 r_3 + \lambda(1 - \lambda)c_2 c_3 r_2 r_3 + (1 - \lambda)^2 c_3^2 r_3^2
\end{aligned} \tag{10}$$

Applying the expectation operator results in

$$\begin{aligned}
E[\alpha^2] &= (1 + w)^2 - \frac{c_1}{2}(1 + w) - \lambda\frac{c_2}{2}(1 + w) - (1 + w)(1 - \lambda)\frac{c_3}{2} \\
&\quad - \frac{c_1}{2}(1 + w) + \frac{c_1^2}{3} + \lambda\frac{c_1 c_2}{4} + (1 - \lambda)\frac{c_1 c_3}{4} \\
&\quad - \lambda\frac{c_2}{2}(1 + w) + \lambda\frac{c_1 c_2}{4} + \lambda^2\frac{c_2^2}{3} + \lambda(1 - \lambda)\frac{c_2 c_3}{4} \\
&\quad - (1 + w)(1 - \lambda)\frac{c_3}{2} + (1 - \lambda)\frac{c_1 c_3}{4} + \lambda(1 - \lambda)\frac{c_2 c_3}{4} + (1 - \lambda)^2\frac{c_3^2}{3}
\end{aligned} \tag{11}$$

Let $c = c_1 = c_2 = c_3$, then after some algebraic manipulation, Eq. (11) becomes

$$\begin{aligned}
E[\alpha^2] &= (1 + w)^2 - c(1 + w) - \lambda c(1 + w) - (1 + w)(1 - \lambda)c \\
&\quad + c^2\left(\frac{1}{3} + \frac{\lambda}{2} + \frac{1 - \lambda}{2} + \frac{\lambda^2}{3} + \frac{\lambda(1 - \lambda)}{2} + \frac{(1 - \lambda)^2}{3}\right) \\
&= (1 + w)((1 + w) - 2c) + \frac{c^2}{6}\left(\lambda^2 - \lambda + 7\right)
\end{aligned}$$

The following expectations are also needed:

$$E[\alpha\beta] = -wE[\alpha] = -w((1+w) - c) \text{ and } E[\beta^2] = w^2 \qquad (12)$$

In order to derive the criteria necessary for order-2 stability, first consider the condition of Eq. (6) in part 3 of Theorem 1:

$$1 + E[\alpha] + E[\beta] \neq 0 \implies c_1 + \lambda c_2 + (1 - \lambda)c_3 \neq 0 \qquad (13)$$

or if $c = c_1 = c_2 = c_3$, simply $c \neq 0$.

Now consider the condition of Eq. (7) in part 3 of Theorem 1:

$$1 - E[\alpha^2] - E[\beta^2] - \left(\frac{2E[\alpha\beta]E[\alpha]}{1 - E[\beta]}\right) > 0$$

$$\implies 2c - 2wc + \left(\frac{2wc^2}{(1+w)}\right) - \frac{c^2}{6}(\lambda^2 - \lambda + 7) > 0 \qquad (14)$$

Solving Eq. (14) as a quadric form equal to 0 leads to

$$c < \frac{12(1 - w^2)}{(\lambda^2 - \lambda + 7)(w + 1) - 12w} \qquad (15)$$

Merging the conditions for order-2 stability in Eqs. (15) and (13) with the conditions for order-1 stability of Eq. (9) leads to the following criteria for order-1 and order-2 stability:

$$0 < c < \frac{12(1 - w^2)}{(\lambda^2 - \lambda + 7)(w + 1) - 12w}, \quad |w| < 1 \qquad (16)$$

This merger is possible because the region defined by Eq. (15) is a subset of the region defined by Eq. (9). The conditions derived for order-2 stability are only the necessary conditions. To verify that they are sufficient, part 2 of Theorem 1 is used: Given the complexity of symbolically solving $\rho(\boldsymbol{B}) < 1$, an empirical approach is utilized in line with that used by Cleghorn and Engelbrecht [6]. The experimental procedure followed is: 10^9 random configurations of the form $\{w, c, \lambda\}$ were generated such that Eqs. (13) and (15) were satisfied. It was then tested if the condition, $\rho(\boldsymbol{B}) < 1$, was satisfied or not. In all of the cases it was found, that if Eqs. (13) and (15) were satisfied, then the condition $\rho(\boldsymbol{B}) < 1$ held. This finding is strong evidence that the criteria of Eq. (16) are both sufficient and necessary for order-1 and order-2 stability.

The manner in which λ affects the stability region is illustrated in Fig. 1. The closer λ gets to 0.5, the more the apex of the stability region extends. As λ approaches either 0 or 1 from 0.5 the stability region reduces in size in a symmetric manner. Given that λ may be initialized to any value in the range $(0, 1)$, selecting coefficients such that Eq. (16) is satisfied for $\lambda = 0$ (or $\lambda = 1$) will ensure that every particle will be both order-1 and order-2 regardless of the particle specific λ.

While the modeling assumption utilized in the section is minimal, it is still required to confirm whether or not the newly derived stability criteria are truly representative of the unsimplified MGPSO's behaviour. This is done in the next section.

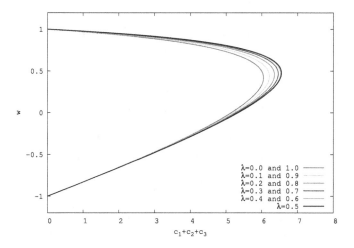

Fig. 1. MGPSO convergent regions for $\lambda = 0$, 0.1, 0.2, 0.3, 0.4, 0.5, 0.6, 0.7, 0.8, 0.9, and 1

4   Empirical Setup

This section utilizes the method for empirically validating the stability region of PSO variants as proposed by Cleghorn and Engelbrecht [1,3].

A swarm size of 64 particles per objective, and 5000 iterations are used in the experiment. Two objective functions where considered. Particle velocities were initialized to **0** and positions were initialized within $(-100, 100)$. The experiment was performed in 50 dimensions. As a result the maximum possible distance between particles in the initial search space is 1414.214. This maximum distance is referred to as Δ_{max} from this point forward. Reported results were capped at Δ_{max} to prevent highly unstable parameter configurations from hindering the presentation of the results.

The measure of stability used in this paper is:

$$\Delta_m (t + 1) = \frac{1}{S_m.n_s} \sum_{i=1}^{S_m.n_s} \| \boldsymbol{x}_i (t + 1) - \boldsymbol{x}_i (t) \|_2. \tag{17}$$

where $S_m.n_s$ is the swarm size for each sub-swarm m. The sum of all $\Delta_{S_m.n_s}(t)$'s is reported as $\Delta(t)$. The objective function used for each objective is

$$CF(\boldsymbol{x}) \in U(-1000, 1000), \tag{18}$$

which was shown to be an effective objective function for stability analysis in [1].

The experiment was conducted over the following parameter region:

$$w \in [-1.1, 1.1], \, c_1 + c_2 + c_3 \in (0, 8], \text{ and } \lambda \in [0, 1], \tag{19}$$

where $c_1 = c_2 = c_3$, with a sample point every 0.1 along w, $c_1 + c_2 + c_3$, and λ. A total of 1840 sample points from the region defined in Eq. (19) were used for each fixed λ. The results reported in Sect. 5 are derived from 50 independent runs for each sample point.

5   Experimental Results and Discussion

This section presents the results of the experiments described in Sect. 4.

A snapshot of all parameter configurations' resulting stability measure values are presented in Figs. 2(a) to 3(e) for the last iteration of MGPSO with λ set to, 0.0, 0.1, 0.2, 0.3, 0.4, 0.5, 0.6, 0.7, 0.8, 0.9, and 1.0. The reported stability measures are the mean derived from the 50 independent runs.

The number of parameter configurations that empirically agree or disagree with the stable/unstable behavior predicted by the theoretically derived stability region of Eq. (16) is presented in Table 1. Eight Δ_m based measurements are presented in Table 1: the number of parameter configurations that are theoretically stable (TS) and unstable (TUS), the number of parameter configurations that where empirically stable (ES) and unstable (EUS), the number of parameter configurations that were found to be empirically stable despite the theory predicting unstable behavior (ES despite TUS), the number of parameter configurations that were found to be empirically unstable despite the theory predicting stable behavior (EUS despite TS), and lastly the percentage error and agreement between the theoretical derivation and the empirical findings. A parameter configuration is classified to be stable if the value of the recorded convergence measure of Eq. (17) is less than $\Delta_{max}(d)$, and unstable if greater than or equal to $\Delta_{max}(d)$, in accordance with the approach of Cleghorn and Engelbrecht [3].

Table 1. Empirical findings versus theoretical prediction

λ	T	TUS	ES	EUS	ES despite TUS	EUS despite TS	Error	Agreement
0	764	1076	759	1081	8	13	1.14%	98.86%
0.1	781	1059	784	1056	12	9	1.14%	98.86%
0.2	796	1044	798	1042	11	9	1.09%	98.91%
0.3	809	1031	809	1031	12	12	1.3%	98.7%
0.4	816	1024	819	1021	11	8	1.03%	98.97%
0.5	817	1023	823	1017	14	8	1.2%	98.8%
0.6	816	1024	816	1024	10	10	1.09%	98.91%
0.7	809	1031	810	1030	9	8	0.92%	99.08%
0.8	796	1044	798	1042	12	10	1.2%	98.8%
0.9	781	1059	781	1059	10	10	1.09%	98.91%
1	764	1076	760	1080	9	13	1.2%	98.8%

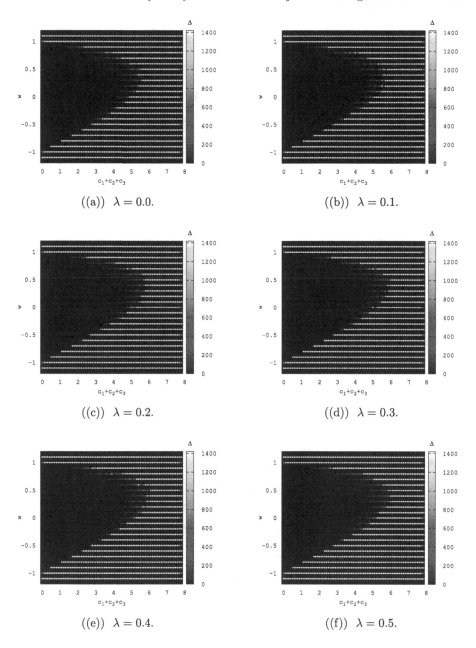

Fig. 2. MGPSO convergence results for $\lambda = 0, 0.1, \ldots, 0.5$

As shown in Figs. 2(a) to 3(e) the shape and size of the regions empirically classified as stable is in-line with the theoretical prediction of Eq. (16). However, the effect of varying λ is harder to see, which is not surprising given how similar

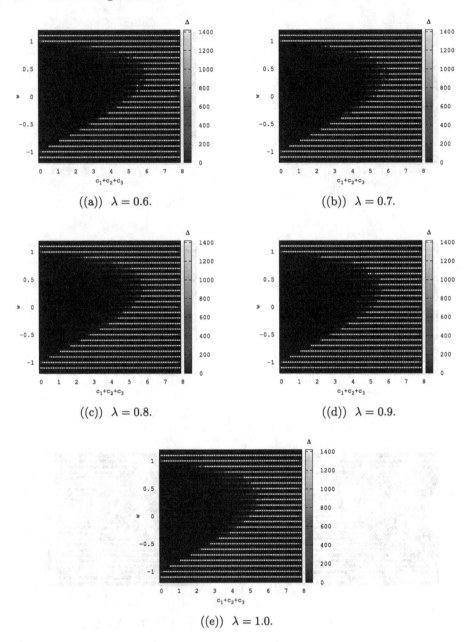

Fig. 3. MGPSO convergence results for $\lambda = 0.6, 0.7, \ldots, 1.0$

the regions that the theoretical derivations predicts are, as illustrated in Fig. 1. The accuracy of the theoretical derivation can be more clearly seen in Table 1. For all the tested λ values, the theoretical prediction had an above 98.7% accuracy.

The accuracy reported was also stable across differing λ values with the largest difference in accuracy reported being only 0.38%.

It is evident from the presented results that the theoretically derived region for particle stability, as defined in Eq. (16) accurately reflect the parameter configurations needed for order-1 and order-2 stability of MGPSO.

6   Conclusion

This paper provided the first theoretical stability analysis of a multi-objective particle swarm optimization (PSO) variant. Specifically, this paper theoretically derived the order-1 and order-2 stable regions for multi-guided PSO (MGPSO) using the minimal required modeling assumptions. The provided order-1 and order-2 stable regions can be utilized by PSO practitioners to make an informed choice when selecting control parameters values for the MGPSO algorithm. Furthermore, the derived criteria for stability were validated, using the empirical method verified by Cleghorn and Engelbrecht [1], under which no simplifying modelling assumptions were placed on the MGPSO algorithm. Given the empirical validation, the theoretical derivation is an accurate representation of MGPSO stability criteria.

References

1. Cleghorn, C.W., Engelbrecht, A.P.: Particle swarm convergence: standardized analysis and topological influence. In: Dorigo, M., et al. (eds.) ANTS 2014. LNCS, vol. 8667, pp. 134–145. Springer, Cham (2014). https://doi.org/10.1007/978-3-319-09952-1_12

2. Cleghorn, C.W., Engelbrecht, A.P.: Fully informed particle swarm optimizer: convergence analysis. In: Proceedings of the IEEE Congress on Evolutionary Computation, pp. 164–170. IEEE Press, Piscataway, NJ (2015)

3. Cleghorn, C.W., Engelbrecht, A.P.: Particle swarm variants: standardized convergence analysis. Swarm Intell. 9(2–3), 177–203 (2015)

4. Cleghorn, C.W., Engelbrecht, A.P.: Particle swarm optimizer: the impact of unstable particles on performance. In: Proceedings of the IEEE Symposium Series on Swarm Intelligence, pp. 1–7. IEEE Press, Piscataway, NJ (2016)

5. Cleghorn, C.W., Engelbrecht, A.P.: Unified particle swarm optimizer: convergence analysis. In: Proceedings of the IEEE Congress on Evolutionary Computation, pp. 448–454. IEEE Press, Piscataway, NJ (2016)

6. Cleghorn, C.W., Engelbrecht, A.P.: Particle swarm stability: a theoretical extension using the non-stagnate distribution assumption. Swarm Intell. 12(1), 1–22 (2018)

7. Corne, D.W., Jerram, N., Knowles, J.D., Oates, M.L.: PESA-II: region-based selection in evolutionary multiobjective optimization. In: Proceedings of the Genetic and Evolutionary Computation Conference, pp. 283–290. ACM Press, New York, NY (2001)

8. Deb, K., Agrawal, S., Pratap, A., Meyarivan, T.: A fast elitist non-dominated sorting genetic algorithm for multi-objective optimization: NSGA-II. In: Schoenauer, M., et al. (eds.) PPSN 2000. LNCS, vol. 1917, pp. 849–858. Springer, Heidelberg (2000). https://doi.org/10.1007/3-540-45356-3_83

9. Kennedy, J., Eberhart, R.C.: Particle swarm optimization. In: Proceedings of the IEEE International Joint Conference on Neural Networks, pp. 1942–1948. IEEE Press, Piscataway, NJ (1995)
10. Nebro, A.J., Durillo, J.J., García-Nieto, J., Coello Coello, C.A., Luna, F., Alba, E.: SMPSO: a new PSO-based metaheuristic for multi-objective optimization. In: Proceedings of the IEEE Symposium on MultiCriteria Decision-Making, pp. 66–73. IEEE Press, Piscataway, NJ (2009)
11. Parsopoulos, K.E., Vrahatis, M.N.: Particle swarm optimization method in multi-objective problems. In: Proceedings of the ACM Symposium on Applied Computing, pp. 603–607 (2002). https://doi.org/10.1145/508791.508907
12. Parsopoulos, K.E., Vrahatis, M.N.: Recent approaches to global optimization problems through particle swarm optimization. Nat. Comput. 1(2), 235–306 (2002)
13. Sierra, M.R., Coello Coello, C.A.: Improving PSO-based multi-objective optimization using crowding, mutation and ∈-dominance. In: Coello Coello, C.A., Hernández Aguirre, A., Zitzler, E. (eds.) EMO 2005. LNCS, vol. 3410, pp. 505–519. Springer, Heidelberg (2005). https://doi.org/10.1007/978-3-540-31880-4_35
14. Scheepers, C.: Multi-guided particle swarm optimization: a multi-objective particle swarm optimizer. Ph.D. thesis, Department of Computer Science, University of Pretoria, Pretoria, South Africa (2018)
15. Scheepers, C., Engelbrecht, A.P.: Multi-guide particle swarm optimization a multi-swarm multi-objective particle swarm optimizer. Swarm Intell. 1–22 (2018, under review)
16. Shi, Y., Eberhart, R.C.: A modified particle swarm optimizer. In: Proceedings of the IEEE Congress on Evolutionary Computation, pp. 69–73. IEEE Press, Piscataway, NJ (1998)
17. Zhang, Q., Li, H.: IEEE transactions on evolutionary computation. Nat. Comput. 11(2), 712–731 (2007)
18. Zitzler, E., Laumanns, M., Thiele, L.: SPEA2: improving the strength pareto evolutionary algorithm. Technical report, Swiss Federal Instituteof Technology (ETH) Zurich (2001)

Swarm Attack: A Self-organized Model to Recover from Malicious Communication Manipulation in a Swarm of Simple Simulated Agents

Giuseppe Primiero[1(✉)] , Elio Tuci[1] , Jacopo Tagliabue[2] ,
and Eliseo Ferrante[3,4(✉)]

[1] The Department of Computer Science, Middlesex University London, London, UK
{G.Primiero,E.Tuci}@mdx.ac.uk
[2] Tooso Inc., San Francisco, USA
tagliabue.jacopo@gmail.com
[3] Laboratory of Socioecology and Social Evolution, KU Leuven, Leuven, Belgium
eliseo.ferrante@bio.kuleuven.be
[4] School of Computer Science, University of Birmingham,
Dubai, United Arab Emirates
e.ferrante@bham.ac.uk

Abstract. Non-centralised behaviour such as those that characterise swarm robotics systems are vulnerable to intentional disruptions from internal or external adversarial sources. Threats in the context of swarm robotics can be executed through goal, behaviour, environment or communication manipulation. Experimental studies in this area are still sparse. We study an attack scenario performed by actively modifying the data between authorised participants. We formulate a robust probabilistic adaptive defence mechanism which does not aim at identifying malicious agents, but to provide the swarm with the means to minimise the consequences of the attack. The mechanism relies on a dynamic modification of the probability of agents to change their current information in view of new contradictory or corroborating incoming data. We investigate several experimental conditions in simulation. The results show that the presence of adversaries in the swarm hinders reaching consensus to the majority opinion when using a baseline method, but that there are several conditions in which our adaptive defence mechanism is highly efficient.

1 Introduction

Swarm robotics is a scientific and engineering field that deals with the design of collective behaviours for a swarm of inexpensive, relatively incapable robots to solve tasks in large and unstructured environments that require scalability, flexibility, and robustness [3]. The robustness of a robot swarm makes it resilient to external random and non-systematic perturbations. However, its flexibility makes it particularly vulnerable to intentional and systematic disruptions from

© Springer Nature Switzerland AG 2018
M. Dorigo et al. (Eds.): ANTS 2018, LNCS 11172, pp. 213–224, 2018.
https://doi.org/10.1007/978-3-030-00533-7_17

an adversarial source. Some of the literature – especially in the military field – has argued that swarm disruption through insertion of adversarial agents is even more cost-effective than traditional weaponised means [17]. This has obvious consequences also for non-military scenarios, with swarm-like technologies increasingly spread in the Internet of Things infrastructures with associated security problems, see e.g. [4].

In the general context of security, many attack scenarios are possible: Denial of Service (DoS), placement of physical barriers, illegitimate impersonation of identity, system penetration, authorisation violation, planting, eavesdropping, modification of data in transit, and so on [14]. Threats to swarm architectures have been recently investigated, see e.g. [7,16]. In [15] four different types of manipulation are illustrated for the specific context of swarm robotics: individual's goal manipulation, individual behaviour manipulation, environment manipulation, and communication manipulation. The possible attack vectors used to carry out such forms of manipulation are: replay, where the attack is performed by playing back past messages to the same recipients [2,6,19,22]; physical Capture and Tampering, where a unit of the swarm is taken over by the attacker and data extracted or added [1]; software or firmware attacks, where code is modified or vulnerabilities exploited [13]; internal or external communications attacks, aiming at eavesdropping, data blocking or data modification; reconnaissance, to prepare for one of the above types of attacks.

Within this extensive taxonomy for swarm attacks, only few scenarios have been so far experimentally studied. The study in [16] considers the cooperative navigation approach presented in [5] and two types of attacks: DoS and communication manipulation of data related to distance to target and freshness of the data transmitted. In their analysis, 10 attackers are statically positioned in the environment to perform communication jamming in 1% of the cases and for a limited time period. The results on the DoS attack show only that the success of the attack is inversely proportional to the number of the non-malicious entities, and that the average time to find the intended target is proportional to the number of attackers who act collaboratively when not attacking. No method is proposed for the swarm to counter the attack. Recently, a simulated model of a trust and reputation protocol has been presented to mitigate black hole style attacks on VANETs, see [10]. These are data modification attacks by members of the swarm who block data transmission or perform false data distribution. This study does not meet the swarm robotics assumptions, as the defined counter attack strategy identifies the best optimal configuration, including network properties (e.g. size and topology), population distribution (e.g. proportion of the attackers) and contextual conditions (e.g. the current content coverage). An interesting recent study has proposed an approach based on the blockchain to achieve collective decision-making in a swarm of robots by considering Byzantine attacks, that is, attacks of any types [18].

In this paper, we focus specifically on a *communication manipulation* using *modification of data in transit* attack scenario, i.e. a threat performed by actively modifying the data sent between authorised participants. We do not aim at the identification of the malicious agents. Instead, we propose a robust probabilistic adaptive defence mechanism to provide the swarm with the means to minimise the consequences of the attack. In the proposed scenario, a given piece of information needs to be propagated from few individuals to the rest of the swarm, while members of another subset of the swarm communicate the wrong piece of information. This scenario can be modelled as the most basic form of the best-of-n problem in the context of collective decision-making [20], where agents do not have a way to evaluate the quality of the two pieces of information (symmetric quality), and the environment is also symmetric with respect to cost of accessing the two options (symmetric costs). In such basic symmetry breaking scenario, the predicted outcome is usually consensus to the piece of information held by the majority [9, 21]. However, in this paper we will show that the simplest addition of adversarial agents will hinder reaching consensus to the majority opinion, and in the case in which the vanilla voter model [21] is used the system will not achieve consensus at all. A recent general review of collective decision making [8] pointed out the interesting fact that, although the swarm always owns more knowledge than the sum of its parts, the emergence of this knowledge depends from the specific mechanism used by each individual to integrate the information of its neighbours. To achieve consensus on the piece of information held by the majority, we propose a new version of the voter model where each member of the swarm has a probability p to change its individual knowledge state based on new incoming messages from the neighbours. In particular, we describe and test a *non-adaptive probabilistic defence mechanism* where $p \leq 1$ is shared by all the agents of the swarm and remains fixed for the entire agent's life; and an *adaptive probabilistic defence mechanism* where p is initially set to 1 for all agents, but then, during the course of the simulation, each agent adjusts this individual probability according to information updating rules based on communication among the agents. In the context of the block-chain decision-making work mentioned above [18], our work differs in that the best collective decision is not determined by constraints in the environment, rather it results from the knowledge held by the majority. The results of our simulations indicate that the adaptive probabilistic defence mechanism can be largely effective in multiple contexts differing for the proportion of malicious agents carrying out the attack.

The rest of the paper is organized as follows: in Sect. 2, we describe the simulation model. In Sect. 3, we provide experimental results where we compare the non-adaptive probabilistic defence mechanism against its adaptive version: here, we analyse the performance in terms of number of attackers, and by varying the parameters of the defence mechanism (the static probability in the non-adaptive version and the update rates for the adaptive version). In Sect. 4, we conclude and we highlight our plan of improvement and future work.

2 The Model

In this section, we describe the type of attack of interest and the probabilistic defence mechanisms deployed to limit its effects. Our model focuses on the communication protocol, while the actual target of the swarm is irrelevant to both the attack performed and the defence mechanism deployed.

Our simulated world is a toroidal grid of dimensions 30×30 in which a swarm of 100 agents moves randomly. At every timestep, each agent occupies a cell. The agents differ in terms of the information they hold at the beginning of the simulation. The *receivers* are agents committed to a piece of information labelled 0. The *discoverers* are agents committed to a piece of information labelled 1. The *attackers* are agents committed to information labelled -1. While the *receivers* can change the content of their information to either 1 or -1 based on the communication that regulates the agents' interaction, discoverers and attackers never change the content of their information during the simulation. The objective of the discoverers is to disseminate the correct piece of information to the receivers. The attackers create a hostile environment since they disseminate a wrong piece of information (i.e. -1) to the receivers, thus preventing the swarm from converging to the desired consensus in which all receivers hold the correct piece of information (i.e. 1).

In our simulation model, all agents move in the same way by randomly selecting one of the eight possible directions from its cell and a random step length between a minimum step of 0 cells to a maximum step of 3 cells. Communication happens by proximity: at every timestep, every receiver checks for the presence on its cell of any other agents currently holding information 1 or -1. With multiple informed agents on the same cell, a receiver selects one randomly among those that are already committed to either 1 or -1. This can be either a discoverer, an attacker, or another receiver already committed to a piece of information different from 0.

This scenario models a communication manipulation attack, since the attackers disseminate information labelled -1 knowing that the correct piece of information is 1. Note that for the receivers both 1 and -1 are equally likely to represent the correct piece of information. A receiver committed to either 1 or -1 never gets back to an uncommitted state (i.e. holding information labelled 0). The objective of our study is to test the effectiveness of a probabilistic defence mechanism that allows the swarm to mitigate the disruptive effects of the communication manipulation attack (i.e. receivers committed to -1). We investigate two versions of this defence mechanism: an adaptive and a non-adaptive one. In both mechanisms, when a receiver is uncommitted, it has 100% probability to accept the first information which becomes available, either 1 or -1. In the non-adaptive version, at every next stage of the communication protocol (i.e. every timestep in which a receiver shares the same position with at least one agent committed to either 1 or -1), receivers have a fixed probability p to change their opinion when receiving new information. We investigate experimental conditions in which p is set to 1, 0.5, and 0.001. In the adaptive version of the defence mechanism, p is set to 1 for every receiver at the beginning of the simulation (i.e. timestep 0). Receivers individually change p during their lifetime according to the following: for every confirmation of information received (i.e. receiver holding 1 and receiving 1, or receiver holding -1 and receiving -1), p decreases as $p = p * k$, with $k = [0.4, 0.6, 0.8]$, and for every information received which contradicts the currently held one (i.e. receiver holding -1 and receiving 1, or receiver holding 1 and receiving -1), p increases as $p = \frac{p}{z}$, with $z = [0.4, 0.6, 0.8, 1.0]$.

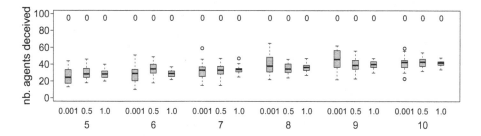

Fig. 1. Graphs showing the number of deceived agents for all experimental conditions with receivers using the non-adaptive defence mechanism. The first row of labels on the x-axis refers to the values of p; the second refers to the number of attackers. Each box is made of 20 points (i.e. 20 runs), with each point referring to the number of deceived agents after 50.000 timesteps. The number of runs (out of 20) that reached at least once the desired consensus is indicated above each box.

3   Results

For the non-adaptive defence mechanism, we collected data on a set of 18 experimental conditions corresponding to all the possible combinations given by three values for the parameter $p = [0.001, 0.5, 1.0]$, and by the use of six swarms with different initial number of attackers (i.e. $5, 6, 7, 8, 9, 10$). For the adaptive defence mechanism, we collected data on a set of 72 experimental conditions corresponding to all the possible combinations given by three values for the parameter $k = [0.8, 0.6, 0.4]$, four values for the parameters $z = [1.0, 0.8, 0.6, 0.4]$, and by the use of six swarms with different initial number of attackers (i.e. $5, 6, 7, 8, 9, 10$). In all experimental conditions, the swarm's size is fixed at 100 agents, and the number of discoverers is fixed at 10. Each simulation run lasts 50.000 timesteps. 20 differently seeded simulation runs are repeated for each experimental condition. At every timestep, the position of each agent is updated according to the navigation mechanism, and the information held by each receiver is updated according to the communication protocol, both explained in Sect. 2. All experiments reported below have been executed on a machine with 64 bit Ubuntu 16.04 system, 40 GB RAM, 4 2.8 GHz cores and using NetLogo 6.0.[1]

3.1   Non-adaptive Probabilistic Defence Mechanism

In this section, we describe the results of the simulations with the non-adaptive probabilistic defence mechanism. We recall that, with this mechanism, uncommitted receivers always accept the first information passed by either a discoverer or an attacker; they have a fixed probability $p = [0.001, 0.5, 1]$ to change their current information if the randomly selected sender among possibly many in the same cell happens to send contradictory data (note that special case $p = 1$

---

[1] The NetLogo code for this model and a C translation used to verify results are both available at https://github.com/gprimiero/swarmattack.

corresponds to the vanilla Voter model, as explained in Sect. 1). We show results which indicate how this defence mechanism is largely non effective in mitigating the distribution of false information by attackers. Results are reported in Fig. 1 where the graph shows the number of deceived agents at the end of each simulation run. Deceived agents are receivers committed to -1.

First of all we note that, as expected, a larger deception diffusion is correlated to a higher number of attackers. The minimum median value is slightly above 20 agents deceived with 5 attackers and $p = 0.001$; the absolute minimum of deceived agents is just above 10 with 6 attackers and $p = 0.001$. The absolute maximum value of deception is above 60 with 8 attackers; the highest median value is above 40 with 9 attackers and $p = 0.001$. On top of being non effective in mitigating the attack, the non-adaptive mechanism, for all three values of p, generates simulation dynamics by which the swarms fail to converge on any of the two consensus points (i.e. either all receivers committed to 1, or all receivers committed to -1, see the numbers on top of each box in Fig. 1). This result has been confirmed by longer simulations, in which each run is executed for 2 millions timesteps (data not shown). We also note that the interquartile range is always at most within 20 points percentage and it tends to decrease with the increase of the value of p. These results suggest that, independently of the number of attackers, facilitating information update (in terms of increasing the probability of changing opinions) stabilizes the infection range. In general, the non-adaptive defence mechanism does not allow to stop the deception perpetrated by the attackers.

3.2 Adaptive Probabilistic Defence Mechanism

Contrary to the non-adaptive defence mechanisms, the adaptive defence mechanism appears to be largely effective in various experimental conditions, particularly when the number of attackers is less than 8. The graphs in Fig. 2 show the number of agents deceived for each value of k, and for each different number of initial attackers. For all values of k, the level of deception tends to decrease with the decrement of the number of initial attackers. However, for $k = 0.8$ more than 60% of the simulation runs (i.e. 49/80, see numbers on top of each box in Fig. 2) managed to converge at least once on the desired consensus point even with 9 initial attackers. Since the experimental conditions with $k = 0.8$ returned the highest number of runs that reached the desired consensus compared to those with $k = 0.4$ and $k = 0.6$, in the following we show the results of further analysis for the adaptive defence mechanisms with $k = 0.8$ only.

With $k = 0.8$ the mechanism is most successful with deception entirely contained when z is set to any value below 1.0 with up to 7 attackers, with $z = [0.4, 0.6]$ with 8 attackers, and also with $z = 0.4$ with 9 attackers (see Fig. 3a). Generally speaking, the results suggest that for $k = 0.8$, lower values of z help to constraint the deception for a larger set of swarms that differ in the initial number of attackers. In other words, making p to increase quicker is a better strategy to prevent deception diffusion in a larger set of swarm differing for the initial number of attackers. For $z = 0.8$ with 8 attackers, and $z = 0.6$ and $z = 0.8$ with 9 attackers, the efficacy of our defence mechanism starts reducing

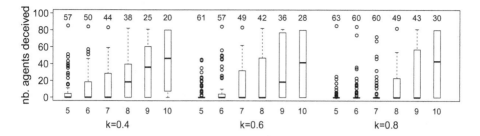

Fig. 2. Graphs showing the number of deceived agents for all experimental conditions with $k = [0.4, 0.6, 0.8]$. The first row of labels on the x-axis refers the number of attackers. Each box is made of 80 points (i.e. 4 values of z times 20 runs for each value), with each point referring to the number of deceived agents after 50.000 timesteps. The number of runs (out of 80) that reached at least once the desired consensus is indicated above each box.

and the swarm converges to the desired consensus point in fewer cases. In particular, for $z = 0.8$ with 8 attackers, and $z = 0.6$ with 9 attackers the swarms tend to converge with almost equal frequency to both consensus points. Distinctive dynamics are generated by the conditions with $z = 1$. As soon as the increasing factor used returns just p, deception does no longer disappear, albeit with values sensibly lower than the swarms with a corresponding number of attackers and any non-adaptive defence mechanism (see Fig. 1). Note also that the number of runs that reached the desired consensus at least once is 0 or very close to 0 in all the conditions with $k = 0.8$ and $z = 1.0$ (see Fig. 3a, numbers above boxes). This suggests that the adaptive defence mechanism can converge to any consensus point, and also effectively operate only when the adaptiveness allow p to decrease in response to evidence matching currently held information, as well as to increase in response to contradictory information.

In Fig. 3b we illustrate for each experimental condition the number of value-change events; that is, the number of times a receiver changes its opinion from 0 to -1 or 1, from 1 to -1, or from -1 to 1. In each run, the number of value-change events is computed from the run start to the timestep at which the first desired consensus is achieved (i.e. all receivers holding 1). This measure is significant because it represents the actual number of interactions in the model and it can be used as a proxy to evaluate the time required by the swarm to reach consensus. Given that our message-passing model is purely symbolic, it does not account for any form of noise. By referring to the value-change events, we can directly account for the number of effectively occurring message passing operations: in a real environment, it is more probable that a given interaction will not occur at all, rather than the operation being nullified by a non-readable, partial or scrambled message. In the limit case, one can consider the latter situation as a non-occurred event, so our measure is a sensible way to express model dynamics even in comparison with non-symbolic implementations. Note that number of value-change events increases with the number of attackers. The absolute minimum value is above 200 with 5 attackers and $z = 0.8$, the absolute maximum

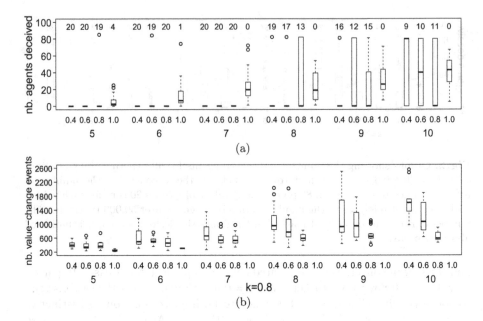

Fig. 3. Graphs showing (a) the number of deceived agents; (b) the number of value-change events before the first desired consensus is reached, for all experimental conditions with $k = 0.8$. The first row of labels on the x-axis refers to the values of z; the second refers to the number of attackers. Each box is made of 20 points (i.e. 20 runs), with each point referring to the number of deceived agents after 50.000 timesteps. In (a), the number of runs (out of 20) that reached at least once the desired consensus is indicated above each box.

value is around 1600 with 9 attackers and $z = 0.4$. This results from the obvious fact that approximating a 1:1 proportion between attackers and discoverers there is a higher dynamics of changes to take into account. Moreover, when p tends to return more quickly to 1 in response to the perception of contradictory information (i.e. for lower values of z), the number of value-change events tends to increase (see Fig. 3b).

Figure 4 presents the historical evolution of the simulation dynamics for the experimental condition in which the number of attackers is 7 and the number of discoverers is 10 (see Fig. 4a and c), and for the experimental condition in which the number of attackers and the number of discoverers is 10 (see Fig. 4b and d). In Fig. 4a and in Fig. 4b, white boxes show how the number of deceived agents changes when the receivers exploit the adaptive defence mechanism with $z = k = 0.8$. The grey boxes show how the number of deceived agents changes when the receivers have a non-adaptive defence mechanism with $p = 0.001$[2].

---

[2] We chose to illustrate the dynamics of the non-adaptive defence mechanism for $p = 0.001$ instead of those generated by $p = 0.5$ and $p = 1.0$ because as shown in Fig. 4c and d, the values of p in the adaptive probabilistic defence mechanism tend to converge to 0.

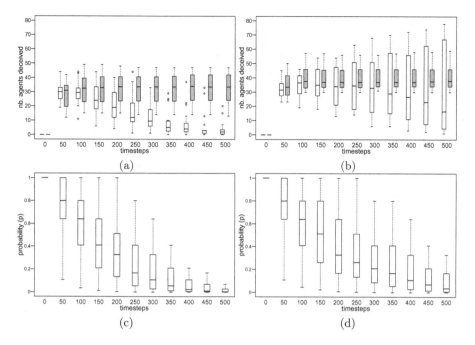

Fig. 4. Graphs showing the number of deceived agents and the probability p of every agent for the condition with 7 attackers (see a and c), and with 10 attackers (see b and d), every 50 timesteps, for 20 runs. In all graphs, white boxes refers to conditions in which the adaptive defence mechanisms operates with $k = 0.8$ and $z = 0.8$. In (a) and (b), the grey boxes refer to the condition in which the probability $p = 0.001$ and receiver exploit the non-adaptive defence mechanism (i.e. $k = 1$ and $z = 1$).

Looking at Fig. 4a, we can notice that there is an initial increase in the number of deceived agents, followed by a quick decrease from step 200 which progresses towards convergence to fully contained deception (see also Fig. 3 for $z = 0.8$ and 7 attackers). The initial increase is to be attributed to the fact that uncommitted receivers commit to any information different from 0, and by the fact that the probability p to change their mind remains for quite few timesteps relatively high for all agents since the adaptive defence mechanism operates with a low decreasing and increasing rate ($z = k = 0.8$). However, the adaptive defence mechanism exploits the initial unbalance between number of attackers and number of discoverers to contain and subsequently to reduce the level of deception into the population, until the desired consensus point is reached (see also Fig. 3a for $z = 0.8$ and 7 attackers). With initial probability value $p = 0.001$ (see Fig. 4a grey boxes), there is an extremely low probability for the receivers to change their information after a first encounter with either a discoverer or an attacker: hence the first information received has a higher effect on the final distribution of deceived agents, which justifies lack of convergence on any consensus point and also the higher variability of level of deception between the runs. In Fig. 4b, the same dynamics is analysed for a scenario with 10 attackers, where one should not expect a high success of the protocol due to the high number of deceiving agents. It clearly appears how the 1:1 proportion between discoverers and attackers at the beginning pushes the deception rate up and it cannot be recovered, with the swarm equally likely to converge on either of the two consensus points (see also Fig. 3a for $z = 0.8$ and 10 attackers). With initial probability value $p = 0.001$ (see Fig. 4b grey boxes), the dynamics remain substantially similar to the condition with 7 attackers, with the swarm unable to converge on any consensus point.

Figure 4c and d present the historical evolution of p (i.e. the probability to change information) for receivers in the same configurations as above: each white box represents therefore the probability of a total of 83×20 receivers in Fig. 4c (i.e. 100 agents minus 7 attackers and 10 discoverers, times 20 runs), and 80×20 agents in Fig. 4d (i.e. 100 agents minus 10 attackers and 10 discoverers, times 20 runs) every 50 timesteps from timestep 0 to timestep 500. For both configurations, it is shown how the p progressively decreases: in Fig. 4c this is due to the progressively high probability of agents to meet other agents with the same opinion and the overall convergence of the swarm towards information labelled 1; in Fig. 4d, a similar progression is happening: this time, due to the balance between attackers and discoverers, swarm's convergence is eventually reached but it can be in any of the two consensus points, as already mentioned above.

As mentioned in the introduction, the non-adaptive mechanism closely resembles the voter model and we have illustrated how the introduction of adversarial agents makes consensus unreachable. The results for the adaptive mechanism introduced in this section show how making opinion change more and more unlikely until unnecessary prevents the diffusion of the attack under appropriate conditions.

4   Conclusions

In this paper we have presented a non-adaptive and an adaptive defence mechanism to mitigate malicious manipulation of communications in swarms of simulated agents. The results indicate that, contrary to its non-adaptive counterpart, the adaptive defence mechanism managed to contain and, in several experimental conditions to suppress, the dissemination of wrong information. Its effectiveness progressively vanishes with the number of malicious agents approaching that of legitimate agents (i.e., the discoverers, see Sect. 2). Nevertheless, by exploring the parameter space, we found values in the adaptive defence mechanisms that proved to be extremely effective in suppressing the deception even in swarms where the number of attackers is very close to the number of discoverers. In all runs in which the swarm managed to reach the desired consensus, the adaptive mechanism exploits even the slighter asymmetries between the number of attackers and the number of discoverers to generate virtuous dynamics that eventually lead to the suppression of the deception in the population.

Given the promising results obtained in this initial set of experiments, various lines for future research work are worth pursuing. It is interesting to explore the effectiveness of the adaptive mechanism in different operating conditions, as generated by varying the proportion of discoverers with respect to the swarm size, or the time when the attack starts with respect to the level of diffusion of the correct information. This last parameter is called *network coverage* and it is investigated for the networked case in [10]. Another possible variation on the present scenario to test the efficacy of the mechanism is given by a different attack vector. A viable possibility is to implement a version of a Gray Hole attack, with the malicious agents performing the data manipulation only for a particular period of time, similarly to what investigated in [16]. It is also our plan to test the adaptive probabilistic defence mechanism with physical robots communicating with different protocols to see how the inherent noise of physical systems and type of communication devices influence its effectiveness. Finally, we plan to situate more formally these models also in the framework of collective decision-making, considering both the case with and without uncommitted agents, see respectively [21] and [11,12].

Acknowledgments. The authors wish to thank Prof. Franco Raimondi for support in setting up the computing cluster required by the experiments in this paper.

References

1. Akdemir, K.D., Karakoyunlu, D., Padir, T., Sunar, B.: An emerging threat: eve meets a robot. In: Chen, L., Yung, M. (eds.) INTRUST 2010. LNCS, vol. 6802, pp. 271–289. Springer, Heidelberg (2011). https://doi.org/10.1007/978-3-642-25283-9_18

2. Aura, T.: Strategies against replay attacks. In: 10th Computer Security Foundations Workshop (CSFW 1997), Rockport, Massachusetts, USA, 10-12 June 1997, pp. 59–69 (1997). https://doi.org/10.1109/CSFW.1997.596787

3. Brambilla, M., Ferrante, E., Birattari, M., Dorigo, M.: Swarm robotics: a review from the swarm engineering perspective. Swarm Intell. **7**(1), 1–41 (2013)

4. Chamoso, P., De la Prieta, F., De Paz, F., Corchado, J.M.: Swarm agent-based architecture suitable for internet of things and smartcities. In: Omatu, S., et al. (eds.) Distributed Computing and Artificial Intelligence, 12th International Conference. AISC, vol. 373, pp. 21–29. Springer, Cham (2015). https://doi.org/10.1007/978-3-319-19638-1_3

5. Ducatelle, F., et al.: Cooperative navigation in robotic swarms. Swarm Intell. **8**(1), 1–33 (2014)

6. Gong, L.: A variation on the themes of message freshness and replay or, the difficulty in devising formal methods to analyze cryptographic protocols. In: Proceedings of the 6th IEEE Computer Security Foundations Workshop - CSFW 1993, Franconia, New Hampshire, USA, 15-17 June 1993, pp. 131–136 (1993). https://doi.org/10.1109/CSFW.1993.246633

7. Higgins, F., Tomlinson, A., Martin, K.: Survey on security challenges for swarm robotics. In: Fifth International Conference on Autonomic and Autonomous Systems (ICAS), pp. 307–312 (2009)

8. Laan, A., Madirolas, G., de Polavieja, G.: Rescuing collective wisdom when the average group opinion is wrong. Front. Robot. AI 4, 1–21 (2017)
9. Montes de Oca, M., Ferrante, E., Scheidler, A., Pinciroli, C., Birattari, M., Dorigo, M.: Majority-rule opinion dynamics with differential latency: a mechanism for self-organized collective decision-making. Swarm Intell. 5, 305–327 (2011)
10. Primiero, G., Martorana, A., Tagliabue, J.: Simulation of a trust and reputation based mitigation protocol for a black hole style attack on VANETs. In: 2018 IEEE European Symposium on Security and Privacy Workshops, EuroS&P Workshops 2018, London, United Kingdom, 23-27 April 2018, pp. 127–135 (2018). https://doi.org/10.1109/EuroSPW.2018.00025
11. Reina, A., Marshall, J.A.R., Trianni, V., Bose, T.: Model of the best-of-N nest-site selection process in honeybees. Phys. Rev. E 95(5), 052411 (2017). https://doi.org/10.1103/PhysRevE.95.052411
12. Reina, A., Valentini, G., Fernández-Oto, C., Dorigo, M., Trianni, V.: A design pattern for decentralised decision making. PLoS ONE 10(10), e0140950 (2015). https://doi.org/10.1371/journal.pone.0140950
13. Roosta, T., Shieh, S., Sastry, S.: Taxonomy of security attacks in sensor networks and countermeasures. In: IEEE International Conference on System Integration and Reliability Improvements, pp. 13–15 (2006)
14. Saljooghinejad, H., Bhukya, W.N.: Layered security architecture for masquerade attack detection. In: Cuppens-Boulahia, N., Cuppens, F., Garcia-Alfaro, J. (eds.) DBSec 2012. LNCS, vol. 7371, pp. 255–262. Springer, Heidelberg (2012). https://doi.org/10.1007/978-3-642-31540-4_19
15. Sargeant, I., Tomlinson, A.: Review of potential attacks on robotic swarms. In: Bi, Y., Kapoor, S., Bhatia, R. (eds.) IntelliSys 2016. LNNS, vol. 16, pp. 628–646. Springer, Cham (2018). https://doi.org/10.1007/978-3-319-56991-8_46
16. Sargeant, I., Tomlinson, A.: Maliciously manipulating a robotic swarm. In: Proceedings of the International Conference Embedded Systems, Cyber-Physical Systems and Applications (ESCS), pp. 122–128 (2016)
17. Scharre, P.: Robotics on the battlefield part II: the coming swarm. Technical report, Centre for a New American Security (2014)
18. Strobel, V., Castello, F., Dorigo, M.: Managing byzantine robots via blockchain technology in a swarm robotics collective decision making scenario. Technical report TR/IRIDIA/2017-013, IRIDIA, Université Libre de Bruxelles, Brussels, Belgium (2017)
19. Syverson, P.F.: A taxonomy of replay attacks. In: Proceedings of the Seventh IEEE Computer Security Foundations Workshop - CSFW 1994, Franconia, New Hampshire, USA, 14-16 June 1994, pp. 187–191 (1994). https://doi.org/10.1109/CSFW.1994.315935
20. Valentini, G., Ferrante, E., Dorigo, M.: The best-of-n problem in robot swarms: formalization, state of the art, and novel perspectives. Front. Robot. AI 4, 9 (2017). https://doi.org/10.3389/frobt.2017.00009. https://www.frontiersin.org/article/10.3389/frobt.2017.00009
21. Valentini, G., Ferrante, E., Hamann, H., Dorigo, M.: Collective decision with 100 Kilobots: speed versus accuracy in binary discrimination problems. Auton. Agents Multi-Agent Syst. 30(3), 553–580 (2016)
22. van Tilborg, H., Jajodia, S. (eds.): Encyclopedia of Cryptography and Security. Springer, Heidelberg (2011)

Task-Agnostic Evolution of Diverse Repertoires of Swarm Behaviours

Jorge Gomes[1,2,6(✉)] and Anders Lyhne Christensen[2,3,4,5]

[1] BioISI, Faculdade de Ciências da Universidade de Lisboa, Lisbon, Portugal
jmgomes@fc.ul.pt
[2] BioMachines Lab, Lisbon, Portugal
[3] Instituto Universitário de Lisboa (ISCTE-IUL), Lisbon, Portugal
[4] Instituto de Telecomunicações, Lisbon, Portugal
[5] Maersk McKinney Moller Institute,
University of Southern Denmark, Odense, Denmark
[6] Sonodot Ltd., London, UK

Abstract. Quality diversity algorithms are evolutionary algorithms that aim to evolve diverse repertoires of high-quality solutions. Quality diversity has recently been used with considerable success to evolve repertoires of single-robot controllers in a wide range of applications. In this paper, we propose a methodology for the evolution of repertoires of general swarm behaviours. We use a quality diversity algorithm that relies on a behaviour characterisation and a quality metric that are task-agnostic, meaning that the repertoire evolution is not driven towards solving any specific task. We use a total of eight swarm robotics tasks to evaluate the behaviours contained in the evolved repertoires a-posteriori, and compare their performance with direct task-specific evolution. We show that the repertoires contain a wide diversity of swarm behaviours, and for most of the tasks, the behaviours in the repertoire have a performance close to the performance achieved by task-specific evolution.

1 Introduction

Swarm robotics systems (SRS) represent an approach to collective robotics, in which large groups of relatively simple and autonomous robots display collectively intelligent behaviours [5]. Control in a SRS is decentralised, meaning that each individual robot operates based on its local observations of the environment and coordination with neighbouring robots. During task execution, the swarm-level behaviour emerges from the interactions between neighbouring robots, and from the interactions between robots and the environment. SRS can display a number of desirable properties, such as robustness, flexibility, and scalability, and thus have a considerable potential in several real-world domains, such as search and rescue, exploration, surveillance, and clean up [1,3,12].

A key challenge in designing SRS is the synthesis of behavioural control [19]. Manual design of control for each robot requires the decomposition of the swarm-level behaviour into individual behavioural rules that lead to the

© Springer Nature Switzerland AG 2018
M. Dorigo et al. (Eds.): ANTS 2018, LNCS 11172, pp. 225–238, 2018.
https://doi.org/10.1007/978-3-030-00533-7_18

desired self-organized behaviour. As an alternative, a number of automatic and semi-automatic design methods have been proposed. In AutoMoDe [18,19], for instance, robot control is produced by selecting, instantiating, and combining pre-existing parametric modules, through an optimisation process. Gauci et al. [20,21], on the other hand, have shown how to produce minimalistic controllers by optimising a small number of selected control parameters. The majority of works in automatic design have, however, relied on the evolution of neural network controllers, an approach known as evolutionary swarm robotics [38,39].

Evolutionary algorithms can exploit the intricate dynamics of self-organised behaviours [28], and have thus been used to produce control for a wide variety of swarm robotics tasks [3,38]. While, traditionally, evolutionary algorithms are driven towards solutions for a specific task, based on a task-specific fitness function, more recent works have shown that novelty-driven evolutionary algorithms are a valuable tool for evolutionary robotics [9,11,35], including swarm robotics applications [4,22,27]. Algorithms such as novelty search [31] and quality diversity [7,37] work by rewarding *behavioural novelty* instead of scoring solutions solely based on task performance. The novelty of a solution corresponds to its behavioural difference with respect to the solutions that have been evolved so far. The behavioural difference is calculated based on a behaviour characterisation that captures how the robots interact with one another and the environment.

Quality diversity algorithms, including Novelty Search with Local Competition (NSLC) [32], MAP-Elites [34,40], and derived techniques [7], try to find the highest-quality solutions for different regions of the behaviour space, meaning they can be used to build repertoires of diverse and high-quality solutions. Quality diversity has been used to evolve repertoires of robot behaviours in a number of domains, including: virtual walking creatures [32]; morphological designs for walking soft robots [34]; locomotion behaviours for legged robots [6–8,15,36,40] and four-wheeled steering robots [14,15]; robotic arm behaviours [7,30,34]; and controllers for maze-navigation tasks [37,40,41]. Studies have shown diverse repertoires to be valuable for online behaviour adaptation [6], hierarchical control [15], among other uses [17].

In this paper, we propose an approach for the evolution of repertoires of *general swarm behaviours* using a quality diversity evolutionary algorithm. The behaviour characterisation and the quality metric used in the repertoire evolution are completely task-agnostic, meaning that the repertoire evolution process is not driven towards solving any specific task or set of tasks. The evolutionary process is driven towards the generation of arbitrary but distinct swarm behaviours, thus departing from the vast majority of previous works where automatic design methods are used as means to solve a specific task or achieve a specific swarm behaviour. Automatically generating a diverse set of basic and general swarm behaviours, as we propose in this paper, can facilitate the synthesis of controllers for complex tasks. Complex swarm robotics behaviours are often achieved by combining several simpler swarm behaviours, such as aggregation and navigation [1,3,13,19]. Having a large repertoire of such basic behaviours allows the achievement of more capable controllers by selecting and combining behaviours from that repertoire [17,23,26].

In a recent study, Engebråten et al. [17] evolved repertoires of swarm behaviours using the MAP-Elites algorithm [34]. The behaviour of each controllers is, however, characterised according to its performance in two pre-defined tasks. This means that the repertoire contains diverse solutions for those specific tasks, not a diverse set of general swarm behaviours as we propose in this paper. In another recent study, Brown et al. [4] use novelty search [31] to discover the behavioural possibilities for a swarm of robots with extremely limited capabilities. The robots used in our study are considerably more capable, and we are concerned with evolving repertoires of useful and high-quality behaviours, rather than exclusively unveiling the behavioural capabilities of the swarm.

2   Methodology

In our approach, each candidate solution in the evolutionary process corresponds to a robot controller that is copied to all robots in a swarm. A robot controller is a neural network that continually receives the current sensory inputs of a robot and outputs the robot's actuator values. During evolution, each controller is evaluated in simulation by executing the controller on a swarm for a fixed amount of time, in a number of different simulations. In these simulations, the behaviour of the swarm as a whole is characterised using a task-agnostic behaviour characterisation [10]. This behaviour characterisation, together with a task-agnostic quality metric, is used to drive the quality diversity evolutionary process. This means that the evolutionary process is driven towards novel high-quality swarm behaviours, without regard to any specific task.

We follow the quality diversity framework proposed by Cully et al. [7], that treats the selection of novel and promising solutions (*selection*) and the construction of the repertoire (*container*) as two independent steps. Our evolutionary process is based on Novelty Search with Local Competition (NSLC) [32], which has shown promising results as a quality diversity technique [7,8,32,37]. The procedure used for the generation of a repertoire is summarised in Algorithm 1.

2.1   Novelty Search with Local Competition

NSLC [32] is an extension of novelty search [31], in which both the quality and the behavioural novelty of the individuals is rewarded. The evolutionary process is a Pareto-based multi-objective evolutionary algorithm, where the two objectives are respectively, the behavioural novelty score and the local competition score. The *novelty score* is calculated in the same way as in novelty search [25,31], with the novelty score corresponding to mean behaviour distance to the k-nearest neighbours (Algorithm 1, step 14), encompassing both the individuals from the current population and individuals from an archive (step 13). The archive is composed of individuals randomly selected during evolution [25] (steps 16–18), and represents a sample of what has been evolved so far. The *local competition score*, on the other hand, corresponds to the number of individuals in the same k-nearest neighbours that are outperformed by the individual currently under evaluation, with respect to a provided quality metric (step 15).

Algorithm 1 Repertoire evolution with NSLC and a separate container.

1: Let S be the maximum archive size, and s the generational growth.
2: Let ℓ be the container distance threshold, and ϵ a constant in $[0, 1]$.
3: Let E be a set of randomly generated environments.
4: $\mathcal{A} \leftarrow \emptyset$, $\mathcal{C} \leftarrow \emptyset$ ▷ *Novelty archive, Container*
5: $\mathcal{P} \leftarrow \texttt{RandomInitialPopulation}()$ ▷ *Population*
6: **for each** generation **do**
7:     **for each** individual $i \in \mathcal{P}$ **do**   ▷ *Evaluate all the individuals in simulation*
8:         **for each** environment $e \in E$ **do**
9:             $q_e, \mathbf{b}_e \leftarrow \texttt{Evaluate}(i, e)$ ▷ *Simulate i in the environment e*
10:         $B(i) \leftarrow \texttt{GeometricMedian}(\{\mathbf{b}_e : e \in E\})$ ▷ *Median behaviour characterisation*
11:         $Q(i) \leftarrow \sum_{e \in E} \frac{1}{|E|} \cdot q_e$ ▷ *Mean quality score*
12:     **for each** individual $i \in \mathcal{P}$ **do**   ▷ *Calculate novelty and LC scores*
13:         $\mathcal{X} \leftarrow \texttt{NearestNeighbours}(i, \mathcal{A} \cup \mathcal{P}, k)$ ▷ *k-nearest neighbours in archive and pop.*
14:         $N(i) \leftarrow \sum_{x \in \mathcal{X}} \frac{1}{|\mathcal{X}|} \cdot \text{dist}(B(i), B(x))$ ▷ *Novelty score*
15:         $LC(i) \leftarrow \sum_{x \in \mathcal{X}} [Q(i) > Q(x)]$ ▷ *Local competition score*
16:     **if** $|\mathcal{A}| > S - s$ **then**   ▷ *If the archive exceeds its capacity*
17:         $\mathcal{A} \leftarrow \mathcal{A} \setminus \texttt{SelectRandom}(\mathcal{A}, |\mathcal{A}| + s - S)$ ▷ *Remove some individuals*
18:     $\mathcal{A} \leftarrow \mathcal{A} \cup \texttt{SelectRandom}(\mathcal{P}, s)$ ▷ *Update archive with new individuals*
19:     **for each** individual $i \in \mathcal{P}$ **do**   ▷ *Update the container*
20:         $\mathcal{X} \leftarrow \texttt{NearestNeighbours}(i, \mathcal{C}, 2)$ ▷ *2 nearest neighbours in container*
21:         **if** $|\mathcal{C}| > 0 \wedge \text{dist}(B(i), B(\mathcal{X}_1)) > \ell$ **then**   ▷ *Different from most similar in container*
22:             $\mathcal{C} \leftarrow \mathcal{C} \cup \{i\}$ ▷ *Add new to container*
23:         **else if** $|\mathcal{C}| > 1 \wedge Q(i) > Q(\mathcal{X}_1) \wedge \text{dist}(B(i), B(\mathcal{X}_2)) > \ell \times (1 - \epsilon)$ **then**
24:             $\mathcal{C} \leftarrow (\mathcal{C} \setminus \{\mathcal{X}_1\}) \cup \{i\}$ ▷ *Better than a similar one in container, replace*
25:     $\mathcal{P} \leftarrow \texttt{Breed}(\mathcal{P})$ based on the scores $N(i)$ and $Q(i)$ ▷ *Create next generation*
26: **return** $\mathcal{C}$

2.2   Container

The use of a novelty archive composed of randomly selected individuals (steps 16–18) encourages uniform behaviour exploration [7,25]. This archive construction process, however, does not ensure that the highest quality solutions for each behaviour region will be present in the archive. To circumvent this limitation, we use a separate *container* to collect the highest-performing solutions during the evolutionary process, as proposed by Cully et al. [7]. The container is composed of the highest-quality individuals for each behaviour region.

In our implementation, the container does not interfere with the evolutionary process, as suggested by Cully et al. [7]. The archive composed of randomly selected individuals is used to calculate the novelty and local competition scores, while the container merely gathers the most promising individuals. If an individual is significantly different from the others already in the container, it is added to the container (steps 21–22). Otherwise, the individual is compared to the two nearest neighbours in the container. If it has a higher quality than the nearest neighbour, and it is significantly different from the second nearest neighbour, it replaces the nearest neighbour in the container (steps 23–24). This approach is similar to the container implementation proposed in [7], with the difference that we use simple quality domination instead of the proposed ϵ-domination based on both quality and novelty scores.[1]

---

[1] Preliminary experiments revealed that quality domination yielded repertoires composed of higher-quality solutions, and with a similar behaviour space coverage.

2.3 Behaviour Characterisation

A number of task-agnostic behaviour characterisations for multirobot systems and swarm robotic systems have been proposed in previous works [4,22,24]. In this study, we adopt *Systematically Derived Behaviour Characterisations* (SDBC) [24]. With the SDBC approach, the characterisation is automatically derived from the environment's elements, measuring spatial distances between the different elements of the environment, and the state of the robots over time.

The evaluation of each controller (Algorithm 1, step 9) has the objective of characterising the respective general swarm behaviour, that is, how the swarm performs in an arbitrary environment. To this end, each controller is evaluated in a large number of independent trials (50 in our experiments), where the swarm size is varied, as well as the initial positions of the robots and points-of-interest in the environment. In each trial, the robots in the swarm operate for a fixed amount of time (100 s in our experiments) using copies of the controller under evaluation, and the behaviour of the swarm is characterised. The behaviour characterisations obtained in all trials are then combined to obtain a single characterisation that is representative of the swarm's typical behaviour. We resort to the geometric median for this combination (step 10), a robust estimator of location [2].

In our experiments, considering that the environment contains walls, robots, and one point of interest (POI) (see Sect. 3.1), the following behaviour features were derived: mean distance from a robot to (i) all other robots; (ii) the closest robot; (iii) the walls; (iv) and the POI; (v) robot's mean linear speed; and (vi) robot's mean turn speed. Each of these features is measured taking into account all robots of the swarm, during the entire simulation. The behaviour characterisation is then composed of the mean value of each feature, as well as the standard deviation of the feature [24], which indicates how much the feature varied during the simulation trial. Each behaviour characterisation is thus a real-valued vector of length 12. All features were scaled to approximately $[-1, 1]$, using scaling factors obtained beforehand through the generation of a large number of random environment instances.

2.4 Quality Metric

The quality metric is used as a local objective in the quality diversity evolutionary process. That is, given two candidate solutions with a similar behaviour characterisation, solution quality dictates which one is preferred. In our approach, the quality metric is not tied to any specific task. Instead, the quality metric should consider transversal behavioural traits, such as: the number of collisions while performing a behaviour, the energy efficiency of the given behaviour, its transferability to physical robots [8], scalability, robustness, and so on. For this study, we chose the number of collisions as the quality metric, considering both robot-robot collisions and robot-obstacle collisions. A quality metric based on the number of collisions is straightforward to implement, and represents a behavioural trait that is generally desirable in any swarm behaviour, due to practical concerns and safety. The quality of each controller is obtained by averaging the quality score obtained in each of the simulation trials (Algorithm 1, step 11). Considering a swarm of size S, a simulation trial of length T, and c as the total number of collisions, the quality score in a given trial is given by:

$$Q = 1 - c/(T \times S) \tag{1}$$

3   Experimental Setup

3.1   Domain

We defined a simulation-based setup that allowed us to evolve the repertoire of behaviours and to implement a broad range of evaluation tasks, see Fig. 1. Each robot of the swarm is modelled as a circular object, and is equipped with the following sensors and actuators (see Table 1 for parameters): (i) *Obstacle sensors*: Six ray-based sensors that return the distance to the nearest intersection with a wall (if any); (ii) *Agent sensors*: Eight cone-based sensors evenly distributed around the robot, each returning the distance to the closest robot in the respective circular sector (if any); (iii) *POI sensors*: Four cone-based sensors evenly distributed around the robot, each returning the distance to the POI in the respective circular sector (if any); and (iv) *Differential drive*: The movement of the robot is controlled by two actuators that dictate the left wheel and right wheel speed, respectively. Each wheel can move both backwards and forwards.

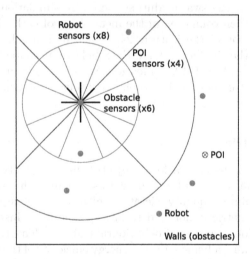

Fig. 1. The experiments are conducted in a 2D environment modelled in the MASON simulator [33]. The environment has a fixed size of 200 × 200 cm and is bounded by impassable walls. The environment can also contain a single *Point-of-Interest* (POI), a non-collidable entity that can be sensed by the robots. The swarm size varies from 5 to 10 robots. The figure depicts an example of an environment, with seven robots and the POI. The sensor ranges are depicted for one of the robots.

Table 1. Experimental parameters. The source code used for the experiments is available at https://github.com/jorgemcgomes/mase/releases/tag/ants18.

Parameter	Value	Parameter	Value	Parameter	Value
Robot and common task parameters					
Environment size	200×200 cm	Obstacle sensor range	15 cm	POI sensors	4 cones
POI sensor range	100 cm	Agent sensors	8 cones	Agent sensor range	50 cm
Max. wheel speed	±5 cm/s	Max. wheel accel.	±5 cm/s^2	Wheel axle length	5 cm
Max. turning speed	≈ 14.3°/s	Control cycle	5 steps/s		
Repertoire evolution: NEAT + NSLC + Container					
Novelty k-nearest	15	Add archive criterion	random	Archive growth (s)	25 inds
Max. archive size (S)	2000	Container threshold ℓ	1	Container ϵ	0.25
Population size	500	Generations	1000	Target species count	10
Recurrency allowed	yes	Survival threshold	20%	Crossover probability	100%
Weight mutation prob.	25%	Add link prob.	5%	Add node prob.	3%
Evaluation samples	50	Evaluation length	100 s		
Direct evolution: NEAT (only different)					
Population size	200	Target species count	5	Evaluation samples	10

3.2   Evaluation Tasks

We defined a number of different tasks based on canonical swarm robotics tasks [1,3,12] to assess the versatility of the evolved repertoires. In all simulation runs of all tasks, the initial positions and orientations of the robots (and POI if present) are random. The number of robots varies randomly from 5 to 10. The tasks and respective fitness functions are described below. Note that these fitness functions have absolutely no influence on the evolution of the repertoires, as they are only used to assess the repertoires a-posteriori.

Aggregation: The swarm has the objective of aggregating anywhere in the arena. The fitness function is inversely proportional to the average distance to the centre of mass over the entire simulation. $T = 150$ s (maximum simulation time).

Clustering: Similar to aggregation, but the fitness function uses the number of clusters instead. Clusters are defined by the single-linkage criterion, with a maximum distance of 15 cm between each two robots. $T = 150$ s.

Coverage: The swarm has the objective of covering the arena as evenly as possible. The arena is discretised into a grid of 10 × 10, and every time a cell is visited by a robot, the value of the cell goes to 1, and then decays constantly at a rate of 0.005/s. The fitness is the average value of all cells over the entire simulation. $T = 200$ s.

Border coverage: Similar to coverage, but only the cells immediately next to a wall are used for the fitness calculation. This means that the swarm is rewarded for good coverage of the space near the walls. $T = 200$ s.

Dispersion: The robots should stay as far as possible from the closest neighbouring robot. The fitness is proportional to the average distance to the nearest neighbour, averaged over the entire simulation. $T = 100$ s.

Phototaxis: The swarm should get as close as possible to the POI present in the arena. The fitness is inversely proportional to the mean distance of the robots to the POI, averaged over the simulation. $T = 100$ s.

Dynamic phototaxis: Similar to Phototaxis, but the POI moves randomly inside the arena, with a speed of 75% the maximum speed of the robots, and changing direction every time it hits a wall. $T = 150\,\mathrm{s}$.

Flocking: The fitness function rewards robots for having an orientation similar to the other robots within a radius of 25 cm (half the robot sensing range), and for moving as fast as possible. The swarm is therefore encouraged to aggregate and keep moving in flocks, the bigger the flocks the better. $T = 200\,\mathrm{s}$.

In all tasks, the swarm is penalised for robot-robot collisions and robot-wall collisions. This is achieved by multiplying the task-specific fitness function (F_T) by a collision coefficient (F_C):

$$F = F_T \times F_C, \quad F_C = \max\left(0, \, 1 - \frac{c}{S \times T \times C_{max}}\right), \tag{2}$$

where c is the total number of collisions, S is the number of robots, T the maximum simulation time for that task, and C_{max} is a factor that controls the maximum number of collisions allowed, set to 0.1 in our experiments.

3.3   Evolutionary Setup

Repertoire Evolution. As described in Sect. 2.1, the repertoire evolution is driven by NSLC, and the repertoire is built using a *container* during the evolutionary process, see Sect. 2.2. The environment used for the repertoire evolution is similar to the environment of the evaluation tasks: the number of robots is randomly varied from 5 to 10; there is always one POI somewhere in the arena; and each of the 50 simulations lasts 100 s. Note that the tasks and fitness functions described in Sect. 3.2 are not used in any way during the repertoire evolution. The parameters of the evolutionary process are listed in Table 1.

Direct Evolution. For establishing a comparison, we evolve controllers specifically for each of the tasks. The performance obtained by the task-specific controllers is used as an upper bound for the performance achievable in each task. These task-specific controllers are evolved using the NEAT neuroevolution algorithm, with parameters similar to the repertoire evolution, but the individuals are scored exclusively by the fitness function of the respective task. For each task, we conduct 30 evolutionary runs, each for 500 generations.

Randomly Generated Repertoire. The quality of the controllers from the evolved repertoires is additionally compared with controllers from randomly generated repertoires. The performance achieved in each task by these repertoires establishes a lower bound of performance that helps gauging the difficulty of each task and the effectiveness of the evolutionary process. This lower bound does not necessarily correspond to the lowest performance obtainable, but rather to the performance that can be trivially obtained with random exploration of the controller parameter space. The random repertoires are generated by creating 1000 three-layer feed-forward neural networks, each with a number of hidden neurons varying from 0 to 20, and with all connections assigned random weights.

4   Results

4.1   Comparison with Direct Evolution

We repeated the repertoire evolution 10 times, thus obtaining 10 evolved repertoires. The mean size of the repertoires was 981 ± 34 controllers, and the mean quality score was 0.99 ± 0.03, meaning that collisions were rare in the vast majority of behaviours. The best-of-generation controllers evolved with direct evolution, and the controllers in all repertoires, were re-evaluated in the tasks using 100 simulation runs per controller, see Fig. 2. In all tasks, the evolved repertoires significantly outperform the randomly generated repertoires, which confirms that the evolutionary process is discovering diverse high-quality solutions, and that none of the tasks are trivial. Comparing with the performance achieved by direct task-oriented evolution, we observe three different scenarios:

Aggregation, Phototaxis: the performance achieved by the evolved repertoires is not significantly different from direct evolution (Mann-Whitney U test, $p < 0.05$). That is, driving evolution towards the solution of these tasks was not more effective than the unrestricted exploration of the behaviour space. This result might be explained by the potential deceptiveness of the tasks. As shown in previous works [27], novelty-driven evolution can yield promising results in the evolution of swarm behaviours.

Coverage, Border coverage, Dispersion: there is a significant performance difference, but the achieved fitness values are close to one another in absolute terms. This means that the repertoires contained solutions for solving these tasks, and the practical differences in terms of solution quality are minor.

Clustering, Dynamic phototaxis, Flocking: the difference to direct evolution is more significant with respect to the highest fitness scores achieved, and corresponds to observable performance differences.

The fact that the controllers in the repertoires could not outperform task-oriented evolution was expected: we are comparing general swarm behaviours with swarm behaviours evolved specifically for that task. Nonetheless, it is noteworthy that in most tasks, the repertoires contained controllers that achieved relatively high performance, and could even match the performance obtained with task-oriented evolution in two tasks. These results also show that the employed quality diversity algorithm was effective in behaviour exploration, as the repertoires contained controllers that performed well in fairly different tasks.

4.2   Repertoire Diversity

To gain insight into the diversity of the repertoires, we measured the fitness achieved in each task by every controller of each repertoire. We then averaged

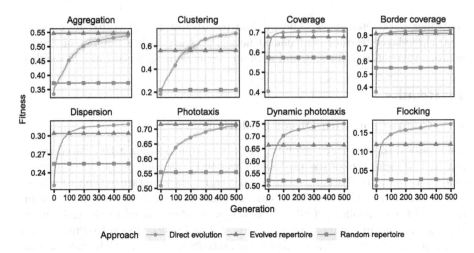

Fig. 2. Fitness achieved in the evaluation tasks with each approach. The fitness values shown for the repertoires correspond to the fitness of the highest-performing controller found in each repertoire for each task.

the fitness scores achieved by the controllers from each behaviour region, in each task. Figure 3 show that most of the regions contain controllers that perform relatively well in some task. This suggests that a significant part of the repertoires is composed of potentially useful controllers. The controllers that work best for

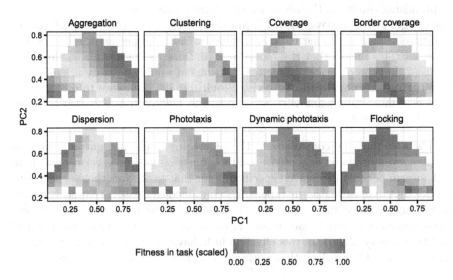

Fig. 3. Mean fitness of the controllers in the evolved repertoires, in each of the evaluation tasks. The behaviour space was reduced from 12 to 2 dimensions using Robust PCA [29] (the two first PCs account for 77% of the variance), and then discretised for calculating the mean fitness of the controllers in each behaviour region.

a given task tend to be focused around a certain region, which varies in size. For instance, only a small subset of controllers perform well in the Flocking task, while a large number of controllers perform well in the Coverage task.

We can also observe that the highest-performing controllers for significantly different tasks belong to significantly different regions. For instance, note that the high-performing regions for Aggregation and Dispersion are disjoint, the high-performing region for Clustering is a subset of the high-performing region of Aggregation, and the high-performing regions for similar tasks overlap: Coverage and Border coverage, Phototaxis and Dynamic phototaxis. These observations suggest that the behaviour characterisation is adequately capturing the controllers' behaviour, and that the behaviour space is at least moderately continuous. That is, controllers close in the behaviour space perform similarly.

5   Conclusions

In this study, we evolved repertoires of general swarm behaviours using a quality diversity algorithm that relied on task-agnostic behaviour characterisations and a task-agnostic quality metric. The evolved repertoires were compared with direct task-oriented evolution in a total of eight swarm robotics tasks. The controllers in the repertoires could closely approximate or match the performance of direct task-oriented evolution in five out of the eight tasks, which is noteworthy given that the repertoire evolution process was conducted without reference to any of the tasks. A single repertoire contained controllers that could solve several different tasks. We showed that the evolved repertoires contained a high diversity of potentially useful controllers. Most of the behaviour regions in the repertoire were found to have controllers that could be useful for solving at least one task.

Diverse repertoires of swarm behaviours have a number of potential applications, which we will explore in future work, alongside validation in real robotic systems. More complex swarm robotics behaviours are commonly built upon elementary swarm behaviours [1,3,12]. Our results suggest that the evolved repertoires contained many of those elementary behaviours, and therefore they could be a good starting point for synthesising hierarchical control [13,16]. EvoRBC [15,26], for instance, is a recently proposed approach for evolving hierarchical control based on repertoires of behaviours. Diverse repertoires can also be used for online adaptation [6] during task execution in response to external factors such as faults and changes in the environment. Additionally, repertoires can also be used to rapidly find solutions for new, unforeseen tasks.

Overall, our study showed that evolving repertoires of general swarm behaviours is possible, and it can yield a wide diversity of high-quality behaviours. Such repertoires can be an important step towards the synthesis of more complex swarm behaviours in a completely automated manner.

Acknowledgments. Work supported by Fundação para a Ciência e a Tecnologia (FCT), Portugal, with grants UID/MULTI/04046/2013 (BioISI), and UID/EEA/50008/2013 (Instituto de Telecomunicações). This work used the EGI infrastructure with the support of NCG-INGRID-PT (Portugal).

References

1. Bayındır, L.: A review of swarm robotics tasks. Neurocomputing **172**, 292–321 (2016)
2. Beck, A., Sabach, S.: Weiszfeld's method: old and new results. J. Optim. Theory Appl. **164**(1), 1–40 (2015)
3. Brambilla, M., Ferrante, E., Birattari, M., Dorigo, M.: Swarm robotics: a review from the swarm engineering perspective. Swarm Intell. **7**(1), 1–41 (2013)
4. Brown, D.S., Turner, R., Hennigh, O., Loscalzo, S.: Discovery and exploration of novel swarm behaviors given limited robot capabilities. In: Groß, R., et al. (eds.) Distributed Autonomous Robotic Systems. SPAR, vol. 6, pp. 447–460. Springer, Cham (2018). https://doi.org/10.1007/978-3-319-73008-0_31
5. Şahin, E.: Swarm robotics: from sources of inspiration to domains of application. In: Şahin, E., Spears, W.M. (eds.) SR 2004. LNCS, vol. 3342, pp. 10–20. Springer, Heidelberg (2005). https://doi.org/10.1007/978-3-540-30552-1_2
6. Cully, A., Clune, J., Tarapore, D., Mouret, J.B.: Robots that can adapt like animals. Nature **521**(7553), 503–507 (2015)
7. Cully, A., Demiris, Y.: Quality and diversity optimization: a unifying modular framework. IEEE Trans. Evol. Comput. **22**(2), 245–259 (2017)
8. Cully, A., Mouret, J.B.: Evolving a behavioral repertoire for a walking robot. Evol. Comput. **24**(1), 59–88 (2016)
9. Doncieux, S., Bredeche, N., Mouret, J.B., Eiben, A.E.G.: Evolutionary robotics: what, why, and where to. Front. Robot. AI **2**, 4 (2015)
10. Doncieux, S., Mouret, J.B.: Behavioral diversity measures for evolutionary robotics. In: IEEE Congress on Evolutionary Computation (CEC), pp. 1–8. IEEE Press (2010)
11. Doncieux, S., Mouret, J.B.: Beyond black-box optimization: a review of selective pressures for evolutionary robotics. Evol. Intell. **7**(2), 71–93 (2014)
12. Duarte, M., et al.: Evolution of collective behaviors for a real swarm of aquatic surface robots. PLoS ONE **11**(3), e0151834 (2016)
13. Duarte, M., Gomes, J., Costa, V., Oliveira, S.M., Christensen, A.L.: Hybrid control for a real swarm robotics system in an intruder detection task. In: Squillero, G., Burelli, P. (eds.) EvoApplications 2016. LNCS, vol. 9598, pp. 213–230. Springer, Cham (2016). https://doi.org/10.1007/978-3-319-31153-1_15
14. Duarte, M., Gomes, J., Oliveira, S.M., Christensen, A.L.: EvoRBC: evolutionary repertoire-based control for robots with arbitrary locomotion complexity. In: Genetic and Evolutionary Computation Conference (GECCO), pp. 93–100. ACM Press (2016)
15. Duarte, M., Gomes, J., Oliveira, S.M., Christensen, A.L.: Evolution of repertoire-based control for robots with complex locomotor systems. IEEE Trans. Evol. Comput. **22**(2), 314–328 (2018)
16. Duarte, M., Oliveira, S.M., Christensen, A.L.: Evolution of hierarchical controllers for multirobot systems. In: International Conference on the Synthesis & Simulation of Living Systems, pp. 657–664. MIT Press (2014)
17. Engebråten, S.A., Moen, J., Yakimenko, O., Glette, K.: Evolving a repertoire of controllers for a multi-function swarm. In: Sim, K., Kaufmann, P. (eds.) EvoApplications 2018. LNCS, vol. 10784, pp. 734–749. Springer, Cham (2018). https://doi.org/10.1007/978-3-319-77538-8_49
18. Francesca, G., et al.: AutoMoDe-Chocolate: automatic design of control software for robot swarms. Swarm Intell. **9**(2–3), 125–152 (2015)

19. Francesca, G., Brambilla, M., Brutschy, A., Trianni, V., Birattari, M.: AutoMoDe: a novel approach to the automatic design of control software for robot swarms. Swarm Intell. **8**(2), 89–112 (2014)
20. Gauci, M., Chen, J., Li, W., Dodd, T.J., Gross, R.: Clustering objects with robots that do not compute. In: Proceedings of the 2014 International Conference on Autonomous Agents and Multi-agent Systems, pp. 421–428. International Foundation for Autonomous Agents and Multiagent Systems (2014)
21. Gauci, M., Chen, J., Li, W., Dodd, T.J., Groß, R.: Self-organized aggregation without computation. Int. J. Robot. Res. **33**(8), 1145–1161 (2014)
22. Gomes, J., Christensen, A.L.: Generic behaviour similarity measures for evolutionary swarm robotics. In: Genetic and Evolutionary Computation Conference (GECCO), pp. 199–206. ACM Press (2013)
23. Gomes, J., Christensen, A.L.: Comparing approaches for evolving high-level robot control based on behaviour repertoires. In: IEEE Congress on Evolutionary Computation (2018). (in Press)
24. Gomes, J., Mariano, P., Christensen, A.L.: Systematic derivation of behaviour characterisations in evolutionary robotics. In: International Conference on the Synthesis and Simulation of Living Systems (ALife), pp. 212–219. MIT Press (2014)
25. Gomes, J., Mariano, P., Christensen, A.L.: Devising effective novelty search algorithms: a comprehensive empirical study. In: Genetic and Evolutionary Computation Conference (GECCO), pp. 943–950. ACM Press (2015)
26. Gomes, J., Oliveira, S.M., Christensen, A.L.: An approach to evolve and exploit repertoires of general robot behaviours. Swarm Evol. Comput. (2018). https://doi.org/10.1016/j.swevo.2018.06.009. (in Press)
27. Gomes, J., Urbano, P., Christensen, A.L.: Evolution of swarm robotics systems with novelty search. Swarm Intell. **7**(2–3), 115–144 (2013)
28. Hamann, H.: Swarm Robotics: A Formal Approach. Springer, Heidelberg (2018). https://doi.org/10.1007/978-3-319-74528-2
29. Hubert, M., Rousseeuw, P.J., Vanden Branden, K.: ROBPCA: a new approach to robust principal component analysis. Technometrics **47**(1), 64–79 (2005)
30. Kim, S., Doncieux, S.: Learning highly diverse robot throwing movements through quality diversity search. In: Proceedings of the Genetic and Evolutionary Computation Conference Companion (GECCO), pp. 1177–1178. ACM Press (2017)
31. Lehman, J., Stanley, K.O.: Abandoning objectives: evolution through the search for novelty alone. Evol. Comput. **19**(2), 189–223 (2011)
32. Lehman, J., Stanley, K.O.: Evolving a diversity of virtual creatures through novelty search and local competition. In: Genetic and Evolutionary Computation Conference (GECCO), pp. 211–218. ACM Press (2011)
33. Luke, S., Cioffi-Revilla, C., Panait, L., Sullivan, K., Balan, G.: MASON: a multi-agent simulation environment. Simulation **81**(7), 517–527 (2005)
34. Mouret, J., Clune, J.: Illuminating search spaces by mapping elites. CoRR abs/1504.04909 (2015). http://arxiv.org/abs/1504.04909
35. Mouret, J.B., Doncieux, S.: Encouraging behavioral diversity in evolutionary robotics: an empirical study. Evol. Comput. **20**(1), 91–133 (2012)
36. Nordmoen, J., Ellefsen, K.O., Glette, K.: Combining MAP-elites and incremental evolution to generate gaits for a mammalian quadruped robot. In: Sim, K., Kaufmann, P. (eds.) EvoApplications 2018. LNCS, vol. 10784, pp. 719–733. Springer, Cham (2018). https://doi.org/10.1007/978-3-319-77538-8_48
37. Pugh, J.K., Soros, L.B., Stanley, K.O.: Quality diversity: a new frontier for evolutionary computation. Front. Robot. AI **3**, 1–40 (2016)

38. Trianni, V.: Evolutionary Swarm Robotics: Evolving Self-Organising Behaviours in Groups of Autonomous Robots, Studies in Computational Intelligence, vol. 108. Springer, Berlin (2008). https://doi.org/10.1007/978-3-540-77612-3
39. Trianni, V., Nolfi, S., Dorigo, M.: Evolution, self-organization and swarm robotics. In: Blum, C., Merkle, D. (eds.) Swarm Intelligence. Natural Computing Series, pp. 163–191. Springer, Heidelberg (2008). https://doi.org/10.1007/978-3-540-74089-6_5
40. Vassiliades, V., Chatzilygeroudis, K., Mouret, J.B.: Using centroidal Voronoi tessellations to scale up the multi-dimensional archive of phenotypic elites algorithm. IEEE Trans. Evol. Comput. (2017). https://doi.org/10.1109/TEVC.2017.2735550. (in Press)
41. Vassiliades, V., Chatzilygeroudis, K., Mouret, J.B.: Comparing multimodal optimization and illumination. In: Genetic and Evolutionary Computation Conference (GECCO) Companion, pp. 97–98. ACM Press (2017)

The Best-of-n Problem with Dynamic Site Qualities: Achieving Adaptability with Stubborn Individuals

Judhi Prasetyo[1]([✉]), Giulia De Masi[2], Pallavi Ranjan[1],
and Eliseo Ferrante[3,4]([✉])

[1] Middlesex University Dubai, Dubai, United Arab Emirates
{j.prasetyo,p.ranjan}@mdx.ac.ae
[2] College of Natural and Health Science, Zayed University,
Dubai, United Arab Emirates
giuliademasi@gmail.com
[3] Laboratory of Socio-Ecology and Social Evolution, KU Leuven,
Leuven, Belgium
eliseo.ferrante@kuleuven.be
[4] School of Computer Science, University of Birmingham,
Dubai, United Arab Emirates
e.ferrante@bham.ac.uk

Abstract. Collective decision-making is one of main building blocks of swarm robotics collective behaviors. It is the ability of individuals to make a collective decision without any centralized leadership, but only via local interaction and communication. The best-of-n problem is a subclass of collective decision-making, whereby the swarm has to select the best option among a set of n possible alternatives. Recently, the best-of-n problems has gathered momentum: a number of decision-making mechanisms have been studied focusing both on cases where there is an explicit measurable difference between the two qualities, as well as on cases when there are only delay costs in the environment driving the consensus to one of the n alternatives. To the best of our knowledge, all the formal studies on the best-of-n problem have considered a site quality distribution that is stationary and does not change over time.

In this paper, we perform a study of the best-of-n problems in a dynamic environment setting. We consider the situation where site qualities can be directly measured by agents, and we introduce abrupt changes to these qualities, whereby the two qualities are swapped at a given time.

Using computer simulations, we show that a vanilla application of one of the most studied decision-making mechanism, the voter model, does not guarantee adaptation of the swarm consensus towards the best option after the swap occurs. Therefore, we introduce the notion of stubborn agents, which are not allowed to change their opinion. We show that the presence of the stubborn agents is enough to achieve adaptability to dynamic environments. We study the performance of the system with respect to a number of key parameters: the swarm size, the difference between the two qualities and the proportion of stubborn individuals.

© Springer Nature Switzerland AG 2018
M. Dorigo et al. (Eds.): ANTS 2018, LNCS 11172, pp. 239–251, 2018.
https://doi.org/10.1007/978-3-030-00533-7_19

1   Introduction

Collective decision-making is a central cornerstone in most natural and artificial collective systems. In the context of artificial systems, collective decision-making is one of the most important building blocks for swarm robotics systems [3]. Many swarm robotics problems such as deciding a common direction for coordinated motion [8], or a common area in the environment to aggregate to [6], can be seen as instances of collective decision-making [28]. When a swarm needs to make a collective decision by choosing among a set of discrete alternatives, the resulting problem is called the best-of-n problem in a robot swarm, and has been thoroughly reviewed in [28].

In this paper, we consider a version of the best-of-n problem whereby a swarm of robots with minimal capabilities has to achieve consensus to one among n options, and in particular to the one with the best quality, while interacting only locally in an environment that is symmetric with respect to the distribution of the n options (that is, all options can be evaluated on average in the same amount of time). The robots do not communicate the perceived option quality. On the contrary, they can only advertise one option at the time, the one corresponding to their current opinion, and they use a *decision mechanism* to change their current opinion after observing their neighbors in local proximity. Many decision mechanisms have been utilized in the past, the most common ones being the voter model [2,30] and the majority rule [14]. Consensus is built over time using a mechanism called positive feedback modulation [11], whereby fluctuations in robot's opinions distributions will eventually produce a bias towards one of the two options, which will make that option more likely to be observed and henceforth reinforcing this bias, until consensus is reached. To date, the literature on the best-of-n for a mobile robot swarm has mainly focused to the static environment case, whereby both environment and option quality do not change over time, with only few exceptions [28].

In this paper, we consider the best-of-n problem in dynamic environments. More precisely, we consider the case where the environment is static and symmetric with respect to option distribution, but the quality is asymmetric and furthermore abruptly changes over time. The goal of the swarm is collectively select the option corresponding to the best quality, and at the same time to adapt this decision and shift its consensus state to another option if this becomes the best one. We consider the voter model as the main decision mechanism, and the positive feedback modulation mechanism first proposed in [30], which consists in robots advertising each option to their neighbors for a time that is proportional, on average, to the quality of each option, an idea that is inspired by the waggle dance behavior exhibited by honeybees [25].

We perform experiments using multi-agent computer simulations where the spatial dimension is taken into account, and each robot is abstracted by an agent. We show that the vanilla application of the voter model does not allow the swarm to adapt to a dynamic change of the options quality. To solve this problem, we introduce the notion of *stubborn* agents, that is, agents that do not apply any decision mechanism and therefore never change their opinion. Stubborn agents can be seen as scouts, constantly exploring their favorite opinion, irrespective of the opinion of others and of the consensus state of the swarm. We perform simulated experiments where we analyze the above idea by studying the effect of the key parameters: swarm size, proportion of stubborn individuals, and ratio of the option quality. We also perform preliminary experiment with another decision mechanism, the majority rule, showing that in this case the dynamics are completely different.

The rest of the paper is organized as follows. In Sect. 2, we analyze the literature in collective decision making and we relate our work considering also the few cases where the environment can be considered dynamic. In Sect. 3, we define the dynamic best-of-n problem, the collective decision-making mechanism, and the idea of having the stubborn individuals. In Sect. 4, we present our experimental setup by explaining the specific task and the parameters that have been studied. In Sect. 5 we present the results, while in Sect. 6 we conclude and we discuss the other possible directions in which this work can be extended.

2   Related Work

The direct biological inspiration of the best-of-n problem and in particular of the scenario chosen here comes from the collective behaviors of social insects such as ants [9] and more specifically bees [13, 25]. The literature on the best-of-n related to swarm robotics will be reviewed using two categories introduced in [28]. We conclude this section by analyzing the few works in the best-of-n that can be considered conducted in a dynamic environment.

In the first category, we find work whereby robots cannot measure directly the quality of the different options. Instead, there are asymmetries in the environment that bias the collective decision towards one of the n options. For example, in [5, 10], a classical aggregation task inspired by cockroaches is presented. Here, the asymmetry in the environment is represented by the different size of a number of shelters, that are perceived by the robots in the exact same way. Thanks to this environmental effect, robots are able to select the right shelter to aggregate in. Another example of environmental asymmetry is shown in [14, 27], whereby the environment is represented by a classical double bridge [7] and robots have to find the shortest path between two bridges connecting the nest to the food source. Differently from our work, here the asymmetries between the two paths induces agents selecting the shortest path to appear more frequently in the nest, therefore biasing the process towards that path. In this work, the majority rule was used as decision mechanism. In a subsequent work [23] performed on the same scenario, another mechanism called the k-unanimity rule (the agent switches opinion only after observing the same option k times in a row) was used.

In the second category, we find works in which the quality can be directly measured, as per our case, but are conducted in a static environment. The baseline studies on direct modulation of positive feedback through quality was performed in [29–31], whereby the authors thoroughly analyzed the voter model and the majority rule through real robot experiments, simulations, ordinary differential equations, and chemical reaction network models, and studied the speed versus accuracy trade-off. The authors in [19–21] developed a decision-making strategy that, differently from our work, includes also an uncommitted opinion (neither of the n alternative), a recruitment mechanism, and an inhibition mechanism, as in honeybees [26]. In a recent follow-up study [18], they have shown how this model can be general to encompass not only decision-making in social insects but also in the human brain [13]. Finally, in [15], the authors considered the best-of-n problem in an aggregation task. Here, agents use a direct recruitment mechanism and are able to commit by using a quorum-based mechanism that makes the swarm aware of the consensus level reached.

In the context of dynamic environment, relatively little effort has been put, however the idea of having the swarm not converging to a full consensus is not new in this paper. For example, biological studies have found that having only a large majority committing to an option rather than the unanimity allow fish schools to swiftly adapt to perturbations [4]. Back to artificial systems, in [16], the authors considered a task-sequencing problem that can be seen as a best-of-2 with two options: task complete and task incomplete. These have dynamic qualities because the task completion level, corresponding to the size of the cleared area, changes over time. The authors of [1] studied a dynamic version of aggregation. Here, each shelter emits a different sound that varies over time, and the swarm has to aggregate in the shelter with the loudest sound. The method is based on a fuzzy version of the original BEECLUST algorithm [12,24]. In the original BEECLUST, after a waiting period, each agent chooses a new direction of motion at random, while in [1] a fuzzy controller that maps the loudness and the bearing of the sound determines the new direction of motion. Differently from all this work that focused on specific application scenarios, in this paper we perform a systematic study of a minimal model of the dynamic best-of-n problem, in order to understand better the effect of the most important key parameters.

3   The Model

In this section, we define the dynamic best-of-n problem (Sect. 3.1) and the collective decision-making model introduced (Sect. 3.2).

3.1   The Dynamic Best-of-n Problem

The best-of-n problem requires a swarm of agents to make a collective decision among n possible alternatives towards the choice that has the best quality. A typical example is the choice of best location for honeybees' swarm foraging. Each of the n options has an intrinsic quality ρ_i with $i \in 1, \ldots, n$. Qualities ρ_i are defined in $[1.05, 1.5, 3]$. A best-of-n problem reaches the optimal solution when the collective decision of the swarm is for the option with maximum quality. That means that a large majority $M \leq N(1 - \delta)$ of agents agrees on the same option, where δ is a small number chosen by the experimenter. In the case where $\delta = 0$ there is *perfect consensus*.

In this paper, as for the majority of the studies [29], we restrict n to 2 options, labeled a and b, having intrinsic quality ρ_a and ρ_b. Without loss of generality, one option quality ρ_a is set to 1 while $\rho_b > 1$. No cost is included in the current model, which means that the time needed to explore and assess the quality of both options is symmetric [28]. Each agent can measure the quality of different options, but cannot communicate it but rather can only advertise the option itself using local communication (see Sect. 3.2). In dynamic environments as introduced here, qualities can change over time: $\rho_a = \rho_a(t)$ and $\rho_b = \rho_b(t)$. In this study, we only consider qualities that are piece-wise constant: at a given time T_C, the two qualities are swapped. Namely $\rho_a(t)$ and $\rho_b(t)$ remains constant for $t < T_C$, they are swapped at T_C ($\rho_a(T_C) = \rho_b(T_C-1)$, $\rho_b(T_C) = \rho_a(T_C-1)$), and again remain constant afterwards.

3.2 The Decision Mechanism and the Stubborn Agents

We consider two kinds of agents: *normal* and *stubborn*. Each agent has an initial opinion, which consists in one of the two options a or b. Normal agents are able to change their opinion by applying a decision mechanism that relies on the observation of other agents in local proximity. Stubborn agents instead never change their opinion and keep the one they have at the very beginning, either a or b.

Initially agents are positioned inside the nest. Then, they move toward the region corresponding to their opinion. They spend there an exponentially-distributed amount of time (sampled independently per agent) that does not depend on the option, during which they measure the quality of that site. Then they go back to the nest, each at a different time, and they start disseminating their opinion. Agents within the nest needs to be well-mixed in order to avoid agents with same opinion clustering near each other. A random walk is implemented in order to meet this well-mixed assumption as much as possible.

The agents controller is represented by the finite state machine in Fig. 1a. Accordingly, agents can have one of the following 4 possible states: dissemination state of opinion a (D_a), dissemination state of opinion b (D_b), exploration state of opinion a (E_a), exploration state of opinion b (E_b). In the figure, solid lines represent deterministic transitions, while dotted lines stochastic transitions. The symbol VR indicates that the voter model is used at the end of the dissemination state (in the case where the majority rule is used, this will be mentioned). In the dissemination state, the agent disseminates his opinion continuously to other agents he meets that are also in the dissemination state. The time spent by the agent disseminating its opinion is randomly sampled from an exponential distribution characterized by a parameter proportional to the quality of the region. As a consequence it is more probable to meet neighbors with the best opinion than meeting those with the worst one. This mechanism is called *modulation of positive feedback* and it is the driving mechanism to make the group converge on the option with the best quality. At the end of dissemination, each agent can change its opinion based on the opinions of other agents and using either the voter model or the majority rule. Both the voter model and the majority

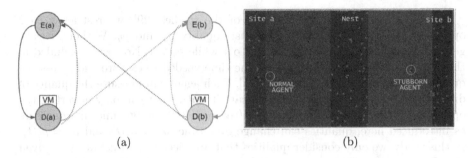

Fig. 1. Panel a: Probabilistic finite state machine. D_a, D_b, E_a and E_b represent the dissemination and exploration state. Solid lines represent deterministic transitions, while dotted lines stochastic transitions. The symbol VR indicates that the voter model is used at the end of the dissemination state. Panel b: Screenshot of the simulation arena. This image is taken from NetLogo software.

rule depend on the opinion of neighbors, that is the agents within a specified spatial radius (in our experiments set to 10 units). In the voter model, the agent switches its opinion to the one of a random neighbors. In the majority model, the agent changes its opinion to the one held by the majority of its neighbors.

4   Experimental Setup

The experiments have been conducted first on NetLogo simulator for fast prototyping. Then, the systematic simulations have been run using the simulator developed in [29].

Agents move on a 2-dimensional arena of size 200 (width) $\times$ 100 (height) units (see Fig. 1 for a screenshot within NetLogo). In the binary model, the arena comprises a central region called the *nest*, where initially all agents are and where they meet to perform the decision-making process. The two external areas are the *sites* and represent the two options: option a on the left and option b on the right.

In order to test the robustness of the model, the most important parameters have been studied. For the voter model, as evident from Table 1, the total number of agents has three different values: 40, 100, 500. Without loss of generality, the interplay between ρ_a and ρ_b can be studied simply by keeping one of them fixed (ρ_a before the environment changes, and ρ_b after it changes) to a value of 1 and by changing the other one. The values of the second quality studied are: 1.05, 1.5, 3, indicating small, medium, and large difference in quality, respectively. The proportion of stubborn individuals have been studied in the range $\{5\%, 10\%, 20\%\}$, equally distributed between the two opinions.

For the majority model only one set of parameters has been run (see Table 1). As initial conditions of each run, 50% of agents have opinion a and 50% of agents have opinion b.

Table 1. Model parameters used in experiments

	Voter	Majority
N	{**40**, **100**, **500**}	100
ρ_a (ρ_b after change)	1	1
ρ_b (ρ_a after change)	{**1.05**, 1.5, **3**}	3
S	{**0.05**, 0.1, **0.2**}	0.1

The dissemination time is exponentially distributed with parameter $\tau_D = g\rho$ with $g = 100$. The time of exploration is also exponentially distributed, with parameter set to $\tau_E = 10$, therefore independent of the site.

In the dynamic environment considered in this paper, a new time parameter T_C is introduced: the time when the value of ρ_a and ρ_b are abruptly changed by swapping their values. In this study $T_C = 2500$, a value empirically chosen as a compromise to allow both consensus to the best option prior to change and reasonably short runs. For each configuration of parameters, an ensemble of simulation has been realized, consisting of $R = 50$ runs.

5   Results

The different configuration sets are compared in terms of temporal evolution of opinions. In particular only the proportion of agents with opinion a (p_a) is monitored, as the percentage of agents with opinion b (p_b) is simply given by $p_b = 1 - p_a$. The plots report all the trajectories of this quantity over time (in simulated seconds, sampled every $\Delta t = 0.1$ steps) for all runs. In the paper are reported only the most relevant plots to discuss the results. All the plots of the full study, together with example videos, are available as Supplementary Material [17]. Table 1 reports in **bold** the parameters whose plots are included in the main text.

5.1   The Vanilla Voter Model

Figure 2 shows the results of runs of voter model without stubborn (also called vanilla voter model) for two different values of quality ratio: 1.05 (low) and 3 (high). It is interesting to note that for a low value of quality ratio the convergence is never reached, while for high value of quality ratio the convergence is reached but there is no adaptation to the environmental change. We will see next how stubborn individuals will play a driving role to get convergence and adaptation in the dynamic best-of-n problem.

5.2   Effect of Quality Ratio and of Proportion of Stubborn Individuals

Figure 3 reports the results of runs for four different cases of systems of 100 agents: across rows, we vary the ratio ρ_a/ρ_b from very low (1.05) to very high

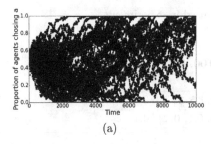 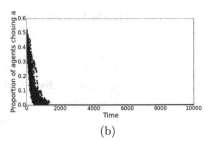

(a)   (b)

Fig. 2. Opinion evolution for a voter model with no stubborn, for two different values of quality ratio: 1.05 (*a*) and 3 (*b*). For low quality ratio there is no convergence. For high quality ratio the convergence to one option is reached but there is no adaptation to the change of opinion quality

(3). Across columns, we vary the stubborn percentage from 5% to 20%. It is evident the large role played by quality ratio value: whatever the values of stubborn presence, in the case of low quality ratio there is no convergence of opinions, neither adaptation. On the other hand, for large quality ratio value (3) there is good convergence and adaptation, irrespective of the proportion of stubborn individuals (which only affects the final value of the consensus state in a decreasingly proportional way). Interestingly, while the presence of stubborn individuals has been shown to be fundamental to have convergence and adaptation, its percentage does not seem to significantly contribute in terms of time nor in terms of variance of number of agents following the opinion.

5.3   Effect of Swarm Size Versus Proportion of Stubborn Individuals

Keeping constant the percentage of stubborn individuals, a big role of the swarm size is disclosed by Fig. 4 (the quality ratio varies across rows, while the swarm size across columns). Increasing the population size decreases the variance of fraction of agents following a certain opinion (here *a*), while the convergence or non-convergence are determined by the value of the quality ratio. In the case of low quality ratio, for small swarm size there is no convergence; increasing the size of the swarm show a certain tendency to convergence. Interestingly, in the case of high quality ratio, increasing the swarm size reduces the variance of adaptation time.

Since the number of stubborn individuals increases with the swarm size, a doubt arises if convergence is observed only for an absolute number of stubborn agents larger than a critical mass. This is denied by the evidence of other configurations. For instance, in the swarm with $N = 100$ agents and stubborn percentage 20% the number of stubborn individuals is 20, which is comparable to the number of stubborns of the swarm with $N = 500$ agents and stubborn percentage 5%, that is 24 stubborns. The first case does not converge while the second case does converge: therefore it can be concluded that there is an intrinsic role of the size of the swarm itself, not related to any critical mass of stubborns.

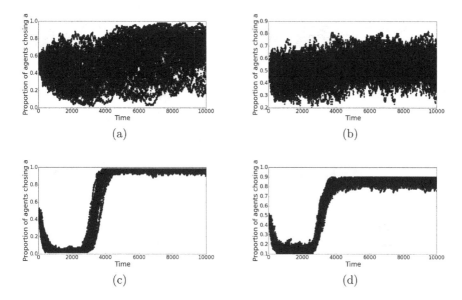

Fig. 3. Different cases of systems of $N = 100$ agents. (a) $S = 5\%$ and $\rho_a/\rho_b = 1.05$, (b) $S = 20\%$ and $\rho_a/\rho_b = 1.05$, (c) $S = 5\%$ and $\rho_a/\rho_b = 3$, and (d) $S = 20\%$ and $\rho_a/\rho_b = 3$. It shows that quality ratio has a higher effect than the percentage of stubborn.

The only thing that could be further argued is that, rather than being an effect of the swarm size, it may be an effect of increased density instead. This will be confirmed or denied in future work.

5.4 The Majority Rule with Stubborn Individuals

Also a majority decision model has been implemented to compare with voter model results. Results show that the majority model never works, but does a sort of spontaneous symmetry breaking (more often biased to the option that is best at the beginning of the experiment, b) and is not sensitive to the presence of stubborn individuals. We will speculate more about this in the conclusions.

6 Conclusion, Discussion, and Future Work

In this work, we have introduced the dynamic best-of-n problem, and we have proposed a solution to this problem when the environment is asymmetric with respect to the option qualities that can be assessed by the robots and symmetric with respect to the time needed to assess each option [28]. The proposed solution consists in a combination of direct modulation of positive feedback coupled with the voter model and with the introduction of stubborn agents, that is, agents that do not change their opinion and stay committed to their initial option.

Through simulation experiments, we have shown that the voter model alone (i.e. without the stubborn agents) cannot make the swarm adapt to abrupt

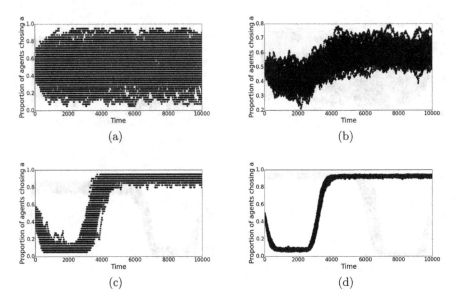

Fig. 4. The effect of the swarm size 40 and 500 for the two quality differences 1.05 and 3: (a) $N = 40$ and $\rho_a/\rho_b = 1.05$, (b) $N = 500$ and $\rho_a/\rho_b = 1.05$, (c) $N = 40$ and $\rho_a/\rho_b = 3$, and (d) $N = 500$ and $\rho_a/\rho_b = 3$. In the case of low quality ratio, for small swarm size there is no convergence; increasing the size of the swarm show a certain tendency to convergence. In the other case (high quality ratio), increasing the swarm size reduces the variance of adaptation time.

changes in the option qualities. After introducing stubborn agents, we have studied the effect of key parameters: the ratio of the two qualities, the proportion of stubborn agents, and the swarm size. Firstly, we reported that, as expected and reported in previous studies [14], the difference in site quality place a crucial role, whereby the ability to adapt to the environmental changes is strongly linked to the system accuracy, and higher level of accuracy and adaptability are observed with increasing ratio between the qualities. Secondly, contrarily to initial expectations, we found that increasing the ratio of stubborn individuals does not have an effect on neither the accuracy nor the adaptability. Finally, and surprisingly, we found that increasing swarm size has a beneficial effect on both consensus and adaptability: when the quality ratio is high (easier problems), the swarm is able to react faster with smaller variations on the reaction times; when the quality ratio is low (harder problems), a small swarm is not able to achieve consensus at all, while a larger swarm shows sign of approaching consensus and at the same time of adaptability to the change in the environment. This trend further confirms our speculation that, at least in this system, consensus-reaching and adaptability are strongly interlinked, as in previous study it was already found that accuracy increases with increasing swarm sizes as this will eventually approach a continuum model that, when studied using ordinary differential equations (ODEs), predicts that the swarm always achieves consensus to the best

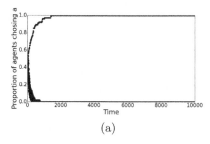
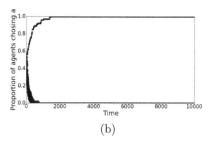

(a)   (b)

Fig. 5. Majority model for a system of $N = 100$ agents, $\rho_a/\rho_b = 3$, without stubborn (a) and with 10% of stubborns (b). Whether there are stubborn individuals or not the majority model never converge and a symmetry breaking is observed.

quality [30]. We concluded the experimental analysis with a preliminary study of the majority rule model, by showing that this model is ineffective in reaching consensus to the right option and at adapting to environmental changes. The latter is due to the effect of spatiality, as stubborn individuals committed to the same options are very unlikely to appear next to each other (Fig. 5).

This present study has revealed new insights but at the same time has raised new questions that we plan to investigate in future studies. Firstly, we would like to study this system using theoretical models such as ODEs and chemical reaction network modeling, as in [29], because this allows to study more broadly the effect of parameters and to have a deeper understanding of the dynamics. To do so, stubborn agents may be replaced with a spontaneous transition rate of all agents to a random opinion, as this can be more easily modeled. However, the equivalence of the two models needs to be tested. Secondly, we plan to systematically study the majority-rule to determine whether there is a variant that can make this mechanism adaptive as well. This study will start by systematically varying the proportion of stubborn agents, because we suspect that there could be a higher value which will cancel out the effects of spatiality and make also this model effective, thus we expect to see a sort of phase transition with respect to this parameter. Finally, provided enough resources, we plan to perform the experiments on real robots, likely kilobots [22], in order to have a proof of concept in the real world and potentially discover new factors influencing adaptability.

Acknowledgments. We would like to thank Gabriele Valentini for the code of his multi-agent simulator. Besides partial support of computational facilities of the KU Leuven, this work has been made possibly purely by the efforts in terms of personal investment of both time and computing resources of all the authors involved.

References

1. Arvin, F., Turgut, A.E., Bazyari, F., Arikan, K.B., Bellotto, N., Yue, S.: Cue-based aggregation with a mobile robot swarm: a novel fuzzy-based method. Adapt. Behav. **22**(3), 189–206 (2014)

2. Baronchelli, A., Díaz-Guilera, A.: Consensus in networks of mobile communicating agents. Phys. Rev. E **85**, 016113 (2012). https://doi.org/10.1103/PhysRevE.85. 016113

3. Brambilla, M., Ferrante, E., Birattari, M., Dorigo, M.: Swarm robotics: a review from the swarm engineering perspective. Swarm Intell. **7**(1), 1–41 (2013)

4. Calovi, D.S., Lopez, U., Schuhmacher, P., Chaté, H., Sire, C., Theraulaz, G.: Collective response to perturbations in a data-driven fish school model. J. Roy. Soc. Interface **12**(104), 20141362 (2015). https://doi.org/10.1098/rsif.2014.1362

5. Campo, A., Garnier, S., Dédriche, O., Zekkri, M., Dorigo, M.: Self-organized discrimination of resources. PLoS ONE **6**(5), e19888 (2010)

6. Correll, N., Martinoli, A.: Modeling and designing self-organized aggregation in a swarm of miniature robots. Int. J. Rob. Res. **30**(5), 615–626 (2011)

7. Deneubourg, J.L., Goss, S.: Collective patterns and decision-making. Ethol. Ecol. Evol. **1**(4), 295–311 (1989)

8. Ferrante, E., Turgut, A.E., Huepe, C., Stranieri, A., Pinciroli, C., Dorigo, M.: Self-organized flocking with a mobile robot swarm: a novel motion control method. Adapt. Behav. **20**(6), 460–477 (2012)

9. Franks, N.R., Pratt, S.C., Mallon, E.B., Britton, N.F., Sumpter, D.J.T.: Information flow, opinion polling and collective intelligence in house-hunting social insects. Philos. Trans. R. Soc. B: Biol. Sci. **357**(1427), 1567–1583 (2002)

10. Garnier, S., Gautrais, J., Asadpour, M., Jost, C., Theraulaz, G.: Self-organized aggregation triggers collective decision making in a group of cockroach-like robots. Adapt. Behav. **17**(2), 109–133 (2009)

11. Garnier, S., Gautrais, J., Theraulaz, G.: The biological principles of swarm intelligence. Swarm Intell. **1**(1), 3–31 (2007)

12. Kernbach, S., Thenius, R., Kernbach, O., Schmickl, T.: Re-embodiment of honeybee aggregation behavior in an artificial micro-robotic system. Adapt. Behav. **17**(3), 237–259 (2009)

13. Marshall, J.A.R., Bogacz, R., Dornhaus, A., Planqué, R., Kovacs, T., Franks, N.R.: On optimal decision-making in brains and social insect colonies. J. R. Soc. Interface **6**(40), 1065–1074 (2009)

14. Montes de Oca, M.A., Ferrante, E., Scheidler, A., Pinciroli, C., Birattari, M., Dorigo, M.: Majority-rule opinion dynamics with differential latency: a mechanism for self-organized collective decision-making. Swarm Intell. **5**, 305–327 (2011)

15. Parker, C.A.C., Zhang, H.: Cooperative decision-making in decentralized multiple-robot systems: the best-of-n problem. IEEE/ASME Trans. Mechatron. **14**(2), 240–251 (2009)

16. Parker, C.A.C., Zhang, H.: Collective unary decision-making by decentralized multiple-robot systems applied to the task-sequencing problem. Swarm Intell. **4**, 199–220 (2010)

17. Prasetyo, J., Masi, G.D., Ranjan, P., Ferrante, E.: The best-of-n problem with dynamic site qualities: achieving adaptability with stubborn individuals (2018). http://bio.kuleuven.be/ento/ferrante/FerranteSupp2018-001/index.html, Supplementary Material. Accessed 30 Apr 2018

18. Reina, A., Bose, T., Trianni, V., Marshall, J.A.R.: Psychophysical laws and the superorganism. Sci. Rep. **8**(4387) (2018). https://doi.org/10.1038/s41598-018-22616-y

19. Reina, A., Marshall, J.A.R., Trianni, V., Bose, T.: Model of the best-of-n nest-site selection process in honeybees. Phys. Rev. E **95**(5), 052411 (2017). https://doi.org/10.1103/PhysRevE.95.052411

20. Reina, A., Miletitch, R., Dorigo, M., Trianni, V.: A quantitative micro-macro link for collective decisions: the shortest path discovery/selection example. Swarm Intell. **9**(2–3), 75–102 (2015)
21. Reina, A., Valentini, G., Fernández-Oto, C., Dorigo, M., Trianni, V.: A design pattern for decentralised decision making. PLoS ONE **10**(10), e0140950 (2015)
22. Rubenstein, M., Ahler, C., Hoff, N., Cabrera, A., Nagpal, R.: Kilobot: a low cost robot with scalable operations designed for collective behaviors. Rob. Auton. Syst. **62**(7), 966–975 (2014)
23. Scheidler, A., Brutschy, A., Ferrante, E., Dorigo, M.: The k-unanimity rule for self-organized decision-making in swarms of robots. IEEE Trans. Cybern. **46**(5), 1175–1188 (2016)
24. Schmickl, T., et al.: Get in touch: cooperative decision making based on robot-to-robot collisions. Auton. Agents Multi-Agent Syst. **18**(1), 133–155 (2009)
25. Seeley, T.D.: Honeybee Democracy. Princeton University Press, Princeton (2010)
26. Seeley, T.D., Visscher, P.K., Schlegel, T., Hogan, P.M., Franks, N.R., Marshall, J.A.R.: Stop signals provide cross inhibition in collective decision-making by honeybee swarms. Science **335**(6064), 108–11 (2012)
27. Valentini, G., Birattari, M., Dorigo, M.: Majority rule with differential latency: an absorbing Markov chain to model consensus. In: Gilbert, T., Kirkilionis, M., Nicolis, G. (eds.) Proceedings of the European Conference on Complex Systems 2012. Springer Proceedings in Complexity, pp. 6651–658. Springer, Cham (2013). https://doi.org/10.1007/978-3-319-00395-5_79
28. Valentini, G., Ferrante, E., Dorigo, M.: The best-of-n problem in robot swarms: formalization, state of the art, and novel perspectives. Front. Rob. AI **4**, 9 (2017). https://doi.org/10.3389/frobt.2017.00009
29. Valentini, G., Ferrante, E., Hamann, H., Dorigo, M.: Collective decision with 100 Kilobots: speed versus accuracy in binary discrimination problems. Auton. Agents Multi-Agent Syst. **30**(3), 553–580 (2016)
30. Valentini, G., Hamann, H., Dorigo, M.: Self-organized collective decision making: the weighted voter model. In: Lomuscio, A., Scerri, P., Bazzan, A., Huhns, M. (eds.) Proceedings of the 13th International Conference on Autonomous Agents and Multiagent Systems, AAMAS 2014, pp. 45–52. IFAAMAS (2014)
31. Valentini, G., Hamann, H., Dorigo, M.: Efficient decision-making in a self-organizing robot swarm: on the speed versus accuracy trade-off. In: Bordini, R., Elkind, E., Weiss, G., Yolum, P. (eds.) Proceedings of the 14th International Conference on Autonomous Agents and Multiagent Systems, AAMAS 2015, pp. 1305–1314. IFAAMAS (2015)

The Impact of Interaction Models on the Coherence of Collective Decision-Making: A Case Study with Simulated Locusts

Yara Khaluf[⊠], Ilja Rausch, and Pieter Simoens

IDLab, INTEC, Ghent University-IMEC, Gent, Belgium
{yara.khaluf,ilja.rausch,pieter.simoens}@ugent.be

Abstract. A key aspect of collective systems resides in their ability to exhibit coherent behaviors, which demonstrate the system as a single unit. Such coherence is assumed to be robust under local interactions and high density of individuals. In this paper, we go beyond the local interactions and we investigate the coherence degree of a collective decision under different interaction models: (i) how this degree may get violated by massive loss of interaction links or high levels of individual noise, and (ii) how efficient each interaction model is in restoring a high degree of coherence. Our findings reveal that some of the interaction models facilitate a significant recovery of the coherence degree because their specific inter-connecting mechanisms lead to a better inference of the swarm opinion. Our results are validated using physics-based simulations of a locust robotic swarm.

1 Introduction

The move towards large-scale distributed systems promotes the field of collective decision-making as a fundamental area of research to address novel distributed control mechanisms. Collective decision-making encompasses (i) the decision mechanisms used at the individual level, and (ii) the emergent behavior at the system level. In this paper, we focus on binary decision-making processes, also known as symmetry-breaking [3], in which two choices of the same quality are available and the system needs to select one in a self-organized manner. For systems comprising only one individual, the solution is rather trivial (i.e., random). In contrast, in collective systems a mutual agreement (i.e., a consensus) needs to be achieved [11]. In order to enable such an agreement, the presence of noise is usually substantial. In particular, symmetry-breaking was mostly used to demonstrate the role of noise (i.e., random choices of the individuals) in pushing the system out of an equilibrium state [13]. Hence, it plays a key role in shifting the system towards one of the two options. This shift is then amplified [22] using interactions of a particular kind, referred to as positive feedback loops. While positive feedback is dominating, more individuals become in favor of the selected option and the coherence degree—represented by the fraction of individuals sharing the same opinion—increases until a consensus is achieved and

© Springer Nature Switzerland AG 2018
M. Dorigo et al. (Eds.): ANTS 2018, LNCS 11172, pp. 252–263, 2018.
https://doi.org/10.1007/978-3-030-00533-7_20

100% of individuals agree on the selected option [8,16]. At this point, negative feedback helps the system to preserve its selected option by damping the noise at the individual level and demotivating the individuals to change their opinion. The specific balance between positive feedback, negative feedback, and noise defines if the system reaches a decided or undecided global state. For example, a high level of noise may exceed the influence of the feedback loops and hence keep the system in an undecided state. Similarly, applying a strong enough positive feedback around a particular option may push the system to decide in favor of that particular option even in cases of low system densities—i.e., low number of feedback loops [17].

In this paper, we show that the interaction model is a key parameter to tune the balance between feedback loops and noise such that the collective system becomes decided even under significant noise levels. The interaction model is exploited by the individuals to exchange their decisions (opinions). Individuals interact locally using proximity models [10,20]. Local interactions allow collective systems to exhibit scalability, since the decisions at the individual level are made based on the personal preferences and on the influence of neighbors located in the immediate proximity. The latter is independent of the system size when the density is constant, hence, the functionality of the system is preserved at any scale (size). However, the implicit assumption of sufficient local interactions, [24], is only valid under moderate noise level e.g., moderate individual deficiency. Very limited research focuses on the impact of the interaction model on the robustness of collective systems against high level of noise [5].

In this study, we use *coherence degree* as a quality measure of collective behavior. Coherence degree is defined as the fraction of individuals adopting a common opinion/committing to a common option. In statistical physics, in a phase transition this measure is referred to as the order parameter. We use this measure to compare the efficiency of different interaction models in preserving the coherence of a collective behavior in a robot swarm under high levels of noise. The noise level is increased using one of the following two mechanisms: (i) reducing the impact of feedback loops through massive break-down of interaction links between the individuals by introducing robot breakdowns, and (ii) increasing the tendency to switch opinion randomly at the individual level. We go beyond the proximity (local) interaction model, and investigate scale-free [1], small-world [23], and regular models. As a case study, we use the locust marching behavior, in which individuals need to decide on their motion direction, and they have only two options of going *right* or *left*. Consequently, a consensus is achieved when all individuals select the same direction of motion. This corresponds to the highest degree of coherence in the collective behavior (100% of the individuals agree).

The paper is organized as follows, in Sect. 2, we specify the decision-making model used by the individuals to select their direction and velocity of motion in a locust marching scenario. The different interaction models, which we investigate in this paper are explained in Sect. 3. In Sect. 4, we present both the robot and environment configurations used in our physics-based simulations. Subsequently, the results obtained from these simulations are demonstrated and discussed in Sect. 5, and the paper is concluded in Sect. 6.

2   Decision-Making Model

Our collective model, as mentioned above, is inspired by one of the prominent natural examples of self-organized behavior: the collective motion (referred to as the marching bands) of desert locust swarms (*Schistocerca gregaria*). It has been previously shown by Buhl et al. that the collective behavior of marching locusts is similar to the behavior of the particles described by the Czirók model [6]. The latter originates from the Vicsek model and describes self-propelled interacting particles moving in $1D$ [9,21]. We adopt the Czirók model to define the individual decision-making processes, except that in our implementation locusts are simulated by a swarm of N homogeneous robots. Similar to Buhl et al., we consider a ring-shaped arena, where robots need to decide to go either left (i.e., the left-goers) or right (i.e., the right-goers), while avoiding collisions with other robots or the arena walls. The initial position and orientation of the robots are sampled from a uniform distribution. Following the discrete Czirók model [4,26], position $x_i(t)$ and velocity $u_i(t) \in \mathbb{R}$ of robot i are updated at every time step $\Delta t = 1$ according to

$$x_i(t+1) = \nu u_i(t) \tag{1}$$

$$u_i(t+1) = \delta_s \left[G(\langle u_i(t) \rangle) + \zeta_i(t) \right], \tag{2}$$

with ν being a speed parameter and $\zeta_i(t) \in [-1.0, 1.0]$ uniformly distributed noise. The propulsion and friction forces are given by the piece-wise continuous function

$$G(\langle u_i(t) \rangle) = \begin{cases} \frac{1}{2}(\langle u_i(t) \rangle + 1), & \langle u_i(t) \rangle > 0 \\ \frac{1}{2}(\langle u_i(t) \rangle - 1), & \langle u_i(t) \rangle < 0 \\ 0, & \text{otherwise,} \end{cases} \tag{3}$$

where $\langle u_i(t) \rangle$ is the average over the set of velocities of i's neighbors. Finally, in Eq. (2) we extended the discrete Czirok model by the factor δ_s, which is -1 with probability p_s, and 1 otherwise. Note that the originial Czirok model does not include δ_s, which we added to introduce the probability p_s to spontaneously switch the heading direction, i.e. the sign of $u_i(t)$. This is the main extension of the Czirok model that allows to account for an additional noise on the individual decision-making process [12], while $\zeta_i(t)$ represents the sensor noise, i.e. the uncertainty in the perception of the neighborhood opinion. From the sign of $u_i(t)$ we can deduce the state of robot i: if $u_i(t) > 0$ ($u_i(t) < 0$) then we denote i as a left-goer robot (right-goer robot), respectively. Therefore, the *collective state* or *collective behavior* of the system is given by:

$$\phi(t) = \frac{1}{N} \sum_i^N \frac{u_i(t)}{|u_i(t)|} \tag{4}$$

The absolute value of this measure represents the degree of *coherence* in the collective behavior (system's decision). When this degree is $|\phi(t)| = 1$, i.e., 100% of the individuals agree on one opinion, the system reaches *consensus*. Fluctuations that occur due to different sources of noise—i.e. sensor noise ζ and spontaneous opinion switching p_s—affect directly the degree of coherence achieved at the system level.

3   Interaction Models

Four different types of interaction (network) models are considered in this paper: (i) proximity network (PN), (ii) regular network (RN), (iii) small-world (SW), and (iv) scale-free (SF). These models differ fundamentally in how individuals connect.

First, PN models describe topologies in which each robot interacts with those robots that are within its proximity communication radius. Therefore, the choice of the communication radius has a significant influence on the communication *degree* of the robot. Two extreme values of the communication radius are $r_c = 0$ and $r_c > d_A$, where d_A denotes the diameter of the arena. In former case, there are no interactions, i.e. the degree of every robot is zero and the collective behavior is the average over the opinions that are purely governed by noise. When $r_c > d_A$, every robot interacts with every other robot in the arena which results in a complete network. For a swarm with an even number of individuals (majority is always available in the individual's neighborhood) this will always lead to consensus. However, complete networks have no locality because every robot knows the state of the entire system at every time step. In contrast, we obtain PN models with local interactions over a limited communication radius r_c and an approximated average of communication degree $\langle k \rangle$.

Second, RN models are topologies in which all robots have equal degree and in which neighbors are selected at random, irrespective of their physical distance. Consequently, robots might be connected to robots at larger physical distances.

Third, SW models originate by randomly rewiring RN networks such that the network distance (hop count) between two randomly chosen nodes grows proportionally to the logarithm of the total number of nodes [15,23]. Following the Watts-Strogatz model [23], we generate SW models by starting from a RN model and replace random links with new links—between any two nodes irrespective of their physical distance—that are sampled from a uniform distribution with probability p_{sw}. Differently from RN models, the latter process introduces a number of *hubs*, i.e. nodes with above-average degree.

Fourth, SF models are a special type of SW models, observed in a large number of natural and biological systems [15,25]. They are characterized with a power-law distributed degree and a very short average distance, making these networks *ultra-small* [7]. On the one hand, this implies the presence of a few extraordinary hubs with degrees that are far higher than the network average. On the other hand, most nodes have a relatively low degree such that the removal of a random node is not likely to affect the system connectivity. Therefore, SF networks are known to be robust against random node failure. Following the Barabasi-Albert model [2], we generate scale-free networks by starting from a small complete network of 10 nodes (robots). Subsequently, each of the remaining $N - 10$ nodes (N denotes the size of the swarm) is iteratively added with a fixed degree. Each of the newly added nodes is connected to a node i with a probability proportional to i's degree k_i, a process also known as *preferential-attachment* [2]. The latter step increases k_i and at the end of the network growth process the resulting average network degree amounts to $\langle k \rangle$ being the same as in the other interaction models.

4 Simulations

In this section we describe the physics-based simulations that we conducted using ARGoS [19] to analyze the global collective behavior of a homogeneous swarm of $N = 500$ simulated Footbots[1]. ARGoS allows us to perform our simulations with taking the robot's physics into consideration, and hence facilitates the generation of more realistic results. In the following, we present the configurations adapted at both individual and environment levels.

4.1 Robot Configuration

In our simulations, robots move randomly with a linear speed of $\nu = 5\,m/ts$, and try to avoid collisions by halting either the left or the right wheel, depending on its orientation relative to the position of the nearby robot or wall. However, in cases when collision avoidance requires to turn more than $90°$, the robot is programmed to maintain the sign of $u_i(t)$. Therefore, collision avoidances do not constitute an additional source of spontaneous direction switching. Nevertheless, the density of robots used in our experiments—i.e., the number of robots over the area of the arena allows to account for a minimized level of spacial interferences [14]. Additionally, the rate of collision avoidances is greatly reduced when the swarm is in consensus because then robots move in the same direction and the possibility of a potential collision becomes negligible. While moving, each robot communicates and exchanges opinions with its neighbors. The opinion of robot i is given by the value of $u_i(t)$ (Eq. (2)), for which the sensor noise is set as $\zeta \in [-1, 1]$, and the spontaneous switching probability p_s. The two types of implemented communications are proximity (short-range) communications and targeted long-range communications. To enable long-range communications over the whole arena area, we assume that the inner walls of the ring-shaped arena are lower than the level of the range-and-bearing sensors and actuators but high enough to be perceived by the robot as physical obstacles. This allows to include all communication models described in the previous section.

---

[1] The large swarm size is chosen for statistical reliability and a sound comparison of the features of the interaction models, which often occur in the limit of large N. http://www.swarmanoid.org/swarmanoid_hardware.php.

4.2   Environment Configuration

An important system property is the density of the swarm inside the arena [6,9]. Therefore, apart from the size of the swarm, the shape and the area of the environment have a significant influence on the system dynamics. In our experiments, swarm robots are confined within a ring-shaped arena with a diameter of 4 m (24 m) for the inner (outer) walls, respectively. The form of the arena encourages the robots to move either clockwise or counter-clockwise, i.e. right or left. Because the arena has a finite radial width, we program the robots to avoid unnecessary radial movement by always maintaining an angle of $(90 \pm 5)°$ to a light beacon located in the center of the arena (i.e. unless collision avoidance is required).

To simulate the event of a break-down, we deactivate the majority of robots at $t = t_d$ randomly. When a robot i is deactivated, it stops moving, i.e. $u_i(t_d) = 0$, and all its communication links are broken. The latter implies that the break-down leads to a substantial decrease of interactions, often followed by a loss of consensus or a significant drop in the degree of coherence. However, as we will demonstrate in the next section, certain interaction models allow the swarm to recover the degree of coherence to a higher level even if more than two thirds of the swarm individuals are deactivated. Table 1 summarizes the parameters settings over the different interaction models implemented in this paper. Note that the interaction models are all generated such that the average robot degree is the same over all models for a fair comparison. This is set to $\langle k \rangle \approx 6$. Finally, the rewiring probability p_{sw} for the generation of the SW models is set to $p_{sw} = 0.5$ to guarantee the occurrence of nodes with above-average degree. In this regard, SW models represent a transition model between RN and SF models.

Table 1. Overview of implemented interaction models and the parameters used to generate them.

Interaction model	Parameter	Value
Proximity	Communication range r_c	1.3 m
Regular	Degree k_r	6
Small-world	Rewiring probability p_{sw}	0.5
Scale-free	Degree k_{sf}	3

Figure 1 depicts a top view of the ring-shaped arena, over which the locust swarm is performing collective marching. The screen-shots are taken after a severe loss of robots (i.e., break-down of 65% of the interaction links) for the three interaction models PN, SW, and SF.

5   Results and Discussion

We launch different sets of physics-based simulations to analyze the influence of the interaction model on the degree of coherence achieved in a locust swarm that

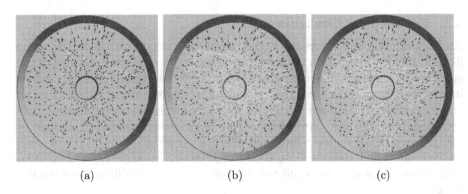

Fig. 1. Top view on the locust swarm. Green (red) robots are left-goers (right-goers), respectively, and black robots are deactivated. (a) PN, (b) SW, and (c) SF, all after break-down. For a clearer picture of the dynamics, see the recordings [18]. (Color figure online)

is exposed to a high level of noise. We first simulate a strong sudden occurrence of noise through a major break-down in the interaction links. The intensity of the break-down—represented by the number of deactivated robots—has the main influence on the achieved degree of coherence. Secondly, we use a new set of experiments to simulate the high level of noise through increasing the probability of the individual spontaneous switch, together with the incidence of robot break-down events.

We start with a robot locust swarm that suffers a major break-down after reaching a high degree of coherence (i.e. over 75% of the robots agree on the same direction). The individuals in this set of experiments are characterized with relatively low level of individual noise (i.e., spontaneous switch probability $p_s = 0.02$). The individual noise is set low in order to focus on the robustness of the different interaction models against random break-down. The coherence degree is measured using the absolute value $|\phi(t)|$—i.e., the fraction of robots agreeing on the same opinion. We analyze the corresponding time average $|\bar{\phi}|$ as a function of the percentage of deactivated robots to reveal the efficiency of each interaction model in preserving a high coherence degree.

In each experiment, and for all interaction models, we first let the swarm achieve a high coherence degree (over 75% of the robots). Subsequently, we deactivate a certain percentage of the robots. As mentioned above, for a fair comparison of the different interaction models, all simulations are configured such that the average robot degree before the break-down is the same, here set to $\langle k \rangle \approx 6$. We analyze the time evolution of the coherence degree $|\phi(t)|$ to study the efficiency of the applied interaction model in preserving a high $|\phi(t)|$. The left part of Fig. 2 demonstrates the results, in which every data point represents the coherence degree $|\phi(t)|$ averaged over the post break-down time—i.e., between the break-down event and the end of simulation at $T = 5000ts$. The time evolution of $|\phi(t)|$ is illustrated in the inset, for the PN model and the deactivation of 65% of robots. The resulting value of $|\bar{\phi}|$ shown in the inset corresponds to one data point in the major plot, other data points were obtained accordingly. As shown, $|\phi(t)|$ drops for all interaction models with increasing percentages of deactivated robots. Nevertheless, there are significant differences among the interaction models. We can notice that all models except of the PN model were able to preserve a high coherence degree ($|\bar{\phi}| > 0.75$) up to 50% of deactivation[2]. Furthermore, the SF interaction model shows a pronounced tendency of achieving higher $|\bar{\phi}|$ than the other models starting from 65% of deactivation. At this point, the performance of the SW and RN models demonstrate a clear drop. However, under the RN model $|\bar{\phi}|$ drops faster than it does under the SW, as can be seen between the deactivation of 65% and 75%. This can be explained by the fact that, for the same average degree, the SW model has a higher clustering coefficient than the RN model: it has a number of robots with higher connectivity degree. Beyond 75% deactivation, SW and RN models start to behave similarly and in agreement with the PN model. SF preserves its superior performance up to 85% of deactivation, due to the high robustness of its hubs.

We continue our experiments with the same low level of individual noise ($p_s = 0.02$) and we fix the deactivating percentage to 65%, at which the coherence degree drops for all models below $|\bar{\phi}| < 0.5$. For this setting, we investigate the efficiency of each interaction model in recovering the coherence degree to a higher level. We analyze this efficiency for different robot average communication degree. For this purpose, we define $|\bar{\phi}|$ as a function of the robot's average communication degree $\langle k \rangle$, aiming to determine the minimum rewiring threshold needed for the influence of the feedback loops to overcome the noise influence and hence increasing the degree of coherence in the collective behavior. In this set of simulations, the system starts with the PN model, followed by a break-down of 65% of interaction links at $t_d = 10000ts$. Next, we use targeted rewiring at $t = 12500ts$, to generate the different interaction models with the same average robot degree. The right side of Fig. 2 demonstrates the results of these experiments. The inset shows an example for the time evolution of $\phi(t)$ for the PN model before the break-down event (the black dashed line), and PN (blue) or SF (orange) interactions for $t > 12500ts$. The latter period of time is used to compute $|\bar{\phi}|$ which corresponds to one point in the major plot. This plot shows that SF, SW, and RN models are able to restore a high coherence degree $|\bar{\phi}| > 80\%$ with an average robot degree of $\langle k \rangle = 4$, while the PN model requires $\langle k \rangle \approx 7.5$ to restore a similar degree of coherence.

As mentioned above, after examining the robustness of the interaction models against mere break-down events, we continue investigating their robustness against increasing the level of individual noise (i.e., the spontaneous switch p_s). For this purpose, we run a new set of experiments, in which we increase the level of individual noise, first, to $p_s = 0.05$. Results are illustrated on the left side of Fig. 3. In this figure, we can notice that despite low coherence degree $|\bar{\phi}| < 0.5$, the SF model outperforms other models significantly starting from the deactiva-

---

[2] The results for the deactivation of $<50\%$ of the robots are shown in the supplementary materials [18].

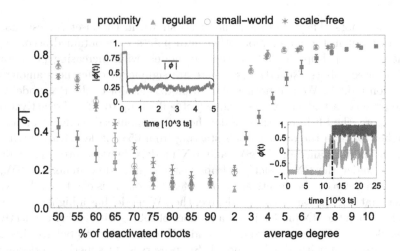

Fig. 2. Comparison of the average collective state $|\bar{\phi}|$ as a function of the percentage of deactivated robots (left) and the average communication degree (right). Left inset: time evolution of $|\phi(t)|$ after the break-down. Right inset: Time evolution of $\phi(t)$ starting with PN model, break-down, targeted rewiring of interactions to a SF model (orange) or no rewiring (blue, continued). Data points were averaged over 30 runs. (Color figure online)

tion percentage of 55%. Below this (critical) percentage of deactivations, all SF, SW and RN models were able to generate similar degree of coherence, which is significantly better compared to the PN model. A behavior similar to the one shown in Fig. 2 is observed, that is the SW model generating higher $|\bar{\phi}|$ than the RN model for specific deactivation percentages before converging to a similar behavior that approaches the PN model. To better investigate the specific role of the individual noise, we start with a swarm of 175—i.e., 35% (65% deacti- vation) of the total $N = 500$—and with an average robot degree of 4—i.e., the average degree at which all models (except for the PN model) are able to achieve $|\bar{\phi}| > 80\%$ with a swarm of 175 in the previous experiments, see the right side of Fig. 2—and we analyze $|\bar{\phi}|$ for different values of p_s. Results are illustrated on the right side of Fig. 3. In this figure, we can notice the significantly higher robustness of all interaction models in comparison to the PN model in terms of the obtained $|\bar{\phi}|$. The behavior of SF, SW, and RN models seems similar up to a certain noise level (here $p_s = 0.08$), up which the SF model starts to clearly out- perform other models (i.e., demonstrating higher coherence degree). This can be also deduced from the inset on the right side of Fig. 3, which corresponds to one data point and shows the phase transition of the coherent behavior generated by the SF model. Moreover, both SW and RN models demonstrate a behavior that is initially similar to the SF model and later approaches the PN model.

In general, the data points at which the behavior of SW and RN models aligns with the behavior of the PN model (see Figs. 2 and 3) indicate that it is not merely the presence of long-range interactions that contributes to the increased

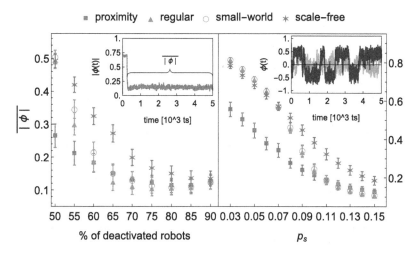

Fig. 3. Comparison of the average collective state $|\bar{\phi}|$ as a function of the percentage of deactivated robots (left) and the spontaneous switching probability p_s (right). Left inset: time evolution of $|\phi(t)|$ after the break-down. Right inset: Time evolution of $\phi(t)$ at $p_s = 0.08$, after the break-down, with alternative targeted rewiring of interactions to a SF topology (orange) or no rewiring (blue, continued) and an average degree of $\langle k \rangle \approx 4$. Data points were averaged over 30 runs. (Color figure online)

degree of coherence. Nor is it the value of the network distance, as RN and SW models behave similarly in our experiments. Instead, our conjecture is that it is the fraction of well-connected individuals that significantly influences coherence and robustness of the collective system affected by severe noise levels. On the one hand, these well-connected individuals have access to a sufficiently large sample of the swarm to reliably estimate its collective state. On the other hand, they can reach and influence the opinion of a significant number of individuals. Both features allow the system to preserve sufficient feedback loops that counteract the effects of noise.

Finally, the ratio of positive feedback to noise also defines the level of *adaptivity* of a collective system. When this ratio is critical, phase transitions occur [6,21] and the swarm is able to explore different options (in our case two options), making the swarm more adaptive to environmental changes. These phase transitions were observed in our systems as well, an example is shown in the right inset of Fig. 2 before the break-down (i.e., for $t < 10000\,\text{ts}$). However, in the same inset one can see that after rewiring the interactions to a SF model (orange data points at $t > 12500\,\text{ts}$), the presence of positive feedback outweighs the effects of noise. Thus, the increased coherence degree stabilized and the adaptivity decreased significantly such that no phase transitions occur until the end of the experiment. However, adaptivity can be restored by increasing the level of noise, as we can see by considering the right inset of Fig. 3 for comparison, demonstrating that the balance between positive feedback and noise is crucial to the collective system performance.

6   Conclusion

In this paper, we have investigated the impact of the interaction model used in collective system on the coherence degree of its decisions under high levels of noise. Beyond local interactions, we have investigated the coherence degree of a collective behavior under: SF, SW, and RN interaction models. The interaction models were analyzed using the case study of locust marching, which represents a symmetry-breaking decision-making problem. Our results have revealed a clear evidence of the significant role the interaction model plays in defining the coherence degree under different noise sources. SF has shown an outstanding performance over other models when the level of noise in the system exceeds a particular threshold. The influence of noise was increased either though introducing a break-down of a particular percentage of the interaction links or through increasing the probability to spontaneously switch of the opinion at the individual level (i.e., the individual noise). SW and RN models act similar to SF up to a specific level of noise, after which both demonstrate a drop in performance, however, a smaller drop of the SW model. Starting from a particular noise level, SW and RN show similar behavior that approaches the behavior of the PN model. Our findings can help as a preliminary step on the way to engineering artificial swarms with a robust coherence degree against high levels of noise from different sources.

In future work, we plan to compare the different interaction models in terms of the mean time required to achieve high coherence levels or even consensus. Furthermore, we plan to investigate the exact relation between the drop in the coherence degree and the connectivity measures of the collective system, such as the clustering coefficient and the fraction of hubs. Finally, it is worthwhile to examine the combination of PN and SF models that could additionally improve coherence degree of collective behavior as well as collective response to localized external stimuli.

References

1. Albert, R.: Scale-free networks in cell biology. J. Cell Sci. **118**(21), 4947–4957 (2005)
2. Albert, R., Barabási, A.L.: Statistical mechanics of complex networks. Rev. Mod. Phys. **74**, 47–97 (2002)
3. Anderson, P.W., et al.: More is different. Science **177**(4047), 393–396 (1972)
4. Ariel, G., Ayali, A.: Locust collective motion and its modeling. PLOS Comput. Biol. **11**(12), 1–25 (2015). https://doi.org/10.1371/journal.pcbi.1004522
5. Ballerini, M., et al.: Interaction ruling animal collective behavior depends on topological rather than metric distance: evidence from a field study. Proc. Natl. Acad. Sci. **105**(4), 1232–1237 (2008)
6. Buhl, J., et al.: From disorder to order in marching locusts. Science **312**(5778), 1402–1406 (2006)
7. Cohen, R., Havlin, S.: Scale-free networks are ultrasmall. Phys. Rev. Lett. **90**, 058701 (2003)

8. Corning, P.A.: Synergy and self-organization in the evolution of complex systems. Syst. Res. Behav. Sci. **12**(2), 89–121 (1995)
9. Czirók, A., Barabási, A.L., Vicsek, T.: Collective motion of self-propelled particles: Kinetic phase transition in one dimension. Phys. Rev. Lett. **82**, 209–212 (1999). https://doi.org/10.1103/PhysRevLett.82.209
10. Dorigo, M., Birattari, M., Brambilla, M.: Swarm robotics. Scholarpedia **9**(1), 1463 (2014)
11. Hamann, H.: Swarm Robotics: A Formal Approach. Springer, Heidelberg (2018). https://doi.org/10.1007/978-3-319-74528-2
12. Huepe, C., Zschaler, G., Do, A.L., Gross, T.: Adaptive-network models of swarm dynamics. New J. Phys. **13**(7), 073022 (2011)
13. Hurtado, P.I., Garrido, P.L.: Spontaneous symmetry breaking at the fluctuating level. Phys. Rev. Lett. **107**(18), 180601 (2011)
14. Khaluf, Y., Birattari, M., Rammig, F.: Analysis of long-term swarm performance based on short-term experiments. Soft Comput. **20**(1), 37–48 (2016)
15. Khaluf, Y., Ferrante, E., Simoens, P., Huepe, C.: Scale invariance in natural and artificial collective systems: a review. J. R. Soc. Interface **14**(136), 20170662 (2017)
16. Khaluf, Y., Hamann, H.: On the definition of self-organizing systems: relevance of positive/negative feedback and fluctuations. In: ANTS 2016. LNCS, vol. 9882, p. 298. Springer, Heidelberg (2016). [extended abstract]
17. Khaluf, Y., Pinciroli, C., Valentini, G., Hamann, H.: The impact of agent density on scalability in collective systems: noise-induced versus majority-based bistability. Swarm Intell. **11**(2), 155–179 (2017)
18. Khaluf, Y., Rausch, I., Simoens, P.: Supplementary materials for "impact of interaction models on the coherence of collective behavior: a case study with robot locusts" (2018). https://drive.google.com/file/d/1ye5_uqY9Y94x6Rsb OEV0kvPUis3NpFSF/view?usp=sharing. Accessed 16 Apr 2018
19. Pinciroli, C., et al.: Argos: a modular, parallel, multi-engine simulator for multi-robot systems. Swarm Intell. **6**(4), 271–295 (2012)
20. Valentini, G., Hamann, H., Dorigo, M.: Efficient decision-making in a self-organizing robot swarm: on the speed versus accuracy trade-off. In: Proceedings of the 2015 International Conference on Autonomous Agents and Multiagent Systems, pp. 1305–1314. International Foundation for Autonomous Agents and Multiagent Systems (2015)
21. Vicsek, T., Czirók, A., Ben-Jacob, E., Cohen, I., Shochet, O.: Novel type of phase transition in a system of self-driven particles. Phys. Rev. Lett. **75**, 1226–1229 (1995). https://doi.org/10.1103/PhysRevLett.75.1226
22. de Vries, H., Biesmeijer, J.C.: Self-organization in collective honeybee foraging: emergence of symmetry breaking, cross inhibition and equal harvest-rate distribution. Behav. Ecol. Sociobiol. **51**(6), 557–569 (2002)
23. Watts, D.J., Strogatz, S.H.: Collective dynamics of 'small-world' networks. Nature **393**(6684), 440 (1998)
24. Winfield, A.F., Nembrini, J.: Safety in numbers: fault-tolerance in robot swarms. Int. J. Model. Identif. Control **1**(1), 30–37 (2006)
25. Wolf, Y.I., Karev, G., Koonin, E.V.: Scale-free networks in biology: new insights into the fundamentals of evolution? Bioessays **24**(2), 105–109 (2002)
26. Yates, C.A.: Inherent noise can facilitate coherence in collective swarm motion. Proceedings of the National Academy of Sciences **106**(14), 5464–5469 (2009). https://doi.org/10.1073/pnas.0811195106

The Importance of Component-Wise Stochasticity in Particle Swarm Optimization

Elre T. Oldewage[1,2]([envelope]) [ORCID], Andries P. Engelbrecht[1,3] [ORCID],
and Christopher W. Cleghorn[1] [ORCID]

[1] Department of Computer Science, University of Pretoria, Pretoria, South Africa
vze.ezv@gmail.com, {engel,ccleghorn}@cs.up.ac.za
[2] Council for Scientific and Industrial Research, Pretoria, South Africa
[3] Institute for Big Data and Data Science, Pretoria, South Africa

Abstract. This paper illustrates the importance of independent, component-wise stochastic scaling values, from both a theoretical and empirical perspective. It is shown that a swarm employing scalar stochasticity is unable to express every point in the search space if the problem dimensionality is sufficiently large in comparison to the swarm size. The theoretical result is emphasized by an empirical experiment, comparing the performance of a scalar swarm on benchmarks with reachable and unreachable optima. It is shown that a swarm using scalar stochasticity performs significantly worse when the optimum is not in the span of its initial positions. Lastly, it is demonstrated that a scalar swarm performs significantly worse than a swarm with component-wise stochasticity on a large range of benchmark functions, even when the problem dimensionality allows the scalar swarm to reach the optima.

1 Introduction

The particle swarm optimization (PSO) algorithm employs stochasticity as an important mechanism to avoid premature convergence to local optima. The stochasticity should (usually) be applied in every dimension (i.e. component-wise) to ensure independence between position updates in each dimension. However, it is a common mistake for scalar stochastic values to be used instead, which restricts the swarm's movement and degrades performance [7,13,18,19].

This paper investigates the effect of using scalar stochasticity, both theoretically and empirically. The paper begins by introducing the PSO algorithm and briefly discussing the importance of component-wise stochasticity in Sect. 2. Section 3 provides theoretical results to formalize the restriction on the swarm's movement caused by scalar stochasticity. It is shown that there is a problem of "reachability", i.e. a swarm with scalar stochasticity will not be able to reach the optimum if the problem dimensionality is higher than the size of the swarm.

Section 4 examines the empirical effects of reachability on the performance of a swarm employing scalar stochasticity. The section compares a scalar swarm's

M. Dorigo et al. (Eds.): ANTS 2018, LNCS 11172, pp. 264–276, 2018.
https://doi.org/10.1007/978-3-030-00533-7_21

performance on a number of constructed benchmark functions with optima defined to be reachable or unreachable by a scalar swarm. Section 5 goes on to compare the performance of a swarm with scalar stochasticity to a swarm with component-wise stochasticity for a wide range of benchmark functions that are not biased towards or against the scalar swarm. Section 6 concludes the paper.

2   Background

This section briefly discusses the PSO algorithm and introduces relevant concepts regarding the importance of component-wise stochasticity.

PSO is a stochastic, population-based optimization algorithm [4] that does not require gradient information and may thus be applied to black box optimization problems. A swarm consists of a number of particles. The position of a particle in the search space represents a potential solution to the optimization problem. The particle moves through the search space for a number of iterations, using local information (the best position encountered by the particle thus far, called the personal best position) and global information (the best position encountered by all the particles within the given particle's logical neighbourhood, called the global or local best position, depending on how the neighbourhood is defined). This paper considers the global best topology, but the findings presented are applicable to arbitrary topologies. Each particle i's position is updated at iteration t according to:

$$\mathbf{x}_i^{t+1} = \mathbf{x}_i^t + \mathbf{v}_i^{t+1} \tag{1}$$

where $\mathbf{x}_i^{t+1}$ denotes the position of particle i at iteration $t+1$ and $\mathbf{v}_i^{t+1}$ denotes its velocity at iteration $t+1$. The particle's initial position, $\mathbf{x}_i^0$, is usually drawn from a uniform random distribution over the search space boundaries (in every dimension). PSO with inertia weight, as introduced in [17], updates particle i's velocity at iteration t in every dimension j as below:

$$v_{ij}^{t+1} = wv_{ij}^t + c_1 r_{1j}(y_{ij}^t - x_{ij}^t) + c_2 r_{2j}(\hat{y}_{ij} - x_{ij}^t) \tag{2}$$

where v_{ij}^{t+1} denotes particle i's velocity in dimension j, w denotes the inertia weight, and c_1 and c_2 denote the cognitive and social acceleration constants respectively. $r_{1j}, r_{2j} \sim U(0,1)$ are random numbers sampled between 0 and 1 at every iteration. y_{ij}^t denotes the personal best position of particle i in the jth dimension and $\hat{y}_{ij}$ denotes jth dimension of the best position found by all the particles in the neighbourhood of particle i. Particle neighbourhoods are usually defined by logical indexing of the swarm. When neighbourhoods are strict subsets of the entire swarm, the algorithm is referred to as a local best PSO. If every particle's neighbourhood consists of the entire swarm, the algorithm is referred to as a global best PSO and $\hat{y}_i = \hat{y}$ is called the global best position. Except where otherwise specified, this paper considers a global best PSO.

The stochastic scaling components, $\mathbf{r}_1$ and $\mathbf{r}_2$ can also be expressed as diagonal matrices, $\mathbf{R}_1$ and $\mathbf{R}_2$, as illustrated below:

$$\mathbf{v}_i^{t+1} = w\mathbf{v}_i^t + c_1 \mathbf{R}_1(\mathbf{y}_i^t - \mathbf{x}_{ij}^t) + c_2 \mathbf{R}_2(\hat{\mathbf{y}} - \mathbf{x}_i^t) \tag{3}$$

where $\mathbf{R}_1$ and $\mathbf{R}_2$ are diagonal matrices, with $\mathbf{r}_1$ and $\mathbf{r}_2$ forming their diagonals.

This paper emphasizes that if r_1 and r_2 are random scalars, then the swarm's movement becomes entirely linear. Every position investigated by the swarm will necessarily be a linear combination of the swarm's initial positions, velocities and personal best positions. If the swarm size is too small relative to the problem dimensionality, then the swarm can only reach a subspace within the larger search space (as is proved in Sect. 3). Since the swarm is typically initialised randomly, there is a possibility that the optimum can not be expressed as a linear combination of the swarm's initial positions, i.e. the swarm will never be able to find the optimum.

The effect of using scalar r_1 and r_2 values has been discussed in literature [14,20]. Paquet and Engelbrecht introduced a Linear PSO [14], in order to solve constrained linear optimization problems. If the swarm is initialised so that all positions are within the problem constraints, then using scalar r_1 and r_2 values ensures that the swarm can never leave the feasible space, which forms a subspace of the search space. In [20], purposeful dimensional coupling via shared $\mathbf{r}_1$ and $\mathbf{r}_2$ components was suggested to reduce the unwanted roaming exhibited by PSO in high dimensional problem spaces.

Throughout the remainder of the paper, a swarm that uses scalar values for r_1 and r_2 will be called a "scalar swarm". A swarm that uses vectors for $\mathbf{r}_1$ and $\mathbf{r}_2$ which are multiplied with the cognitive and social components in every dimension (or, alternatively, are diagonal matrices) will be called a "vector swarm".

3 Theoretical Results

This section provides a theoretical discussion regarding the consequences of using a scalar PSO. A number of definitions and key concepts are introduced first [15].

Definition 1. *A set of vectors $\mathcal{I} = \{\mathbf{z}_1, \mathbf{z}_2, \ldots, \mathbf{z}_M\} \subset \mathbb{R}^n$ is linearly dependent if there exists a finite number of distinct vectors, $\mathbf{z}_1, \mathbf{z}_2, \ldots \mathbf{z}_K \in \mathcal{I}$, and scalars, $a_1, a_2, \ldots, a_K \in \mathbb{R}$, not all zero, such that*

$$a_1\mathbf{z}_1 + a_2\mathbf{z}_2 + \ldots + a_K\mathbf{z}_K = \mathbf{0} \tag{4}$$

Since at least one scalar is non-zero, say $a_1 \neq 0$, the vector $\mathbf{z}_1$ can be expressed as a linear combination of the other vectors:

$$\mathbf{z}_1 = -\frac{a_2}{a_1}\mathbf{z}_2 - \ldots - \frac{a_K}{a_1}\mathbf{z}_K \tag{5}$$

Thus, the set $\mathcal{I}$ is linearly dependent if and only if at least one element in $\mathcal{I}$ can be written as a linear combination of the other elements in $\mathcal{I}$.

Definition 2. *A set of vectors $\mathcal{I} = \{\mathbf{z}_1, \ldots, \mathbf{z}_M\} \subset \mathbb{R}^n$ is linearly independent if*

$$a_1\mathbf{z}_1 + a_2\mathbf{z}_2 + \ldots + a_M\mathbf{z}_M = \mathbf{0} \tag{6}$$

can only be satisfied by $a_1 = a_2 = \ldots = a_M = 0$. Thus no element in $\mathcal{I}$ can be written as a linear combination of other elements from $\mathcal{I}$.

Definition 3. *Let $\mathcal{I}$ be a non-empty set of vectors from $\mathbb{R}^n$ (i.e. $\emptyset \neq \mathcal{I} \subset \mathbb{R}^n$). Then the span of $\mathcal{I}$ is the smallest subspace $W \subseteq \mathbb{R}^n$ that contains $\mathcal{I}$. Thus $span(\mathcal{I}) = W$. The subspace W consists of all linear combinations of elements of $\mathcal{I}$, given below (where $|.|$ denotes set cardinality):*

$$span(\mathcal{I}) = \left\{ \sum_{k=1}^{|\mathcal{I}|} a_k \mathbf{z}_k \;\middle|\; \mathbf{z}_k \in \mathcal{I}, a_k \in \mathbb{R}, k \in \mathbb{N} \right\} \tag{7}$$

Definition 4. *A non-empty set of vectors, $\mathcal{I} = \{\mathbf{z}_1, \ldots, \mathbf{z}_M\}$, is a spanning set for a subspace $W \subseteq \mathbb{R}^n$ if and only if any element in W can be expressed as a linear combination of elements in $\mathcal{I}$. In other words, for any non-zero $\mathbf{z} \in W$, there exist scalars $a_1, a_2, \ldots, a_M$ with at least one $a_i \neq 0$ such that*

$$\mathbf{z} = a_1 \mathbf{z}_1 + a_2 \mathbf{z}_2 + \ldots + a_M \mathbf{z}_M \tag{8}$$

Definition 5. *A basis for a subspace S of $\mathbb{R}^n$ is a set of vectors $\mathcal{B}$ in S where $\mathcal{B}$ is a spanning set for S, and $\mathcal{B}$ is linearly independent.*

Armed with these definitions, it is proven below that if r_1 and r_2 are scalars, then the positions of any particle at any iteration of the search space must be a linear combination of their initial positions, personal best positions, and velocities. The theorem below is for a local best PSO, since global best PSO can be considered as a special case of local best PSO where the neighbourhood is the entire swarm. For the sake of generality, the theorem makes no assumptions regarding the initial particle velocities or personal best positions.

Theorem 1. *For a particle swarm governed by the movement update equations in Eqs. (1) and (2), at any iteration $t \geq 0$, the position $\mathbf{x}_i^t$ of any particle i is in the span of $\mathcal{I}$ where $\mathcal{I} = \{\mathbf{x}_1^0, \mathbf{y}_1^0, \mathbf{v}_1^0, \ldots, \mathbf{x}_m^0, \mathbf{y}_m^0, \mathbf{v}_m^0\}$.*

Proof. Suppose that particle velocities, positions and personal best positions are initialised randomly within the search space. Let the set of all these initial points be given by $\mathcal{I} = \{\mathbf{x}_1^0, \mathbf{y}_1^0, \mathbf{v}_1^0, \ldots, \mathbf{x}_{n_s}^0, \mathbf{y}_{n_s}^0, \mathbf{v}_{n_s}^0\}$. Assume that all the elements in $\mathcal{I}$ are unique and non-zero. These assumptions are made without loss of generality: the probability of obtaining a zero vector from a uniform initialisation is zero, since the probability of a continuous random variable being a particular constant is zero. Similarly, the probability of sampling two equal vectors is zero because the set of such points have zero measure.

The position of any particle at $t = 0$ is in $\mathcal{I}$ by the definition of $\mathcal{I}$. Thus, the hypothesis holds for the case $t = 0$.

At iteration $t = 1$, the position of any particle i is given by

$$\mathbf{x}_i^1 = \mathbf{x}_i^0 + \mathbf{v}_i^1 \tag{9}$$

Since $\mathbf{x}_i^0 \in \mathcal{I}$, it is only necessary to prove that $\mathbf{v}_i^1 \in span(\mathcal{I})$ for $\mathbf{x}_i^1$ to be in the span of $\mathcal{I}$. According to the velocity update equation,

$$\mathbf{v}_i^1 = w\mathbf{v}_i^0 + c_1 r_1(\mathbf{y}_i^0 - \mathbf{x}_i^0) + c_2 r_2(\hat{\mathbf{y}}_i^0 - \mathbf{x}_i^0) \tag{10}$$

$$= w\mathbf{v}_i^0 + c_1 r_1 \mathbf{y}_i^0 + c_2 r_2 \hat{\mathbf{y}}_i^0 - (r_1 c_1 + r_2 c_2)\mathbf{x}_i^0 \tag{11}$$

By definition, $\mathbf{v}_i^0, \mathbf{y}_i^0, \mathbf{x}_i^0 \in \mathcal{I}$. Additionally, the neighbourhood best position is chosen from among the personal best positions of the other particles in the neighbourhood, so that $\hat{\mathbf{y}}_i^0 \in \mathcal{I}$. Thus, if particle i is not the neighbourhood best, then $\mathbf{v}_i^1$ is a linear combination of four distinct elements from $\mathcal{I}$. If particle i is the neighbourhood best, then $\mathbf{y}_i^0 = \hat{\mathbf{y}}_i^0$ and $\mathbf{v}_i^1$ is a linear combination of three distinct elements from $\mathcal{I}$. In either case, $\mathbf{v}_i^1 \in span(\mathcal{I})$ by Definition 3. The fact that $\mathbf{v}_i^1 \in span(\mathcal{I})$ will be referred to as (*). Therefore, since $\mathbf{x}_i^1$ is the sum of two elements in $span(\mathcal{I})$, $\mathbf{x}_i^1$ is in the span of $\mathcal{I}$. Since this is true for any particle i, all the particles' positions at iteration i must be in the span of $\mathcal{I}$ - this fact will be referred to as (**).

Suppose for all iterations $s \leq t$, that the positions of all the particles are in the span of $\mathcal{I}$. It will now be proved that the positions of all the particles must still be in the span of $\mathcal{I}$ at iteration $t+1$. The position of any particle i is given by the position update equation:

$$\mathbf{x}_i^{t+1} = \mathbf{x}_i^t + \mathbf{v}_i^{t+1} \tag{12}$$

where $\mathbf{x}_i^t \in span(\mathcal{I})$ by virtue of the inductive assumption. It thus remains to prove that $\mathbf{v}_i^{t+1}$ is in the span of $\mathcal{I}$:

$$\mathbf{v}_i^{t+1} = w\mathbf{v}_i^t + c_1 r_1(\mathbf{y}_i^t - \mathbf{x}_i^t) + c_2 r_2(\hat{\mathbf{y}}_i^t - \mathbf{x}_i^t) \tag{13}$$

$$= w\mathbf{v}_i^t + c_1 r_1 \mathbf{y}_i^t + c_2 r_2 \hat{\mathbf{y}}_i^t - (r_1 c_1 + r_2 c_2)\mathbf{x}_i^t \tag{14}$$

where $\mathbf{x}_i^t \in span(\mathcal{I})$ by the inductive assumption. It remains to prove that $\mathbf{v}_i^t$, $\mathbf{y}_i^t$ and $\hat{\mathbf{y}}_i^t$ are in the span of $\mathcal{I}$. According to the position update equation,

$$\mathbf{x}_i^t = \mathbf{x}_i^{t-1} + \mathbf{v}_i^t \tag{15}$$

$$\implies \mathbf{v}_i^t = \mathbf{x}_i^t - \mathbf{x}_i^{t-1} \tag{16}$$

In other words, $\mathbf{v}_i^t$ is a linear combination of $\mathbf{x}_i^t$ and $\mathbf{x}_i^{t-1}$, both of which are elements in the span of $\mathcal{I}$ by the inductive assumption. Thus, $\mathbf{v}_i^t$ is also in the span of $\mathcal{I}$. The personal best position of any particle i can only be equal to one of the particle's previous positions. But, by the inductive assumption, all the particle's previous positions were in the span of $\mathcal{I}$. Thus, $\mathbf{y}_i^t$ must be in the span of $\mathcal{I}$. Similarly, the neighbourhood best position must be equal to a previous position of some particle in i's neighbourhood, all of which are in the span of $\mathcal{I}$ by the inductive assumption. Therefore, the velocity and also the position of particle i at iteration $t+1$ must be in $span(\mathcal{I})$.   □

Theorem 1 implies that, when r_1 and r_2 are random scalar values, the positions of the particles are limited to be in the span of their initial velocities, positions and personal best positions. If either of the assumptions on $\mathcal{I}$ does not hold (e.g. some vectors are multiples of another or some are zero), then the positions of the particles are limited further to being linear combinations of all non-zero, linearly independent initial velocities, positions and personal best positions. The question arises whether any point in the search space can be expressed in terms of such linear combinations. This question is answered by the theorem below:

Theorem 2. *Suppose $\mathcal{I}$ contains m linearly independent vectors and $S = [L, U]^n$. If $m < n$ then $span(\mathcal{I}) \cap S \subsetneq S$. Thus $\mathcal{I}$ can only be a spanning set of S if it contains at least n linearly independent elements.*

Theorem 2 follows from the fundamental theorem of invertible matrices (as given in [15]), the exact proof as given in [10] is not reproduced here. If $\mathcal{I}$ constitutes a spanning set for the search space (i.e. the span of $\mathcal{I}$ is larger than the search space or equal to), then any point in the search space can theoretically be reached by the particles. However, if $\mathcal{I}$ is not a spanning set of the search space (i.e. the span of $\mathcal{I}$ is a strict subspace within the search space), then the particles can not reach every position in the search space. If the global optimum happens to be outside the span of $\mathcal{I}$, then the particles will never be able to find it. Since initial positions and personal best positions are typically generated randomly, this is a realistic scenario in high dimensional spaces.

If the swarm's velocities are initialised to zero and the initial personal best positions are set equal to the initial positions, then the portion of the search space that can be reached by the particles is even smaller. Additionally, the swarm may lose degrees of freedom throughout the search. Since the algorithm is executed on a computer with limited precision, some of the vectors in $\mathcal{I}$ may be cancelled out further in the search. Though unlikely, the span of the swarm may in fact decrease as the search progresses. Thus, if the size of the swarm is much smaller than the dimensionality of the search space, then the swarm will be unable to reach a large part of the search space. Unfortunately, simply increasing the number of particles in the swarm is not an adequate solution, because it greatly increases the computational cost. Additionally, the swarm size parameter influences the swarm's searching behaviour in other ways, so changing the swarm size drastically may have unintended consequences [5,9].

4   Illustration of Reachability

This section aims to illustrate the importance of reachability, i.e. that the scalar swarm's performance is severely penalized if the optimum is not in the span of the swarm's initial positions. Section 4.1 describes the experiment's empirical method and Sect. 4.2 summarizes the results.

4.1   Empirical Method - Reachability

The optimum can be placed inside or outside the swarm's initial span by shifting: $f(\mathbf{x})_{Sh} = f((\mathbf{x} - \boldsymbol{\gamma}))$, where $\mathbf{x}$ denotes the position vector to be evaluated, f denotes the objective function and $\boldsymbol{\gamma}$ denotes the shift vector. If the scalar swarm performs well when the shift places the optimum within its reach, but poorly when the optimum is outside the swarm's span, then the importance of reachability will be empirically justified.

Unfortunately, the performance of a swarm on a function with a reachable optimum can not be compared directly with its performance on an unreachable version of that same function. The swarm would essentially be optimizing different functions, since applying a particular shift may change the problem's difficulty. Thus, in order to compare the influence of reachability on the scalar swarm's performance, a total of 30 different shifts were generated for each benchmark problem: 15 placed the optimum within the scalar swarm's span and 15 moved the optimum to an unreachable region. The swarm was run 30 times on all 30 versions of each benchmark function. The suite of benchmark functions consisted of Ackley, Absolute Value, Elliptic, Griewank, Quartic, Rastrigin, Rosenbrock, Schwefel 1.20, Schwefel 2.21, Spherical and Weierstrass (as defined in [6]). A total of 30×11 functions were thus under consideration.

The process for generating the shifts is described below. First, the Modified Gram-Schmidt method [12] was applied to the particles' initial positions to produce a basis $\mathcal{B}$ containing m-many vectors (where m is the swarm size, as before). For the shift to be reachable, a direction vector $\mathbf{d}$ was generated by a random linear combination of the vectors in $\mathcal{B}$. The direction vector was normalized and used to define a line passing through the search space center. A random point on that line, γ was then chosen as the shift. To produce an unreachable shift, a new random vector $\mathbf{s}$ was chosen (distributed uniformly over the search space in each dimension, like the particle positions). The vector $\mathbf{s}$ was then orthogonalized relative to $\mathcal{B}$, producing $\tilde{\mathbf{d}}$, a direction vector orthogonal to the swarm's span. As before, γ was chosen to be a random point on the line passing through the center of the search space with direction $\tilde{\mathbf{d}}$.

The experiments used PSO with inertia weight as introduced in [17] with the global best topology. The selected inertia weight, $w = 0.7298$ and the acceleration coefficients $c_1 = c_2 = 1.49618$ are known good values suggested by Clerc [3] that guarantee convergence of the swarm (in terms of expectation and variance of particle positions [2]). As suggested by [1], all personal and global best positions were restricted to be within the search space. Each swarm consisted of 10 particles ($m = 10$), so that m is low enough to test problems of dimensionality 5 times larger than m without venturing into large scale optimization). The particles' initial positions were initialised uniform randomly throughout the search space. Each particle's initial personal best position was set equal to its initial position, and its velocity was initialised to zero. Thus, the scalar swarm was limited to the span of its initial positions. The experiments were repeated for dimensions $n = \{15, 20, 25, 50\}$. Every simulation ran 2000 iterations to allow sufficient time for the swarm to converge.

4.2   Results - Reachability

Table 1 compares the scalar swarm's performance on the 165 (11 functions $\times$ 15) problems with reachable optima against the 165 problems with unreachable optima. As mentioned in Sect. 4.1, 30 runs were performed on each problem for statistical significance. Every row of the table corresponds to the results for a given problem dimensionality. Friedman tests with a p-value of 0.05 were used to detect statistically significant differences between the scalar and vector PSO's performance (in terms of the best scores attained over all runs on a given function). If the Friedman test indicated a significant difference, pairwise comparisons were done by Mann-Whitney U tests with a p-value of 0.05. If no statistically significant different was found, the result was recorded as a draw.

Table 1. Scalar Swarm's performance on reachable and unreachable problems

Dimensionality	Reachable wins	Draws	Unreachable wins
$n = 15$	104	26	35
$n = 20$	117	21	27
$n = 25$	120	22	23
$n = 50$	93	28	44

As expected from the theoretical discussion, Table 1 shows that the scalar swarm performed significantly better on the problems with reachable optima. It may still be possible for the swarm to perform better on the benchmark functions with theoretically unattainable optima if the swarm's initial subspace is "sufficiently" close to the shifted optimum. Additionally, a given problem with an unreachable problem may still be easier than a problem with a reachable optimum, as discussed in Sect. 4.1, resulting in a few wins for the swarm optimizing the unreachable problems. However, the general trend is that the scalar swarm preform better on a given benchmark function when the optimum is within the span of its initial positions, as would be expected from the theory.

5 Extensive Performance Comparison

As seen in the previous section, it may be that the swarm's reachable subspace may lie in a region sufficiently close to the optimum for it to be a mere technicality that the optimum is unreachable. Since the benchmark suites from the previous section were designed either in favour or against the scalar PSOs, this section compares the performance of the vector and scalar swarms on a large suite of unbiased benchmark functions. The optima for these functions are either shifted by a predefined constant (as specified in the corresponding technical papers) or by a random vector, distributed uniformly over the search space. Section 5.1 details the empirical method and Sect. 5.2 discusses the results.

5.1 Empirical Method - Performance Comparison

The benchmark suite consisted of 28 base functions which are listed in the "Function Name" column of Table 2. A given function f was shifted and rotated to produce f_{ShRot} according to

Table 2. Benchmark functions

Function name	Src	γ	β	Rot	Function name	Src	γ	β	Rot
Absolute Value	f_1	Rand	0.0	No	Rastrigin	f_{12}	Rand	0.0	No
Ackley	f_2	Rand	0.0	No	Rastrigin Rot	f_{12}	2.0	−330	No
Ackley Sh	f_2	10.0	−140	No	Rastrigin Sh	f_{12}	0.0	0.0	Yes
Ackley Rot	f_2	0.0	0.0	Yes	Rastrigin ShRot	f_{12}	1.0	−330	Yes
Ackley ShRot	f_2	−32.0	−140	No	Rosenbrock	f_{13}	Rand	0.0	No
Alpine	F_7	Rand	0.0	No	Rosenbrock Sh	f_{13}	10.0	390	No
Brown	F_{25}	Rand	0.0	No	Rosenbrock Rot	f_{13}	0.0	0.0	Yes
Dixon-Price	F_{48}	Rand	0.0	No	Salomon	f_{14}	Rand	0.0	No
Egg Holder	f_4	Rand	0.0	No	Schaffer 6	f_{15}	Rand	0.0	No
Elliptic	f_5	Rand	0.0	No	Schaffer 6 ShRot	f_{15}	20.0	−300	Yes
Elliptic Sh	f_5	10.0	−450	No	Schwefel	G_5	0.0	0.0	No
Elliptic Rot	f_5	0.0	0.0	Yes	Schwefel 1.2	f_{16}	Rand	0.0	No
Elliptic ShRot	f_5	10.0	−450	Yes	Schwefel 1.2 Sh	f_{16}	10.0	−450	No
Griewank	f_6	Rand	0.0	No	Schwefel 1.2 Rot	f_{16}	0.0	0.0	Yes
Griewank Sh	f_6	10.0	−180	No	Schwefel 2.21	f_{19}	Rand	0.0	No
Griewank Rot	f_6	0.0	0.0	Yes	Schwefel 2.22	f_{20}	Rand	0.0	No
Griewank ShRot	f_6	−60.0	−180	Yes	Shubert	f_{21}	Rand	0.0	No
HyperEllipsoid	f_7	Rand	0.0	No	Spherical	f_{22}	Rand	0.0	No
Michalewicz	f_8	Rand	0.0	No	Spherical Sh	f_{22}	10.0	−450	No
Norwegian	f_9	Rand	0.0	No	Step	f_{23}	Rand	0.0	No
Powell Singular 2	F_{92}	Rand	0.0	No	Vincent	f_{24}	Rand	0.0	No
Quadric	f_{10}	Rand	0.0	No	Weierstrauss	f_{25}	Rand	0.0	No
Quartic	f_{11}	Rand	0.0	No	Weierstrauss Sh	f_{25}	1.0	−130	No

$$f(\mathbf{x})_{ShRot} = f(Q(\mathbf{x} - \boldsymbol{\gamma})) + \beta \qquad (17)$$

where β is a constant scalar, $\boldsymbol{\gamma}$ is either constant or uniform random over the search space in each dimension, and Q is a randomly generated orthogonal matrix. The constants are specified in Table 2.

The "Rot" column in Table 2 indicates whether the function was rotated or not. The transformations provided a total of 46 benchmark functions. The benchmark suite contains uni- and multi-modal functions that are both separable and non-separable. The definitions of the functions and the corresponding bounds were used as in [6], [8] and [16]. The "Src" column of Table 2 lists the identifier of each function according to its source. Function i from [6] is denoted by f_i; function i from [8] is denoted by F_i and function i from [16] is denoted by G_i. The vector and scalar swarms were run on each of the benchmark problems 30 times for statistical significance. Each simulation ran for 2000 iterations.

All of the functions were minimized in 5, 10, 15, 20, and 25 dimensions. As before, the swarm size was set to 10. If the hypothesis proved in the previous section holds, then it is expected for the scalar swarm's performance to dete-

Table 3. Comparison of vector and scalar swarms across dimensionality

Dimensionality	Scalar wins	Draws	Vector wins
$n = 5$	0	2	44
$n = 10$	1	0	45
$n = 15$	1	0	45
$n = 20$	1	2	43
$n = 25$	1	1	44

riorate as the problem dimensionality exceeds the number of particles in the swarm.

5.2   Results - Performance Comparison

Table 3 summarizes the results of the wide performance comparison. The scalar swarm consistently performed better than the vector swarm on the Quadric function (f_{10}) for $n > 5$. However, the vector swarm outperforms the scalar swarm on nearly all of the benchmark functions, even when the problem dimensionality is low enough for the scalar swarm to reach the optimum. Although the reachability of the optimum also plays a role (as shown previously), the scalar swarm's linear movement prevents the swarm from finding good solutions even inside the swarm's initial subspace.

The strong restriction imposed on the scalar swarm becomes apparent in Figs. 1, 2, 3 and 4, which plot typical profiles of the swarm diversity (as defined in [11]), averaged over all runs for $n = 5$ and $n = 25$. As the problem dimensionality increases, the vector PSO's swarm diversity also increases. In contrast, the scalar PSO's diversity profile remains unchanged even as the dimensionality increases.

As shown before [20], restricting the swarm's movement may be beneficial in high dimensional spaces for the very reason that the initial velocity explosion

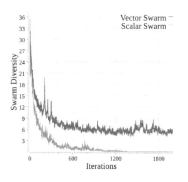

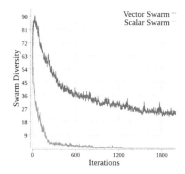

Fig. 1. Diversity, Ackley Shr ($n = 5$)   **Fig. 2.** Diversity, Ackley Shr ($n = 25$)

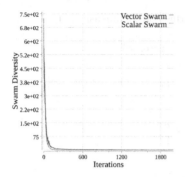

Fig. 3. Diversity, Griewank ($n = 5$)

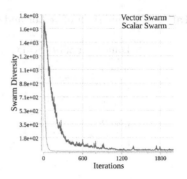

Fig. 4. Diversity, Griewank ($n = 25$)

is mitigated, causing the unchanged diversity profile observed here. However, in low dimensional spaces, the vector swarm outperforms the scalar swarm.

6   Conclusion

This paper demonstrated the importance of employing component-wise stochasticity both theoretically and empirically. Section 3 showed that a swarm's movement is severely restricted by using scalar values for r_1 and r_2. In particular, it is emphasized that a scalar swarm is limited to the span of its initial particle positions, personal best positions and velocities. Thus, the swarm may not be able to reach the optimum. Section 4 shows that reachability is not merely a theoretical problem, but can also be illustrated empirically. The section constructs benchmark functions with optima explicitly defined to be reachable or unreachable by a scalar swarm. The scalar swarm is shown to perform significantly better on benchmark problems with reachable optima, as expected from the theory.

Since the benchmarks in Sect. 4 were biased in favour of, or against the scalar swarms, the artificial benchmarks would not provide a fair comparison between scalar and vector swarms. Towards this end, Sect. 5 demonstrated the performance difference between scalar and vector swarms on an extensive range of benchmarks. It was shown that the vector swarm performs significantly better on almost all of the benchmark functions than the scalar swarm, even when the dimensionality is low enough for the scalar swarm to reach the optimum.

Acknowledgments. This work is based on the research supported by the National Research Foundation (NRF) of South Africa (Grant Number 46712). The opinions, findings and conclusions or recommendations expressed in this article is that of the author(s) alone, and not that of the NRF. The NRF accepts no liability whatsoever in this regard.

References

1. Bratton, D., Kennedy, J.: Defining a standard for particle swarm optimization. In: Proceedings of the IEEE Swarm Intelligence Symposium, pp. 120–127. IEEE Computer Society (2007). https://doi.org/10.1109/SIS.2007.368035
2. Cleghorn, C.W., Engelbrecht, A.P.: Particle swarm stability: a theoretical extension using the non-stagnate distribution assumption. Swarm Intell. **12**, 1–22 (2017). https://doi.org/10.1007/s11721-017-0141-x
3. Clerc, M., Kennedy, J.: The particle swarm - explosion, stability, and convergence in a multidimensional complex space. IEEE Trans. Evol. Comput. **6**(1), 58–73 (2002). https://doi.org/10.1109/4235.985692
4. Eberhart, R., Kennedy, J.: A new optimizer using particle swarm theory. In: Proceedings of the Sixth International Symposium on Micro Machine and Human Science, pp. 39–43, October 1995. https://doi.org/10.1109/MHS.1995.494215
5. Engelbrecht, A.P.: Fitness function evaluations: a fair stopping condition? In: Proceedings of the IEEE Symposium on Swarm Intelligence, pp. 1–8, December 2014. https://doi.org/10.1109/SIS.2014.7011793
6. Engelbrecht, A.: Particle swarm optimization: global best or local best? In: Proceedings of the BRICS Congress on Computational Intelligence and 11th Brazilian Congress on Computational Intelligence (BRICS-CCI CBIC), pp. 124–135, September 2013. https://doi.org/10.1109/BRICS-CCI-CBIC.2013.31
7. Han, F., Liu, Q.: A diversity-guided hybrid particle swarm optimization based on gradient search. Neurocomputing **137**, 234–240 (2014). https://doi.org/10.1016/j.neucom.2013.03.074. Advanced Intelligent Computing Theories and Methodologies
8. Jamil, M., Yang, X.S.: A literature survey of benchmark functions for global optimization problems. Int. J. Math. Model. Numer. Optim. **4**(2), 150–194 (2013)
9. Malan, K., Engelbrecht, A.P.: Algorithm comparisons and the significance of population size. In: Proceedings of the IEEE Congress on Evolutionary Computation (IEEE World Congress on Computational Intelligence), pp. 914–920 (2008)
10. Oldewage, E.: The perils of particle swarm optimisation in high dimensional problem spaces. Master's thesis, University of Pretoria, Pretoria, South Africa (2018)
11. Olorunda, O., Engelbrecht, A.P.: Measuring exploration/exploitation in particle swarms using swarm diversity. In: Proceedings of the IEEE Congress on Evolutionary Computation, pp. 1128–1134, June 2008. https://doi.org/10.1109/CEC.2008.4630938
12. Paige, C.C., Rozložník, M., Strakos, Z.: Modified Gram-Schmidt (MGS), least squares, and backward stability of MGS-GMRES. Soc. Ind. Appl. Math. J. Matrix Anal. Appl. **28**(1), 264–284 (2006). https://doi.org/10.1137/050630416
13. Pandey, S., Wu, L., Guru, S.M., Buyya, R.: A particle swarm optimization-based heuristic for scheduling workflow applications in cloud computing environments. In: 2010 24th IEEE International Conference on Advanced Information Networking and Applications, pp. 400–407, April 2010. https://doi.org/10.1109/AINA.2010.31
14. Paquet, U., Engelbrecht, A.P.: Particle swarms for linearly constrained optimisation. Fundam. Inform. **76**(1–2), 147–170 (2007). http://dl.acm.org/citation.cfm?id=1232695.1232705
15. Poole, D.: Linear Algebra: A Modern Introduction, 3rd edn. Cengage Learning, Canada (2011)
16. Ramezani, F., Lotfi, S.: The modified differential evolution algorithm (MDEA). In: Pan, J.S., Chen, S.M., Nguyen, N.T. (eds.) ACIIDS 2012. LNCS, vol. 7198, pp. 109–118. Springer, Heidelberg (2012). https://doi.org/10.1007/978-3-642-28493-9_13

17. Shi, Y., Eberhart, R.: A modified particle swarm optimizer. In: Proceedings of the IEEE International Conference on Evolutionary Computation, pp. 69–73, May 1998. https://doi.org/10.1109/ICEC.1998.699146

18. Yoshida, H., Kawata, K., Fukuyama, Y., Takayama, S., Nakanishi, Y.: A particle swarm optimization for reactive power and voltage control considering voltage security assessment. IEEE Trans. Power Syst. **15**(4), 1232–1239 (2000). https://doi.org/10.1109/59.898095

19. Zahara, E., Kao, Y.T., Su, J.R.: Enhancing particle swarm optimization with gradient information. In: 2009 Fifth International Conference on Natural Computation, vol. 3, pp. 251–254, August 2009. https://doi.org/10.1109/ICNC.2009.711

20. van Zyl, E., Engelbrecht, A.: Group-based stochastic scaling for PSO velocities. In: Proceedings of the IEEE Congress on Evolutionary Computation (CEC), pp. 1862–1868, July 2016

The Importance of Information Flow Regulation in Preferentially Foraging Robot Swarms

Lenka Pitonakova[1(✉)], Richard Crowder[2], and Seth Bullock[1]

[1] Department of Computer Science, Faculty of Engineering, University of Bristol, Bristol, UK
contact@lenkaspace.net, seth.bullock@bristol.ac.uk
[2] Department of Electronics and Computer Science, Faculty of Physical and Applied Sciences, University of Southampton, Southampton, UK
rmc@ecs.soton.ac.uk

Abstract. Instead of committing to the first source of reward that it discovers, an agent engaged in "preferential foraging" continues to choose between different reward sources in order to maximise its foraging efficiency. In this paper, the effect of preferential source selection on the performance of robot swarms with different recruitment strategies is studied. The swarms are tasked with foraging from multiple sources in dynamic environments where worksite locations change periodically and thus need to be re-discovered. Analysis indicates that preferential foraging leads to a more even exploitation of resources and a more efficient exploration of the environment provided that information flow among robots, that results from recruitment, is regulated. On the other hand, preferential selection acts as a strong positive feedback mechanism for favouring the most popular reward source when robots exchange information rapidly in a small designated area, preventing the swarm from foraging efficiently and from responding to changes.

1 Introduction

Instead of committing to the first source of reward that it discovers, an agent engaged in "preferential foraging" continues to choose between different reward sources in order to maximise its foraging efficiency [13]. This foraging behaviour appears in nature at the level of individual creatures, for example in fish [13] and birds [5], as well as at the collective level, for instance in honey bee [2] and ant [28] colonies. While numerous studies have shown that preferential foraging is advantageous for animals, the conditions under which these advantages transfer to biologically-inspired robot swarms are currently unclear.

During robot swarm foraging, individuals are required to search an unknown environment for *worksites* that contain reward. The robots may perform work directly at the worksite locations during *general foraging* (e.g., [16,24]) or transport resource collected at worksites to a designated deposition area in *central-place foraging* (e.g., [4,14]). Robots may communicate information about worksites, such as their locations, to other members of the swarm in order to facilitate

© Springer Nature Switzerland AG 2018
M. Dorigo et al. (Eds.): ANTS 2018, LNCS 11172, pp. 277–289, 2018.
https://doi.org/10.1007/978-3-030-00533-7_22

faster exploitation of the environment (e.g., [3,23]). In this paper, the effect of robots preferring worksites with higher utilities, i.e., higher reward returns, is explored in general and central-place foraging tasks, where worksite locations are not known in advance and change over time. The foraging tasks presented here are a paradigm for a number of real-world robot swarm tasks, such as package delivery, environment sampling and resource collection.

Two robot swarms with different communication strategies are studied. In Broadcaster swarms, individuals advertise worksite information to nearby robots while they are near worksites. In Bee swarms, robots exchange information in a designated area, that they return to periodically. Our previous work [20,22] suggests that Broadcaster swarms outperform Bee swarms in many foraging environments due to their ability to respond to environmental changes faster, but that Bee swarms are more suitable during central-place foraging in environments with a low worksite density, since their recruitment strategy allows the robots to share information with each other relatively quickly. Here we show that the rapid spread of information through Bee swarms damages their performance when individuals preferentially choose worksites with higher utility, since most of the swarm may tend to concentrate on a single worksite, increasing the negative effects of congestion. On the other hand, Broadcasters form small sub-groups that can use preferential foraging to choose non-congested worksites, because the information spread in these swarms is regulated by the limited communication range of robots.

2 Methods

2.1 Simulated Environment

A continuous-space experimental arena, identical to that in [22] and containing a central circular base surrounded by circular worksites, was created in the ARGoS simulator [18]. An experimental environment was characterised by the number of worksites, N_W, and worksite distance from the base, $D \in \{5, 9, 13, 17\}$ m. There were two types of environment (Fig. 1a and b):

- **HeapN_W**: $N_W \in \{1, 2, 4\}$ high-volume worksites evenly distributed around the base at a distance D from the base edge.
- **Scatter25**: $N_W = 25$ worksites randomly distributed between distance D and $D - 5$ m from the base edge.

The total amount of resource in each environment was set to 100 units and the amount of resource units per worksite, $V_W = 100/N_W$. Each worksite had a 0.1 m radius and there was a colour gradient with 1 m radius around it, that the robots could use to "sense" and navigate towards the worksite (Fig. 1c). The base had a radius of 3 m and featured a light source above its centre that the robots could use to navigate towards the base.

At the beginning of each experiment, a number, $N_R \in \{10, 25, 50\}$ of robots were placed at random positions and with random orientations in the base.

Two types of foraging tasks were investigated, as in [22]:

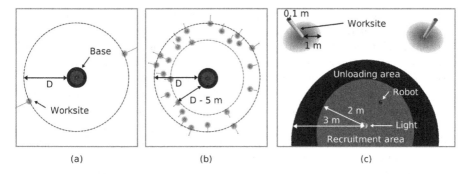

Fig. 1. The (a) Heap2 and (b) Scatter25 environments with worksite distance $D =$ 13 m. (c) The base and nearby worksites. Figure reproduced from [22]

- **Consumption**: Worksites represented "jobs" that could be completed at the worksite locations. A robot that was at a worksite gradually depleted its volume, increasing the swarm's total reward at the rate of 1/400 units per second. Similar tasks were explored in, e.g., [16,24].
- **Collection**: Worksites represented resource deposits. A robot could collect one unit of resource at a time, after which the resource had to be deposited in the base. Similar tasks were explored in, e.g., [4,14].

Additionally, each task had two variants, *slow* and *fast*, that represented different degrees of challenge. In each variant, worksite locations changed every T_C seconds and were chosen randomly according to the environment type. For example, in the Heap2 environment with $D = 5$ m, the two worksites were relocated every T_C seconds, remaining 5 m from the edge of the base. Worksite volumes were replenished after each change. The value of T_C, as well as the total simulation time, T, were set as in [22], so that the environment changed 10 times in the slow variant and 20 times in the fast variant and so that a swarm could deplete around 50% of total worksite volume in each slow change interval. For example, $T_C = 45$ min and $T = 7.5$ h for 50-robot swarms.

2.2 Robots

The simulated MarXbots [1] were differentially steered circular robots with a radius of 8.5 cm. The robots could communicate with each other using a range and bearing module with a signal range of 5 m. We have previously described the robot model in [20]. There were two types of robot swarm, that we parametrised for the best performance in a series of environments [22]:

- **Broadcaster** (Fig. 2a): Robots left the base immediately at the beginning of an experiment to start scouting for worksites. Upon discovering a worksite, a scout started working on it, meaning that it either started depleting its volume (in the Consumption task) or started travelling between the worksite and the base to gradually deposit resource (in the Collection task). The robot kept track of worksite location by using odometry. A robot that was located on the gradient surrounding a worksite also broadcasted the worksite location to other robots that were within the communication range. A scout that received the message was recruited to the worksite and started working from it.

– **Bee** (Fig. 2b): Robots left the base with probability $p(S) = 10^{-3}$ to start scouting for worksites. In the Consumption task, a robot that discovered a worksite first returned to the recruitment area of the base (Fig. 1c) in order to recruit any "observing" robots for $T_R = 120$ s and then resumed working. In the Collection task, a robot recruited after each time that it deposited resource in the base. A scout that could not find any worksites for $T_S = 18$ min returned to the base in order to start observing recruitment signals.

The Broadcaster control strategy was inspired by the behaviour of animals such as sheep [17] and fish [15], where individuals observe each other during foraging and are attracted to locations that others forage from. A similar strategy has also been implemented in robot swarms, e.g. in [3,8,30]. The Bee control strategy was inspired by the foraging behaviour of honey bees [27] and has also been implemented in robot swarms, e.g. in [9,14,23].

Differential steering sensors and motors of the robots were subject to minor noise, which could result in accumulation of errors in the relative vector to a robot's worksite. Therefore, upon arriving to a supposed worksite location that was empty, a robot performed neighbourhood search around the location that lasted for 5 min.

Experiments were performed with *Committed* and *Preferential* swarms. Robots in the Committed swarms remained foraging from the same worksite that they discovered or were recruited to until the worksite was depleted. Robots in Preferential swarms exchanged both worksite locations and worksite utilities and always preferred to forage from a worksite with a higher utility. In the Consumption task, worksite utility, U_W, was equal to its current volume, V_W:

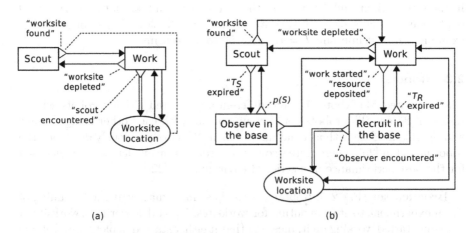

Fig. 2. BDRML [21] representations of the (a) Broadcaster and (b) Bee robot controllers.

$$U_W = V_W \tag{1}$$

In the Collection task, the distance, D_W, between a worksite and the base was also considered, so that worksites that were further away from the base had a lower utility:

$$U_W = V_W/D_W \tag{2}$$

A preferentially foraging robot switched to a worksite with a higher utility if it found such a worksite during its journey to its current worksite location. Additionally, in Preferential Broadcaster swarms, a scout could receive recruitment signals from multiple directions and always chose a worksite with the highest advertised U_W. Furthermore, if the distance between worksites was smaller than the communication range of robots, the robots exchanged information about U_W each second while they were working. If worksite A was being depleted faster than worksite B (because more robots were working at A), its U_W decreased faster, meaning that robots from A were eventually recruited to B. In Preferential Bee swarms, robots exchanged information about U_W while they were recruiting in the base. If there were multiple worksites advertised at the same time, all observing robots and recruiters adopted the worksite with the highest advertised utility.

2.3   Terminology and Data Visualisation

Swarm performance analysis is conducted within the Information-Cost-Reward framework [22]. The framework allows us to identify various costs that the robots incur each second during foraging instead of obtaining reward. The *uncertainty cost*, C_U, is incurred by robots that do not know about worksites. The *displacement cost*, C_D is incurred by robots that know where worksites are located, but are currently not at their worksites, unable to receive reward from them. For example, Bee swarm robots usually incur high amounts of C_D, because they are recruited in the base, i.e., far away from worksites. The displacement cost coefficient, d, represents a ratio between the amount of C_D and the decrease in C_U paid at a given time. When $d = 1$, all robots that know about worksites are displaced from them and no reward is obtained. Intermediate values of $0 < d < 1$ indicate that some robots are displaced and some are receiving reward. Finally, the *misinformation cost*, C_M, is incurred by robots that are away from their worksites and do not know that the worksites have already been depleted by other robots. There is a high potential for a robot to incur C_M during Collection, since it periodically returns to the base to deposit resource.

In the following two sections, swarm performance and ICR metrics are presented in the form of box plots. Each data point represents a median result of 50 independent simulation runs. The surrounding boxes represent the inter-quartile range of the result set, and whiskers represent data in the range of 1.5 times the inter-quartile. Outliers outside this range are shown as plus signs.

3 The Performance of Committed Swarms

In each foraging task, the performance of the Committed swarms depended on the number of worksites, N_W, and the distance of worksites from the base, D (Fig. 3). The performance was generally better in environments with a high worksite density, i.e., when N_W was high, or when D was low. However, note that in the slow Consumption task (Fig. 3a), the Bee swarms experienced a high amount of congestion around worksites in Scatter25 environments when $D \leq$ 9 m, causing their performance to be lower than that in Heap4 environments. The nature of their recruitment strategy in this particular task and environment caused many Bee robots to concentrate on a small number worksites, which was disadvantageous when many worksites needed to be found and exploited.

The Broadcasters outperformed the Bee swarms in the Consumption task, because they did not return to the base in order to recruit, which allowed them to spend more time working. However, in the Collection task, the displacement and misinformation costs associated with central-place recruitment were ameliorated by the fact that all robots had to return to the base periodically in order to deposit resource. This allowed the Bee swarms to surpass or match the performance of Broadcasters in many low-density environments, such as Heap1 and

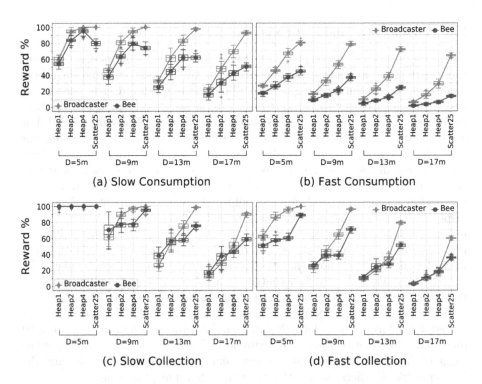

Fig. 3. The amount of reward collected by 50-robot Committed swarms in various tasks and environments.

Heap2, especially in the slow Collection task. Similar trends in absolute and relative swarm performance were discovered for a number of explored swarm sizes ($N_R \in \{10, 25, 50\}$). However, the largest swarms showed the largest differences in relative performance to each other.

4   The Performance of Preferential Swarms

The impact of preferential foraging on swarm performance depended on the environment and task type, as well as on the control strategy of robots. In general, swarm performance was affected more strongly when worksites were closer to the base, i.e., when robots could find worksite information faster. Similarly, larger swarms were affected in a larger number of environments, since the robots could receive information from a larger number of recruiters. Broadcasters were able to improve their performance in the Consumption task, while Bee swarms were negatively affected in both tasks (Fig. 4).

There were two different ways in which Preferential Broadcasters improved their performance in the Consumption task. In Scatter25 environments with a short D, where it was possible for multiple worksites to be located within the communication range of robots, a recruited robot was able to find worksites with larger volumes, i.e. a larger utility, on its way to the location that it was originally recruited to. This allowed the Preferential Broadcaster swarms to spread their foraging effort across multiple worksites better and prevent congestion, which in many environments decreased their displacement cost coefficient compared to the Committed swarms (Fig. 5a and c). This was especially advantageous when the environment changed quickly, i.e., when it was more important to exploit as many worksites as possible in a relatively short amount of time. For example, 50-robot Preferential broadcaster swarms obtained around 12% more reward in the Scatter25 environment when $D = 5\,\mathrm{m}$ (Fig. 4b). Secondly, in Heap environments with a small D, recruits often could not reach an advertised worksite due to congestion around it. Robots in congested areas often made a lot of turns while avoiding each other, causing their odometry-based vector to the worksite to become increasingly incorrect due to the cumulative effect of sensory-motor noise. Preferential Broadcasters communicated about the worksite utility and location periodically, meaning that some recruits were sent to incorrect locations. This cleared congestion and allowed the recruits to explore a new area after they could not find the advertised worksite.

On the other hand, in the Collection task, where robots deposited resource in the base and where congestion around worksites was thus cleared periodically, recruitment to incorrect worksite locations decreased the performance of Preferential Broadcasters in Heap environments with a small D (Fig. 4c and d). In these cases, it was more advantageous for robots to wait until the path to their worksite became less congested, rather than to travel away and search for new worksites. Secondly, in Scatter25 environments, robots could not share information about worksite utilities with each other as frequently as in the Consumption task, since they spent most of the time travelling between the base and worksites, rather than recruiting near worksites.

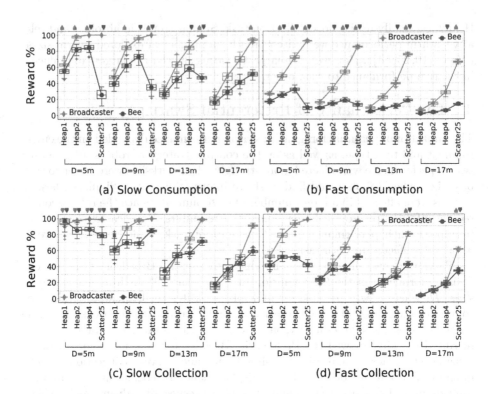

Fig. 4. The amount of reward collected by 50-robot Preferential swarms in various tasks and environments. The up and down arrows above the data points for each environment indicate statistically significant (ANOVA, $p = 0.01$) increase and decrease in performance of the Broadcaster (orange) and Bee (green) Preferential swarms, when compared to the corresponding Committed swarms from Fig. 3 in the same environment. When no arrow is shown for a given environment and swarm type, the Preferential and the Committed swarms of that type performed similarly. (Color figure online)

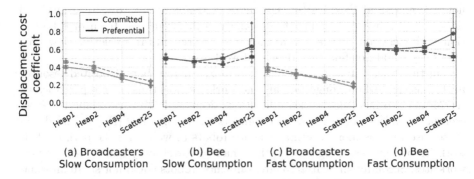

Fig. 5. The displacement cost coefficient, d, of the Committed and Preferential swarms in various tasks and environments with $D = 5$ m.

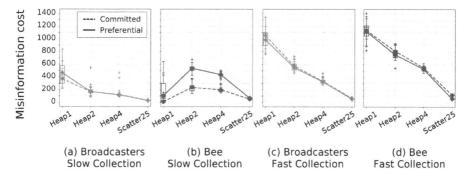

Fig. 6. The misinformation cost, C_M, paid per minute by the Committed and Preferential swarms in various tasks and environments with $D = 5\,\text{m}$.

Preferential Bee swarms exploited the environment less efficiently than the Committed Bee swarms. Since Bee swarm robots utilised central-place recruitment, the preference for a single worksite with the highest utility often spread to the majority of robots. This increased the amount of congestion around that worksite, preventing the robots from exploiting it, as well as from exploring the rest of the arena. In the Consumption task, the swarm's displacement cost coefficient, d, increased as a result of congestion in environments with multiple worksites (Fig. 5b and d), while the d of Broadcasters remained similar or decreased (Fig. 5a and c). Additionally, in the slow Collection task, where recruitment to a single worksite in Bee swarms was much stronger as the robots exchanged information every time they returned to the base, the amount of misinformation cost that the robots incurred increased (Fig. 6b).

As a result of these effects, Preferential foraging decreased the performance of Bee swarms in both tasks, and its impact was significantly stronger than in the Broadcaster swarms (Fig. 4). Consequently, the differences between the performance of Preferential Broadcaster and Bee swarms were stronger than between the Committed swarms. Most notably, Preferential Bee swarms performed significantly worse than Broadcasters in environments with a high worksite density due to the increased amount of costs that they incurred.

5   Discussion

While preferential foraging may be beneficial to a foraging swarm, the results presented here indicate that the spread of preference for a selected source needs to be regulated via a negative feedback mechanism, especially when the environment is dynamic and changes over time.

For instance, bee-inspired swarms in [29] were successfully able to distribute their foraging effort between worksites based on worksite utility. Better worksites were advertised in the base for a longer amount of time, while the regulation of information flow was achieved by allowing agents to choose randomly between advertised worksites, preventing all robots from adopting the same choice. In [20], we explored preferentially foraging bee-inspired swarms in dynamic environments where worksite locations remained the same, but where worksite utilities changed over time. The task of the swarms was to collect resource from the worksites into the base and to switch to a worksite with a higher utility when the environment changed. In line with the results presented here, it was shown that bee swarms employing preferential foraging performed worse than other swarms due to their strong commitment to a small portion of worksites. However, presence of negative feedback in the form of utility-dependent worksite abandonment significantly improved the swarm performance.

Studies with other control strategies also showed that maximising information spread within the swarm is not appropriate when a swarm needs to react to a randomly changing environment. In [24], foraging swarms with localised communication outperformed those with a global communication strategy, where all robots were informed about worksite location and urgencies (analogous to our worksite utilities). Localised communication prevented robots from costly travel to distant worksite locations and from interfering with each other near worksites. In ant-inspired foraging swarms, where robots dropped beacons into the environment in order to form virtual trails to worksites, a disrupted trail could be re-established provided that robots sometimes stopped following social information stored in the beacons and started exploring the environment instead [11]. The importance of maintaining a balance between exploration, exploitation and information sharing has also been demonstrated for other collective tasks, such as area clearing [25] and labour division [12].

It is important to note that other aspects of information exchange, for instance, the granularity of data, also impact foraging performance. In [10], ant-inspired swarms formed virtual pheromone trails using beacons between a base and food sources. It was shown that more efficient foraging routes were formed when the beacons stored simple "hop count" integers rather than fine-grained floating point numbers.

Finally, it is interesting that studies of honey bee foraging, contrary to the results presented here, showed that direct exchange of information about nectar profitability among worker bees *improves* the ability of bee colonies to forage efficiently [2]. Workers sample the nectar that other bees bring into the nest, which helps the colony as a whole to react to rapid changes in nectar quality of different flower patches. However, it is important to point out that bees exhibit preferential nectar source selection in addition to a large repertoire of other self-regulatory and communication behaviours. For example, bees also regularly scout the environment in order to discover new flower patches [26] and they periodically check the profitability of old abandoned patches [6]. The colony also monitors and maintains a healthy nectar intake in order to prevent energy wastage resulting from congestion in the nest [7]. Similarly, ants prefer to follow stronger pheromone trails, but pheromone evaporation [28], and in some ant species, time-dependent decrease of interest of free workers in foraging [19], represent negative feedback mechanisms that regulate the colony's responses.

6   Conclusion

In order for preferential source selection to improve swarm performance, the flow of information about worksite utility should be sufficient, but also regulated through a negative feedback mechanism. Sufficient information flow can occur, for example, in environments with an adequate worksite density, or in swarms where robots can meet each other frequently and thus evaluate different sources of information at the same time. Regulation of information flow can be achieved, e.g., by using the Broadcaster recruitment strategy, where recruitment is limited by the communication range of robots. On the other hand, if many robots are allowed to communicate their preferences at the same time, and when robots adopt the best option with a 100% probability, as was the case in our Bee swarms, the environment is exploited and explored inefficiently, preventing the swarm from responding to changes.

These results point to the importance of studying the context in which a particular behaviour is used in a swarm, especially when nature-inspired behaviours are taken out of their biological context and are applied in an engineered robot control algorithm. A new behaviour, such as preferential foraging, adds a new feedback mechanism that interacts with other feedback mechanisms in unique ways. In the experiments presented here, preferential foraging acted as a strong positive feedback mechanism for the most popular worksite choice when robots exchanged information in a small designated area. On the other hand, the same behaviour facilitated a more even exploitation of resources and a more efficient exploration when recruitment was regulated through negative feedback mechanisms.

Acknowledgments. This work was supported by EPSRC grants EP/G03690X/1, EP/N509747/1 and EP/R0047571.

References

1. Bonani, M., et al.: The MarXbot, a miniature mobile robot opening new perspectives for the collective-robotic research. In: Proceedings of 2010 IEEE/RSJ International Conference on Intelligent Robots and Systems (IROS 2010), pp. 4187–4193. IEEE, Piscataway (2010)
2. De Marco, R., Farina, W.M.: Trophallaxis in forager honeybees Apis mellifera: Resource uncertainty enhances begging contacts? J. Comp. Physiol. A **189**, 125–134 (2003). https://doi.org/10.1007/s00359-002-0382-y
3. Ducatelle, F., et al.: Cooperative navigation in robotic swarms. Swarm Intell. **8**(1), 1–33 (2014)
4. Ducatelle, F., Di Caro, G.A., Pinciroli, C., Gambardella, L.M.: Self-organized cooperation between robotic swarms. Swarm Intell. **5**(2), 73–96 (2011)
5. Gill, F.B., Wolf, L.L.: Nonrandom foraging by sunbirds in a patchy environment. Ecology **58**(6), 1284–1296 (1997)
6. Granovskiy, B., Latty, T., Duncan, M., Sumpter, D.J.T., Beekman, M.: How dancing honey bees keep track of changes: The role of inspector bees. Behav. Ecol. **23**(3), 588–596 (2012). https://doi.org/10.1093/beheco/ars002

7. Gregson, A.M., Hart, A.G., Holcombe, M., Ratnieks, F.L.: Partial nectar loads as a cause of multiple nectar transfer in the honey bee (Apis mellifera): a simulation model. J. Theor. Biol. **222**(1), 1–8 (2003). https://doi.org/10.1016/S0022-5193(02)00487-3

8. Gutiérrez, Á., Campo, A., Monasterio-Huelin, F., Magdalena, L., Dorigo, M.: Collective decision-making based on social odometry. Neural Comput. Appl. **19**(6), 807–823 (2010)

9. Hecker, J.P., Moses, M.E.: Beyond pheromones: evolving error-tolerant, flexible, and scalable ant-inspired robot swarms. Swarm Intell. **9**, 43–70 (2015)

10. Hoff, N., Sagoff, A., Wood, R.J., Nagpal, R.: Two foraging algorithms for robot swarms using only local communication. In: Proceedings of the 2010 IEEE International Conference on Robotics and Biomimetics (ROBIO 2010), pp. 123–130. IEEE, Piscataway (2010)

11. Hrolenok, B., Luke, S., Sullivan, K., Vo, C.: Collaborative foraging using beacons. In: van der Hoek, W., Kaminka, G.A., Lesperance, Y., Luck, M., Sen, S. (eds.) Proceedings of 9th International Conference on Autonomous Agents and Multiagent Systems (AAMAS 2010), pp. 1197–1204. IFAAMAS, Richland (2010)

12. Jones, C., Mataric, M.J.: Adaptive division of labor in large-scale minimalist multi-robot systems. In: Proceedings 2003 IEEE/RSJ International Conference on Intelligent Robots and Systems (IROS 2003), vol. 2, pp. 1969–1974. IEEE, Piscataway (2003)

13. Krause, J., Godin, J.G.J.: Influence of prey foraging posture on flight behavior and predation risk: predators take advantage of unwary prey. Behav. Ecol. **7**(3), 264–271 (1996)

14. Krieger, M.J.B., Billeter, J.B.: The call of duty: self-organised task allocation in a population of up to twelve mobile robots. Rob. Auton. Syst. **30**(1–2), 65–84 (2000)

15. Lachlan, R., Crooks, L., Laland, K.: Who follows whom? Shoaling preferences and social learning of foraging information in guppies. Anim. Behav. **56**(1), 181–190 (1998). https://doi.org/10.1006/anbe.1998.0760

16. Lerman, K., Jones, C., Galstyan, A., Mataric, M.J.: Analysis of dynamic task allocation in multi-robot systems. Int. J. Rob. Res. **25**, 225–242 (2006)

17. Michelena, P., Jeanson, R., Deneubourg, J.L., Sibbald, A.M.: Personality and collective decision-making in foraging herbivores. Philos. Trans. R. Soc. Lond. B Biol. Sci. **277**(1684), 1093–1099 (2010). https://doi.org/10.1098/rspb.2009.1926

18. Pinciroli, C., et al.: ARGoS: a modular, parallel, multi-engine simulator for multi-robot systems. Swarm Intell. **6**(4), 271–295 (2012)

19. Pinter-Wollman, N., et al.: Harvester ants use interactions to regulate forager activation and availability. Anim. Behav. **86**(1), 197–207 (2013)

20. Pitonakova, L., Crowder, R., Bullock, S.: Information flow principles for plasticity in foraging robot swarms. Swarm Intell. **10**(1), 33–63 (2016)

21. Pitonakova, L., Crowder, R., Bullock, S.: Behaviour-data relations modelling language for multi-robot control algorithms. In: Proceedings of 2017 IEEE/RSJ International Conference on Intelligent Robots and Systems (IROS 2017), pp. 727–732. IEEE, Piscataway (2017)

22. Pitonakova, L., Crowder, R., Bullock, S.: The Information-Cost-Reward framework for understanding robot swarm foraging. Swarm Intell. **12**(1), 71–96 (2018). https://doi.org/10.1007/s11721-017-0148-3

23. Reina, A., Miletitch, R., Dorigo, M., Trianni, V.: A quantitative micro-macro link for collective decisions: the shortest path discovery/selection example. Swarm Intell. **9**(2), 75–102 (2015)

24. Sarker, M.O.F., Dahl, T.S.: Bio-Inspired communication for self-regulated multi-robot systems. In: Yasuda, T. (ed.) Multi-Robot Systems, Trends and Development, pp. 367–392. InTech (2011)
25. Schmickl, T., Crailsheim, K.: Throphallaxis within a robotic swarm: bio-inspired communication among robots in a swarm. Auton. Robots **25**(1), 171–188 (2008)
26. Seeley, T.D.: Honey bee foragers as sensory units of their colonies. Behav. Ecol. Sociobiol. **34**(1), 51–62 (1994). https://doi.org/10.1007/BF00175458
27. Seeley, T.D., Camazine, S., Sneyd, J.: Collective decision-making in honey bees: how colonles choose among nectar sources. Behav. Ecol. Sociobiol. **28**, 277–290 (1991)
28. Sumpter, D.J.T., Beekman, M.: From nonlinearity to optimality: pheromone trail foraging by ants. Anim. Behav. **66**(2), 273–280 (2003). https://doi.org/10.1006/anbe.2003.2224
29. Valentini, G., Hamann, H., Dorigo, M.: Self-organized collective decision making: the weighted voter model. In: Proceedings of 13th International Conference on Autonomous Agents and Multiagent Systems (AAMAS 2014), pp. 45–52. ACM, New York (2014)
30. Wawerla, J., Vaughan, R.T.: A fast and frugal method for team-task allocation in a multi-robot transportation system. In: Proceedings of 2010 IEEE International Conference on Robotics and Automation (ICRA 2010), pp. 1432–1437. IEEE, Piscataway (2010)

The Role of Largest Connected Components in Collective Motion

Heiko Hamann(✉) ⓘ

Institute of Computer Engineering, University of Lübeck, Lübeck, Germany
hamann@iti.uni-luebeck.de

Abstract. Systems showing collective motion are partially described by a distribution of positions and a distribution of velocities. While models of collective motion often focus on system features governed mostly by velocity distributions, the model presented in this paper also incorporates features influenced by positional distributions. A significant feature, the size of the largest connected component of the graph induced by the particle positions and their perception range, is identified using a 1-d self-propelled particle model (SPP). Based on largest connected components, properties of the system dynamics are found that are time-invariant. A simplified macroscopic model can be defined based on this time-invariance, which may allow for simple, concise, and precise predictions of systems showing collective motion.

1 Introduction

Collective behaviors of animals and other agents can be described by general principles based on a few typical classes despite their variety and richness of details [25]. At large there are two methods of modeling. Microscopic models abstract individual behaviors but represent properties of each member of the animal group, such as position, direction, and internal states. Examples are swarm models from physics, such as the Vicsek model and similar ones [3,15,24], and models focusing on the agents' energy reserve, their random walks, and more complex behaviors (e.g., taxis behaviors), such as so-called Brownian agents [12, 21,22]. Macroscopic models abstract away individual agents and reduce the state space to only a few variables, such as densities or sizes of groups that are in a certain state (cf. population models). Examples are from biology [13,17,19, 26], physics and mathematics [1,3,4], and swarm robotics [5,16,18,20]. In this context, systems showing collective motion are represented to a large extent by only two distributions: a spatial and a directional distribution. Basically these two distributions are equivalent in their significance for effective collective motion because both physical proximity and coordination in velocities are necessary [25]. However, the directional distribution may have an assumed higher significance in the literature, especially in one-dimensional systems [2,25,26].

Examples for microscopic models are models of self-propelled particles (SPP) that define each particle's motion influenced by neighboring particles [2,26].

© Springer Nature Switzerland AG 2018
M. Dorigo et al. (Eds.): ANTS 2018, LNCS 11172, pp. 290–301, 2018.
https://doi.org/10.1007/978-3-030-00533-7_23

Examples for macroscopic models are Fokker–Planck models that describe global features of collective behaviors, such as velocity distributions and spatial distributions [9,11,26]. The model of Yates et al. [26], in particular, models the velocity distribution of particles based on a Fokker–Planck equation with estimated diffusion and drift coefficients based on experimental data and simulation data, whereas "the drift coefficient represents the mean rate of change of the average velocity" [26]. It turns out that such drift coefficients are time-variant and seem to depend on another state variable of the system following an exponential function $(1 - \exp(-t))$ [6,10,23]. We identify a relevant graph-theoretic feature of the graph that is induced by the robots' positions and their perceivable neighborhoods. This main feature is the size of the largest connected component of this graph, which is the largest subgraph in which all pairs of nodes are connected by paths. Time-invariant features are always of interest in mathematical modeling and so are they here, too. These time-invariant features may be used to formulate concise and precise models of collective motion.

2   Microscopic and Macroscopic Models

We introduce two models: a microscopic model of self-propelled particles (SPP) and a macroscopic model of swarm populations.

2.1   Microscopic Model: SPP

We define a time-discrete microscopic model as a 1-d SPP model of N particles moving on a circle of circumference $U = 1$ without units (i.e., periodic boundary conditions). The number of particles N can also be interpreted as particle density $\rho = N/U = N$. Our model is similar to the model defined by Czirók and Vicsek [3,10] but different in a few aspects. First, the allowed velocities are discrete. Second, particles explicitly decide to switch their direction of motion based on majority decisions. Third, noise is implemented as spontaneous switching of direction. A particle i has coordinate x_i and discrete, dimensionless velocity $u_i \in \{-1, 1\}$. The dynamics are defined by

$$x_i(t + 1) = x_i(t) + v_0 u_i(t), \tag{1}$$

$$u_i(t + 1) = \begin{cases} G(L_i(t), R_i(t)), & \text{with probability } P_d \\ -u_i(t), & \text{with probability } P_n \\ u_i(t), & \text{else} \end{cases} \tag{2}$$

where L_i is the number of neighboring particles located on the interval $[x_i - \Delta r, x_i + \Delta r]$ with velocities $u_j = +1$, that is, neighbors moving counterclockwise or 'left'. Similarly, R_i are neighboring particles with velocities $u_j = -1$, that is, neighbors moving clockwise or 'right'. The perception range of a particle is $\Delta r = 0.002$ and a particle's nominal velocity is $v_0 = 0.001$ (see Table 1 for used parameters). $P_d = 0.1$ is the particle's probability of reconsidering its direction of movement (i.e., on average an agent reconsiders its direction of motion every ten time steps). $P_n = 0.015$ is the particle's probability of inverting its direction of movement spontaneously, hence, it implements noise. With G we implement a local majority decision

$$G(L, R) = \begin{cases} +1, & L > R \\ -1, & R > L, \\ u_r, & R = L \end{cases} \tag{3}$$

with $u_r \in \{-1, +1\}$ is a random tie breaker choosing -1 or $+1$ with equal probabilities.

Table 1. Parameter settings used in simulations.

Parameter	Value
Swarm size N	$\{38, 42, 70\}$
Particle nominal velocity v_0	0.001
Perception range Δr	0.002
Prob. reconsidering P_d	0.1
Prob. spontaneous switch P_n	0.015
Simulated time intervals t	$\{2000, 4000\}$
Circumference U	1

The initial condition is a random uniform distribution for both, the particles' coordinates $x_i \in [0, U)$ and their velocities $u_i \in \{-1, +1\}$. Note that despite the particles' discrete velocities u_i the system is different from cellular automata because the particles' (initial) positions x_i are continuous. As an effect, a particle's neighbors are distributed continuously over its whole neighborhood interval $[x_i - \Delta r, x_i + \Delta r]$ but propagate through it in discrete steps.

Our primary interest is the particle distribution among the two states $u_i = \pm 1$, that is, we observe the fraction $s_t \in [0, 1]$ of particles in state $u_i = +1$ (without loss of generality) over time t. With the settings used in this paper, s fluctuates over the whole interval $s \in [0, 1]$ for all time but with a bimodal distribution that is established after a transient (see Fig. 1a). We interpret the system states that correspond to the two peaks of this bimodal distribution as the aligned states although the particles are not completely aligned ($s < 1$ and $s > 0$). The observed bimodal distribution is expected for effective collective motion and was reported before (e.g., see [10, 26]).

2.2 Largest Component

In the following we apply a graph-theoretic interpretation of the particles' coordinates in the SPP model. The particles are interpreted as nodes of an undirected graph. There is an edge (i, j) between particle i and j if $d(x_i, x_j) < \Delta r$,

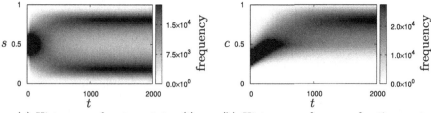

(a) Histogram of system state $s(t)$.   (b) Histogram of swarm fraction part of the largest connected component $c(t)$.

Fig. 1. Distributions of swarm fraction moving left s_t and swarm fraction in largest connected component c_t obtained numerically from the SPP model, $N = 42$, 5×10^5 sample simulation runs.

whereas $d(x_i, x_j)$ gives the distance between the particles. Hence, each agent is connected by an edge to each of its neighbors, who also determine the particle's local majority decision based on $G(L, R)$. The largest connected component (or short: largest component) of a graph is the largest set of nodes M for which it is true that any pair of nodes out of M is connected by a path, which is a sequence of nodes that have an edge from one node to the next in the sequence. For any state of the SPP model at any time the largest component M_t can be calculated. In the following we are interested in the swarm fraction that is in the largest component $c_t = |M_t|/N$ ($|M|$ is the number of nodes in M). We can measure the size of the largest component as the swarm fraction $c_t \in [0, 1]$ that is part of the largest component at time step t (see Fig. 1b). The size of the largest component increases over time because aligned motion typically triggers an aggregation behavior as a side-effect [8]. Later in the simulation runs, the size of the largest component saturates at values of $c \approx 0.8$. An interesting observation is that the swarm is aligned early at $t \approx 500$, while the distribution of largest components stays time-variant until $t > 1000$. Hence, aligning particles is a quick process while forming a big largest component is a slow process for the tested parameter set. The used simulation software and some experiment data is available online[1].

2.3   Macroscopic Model

A reasonable approach is to macroscopically model trajectories s_t as a stochastic process

$$s_{t+1} = s_t + \Delta s(s_t) + \xi_t, \tag{4}$$

for a drift term $\Delta s(s_t)$ and a diffusion term ξ_t with mean $\langle \xi \rangle = 0$. This is similar to approaches based on Langevin equations and the corresponding Fokker–Planck equations [5,18,26]. In the following we focus on the drift term $\Delta s(s)$ only. Within the SPP model we can measure the mean $\langle \Delta s(s) \rangle$ based on averages

---

[1] https://doi.org/10.5281/zenodo.1293372.

over many realizations of the stochastic process due to

$$\langle s_{t+1} - s_t \rangle = \langle \Delta s(s_t) + \xi_t \rangle = \langle \Delta s(s_t) \rangle. \tag{5}$$

The main purpose of this macroscopic model here is only to indicate the meaning of the drift coefficient $\Delta s(s_t)$. To keep such a macroscopic model simple, we prefer time-invariant parameters and features. In particular, the drift coefficient $\Delta s(s_t)$ should preferably be time-invariant. The drift term is the main driver of the system on the global level and determines whether a collective decision-making system is effective. Finding and using time-invariant coefficients could help to formulate a concise and precise model of collective motion.

3   Results

3.1   Time-Invariant Drift Coefficients

Measurements of $\langle \Delta s(s_t) \rangle$ reveal that it is time-variant during a transient (see Fig. 2a). The time-variance of the drift coefficient $\langle \Delta s(s_t) \rangle$ is known and was reported before [23]. In [6,7] empirical evidence is given that $\langle s_t \rangle$ follows an exponential function (symmetry is exploited here and values $s < 0.5$ are mapped by $s' = 1 - s$). This is also what is measured for the size of the largest component $\langle c_t \rangle$ in the SPP model as shown in Fig. 2c (cf. Fig. 1b).

We find that measurements of the drift coefficient $\Delta s(s)$ are time-invariant if we measure them with respect to the size of the current largest component, that is, by defining a two-dimensional function $\Delta s(s, c)$ that takes the size of the largest component as an argument, too. Measurements are shown in Fig. 3. These are measurements at nine different times every 200 time steps and different sizes of the observed largest component. The noise of these measurements is because the 5×10^5 samples are now distributed over all different bins for the largest component sizes and certain component sizes are infrequently observed at certain times. Besides a few outliers the measurements indicate time-invariant behavior of $\Delta s(s, c)$, especially in comparison to Fig. 2a. Also see Fig. 2b showing the drift coefficient for $\Delta s(s, c = 0.5)$, which indicates features hardly observed when measuring only $\Delta s(s)$. $\Delta s(s, c = 0.5)$ has five zeros with three stable fixed points: $s_1^* \approx 0.17$, $s_2^* = 0.5$, and $s_3^* \approx 0.83$.

This approach of using a two-dimensional function $\Delta s(s, c)$ has two main advantages. First, a model of collective motion, for example similar to the Fokker–Planck model of Yates et al. [26], based on time-invariant functions could also give valid predictions for the transient if it is combined with a model for the increase of the largest component c_t over time, which can be simple (cf. Fig. 2c). Second, measurements of $\Delta s(s, c)$ also reveal otherwise hidden system properties that we discuss in the following.

For certain N and c, $\Delta s(s, c)$ has not only three but five zeros (see Fig. 2b) generating a stable fixed point at $s = 0.5$. Hence, the system may travel through state space, that drives the system away from a collective decision towards indecisiveness. This may happen not only early in the transient, as seen in Fig. 2a,

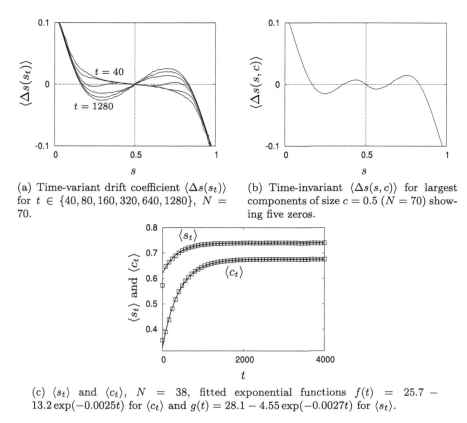

(a) Time-variant drift coefficient $\langle \Delta s(s_t) \rangle$ for $t \in \{40, 80, 160, 320, 640, 1280\}$, $N = 70$.

(b) Time-invariant $\langle \Delta s(s, c) \rangle$ for largest components of size $c = 0.5$ ($N = 70$) showing five zeros.

(c) $\langle s_t \rangle$ and $\langle c_t \rangle$, $N = 38$, fitted exponential functions $f(t) = 25.7 - 13.2 \exp(-0.0025t)$ for $\langle c_t \rangle$ and $g(t) = 28.1 - 4.55 \exp(-0.0027t)$ for $\langle s_t \rangle$.

Fig. 2. Drift coefficients and growth of synchronized group and largest component measured in the SPP model.

but possibly also even later in the transient whenever the largest component has a relatively small size of about $c \approx 0.5$.

Measured time-invariant $\Delta s(s, c)$ based on the sizes of largest components $c \in \{0.29, 0.48, 0.83\}$ for $N = 38$ are shown in Fig. 4a. For a big largest component of $c = 0.83$ the system is sensitive to fluctuations around $s = 0.5$ because they are reinforced with positive feedback (i.e., on average a small majority grows). In contrast the system diminishes fluctuations for $c = 0.29$. This is probably because fluctuations affect almost only the connected component in which they occur, hence, effects in components due to fluctuations are almost independent from each other. With a largest component of $c < 0.5$, even the largest component is only a minority and consequently has no global effect. This is also seen in Figs. 4b and c, which give values of $\Delta s(s, c)$ for most measurable largest component sizes c as a map for bigger densities ($N = 42$ and $N = 70$). Note that the tested swarm sizes of $N \in \{38, 42, 70\}$ are chosen arbitrarily and represent different swarm densities (cf. [14]). Higher densities turn out to show more complexity in their system dynamics but also reduce the observed occurrence

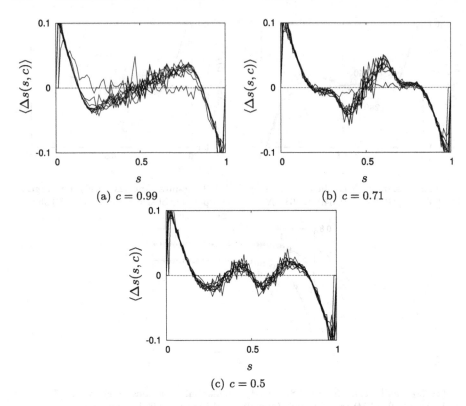

Fig. 3. Time-invariant $\langle \Delta s(s,c) \rangle$ for $N = 70$ and $t \in \{200, 400, 600, 800, 1000, 1200, 1400, 1600, 1800\}$, measured in the SPP model.

of small largest connected components (e.g., $c < 0.4$ for $N = 70$). Figures 4b and c clearly indicate the underlying complex system dynamics of the collective motion process in contrast to mere time-averaged or steady-state measurements of $\Delta s(s)$. The stable fixed point at $s = 0.5$ for $c < 0.5$ is seen as well as the small interval around $c \approx 0.5$ for which we have three stable fixed points (more clearly for $N = 70$ in panel c). For $c > 0.5$ the two remaining stable fixed points get closer to $s = 0$ and $s = 1$ respectively with increasing largest component size.

The unexpected complexity of the underlying system dynamics is prominent in Fig. 4c ($N = 70$). It reminds one of so-called bifurcation scenarios from nonlinear dynamics. For $0.36 \leqslant c \leqslant 0.41$ the system has four stable fixed points, which then collapse for $c \approx 0.41$. For $0.44 \leqslant c \leqslant 0.54$ the system has three stable fixed points, which then collapse into a situation with four stable fixed points again for $0.55 \leqslant c \leqslant 0.59$. Starting from $c \approx 0.6$ the system has increasingly more positive feedback and only two stable fixed points.

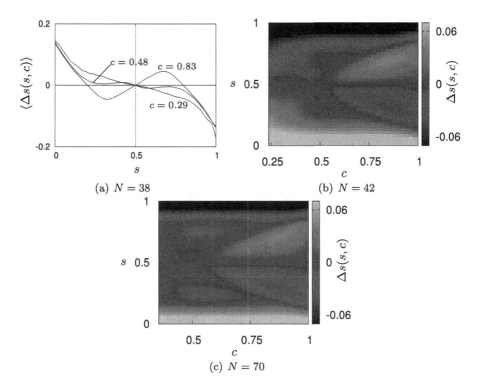

Fig. 4. Drift coefficient $\langle \Delta s(s,c) \rangle$ depending on the largest component for $N \in \{38, 42, 70\}$.

3.2   On the Effectivity of Collective Motion

We are interested in the necessary conditions for effective synchronization in collective motion, which we define by the existence of two stable fixed points at $s = 0.5 \pm x$ for $x > 0$.

Based on the measurements of $\Delta s(s,c)$ we determine bifurcation diagrams depending on the largest component and density (similar to what is shown in Figs. 4b and c). In Fig. 5a we give the bifurcation diagram for a given density $\rho = N = 42$. Two stable fixed points are found for $c \geq 0.54$.

We can investigate the influence of particle density ρ, if we specify which largest component size we associate with each considered density (otherwise we would need to investigate a 3-d system). For this purpose we measure the largest component size in the SPP model in the steady state for a given density (data not shown). Following these measurements, we then define a simple ad-hoc model of the observed effective largest component size by $c^*(\rho) = 1/\rho(-20.7/(1 + \exp(-0.1\rho + 0.9)) + 1.05\rho + 6.2)$. We find a critical density of $\rho_{\text{crit}} = 28$ (Fig. 5b) for the selected parameters. For too sparse densities ($\rho < \rho_{\text{crit}}$) no big enough largest component forms that could trigger global alignment. Once the critical

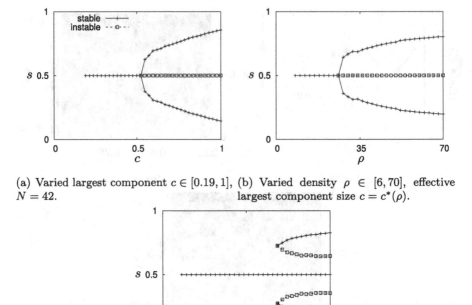

(a) Varied largest component $c \in [0.19, 1]$, (b) Varied density $\rho \in [6, 70]$, effective
$N = 42$.   largest component size $c = c^*(\rho)$.

(c) Varied density $\rho \in [6, 70]$, largest component size $c = 0.5$.

Fig. 5. Bifurcation diagrams obtained using the largest-component-based drift coefficients $\Delta s(s, c)$ measured in the SPP model.

density ρ_{crit} is reached, appropriate largest components form and the collective motion is effective.

Hence, we observe bifurcations in two dimensions (largest component and density) and consequently two conditions have to be met for a collective decision to emerge. A minimum density is needed and at least the majority of the swarm has to be positioned within the largest component. These two conditions are distinguished by their priorities. With too low density, no stable largest cluster of size $c > 0.5$ forms. However, once this critical density is reached, a stable largest cluster of size $c > 0.5$ forms. Moreover, the clearness of the decision (s close to zero or one respectively) increases with increasing largest component c (see also Figs. 4b and c) and with increasing density ρ. In addition, the bifurcation diagram for the borderline situation $c = 0.5$ (Fig. 5c) shows that even above the critical density $\rho_{\text{crit}} = 28$ but with a relatively small largest component the fixed point at $s = 0.5$ stays stable (this is observed on an interval of about $c \in [0.43, 0.56]$) and interestingly two new unstable fixed points emerge which shows the particularity of the $c = 0.5$-situations. Hence, we summarize that a critical density is necessary to observe a collective decision but during the transient, situations may emerge that stabilize the indecisive state $s = 0.5$ temporarily.

4 Conclusion

Based on a simple 1-d SPP model, we have investigated collective motion. Our focus was on measurements of the drift coefficient $\Delta s(s)$, which is time-variant but turns out to be time-invariant once we measure it relative to the size of the largest connected component. This feature of time-invariance may be used to formulate concise and precise models of collective motion. As discussed, a simple temporal model of the increase in size of the largest component can be formulated (exponential function) and can then be used to create a macroscopic model. This macroscopic model then exclusively uses time-invariant coefficients and allows to model the transient behavior of collective motion systems.

Note that the spatial distribution of particles seems to have more impact on the dynamics of collective decisions as modeled in this paper than the current state of majority stake s. Previously reported Fokker–Planck models [11,18,26] based on time-averaged drift coefficients may give good predictions for steady states but are of limited use for investigations of the temporal evolution of collective decisions and may also be of limited use to understand the underlying dynamics.

The drift coefficient of the velocity distribution based on sizes of the largest components is constant over time. Consequently the system dynamics of collective motion is completely described by positive feedback depending on largest components, the dynamics of largest components, and additional noise features (diffusion coefficient).

Future work includes investigating the generalization from 1-d space to 2-d, which could bring a qualitative change in the observed dynamics. Models and findings as reported here can help in swarm engineering to provoke advantageous system states, such as big largest connected components, and to increase performance of an engineered collective system. A key challenge for modeling is trying to answer the question of how the system behaviors that were measured here (e.g., drift coefficient depending on fraction of swarm within largest connected component, Fig. 4) could be predicted by an appropriate model.

References

1. Chazelle, B.: An algorithmic approach to collective behavior. J. Stat. Phys. **158**(3), 514–548 (2015)
2. Czirók, A., Barabási, A.L., Vicsek, T.: Collective motion of self-propelled particles: kinetic phase transition in one dimension. Phys. Rev. Lett. **82**(1), 209–212 (1999)
3. Czirók, A., Vicsek, T.: Collective behavior of interacting self-propelled particles. Physica A **281**, 17–29 (2000)
4. Degond, P., Yang, T.: Diffusion in a continuum model of self-propelled particles with alignment interaction. Math. Models Methods Appl. Sci. **20**, 1459–1490 (2010)

5. Hamann, H.: Space-Time Continuous Models of Swarm Robotics Systems: Supporting Global-to-Local Programming. Springer, Berlin (2010). https://doi.org/10.1007/978-3-642-13377-0

6. Hamann, H.: Towards swarm calculus: universal properties of swarm performance and collective decisions. In: Dorigo, M., Birattari, M., Blum, C., Christensen, A.L., Engelbrecht, A.P., Groß, R., Stützle, T. (eds.) ANTS 2012. LNCS, vol. 7461, pp. 168–179. Springer, Heidelberg (2012). https://doi.org/10.1007/978-3-642-32650-9_15

7. Hamann, H.: Towards swarm calculus: urn models of collective decisions and universal properties of swarm performance. Swarm Intell. **7**(2–3), 145–172 (2013). https://doi.org/10.1007/s11721-013-0080-0

8. Hamann, H.: Swarm Robotics: A Formal Approach. Springer, Cham (2018). https://doi.org/10.1007/978-3-319-74528-2

9. Hamann, H., Meyer, B., Schmickl, T., Crailsheim, K.: A model of symmetry breaking in collective decision-making. In: Doncieux, S., Girard, B., Guillot, A., Hallam, J., Meyer, J.-A., Mouret, J.-B. (eds.) SAB 2010. LNCS (LNAI), vol. 6226, pp. 639–648. Springer, Heidelberg (2010). https://doi.org/10.1007/978-3-642-15193-4_60

10. Hamann, H., Valentini, G.: Swarm in a fly bottle: feedback-based analysis of self-organizing temporary lock-ins. In: Dorigo, M., Birattari, M., Garnier, S., Hamann, H., Montes de Oca, M., Solnon, C., Stützle, T. (eds.) ANTS 2014. LNCS, vol. 8667, pp. 170–181. Springer, Cham (2014). https://doi.org/10.1007/978-3-319-09952-1_15

11. Hamann, H., Wörn, H.: A framework of space-time continuous models for algorithm design in swarm robotics. Swarm Intell. **2**(2–4), 209–239 (2008). https://doi.org/10.1007/s11721-008-0015-3

12. Helbing, D., Schweitzer, F., Keltsch, J., Molnár, P.: Active walker model for the formation of human and animal trail systems. Physical Review E **56**(3), 2527–2539 (1997)

13. Hillen, T., Painter, K.J.: A user's guide to PDE models for chemotaxis. Math. Biol. **58**, 183–217 (2009). https://doi.org/10.1007/s00285-008-0201-3

14. Khaluf, Y., Pinciroli, C., Valentini, G., Hamann, H.: The impact of agent density on scalability in collective systems: noise-induced versus majority-based bistability. Swarm Intell. **11**(2), 155–179 (2017). https://doi.org/10.1007/s11721-017-0137-6

15. Levine, H., Rappel, W.J., Cohen, I.: Self-organization in systems of self-propelled particles. Phys. Rev. E **63**(1), 17101 (2000)

16. Milutinovic, D., Lima, P.: Cells and Robots: Modeling and Control of Large-Size Agent Populations. Springer, Berlin (2007). https://doi.org/10.1007/978-3-540-71982-3

17. Okubo, A.: Dynamical aspects of animal grouping: swarms, schools, flocks, and herds. Adv. Biophys. **22**, 1–94 (1986)

18. Prorok, A., Correll, N., Martinoli, A.: Multi-level spatial models for swarm-robotic systems. Int. J. Robot. Res. **30**(5), 574–589 (2011)

19. Reina, A., Marshall, J.A.R., Trianni, V., Bose, T.: Model of the best-of-n nest-site selection process in honeybees. Phys. Rev. E: **95**, 052411 (2017). https://doi.org/10.1103/PhysRevE.95.052411

20. Reina, A., Valentini, G., Fernández-Oto, C., Dorigo, M., Trianni, V.: A design pattern for decentralised decision making. PLOS ONE **10**(10), 1–18 (2015). https://doi.org/10.1371/journal.pone.0140950

21. Schimansky-Geier, L., Mieth, M., Rosé, H., Malchow, H.: Structure formation by active Brownian particles. Phys. Lett. A **207**, 140–146 (1995)

22. Schweitzer, F.: Brownian Agents and Active Particles: On the Emergence of Complex Behavior in the Natural and Social Sciences. Springer, Berlin (2003)
23. Valentini, G., Hamann, H.: Time-variant feedback processes in collective decision-making systems: influence and effect of dynamic neighborhood sizes. Swarm Intelligence 9(2–3), 153–176 (2015). https://doi.org/10.1007/s11721-015-0108-8
24. Vicsek, T., Czirók, A., Ben-Jacob, E., Cohen, I., Shochet, O.: Novel type of phase transition in a system of self-driven particles. Phys. Rev. Lett. 6(75), 1226–1229 (1995)
25. Vicsek, T., Zafeiris, A.: Collective motion. Phys. Rep. 517(3–4), 71–140 (2012)
26. Yates, C.A., et al.: Inherent noise can facilitate coherence in collective swarm motion. Proc. Natl. Acad. Sci. USA 106(14), 5464–5469 (2009). https://doi.org/10.1073/pnas.0811195106. http://www.pnas.org/content/106/14/5464.abstract

Why the Intelligent Water Drops Cannot Be Considered as a Novel Algorithm

Christian Leonardo Camacho-Villalón[✉][iD], Marco Dorigo[iD],
and Thomas Stützle[iD]

IRIDIA, Université Libre de Bruxelles, Brussels, Belgium
{ccamacho,mdorigo,stuetzle}@ulb.ac.be

Abstract. In this paper we show that intelligent water drops (IWD), a swarm intelligence based approach to discrete optimization proposed by Shah-Hosseini in 2007, is a particular instantiation of the ant colony optimization (ACO) metaheuristic. To do so, in the paper, we identify the components of IWD and place them into the ACO metaheuristic framework. We show therefore that there was no need for a new natural metaphor. We also discuss that the proposed metaphor does not bring any novel insight into the algorithmic optimization process used by IWD.

Keywords: Intelligent water drops · Ant colony optimization
Novel algorithm

1 Introduction

Recently, many so-called *novel* approaches to stochastic optimization based on a natural metaphor have been proposed in the literature. Unfortunately, as also discussed in [29], such natural metaphors are often unnecessary or even misleading. For example, stochastic optimization algorithms based on diverse metaphors such as spiders [8], whales [22], grey wolves [23], birds [2], and so on, have been proposed and published in the literature. However, the real value of using a metaphor is often unclear. In some rare cases such as for harmony search [32] and black holes [24], it has been formally shown that the *novel* algorithm is just a re-formulation, using different terms, of an already well-known algorithm. In general, however, it remains challenging to understand whether the *novel* algorithms are indeed new or not.

We believe that the usage of such new metaphors should be limited to the cases in which they are indeed useful to express a new concept. This means that (i) it should not be possible to express the same algorithmic ideas using the terminology of already existing algorithms, and (ii) the inspiring metaphor should bring some new concepts that are related to the optimization process proposed. Unfortunately, this is often not the case. In our research we intend to examine a number of such *novel* nature-inspired algorithms to understand if they meet the two above-mentioned requirements and therefore deserve to be considered

© Springer Nature Switzerland AG 2018
M. Dorigo et al. (Eds.): ANTS 2018, LNCS 11172, pp. 302–314, 2018.
https://doi.org/10.1007/978-3-030-00533-7_24

novel. In this particular paper, we study the intelligent water drops (IWD) algorithm and its relation to the well-known ant colony optimization metaheuristic [16]. To do so, we first briefly present the ACO metaheuristic (Sect. 2) and the IWD algorithm (Sect. 3), highlighting their constituent components. Then, in Sect. 4, we perform a component-by-component comparison between ACO and IWD and show that IWD is indeed a particular case of ACO and that therefore it was not necessary to introduce a new terminology. We also discuss the fact that the inspiring metaphor does not bring any concepts that are related to the optimization process proposed. Therefore, the proposed IWD algorithm does not meet the two conditions set out in points (i) and (ii) above. Accordingly, we conclude that there is no need for an IWD algorithm and that adding it as a new tool to the optimization tool set is unnecessary and misleading.

2   Ant Colony Optimization

Ant colony optimization (ACO) is a metaheuristic that was first proposed in the early '90s [10,13,14]. The original source of inspiration was the foraging behavior of *Argentine* ants as described in a seminal paper by Deneubourg et al. [9]. In [9], it was shown that ants can find a shortest path between their nest and a food source by depositing pheromones on the ground and by choosing their way using a stochastic rule biased by pheromone intensity. In an analogous way, Dorigo et al. [10,13,14] showed that artificial agents, also called *artificial ants*, that

- move on a graph representation of a discrete optimization problem, where a path on the graph corresponds to a problem solution,
- deposit virtual pheromones on the graph edges, and
- use pheromones to bias the construction of random paths on the graph,

can find high quality solutions by letting their stochastic solution construction routine be biased by the value of virtual pheromones.

After the publication of the seminal algorithm in [13–15], many variants and improvements have been proposed [1,3–7,12,15,17,18,20,28,30]. Most of this work has been summarized in a book [16] where ACO is described as a constructive population-based metaheuristic comprising three main algorithmic components: (i) *stochastic solution construction*; (ii) *daemon actions*; and (iii) a *pheromone update procedure*.

One iteration of the ACO metaheuristic can be described as follows. First, every ant constructs a solution using a *stochastic solution construction* mechanism that iteratively selects solution components to add to the partial solution under construction. Once all ants have completed their solutions, an optional procedure called *daemon action* can be applied.[1] Finally, a *pheromone update procedure* modifies the pheromone trails.[2] Several iterations are executed until

---

[1] Daemon actions, for example, perform a local search procedure to improve an ant's solution or deposit an additional amount of pheromone on some solution components.

[2] In some ACO implementations, the *pheromone update procedure* can be interleaved with the solution construction (e.g., [12,17]), an example being the *local pheromone update procedure* that is implemented in ACS [12].

Algorithm 1. ACO metaheuristic

1: Set initial parameters
2: **while** termination condition not met **do**
3:    **repeat**
4:       Apply *stochastic solution construction*
         % solution components are iteratively added to a partial solution using a
         stochastic selection rule biased by artificial pheromones
5:       Apply *local pheromone update procedure* % optional
6:    **until** construction process is completed
7:    Apply *daemon actions* % optional
8:    Apply *pheromone update procedure*
9: **end while**
10: Return best solution

a termination condition is verified. An algorithmic outline of the ACO meta-heuristic is shown in Algorithm 1.

Artificial pheromones, indicated by τ, are numerical values given to each of the solution components in the search space. They are iteratively modified by ants in order to bias the selection of solution components. Pheromone values can increase due to ants depositing pheromones (positive feedback) or decrease through evaporation (negative feedback). ACO algorithms also use heuristic information, indicated by η, to bias the solution construction process.

In Table 1 we summarize all the most important ACO algorithms. They differ in the way in which stochastic solution construction and pheromone update are implemented.

3   The Intelligent Water Drops Algorithm

The intelligent water drops (IWD) algorithm, published first by Shah-Hosseini in 2007 [25], was proposed as a *novel* nature-inspired algorithm for combinatorial optimization problems. This algorithm is explained using a metaphor in which water streams are seen as groups of individual particles (water drops) moving in discrete steps.

In the words of the author:

> *In the water drops of a river, the gravitational force of the earth provides the tendency for flowing toward the destination ... It is assumed that each water drop flowing in a river can carry an amount of soil. The amount of soil of the water drop increases while the soil of the riverbed decreases. In fact, some amount of soil of the river bed is removed by the water drop and is added to the soil of the water drop.*
> [26, pp. 195]
> *A water drop has also a velocity and this velocity plays an important role in the removing of soil from the bed of the rivers ... The faster water drops are assumed to gather more soil than others.*
> [26, pp. 196]

In computational terms, the intelligent water drops:

- move on a graph representation of a discrete optimization problem, where a path on the graph corresponds to a problem solution,
- modify the amount of soil on the graph edges as a function of their velocity,
- use soil amount to bias the construction of random paths on the graph.

Shah-Hosseini [25–27] has described IWD as a constructive population-based algorithm composed of three algorithmic components: (i) *stochastic solution construction*; (ii) *local soil update procedure*; and (iii) *global soil update procedure*.

One iteration of the IWD algorithm consists of the following steps. First, each water drop constructs a solution using a *stochastic solution construction* mechanism biased by the amount of soil associated to the solution components, so that components with lower soil values have a higher probability to be chosen. After a solution component is selected, a *local soil update procedure* performs two actions: (i) it decreases the soil in the solution component, which is, according to the metaphor, removed by the water drop, and (ii) it increases the soil in the water drop, which indicates that it has been loaded into the water drop. For this procedure to take place, each water drop keeps a record of its own velocity and soil gathered during the iteration. After each water drop has built a complete solution, a *global soil update procedure* updates the soil values using the *iteration-best* water drop (i.e., the water drop that built the best solution in the current iteration). Several iterations are performed before a termination criterion is met and the algorithm stops. Algorithm 2 depicts this process.

Algorithm 2. Intelligent water drops algorithm

1: Set initial parameters
2: **while** termination condition not met **do**
3:     **repeat**
4:         Apply *stochastic solution construction*
            % solution components are iteratively added to a partial solution using a stochastic selection rule biased on amount of soil
5:         Apply *local soil update procedure*
6:     **until** construction process is completed
7:     Apply *global soil update procedure*
8: **end while**
9: Return best solution

It is clear that in the IWD algorithm the *soil* variable plays the same role as *pheromone* in ACO: it represents the numerical information given to the solution components in order to bias their selection during the stochastic construction process. Differently from artificial ants in ACO, water drops have associated a velocity variable. The velocity is an independent property of each water drop, that is, for different solutions constructed different velocities are obtained. When one iteration starts, all water drops have the same initial velocity; however, the velocity of a water drop is updated as a function of the soil found in the edges it traverses while building a solution. The value of soil loaded in the water drops is non-linearly proportional to the *heuristic undesirability*,[3] that is, the inverse of the time needed for the water drops to move from one solution component to another.

4 Discussion

Algorithms 1 and 2 show the general structure of the ACO metaheuristic and of the IWD algorithm. Both are composed of the following three main algorithmic components:

- a *stochastic solution construction* mechanism to iteratively construct solutions biased by a quantity (pheromone/soil) associated to solution components,
- a *local update procedure* to improve the search interleaving the construction mechanism with a local update of pheromone/soil,
- a *global update procedure* to give a positive feedback via modifications of the pheromone/soil associated to specific solutions.

In this section, we present a detailed analysis of the two approaches comparing their algorithmic components in order to clarify if IWD is in fact a new algorithm and deserves to be called a novel approach or should rather be considered a variant of ACO. To this purpose, in Table 1 we schematically present the algorithmic components proposed in some of the best-known ACO variants: Ant System (AS) [13–15], Ant System with Q-learning (Ant-Q) [17], MAX-MIN Ant System (MMAS) [31], Ant Colony System (ACS) [12], Approximate Non-deterministic Tree-Search procedure (ANTS) [20]; and in IWD.

One difference between IWD and ACO is that in ACO pheromone values are always positive, while in IWD the value of soil progressively becomes negative. Unlike ACO pheromones, in IWD the soil is gradually removed by the water drops, which implies that additional mechanisms have to be introduced to manage negative and positive soil values as well as to avoid a possible division by zero.

Another difference is that IWD constructs solutions biased solely by the values of *soil*; that is, no problem-specific information is used to bias solution construction, as opposed to what is done in ACO with *heuristic information*.[4]

---

[3] The author calls *heuristic undesirability* the inverse of the *heuristic information* used in ACO. For example, in the travelling salesman problem the ACO *heuristic information* is commonly defined as $\eta_{ij} = 1/d_{ij}$, where d_{ij} refers to the distance between city i and city j. In IWD, the *heuristic undesirability* is, for the same problem, defined as $HUD_{ij} = d_{ij}$.

[4] The usage of heuristic information is a way to integrate problem-specific information in the stochastic solution construction procedure so as to stochastically favor solution components of lower cost.

As it is shown in the following, IWD's *local soil update* and *global soil update* are special cases of the components used to update pheromones in ACO. However, the function of these components in IWD is different from their typical function in ACO. The *local soil update* procedure is the most different due to the introduction of the water drop velocity and the soil removed from the riverbed, this latter computed using the linear motion equations and the heuristic undesirability.

4.1 Stochastic Solution Construction

Ants construct solutions adding new solution components with a probability computed using a *transition rule* (see second column of Table 1), that is, a function of the pheromone values and of the heuristic information. The transition rule not only states which information will be used by ants to choose the next solution component, but also how the relative importance of such information will be weighted.

The *stochastic solution construction* mechanism used in IWD is a particular case of the *random proportional rule* of AS proposed in [15], in which the parameters τ and η are weighted using $\alpha = -1$ and $\beta = 0$.

Equations 1 and 2 show the *transition rules* in AS and IWD respectively:

$$p_j^{ant} = \frac{[\tau_j]^\alpha \cdot [\eta_j]^\beta}{\sum\limits_{h \in N^f} [\tau_h]^\alpha \cdot [\eta_h]^\beta} \tag{1}$$

$$p_j^{iwd} = \frac{\frac{1}{\epsilon + g(soil_j)}}{\sum\limits_{h \in N^f} \left(\frac{1}{\epsilon + g(soil_h)}\right)} \tag{2}$$

where N^f is the set of feasible solution components and j is one solution component in the search space. The parameter ϵ is a small positive constant added to avoid a possible division by zero in Eq. 2.

From the equations, it can be seen that IWD uses a transition rule that includes only the information given by the *soil* (i.e., *heuristic information* is not used) and that $1/soil$ is used so as to favor solution components with a low soil level (as opposed to ACO variants which favor solution components with a high pheromone level).

Additionally, because the value of soil can become negative, IWD applies a function g to the value of soil to keep it positive in Eq. 2:

$$g(soil_j) = \begin{cases} soil_j & \text{if } \min\limits_{h \in N^f} soil_j \geq 0 \\ soil_j - \min\limits_{h \in N^f} soil_j & \text{otherwise} \end{cases} \tag{3}$$

Table 1. Main algorithmic components used in AS, ANT-Q, MMAS, ANTS, ACS and IWD. We do not show here the *daemon actions* component commonly integrated in ACO implementations. *Daemon actions* are optional problem-specific operations; for example, the application of a local search procedure. In the table, s^k is the solution built by ant k, s^{best} is the solution built by the ant that built the best solution, and iwd^{best} is the solution built by the water drop that built the best solution.

Algorithm	Stoch. solution constr.: Transition rule	Local update procedure	Global update procedure
AS	random proportional rule $$\left[\frac{[\tau_j]^\alpha \cdot [\eta_j]^\beta}{\sum_{h \in Nf}[\tau_h]^\alpha \cdot [\eta_h]^\beta}\right],$$ where N^f is the set of feasible component	ant density $[\tau_j + Q1]$, ant quantity $[\tau_j + (Q_2 \cdot \eta_j)]$, where Q_1 and Q_2 are constants	ant cycle $\left[(1-\rho) \cdot \tau_j + \sum_{k=1}^{ants} \Delta\tau_j^k\right]$, where $\Delta\tau_j^k$ is defined as: $\begin{cases} F(k) & \text{if } j \in s^k \\ 0 & \text{otherwise} \end{cases}$
ANT-Q	pseudo-random-random rule $\begin{cases} \arg\max_{h \in Nf} \{\tau_h^\alpha \cdot \eta_h^\beta\} & \text{if } q \leq q0 \\ S & \text{otherwise} \end{cases}$ where S is random value from a probability distribution given by τ_h^α and η_h^β, and q is a random value from a uniform probability distribution	AQ-values learning rule $\left[(1-\alpha) \cdot \tau_j + \alpha \cdot [\Delta\tau_j + \gamma \cdot \max_{h \in Nf} \tau_h]\right]$, where $\Delta\tau_j$ is defined as: $\frac{W}{cost}$ where W is a constant and $cost$ is the solution cost	delayed reinforcement $\left[(1-\alpha) \cdot \tau_j + \alpha \cdot (\Delta\tau_j^{best} + \gamma \cdot \max_{h \in Nf} \tau_h)\right]$, where $\Delta\tau_j^{best}$ is defined as: $\begin{cases} F(s^{best}) & \text{if } j \in s^{best} \\ 0 & \text{otherwise} \end{cases}$
MMAS	random proportional rule [same as AS]		pheromones update rule $\left[\max\{\tau_{min}, \min(\tau_{max}, (1-\rho) \cdot \tau_j + \Delta\tau_j^{best})\}\right]$, where $\Delta\tau_j^{best}$ is defined as: $\begin{cases} F(s^{best}) & \text{if } j \in s^{best} \\ 0 & \text{otherwise} \end{cases}$
ANTS	additive random proportional rule $\left[\frac{\alpha \cdot \tau_j + [1-\alpha] \cdot \eta_j}{\sum_{h \in Nf} \alpha \cdot \tau_h + [1-\alpha] \eta_h}\right]$	—	trail update $[\tau_j + \Delta\tau_j^k]$, where $\Delta\tau_j^k$ is defined as: $\left[\tau_0 \cdot (1 - \frac{z_{curr} - LB}{\bar{z} - LB})\right]$, where $\bar{z}$ is the average cost of the solutions, z_{curr} is the current cost of the solution and LB is a lower bound for the problem
ACS	pseudo-random-random proportional rule $\begin{cases} \arg\max_{h \in Nf} \{\tau_h^\alpha \cdot \eta_h^\beta\} & \text{if } q \leq q0 \\ \text{[same as AS]} & \text{otherwise} \end{cases}$ where q is defined as in Ant-Q	local pheromone update $\left[(1-\varphi) \cdot \tau_j + \varphi \cdot \tau_0\right]$, where τ_0 is the pheromone lower bound	offline pheromones update $\begin{cases} (1-\rho) \cdot \tau_j + \rho \cdot \Delta\tau_j^{best} & \text{if } j \in s^{best} \\ \tau_j & \text{otherwise} \end{cases}$, where $\Delta\tau_j^{best} = F(s^{best})$
IWD	random selection $\left[\frac{\frac{1}{\epsilon + g(soil_j)}}{\sum_{h \in Nf} (\frac{1}{\epsilon + g(soil_h)})}\right]$	local soil update $\left[(1-\varphi) \cdot soil_j - \varphi \cdot \Delta soil_j\right]$, where $\Delta soil_j$ is defined in Equations 6,7 and 8	global soil update $\begin{cases} (1+\rho) \cdot soil_j - \rho \cdot \Delta soil^{best} & \text{if } j \in iwd^{best} \\ soil_j & \text{otherwise} \end{cases}$, where $\Delta soil_j^{best} = F(iwd^{best})$

4.2   Local Update Procedure

The *local pheromone update procedure* allows the artificial ants to give a negative or a positive feedback to other ants while constructing solutions[5] so as to avoid stagnation[6]. ACO variants implementing the idea of local negative feedback are, for example, Ant-Q [17] and ACS [12]. In the Ant-Q algorithm, pheromones are called AQ-values and the goal of the artificial ants is to learn these values (see AQ-values learning rule in Table 1) so that they can probabilistically favor better solution components.

IWD implements a variant of the *AQ-values learning rule* of Ant-Q, where parameter γ is set to $\gamma = 0$, and $\Delta soil_j$ is defined differently from $\Delta \tau_j$ (see Eqs. 6, 7 and 8). In fact, $\Delta soil_j$ is the only real difference between IWD and what had already been proposed in the context of the ACO metaheuristic. The implementation of this component in Ant-Q and IWD is shown in Eqs. 4 and 5, respectively:

$$\tau_j = (1 - \alpha) \cdot \tau_j + \alpha \cdot \left[\Delta_{\tau_j} + \gamma \cdot \max_{h \in N^j} \tau_h \right] \tag{4}$$

$$soil_j = (1 - \varphi) \cdot soil_j - \varphi \cdot \Delta soil_j \tag{5}$$

Equation 4 interpolates between the current pheromone value τ_j and the maximum pheromone over the possible next components; it simulates the change in the amount of pheromone due to evaporation and ant deposit. In Ant-Q, Equation 4 is used for both local and global reinforcement, the former applied after a solution component is selected during the solution construction, and the latter applied after the construction process finishes and all solutions are completed. However, in most Ant-Q implementations Δ_{τ_j} is defined as zero for the local pheromone update and as $1/cost^{best}$ for the global pheromone update [11,17,19].

Parameters α and γ are the learning step and the discount factor, respectively. The values chosen for these two parameters can favor the exploration or the exploitation behavior of the algorithm. The application of Eq. 4 can either enhance or reduce the exploration capabilities of Ant-Q by slightly reducing or increasing (depending on the values of γ and $\max \tau_h$) the pheromones. In a later ACO variant, ACS, a similar idea was proposed where $\gamma \cdot \max \tau_h$ was replaced by a small constant τ_0.

On the other hand, Eq. 5 intends to model the erosion of soil by water drops. In the metaphor of the IWD algorithm, water drops remove part of the soil every time a solution component is added. In practice, the *local soil update* procedure slightly increases the probability of one solution component to be selected by other water drops (in IWD lower soil values are preferred), thus implementing a

---

[5] Note that the idea of giving a positive feedback during the construction process was explored in some of the first ACO variants: *ant quantity* and *ant density* [10,13] However, these variants were abandoned many years ago because of their inferior performance compared with other ACO variants.

[6] Stagnation happens when the pheromones trails converge and all ants construct the same solutions over and over again.

form of positive feedback. The amount of soil removed by a water drop, $\Delta soil_j$, is computed using the *linear motion equations* of physics. As said in Sect. 3, different water drops have different velocities. The initial water drops velocity is a user selected parameter and its value should be selected empirically by running experiments on the considered problem. In fact, its value can greatly vary from problem to problem; for example, in [27], where the traveling salesman problem is considered, the water drops initial velocity is set 200, while in [26], where the problem considered is the multidimensional knapsack problem, the water drops initial velocity is set to 4. Once a solution component j has been added, the velocity of a water drop vel^{iwd} is updated according to

$$vel^{iwd} = vel^{iwd} + \frac{a_v}{b_v + c_v \times [soil_j]^2} \tag{6}$$

where a_v, b_v, c_v and $soil_j$ are also user selected parameters. The *time* required by the water drop to move from the current solution component to the next one is computed dividing the *heuristic undesirability* (HUD_j) by the water drop's new velocity. HUD_j represents the *distance* in the linear motion equation $V = d/t$, which, hence, becomes

$$time^{iwd} = \frac{HUD_j}{vel^{iwd}} \tag{7}$$

Finally, the amount of soil to be removed and to be loaded into the water drop is a function of the *time* taken by the water drop to move between the two solution components:

$$\Delta soil_j = \frac{a_s}{b_s + c_s \times [time^{iwd}]^2} \tag{8}$$

$\Delta soil_j$ tends to be larger for solution components with lower soil values or for those with small *heuristic undesirability*. In Eqs. 6 and 8, parameters b_v and b_s are used to avoid a possible division by zero. Typical values for the user selected parameters in the above equations are $a_v = 1, b_v = 0.01, c_v = 1$, $a_s = 1, b_s = 0.01, c_s = 1$, and for the initial value of soil 10 000 [27].

The velocity can be seen as an indicator of the quality of the partial solution constructed so far, that is, faster water drops have traversed edges with lower soil. However, putting the desirability of a solution component in terms of the velocity (quality of a partial solution) and of the heuristic information, as is defined for $\Delta soil_j$, is rather similar to the abandoned idea of *ant quantity* (see AS local update procedure in Table 1). Moreover, the *local soil update* component cannot be explained in terms of the inspiring metaphor. For example, if soil is removed, it is unclear why then the new amount of soil is computed by an equation such as Eq. 5 that uses a decay factor φ (and not simply by subtracting $\Delta soil_j$ from the current soil value). Additionally, the metaphor of water drops acting as individual particles removing the soil in the riverbeds is unrealistic, as water in a river should rather be seen as a moving fluid.

4.3   Global Update Procedure

The *global pheromone update procedure* in ACO is performed at the end of an iteration once all solutions have been completed. The main goal of this procedure is to give a *positive feedback* to the solution components included in a set of solutions that is used to deposit pheromones; common choices in ACO algorithms are using the *iteration-best* or *global-best* solution, but other options have been examined. Solution components that receive a higher amount of pheromone will have a higher probability of being selected by other ants in the next iterations.

The *global soil update* is a special case of the *offline pheromone update* in ACS, in which the parameter ρ has a range defined in the interval $[-1, 0]$, differently from its typical range defined in $(0, 1]$. Eqs. 9 and 10 show the definition of this component in ACO and IWD[7] respectively:

$$\tau_j = \begin{cases} (1 - \rho) \cdot \tau_j + \rho \cdot \Delta \tau_j^{best} & \text{if } j \in s^{best} \\ \tau_j & \text{otherwise} \end{cases} \qquad (9)$$

$$soil_j = \begin{cases} (1 + \rho) \cdot soil_j - \rho \cdot \Delta soil_j^{best} & \text{if } j \in iwd^{best} \\ soil_j & \text{otherwise} \end{cases} \qquad (10)$$

where the parameter $\Delta \tau_j^{best}$ is commonly defined as the inverse of the total cost of the solution ($1/cost^{best}$), while $\Delta soil_j^{best}$ is proportional to the soil gathered by the best water drop divided by the number of solution components ($soil^{best}/N^{best}$).

The *global soil update procedure*, as defined in [27], has two different outcomes depending on the value of $soil_j$ in the solution component. Let us consider the first summand in the first case in Eq. 10, $(1 + \rho) \cdot soil_j$. It is easy to see that if $soil_j > 0$, the resulting value of the first summand will be positive and therefore it will contribute with a *negative feedback* to the solution component. In the opposite case, when $soil_j < 0$, the product $(1 + \rho) \cdot soil_j$ will be negative and therefore the first summand will contribute with a *positive feedback* to the solution component. In other words, the first summand can either increase the value of soil if $soil_j > 0$, or decrease it if $soil_j < 0$. Regarding the second summand in the first case of Eq. 10, $-\rho \cdot \Delta soil_j^{best}$, the value of $\Delta soil_j^{best}$ is defined as always positive (see Eq. 8) and as we have it multiplied by $-\rho$, the result of this second summand will always be negative.

5   Conclusions

As the IWD algorithm, there are many other algorithms published as *novel* nature-inspired approaches in the metaheuristics literature. In fact, the already large number of these so-called *novel* approaches has made the selection of optimization algorithms troublesome, specially for those who use them for specific application problems and do not necessarily have a deep knowledge in the field

---

[7] There are two versions of this component in IWD. In [25], the first article proposing IWD, ρ was defined in the range $[0, 1]$ (just as in Eq. 9). However, in a later publication [27], the range of ρ was changed to $[-1, 0]$, leading to a somewhat different behavior of the update as explained here.

of metaheuristics. The very few existing, rigorous analyses of *novel* algorithms have shown in some selected cases that either (i) they simply re-use ideas proposed in the past [24,32], or that (ii) the scientific rationale behind the source of inspiration is incongruous and questionable [21,29].

In this paper, we contribute to such rigorous analyses by examining in more detail the Intelligent Water Drops (IWD) algorithm. In particular, we have shown that the algorithmic components proposed in IWD are not new and that they mainly have been proposed in the context of ant colony optimization (ACO) often already one or two decades earlier. More concretely, we found that the *stochastic construction mechanism* of IWD is a special case of the *random proportional rule* proposed in AS, the very first ACO algorithm. The *local soil update* component is a slight variant of the *AQ-values learning rule* that was proposed in the Ant-Q algorithm, a predecessor of ACS. The only, small, difference of IWD with earlier ACO algorithms is the definition of the $\Delta soil_j$ term in the local soil update; unfortunately, the rationale behind the definition of $\Delta soil_j$ and the definition of the *local soil update* component cannot be explained in terms of the source of inspiration of IWD. Finally, the *global update procedure* is a special case of the *offline pheromone update* proposed in ACS.

If we reconsider the two main criteria we have defined in the introduction, namely the fact that (i) it should not be possible to express the same algorithmic ideas using the terminology of already existing algorithms, and (ii) the inspiring metaphor should bring some new concepts that are related to the optimization process proposed, we can summarize the analysis of our article by saying that the IWD algorithm fails on both criteria.

Acknowledgments. Marco Dorigo and Thomas Stützle acknowledge support from the Belgian F.R.S.-FNRS, of which they are Research Directors.

References

1. Alaya, I., Solnon, C., Ghédira, K.: Ant colony optimization for multi-objective optimization problems. In: 19th IEEE International Conference on Tools with Artificial Intelligence (ICTAI 2007), vol. 1, pp. 450–457. IEEE Computer Society Press, Los Alamitos, CA (2007)
2. Askarzadeh, A.: Bird mating optimizer: an optimization algorithm inspired by bird mating strategies. Commun. Nonlinear Sci. Numer. Simul. **19**(4), 1213–1228 (2014)
3. Birattari, M., Balaprakash, P., Dorigo, M.: The ACO/F-RACE algorithm for combinatorial optimization under uncertainty. In: Doerner, K.F., Gendreau, M., Greistorfer, P., Gutjahr, W.J., Hartl, R.F., Reimann, M. (eds.) Metaheuristics - Progress in Complex Systems Optimization, Operations Research/Computer Science Interfaces Series, vol. 39, pp. 189–203. Springer, New York (2006). https://doi.org/10.1007/978-0-387-71921-4
4. Blum, C.: Beam-ACO–hybridizing ant colony optimization with beam search: an application to open shop scheduling. Comput. Oper. Res. **32**(6), 1565–1591 (2005)
5. Blum, C., Dorigo, M.: The hyper-cube framework for ant colony optimization. IEEE Trans. Syst. Man, Cybern. - Part B **34**(2), 1161–1172 (2004)

6. Bullnheimer, B., Hartl, R.F., Strauss, C.: An improved ant system algorithm for the vehicle routing problem. Ann. Oper. Res. **89**, 319–328 (1999)
7. Cordón, O., de Viana, I.F., Herrera, F., Moreno, L.: A new ACO model integrating evolutionary computation concepts: the best-worst ant system. In: Dorigo, M., et al. (eds.) Abstract Proceedings of ANTS 2000 - From Ant Colonies to Artificial Ants: Second International Workshop on Ant Algorithms, pp. 22–29. IRIDIA, Université Libre de Bruxelles, Belgium, 7–9 September 2000
8. Cuevas, E., Miguel, C., Zaldívar, D., Pérez-Cisneros, M.: A swarm optimization algorithm inspired in the behavior of the social-spider. Expert Syst. Appl. **40**(16), 6374–6384 (2013)
9. Deneubourg, J.L., Aron, S., Goss, S., Pasteels, J.M.: The self-organizing exploratory pattern of the Argentine Ant. J. Insect Behav. **3**(2), 159–168 (1990). https://doi.org/10.1007/BF01417909
10. Dorigo, M.: Optimization, Learning and Natural Algorithms. Ph.D. thesis, Dipartimento di Elettronica, Politecnico di Milano, Italy (1992). (in Italian)
11. Dorigo, M., Gambardella, L.M.: A study of some properties of Ant-Q. In: Voigt, H.-M., Ebeling, W., Rechenberg, I., Schwefel, H.-P. (eds.) PPSN 1996. LNCS, vol. 1141, pp. 656–665. Springer, Heidelberg (1996). https://doi.org/10.1007/3-540-61723-X_1029
12. Dorigo, M., Gambardella, L.M.: Ant colony system: a cooperative learning approach to the traveling salesman problem. IEEE Trans. Evol. Comput. **1**(1), 53–66 (1997)
13. Dorigo, M., Maniezzo, V., Colorni, A.: The Ant System: An autocatalytic optimizing process. Technical report, 91–016 Revised, Dipartimento di Elettronica, Politecnico di Milano, Italy (1991)
14. Dorigo, M., Maniezzo, V., Colorni, A.: Positive feedback as a search strategy. Technical report, 91–016, Dipartimento di Elettronica, Politecnico di Milano, Italy (1991)
15. Dorigo, M., Maniezzo, V., Colorni, A.: Ant system: optimization by a colony of cooperating agents. IEEE Trans. Syst. Man Cybern. - Part B **26**(1), 29–41 (1996)
16. Dorigo, M., Stützle, T.: Ant Colony Optimization. MIT Press, Cambridge, MA (2004)
17. Gambardella, L.M., Dorigo, M.: Ant-Q: a reinforcement learning approach to the traveling salesman problem. In: Proceedings of the Twelfth International Conference on Machine Learning, ML 1995, pp. 252–260. Morgan Kaufmann Publishers, Palo Alto (1995)
18. Guntsch, M., Middendorf, M.: A population based approach for ACO. In: Cagnoni, S., Gottlieb, J., Hart, E., Middendorf, M., Raidl, G.R. (eds.) EvoWorkshops 2002. LNCS, vol. 2279, pp. 72–81. Springer, Heidelberg (2002). https://doi.org/10.1007/3-540-46004-7_8
19. Machado, L., Schirru, R.: The Ant-Q algorithm applied to the nuclear reload problem. Ann. Nucl. Energy **29**(12), 1455–1470 (2002)
20. Maniezzo, V.: Exact and approximate nondeterministic tree-search procedures for the quadratic assignment problem. INFORMS J. Comput. **11**(4), 358–369 (1999)
21. Melvin, G., Dodd, T.J., Groß, R.: Why 'GSA: a gravitational search algorithm' is not genuinely based on the law of gravity. Natural Comput. **11**(4), 719–720 (2012)
22. Mirjalili, S., Lewis, A.: The whale optimization algorithm. Adv. Eng. Softw. **95**, 51–67 (2016)
23. Mirjalili, S., Mirjalili, S.M., Lewis, A.: Grey wolf optimizer. Adv. Eng. Softw. **69**, 46–61 (2014)

24. Piotrowski, A.P., Napiorkowski, J.J., Rowinski, P.M.: How novel is the "novel" black hole optimization approach? Inf. Sci. **267**, 191–200 (2014)
25. Shah-Hosseini, H.: Problem solving by intelligent water drops. In: Proceedings of the 2007 Congress on Evolutionary Computation, CEC 2007, pp. 3226–3231. IEEE Press, Piscataway (2007)
26. Shah-Hosseini, H.: Intelligent water drops algorithm: a new optimization method for solving the multiple knapsack problem. Int. J. Intell. Comput. Cybern. **1**(2), 193–212 (2008)
27. Shah-Hosseini, H.: The intelligent water drops algorithm: a nature-inspired swarm-based optimization algorithm. Int. J. Bio-Inspired Comput. **1**(1–2), 71–79 (2009)
28. Socha, K., Dorigo, M.: Ant colony optimization for continuous domains. Eur. J. Oper. Res. **185**(3), 1155–1173 (2008). https://doi.org/10.1016/j.ejor.2006.06.046
29. Sörensen, K.: Metaheuristics–the metaphor exposed. Int. Trans. Oper. Res. **22**(1), 3–18 (2015). https://doi.org/10.1111/itor.12001
30. Stützle, T., Hoos, H.H.: The $\mathcal{MAX} - \mathcal{MIN}$ and local search for the traveling salesman problem. In: Bäck, T., Michalewicz, Z., Yao, X. (eds.) Proceedings of the 1997 IEEE International Conference on Evolutionary Computation, ICEC 1997, pp. 309–314. IEEE Press, Piscataway (1997)
31. Stützle, T., Hoos, H.H.: $\mathcal{MAX} - -\mathcal{MIN}$ Ant system. Future Gener. Comput. Syst. **16**(8), 889–914 (2000)
32. Weyland, D.: A rigorous analysis of the harmony search algorithm: how the research community can be misled by a "novel" methodology. Int. J. Appl. Metaheuristic Comput. **12**(2), 50–60 (2010)

Short Papers

A Cooperative Opposite-Inspired Learning Strategy for Ant-Based Algorithms

Nicolás Rojas-Morales[1]([⊠]), María-Cristina Riff[1], Carlos A. Coello Coello[2], and Elizabeth Montero[1]

[1] Universidad Técnica Federico Santa María, Valparaíso, Chile
nicolasrojas@acm.org
[2] CINVESTAV-IPN (Evolutionary Computation Group), Mexico City, Mexico

Abstract. In recent years, there has been an increasing interest in Opposite Learning strategies. In this work, we propose COISA, a Cooperative Opposite-Inspired Strategy for Ants. Inspired on the concept of anti-pheromone, in this approach, sub-colonies of ants perform different search processes to construct an initial pheromone matrix. We aim to produce a repel effect to (temporarily) avoid components that were related to an undesirable characteristic. To assess the effectiveness of COISA, we selected Ant Knapsack, a well-known ant-based algorithm that efficiently solves the Multidimensional Knapsack Problem. Results in benchmark instances show that the performance of Ant Knapsack is improved considering the opposite information, so that it can reach better solutions than before.

1 Introduction

We propose here an Opposite-Inspired Learning strategy where the search process of an ant-based algorithm is divided into two steps: a *First Step* used to identify a $u\mathcal{D}$-characteristic from complete instantiations and a *Second Step* used to solve the problem of interest. Three sub-colonies of ants cooperate obtaining information during the *First Step*. Such information will be considered in the *Second Step* to change its decisions during the construction process. Each sub-colony performs a search process defined by a *Method*. Here, we propose a collaboration between these *Methods* that were previously proposed in [10,11]. Sections 2 and 3 present details of our proposed strategy.

Opposite Learning (OL) is a search strategy that has been applied for mapping candidate solutions with the objective of increasing the coverage of the solution space [5]. Opposition-Inspired Learning (OIL) [12] was proposed considering that, in some cases, the idea of mapping solutions is not intuitive because of some algorithm-specific properties. Some previous OIL ant-based approaches have been proposed [2,4,6,7]. In our case, the term *opposite* is related to the possible decisions made by ant-based algorithms, that could lead the search process towards poor quality candidate solutions.

© Springer Nature Switzerland AG 2018
M. Dorigo et al. (Eds.): ANTS 2018, LNCS 11172, pp. 317–324, 2018.
https://doi.org/10.1007/978-3-030-00533-7_25

To evaluate our strategy, we selected the well-known Ant Knapsack (AK) algorithm [1] originally proposed for solving the Multidimensional Knapsack Problem (MKP). The implementation in AK is described in Sect. 4. It is important to mention, however, that our objective is not to propose the best algorithm for the MKP. The idea is to evaluate the use of a learning strategy to focus the search process of a specific ant-based algorithm.

2   An OIL Strategy for Ant-Based Algorithms

Let's assume a combinatorial problem $\mathbb{P}$ and an ant-based algorithm A. We are interested in improving the search process of A, in terms of the quality of the solutions that A can build. For this, we are interested in providing useful information to A in order to improve its *intermediate* decisions. Let's assume that each ant k of A incrementally constructs a complete instantiation of a solution I_C^k, making stochastic *intermediate* decisions to include components into a partial instantiation I_P^k.

In most cases, components are included in I_C^k because a certain preference related to their heuristic knowledge (η) and pheromone information (τ) was considered in the *intermediate* decisions of A. As η is particularly defined in A and the information in the pheromone matrix is limited by the vertices which were already visited during the current execution, in some cases, the information provided to perform *intermediate* decisions might be poor. Considering that $\mathbb{P}$ is complex to solve, this information can affect some *intermediate* decisions and lead the construction process to solutions with less quality than expected.

Let's assume that I_C^k has some characteristic w that can be measurable and related to: a structural property of I_C^k, a quality feature of I_C^k, a feature related to the (in)feasibility in I_C^k, and a problem-specific property feature not detectable by $\mathbb{A}$, among others. During the construction process of I_C^k, *intermediate* decisions are biased giving priority to some components that look more promising than others. We name this characteristic as *undesirable* ($u\mathcal{D}$) because these intermediate decisions prefer components that are locally interesting, but finally produced that $F(I_C^k) < F(I_C^*)$.[1] It is important to remark that this characteristic is not inherent to the problem $\mathbb{P}$, but it cannot be perceptible by the current pheromone information and by the heuristic knowledge, as it is specifically defined in A. We propose to learn about this $u\mathcal{D}$−characteristic w in I_C^k to decrease the attraction to components that A considers promising. The objective is to allow A to consider other *intermediate* decisions during its construction process and, finally, obtain better quality solutions.

Let $S_{(A,i)}$ be a set of complete instantiations obtained by A during its ith iteration and w a $u\mathcal{D}$-characteristic. As w is measurable, solutions in $S_{(A,i)}$ can be compared considering the presence of a $u\mathcal{D}$-characteristic. We define $S_{(A,i)}^w$ as the set of complete instantiations that have more presence of w. As the pheromone produces a modification of the way in which the problem is represented and

---

[1] Considering that $\mathbb{P}$ is a maximization problem with an objective function F and I_C^* is an optimal solution.

perceived by artificial ants [3], we decided to use the pheromone to learn about the $u\mathcal{D}$-characteristics. Our hypothesis is the following: if we consume a certain amount of resources in identifying and learning about some $u\mathcal{D}$-characteristic w in $S_{(A,i)}$, the search process could be further focused making decisions using this knowledge so that we can obtain complete instantiations of a better quality. For this, we propose to divide the search process of A into two steps.

2.1 Division of the Process

First, we propose to divide the search process of A into two steps: a *First Step* (*FS*) to learn about w in $S_{(A,i)}$ and a *Second Step* (*SS*) performed by A using the knowledge obtained in the *FS*. Inspired in the concept of *anti-pheromone* [13], the idea is to produce a repellent effect to some pairs of components of solutions in $S_{(A,i)}^w$, allowing A to consider other components that, originally, would not be included. From now on, the pheromone used during the *FS* will be called *anti-pheromone*. As A was designed, normal pheromone is used during the *SS*.

   Let A° be an ant-based algorithm that will perform the *FS* and let's use *anti-pheromone* to decrease the attraction of paths that are related with complete instantiations in $S_{(A,i)}^w$. The definition of the representation for $\mathbb{P}$ and the *state transition rule* of A° is the same as in A. The *FS* is performed by A° consuming an amount $B*maxRes$ of resources[2], where $B \in [0,1]$ is a parameter that defines the budget of resources designed for the learning step. At the end of the *FS*, an initial pheromone matrix will be obtained and used by A. Finally, A performs its search process considering the remaining $(1-B)*maxRes$ resources.

2.2 Methods

In order to explore and compare different possibilities to identify a $u\mathcal{D}$-characteristic, we propose three different *Methods*. Each method will consider a different definition for the heuristic knowledge and anti-pheromone management for A°. The methods are named *Soft Opposite-Learning (SOL)*, *Worst Opposite-Learning (WOL)* and *Half Opposite-Learning (HOL)*.

SOL: This method is focused on identifying a $u\mathcal{D}$-characteristic related to the quality of complete instantiations but trying to perform a similar search process as in A. For this, η of A° will be the same as in A. On the other hand, anti-pheromone will be decreased in edges that are related to the lowest quality solution of each iteration. The information obtained during a *FS* performed by the SOL method will reduce the level of attraction produced by the heuristic knowledge in the corresponding *intermediate* decisions of A.

WOL: This method is focused on evaluating the effect of taking totally opposed decisions to the objective of the problem $\mathbb{P}$. For this, the heuristic knowledge

---

[2] These resources can be execution time, a fixed number of evaluations, and conflict checks, among others. In general, the amount of resources can be defined considering how A was originally evaluated.

should be inverted in each intermediate decision (Eq. 1), where $J_k(i)$ is a list of candidate components, η_{ij}^A is the heuristic knowledge of A, and the maximum and minimum heuristic knowledge of values of the current decision are also considered. Here, the construction process is biased towards actual poor quality solutions by the translated heuristic information and the anti-pheromone. Furthermore, components of the lowest quality solution obtained will be marked with *anti-pheromone* at each iteration.

$$\eta_{ij}^{A^\circ} = \max_{u \in J_k(i)} (\eta_{iu}^A) + \min_{u \in J_k(i)} (\eta_{iu}^A) - \eta_{ij}^A \tag{1}$$

HOL: This method is focused on detecting a problem-specific $u\mathcal{D}$-characteristic w. In order to detect this problem-specific feature, the heuristic knowledge η^{A° should be redefined in A°. The construction process should be guided considering new information from η^{A°, allowing the search process to consider information in the presence of w in the complete instantiations obtained. Anti-pheromone will be used to reinforce the objective of η^{A°.

3   Cooperation Between Sub-colonies

In our proposed approach, three sub-colonies of ants cooperate in the construction of a pheromone matrix. Each sub-colony focuses in obtaining information about a $u\mathcal{D}$-characteristic and is guided by one *Method*. During the *FS*, all ants consider the same *anti-pheromone* matrix M to construct solutions. At the end of each iteration, *anti-pheromone* will be updated by ant^{SOL}, ant^{HOL} and ant^{WOL} considering the following rule:

$$anti\tau_{ij}^{new} = anti\tau_{ij}^{old} - \Delta_{ij}^{SOL} - \Delta_{ij}^{HOL} - \Delta_{ij}^{WOL} \tag{2}$$

where Δ_{ij}^{SOL}, Δ_{ij}^{HOL} and Δ_{ij}^{WOL} are the decreased amounts of antipheromone. As the collaboration of these three sub-colonies can be time consuming, we decided to execute the *FS* in parallel and the *SS* is executed sequentially. *COISA* was implemented in *POSIX Threads*. Two types of threads will be considered: *constructor* or *manager* threads. Considering a total of N threads and m ants, one *manager* thread will be focused on the pheromone management and $(N - 1)$ *constructor* threads are focused in constructing and evaluating solutions. The *manager* thread waits until all constructor threads finish, using a barrier, to construct their tasks to update the pheromone matrix M. For the synchronization of all the threads, a barrier, a conditional variable and a mutex are used.

4   Case Study: Multidimensional Knapsack Problem

Multidimensional Knapsack Problem (MKP) is an *NP-hard* combinatorial optimization problem. It considers a set of objects and a knapsack with T dimensions, each one with a maximium capacity defined (b_t). Each object has a defined profit

p_i and weight w_{it} in each problem dimension t. The idea is to select a subset of objects maximizing the total profit, satisfying each capacity constraint.

Here, we introduce *COISA* into Ant Knapsack [1] (AK), a well-known ACO algorithm designed for solving the MKP. AK is a $\mathcal{MAX} - \mathcal{MIN}$ Ant System [14] that constructs feasible complete instantiations. Pheromone represents the desirability of including pairs of objects simultaneously. The heuristic knowledge is defined as: $\eta_{I_P^k}(o_j) = \frac{p_j}{\sum_{t=1}^{T} \frac{w_{jt}}{CC_t}}$, where CC_t is the Current Capacity in dimension t (defined as $CC_t = b_t - \sum_{o_v \in I_P^k} w_{vt}$). Pheromone is deposited in each pair of objects of the best quality solution found of each iteration (L_{b_i}). Here, an amount of $\Delta \tau = \frac{1}{1+|F(L_{b\_f})-F(L_{b\_i})|}$ is deposited, considering that L_{b_f} is the best solution found during the execution.

4.1   Details of the Implementation

This section presents some details that should be considered before the implementation of *COISA* in AK. First, the amount of *anti-pheromone* $\Delta anti\tau$ is defined similarly as in AK. In this case, the worst solution found in the current iteration (L_{w_i}) and the worst solution found during the execution (L_{w_f}) are considered. Moreover, as in AK, one ant per sub-colony will be allowed to deposit anti-pheromone during the *FS*. In order to obtain information without any perturbation, the evaporation is not considered during the *FS*.

SOL and WOL methods are implemented as was already explained in Sect. 2.2. For the HOL method, it is necessary to define a heuristic knowledge for guide its search process. In this case, we considered the same η used in [11]: $\eta_{I_P^k}(o_j) = \frac{p_j}{\sum_{t=1}^{T} RC_t}$, where RC_t is the *remaining capacity* in the dimension t defined as $RC_t = b_t - w_{jt}$. In this case, the $u\mathcal{D}-$characteristic points to identify the *core* of objects for which it is hard to decide if they will be part of an optimal solution or not [8]. Moreover, *anti-pheromone* will mark the lower quality solution of each iteration.

5   Experiments and Results

We considered two sets of 30 instances from the OR Library proposed by Chu and Beasley: 10×100 (10 dimensions and 100 objects) and 5×100 (5 dimensions and 100 objects). In order to compare the collaboration between the three sub-colonies, we present results by each method independently: SOL-AK, HOL-AK and WOL-AK. For all the executions we considered a number of ten threads. The hardware platform used was a Power Edge R630 server with 2 Intel(R) Xeon(R) CPUE5-2680v3 @ 2.50 GHz, 128 GB of RAM using Ubuntu x64 16.10 distribution. We considered the same parameter values proposed in [1]: $\alpha = 1$, $\beta = 5$, $\rho = 0.01$, $N_{Total} = 30$, $\tau_{max} = 6$ and $\tau_{min} = 0.01$. To determine the parameter values for our approaches we used Evolutionary Calibrator (EVOCA) [9], a parameter tuner algorithm, considering randomly selected instances from both sets. The objective was to obtain the number of ants for each sub-colony and the

Table 1. Results for set 10×100 from OR library

#	BK	AK			COISA-AK			SOL-AK			HOL-AK			WOL-AK		
		AVG	SDV	BEST	AVG	SDV	BEST	AVG	SDV	BEST	AVG	SDV	BEST	AVG	SDV	BEST
1	23064	23016.0	42.2	23064	23014.3	46.4	23064	22998.6	47.5	23057	23008.0	41.0	23064	23006.7	42.7	23064
2	22801	22714.0	67.2	22801	22702.2	83.8	22801	22713.8	66.5	22801	22694.6	58.3	22801	22693.6	69.4	22801
3	22131	22034.0	66.9	22131	22046.6	56.0	22131	22024.4	69.7	22131	22008.2	69.9	22131	22035.5	65.3	22131
4	22772	22634.0	60.6	22717	22613.3	63.9	22763	22623.4	64.0	22772	22598.2	73.7	22772	22601.7	53.8	22709
5	22751	22547.0	66.3	22654	22559.2	47.6	22654	22543.2	70.8	22697	22533.0	66.9	22697	22542.7	51.4	22697
6	22777	22602.0	63.3	22716	22593.4	46.8	22716	22610.3	51.4	22716	22594.7	46.0	22664	22591.9	40.5	22675
7	21875	21777.0	44.9	21875	21790.8	36.7	21875	21773.4	45.5	21875	21780.1	48.6	21875	21774.3	54.2	21875
8	22635	22453.0	89.2	22551	22498.8	54.1	22635	22512.0	40.6	22551	22511.7	57.8	22635	22500.1	57.5	22635
9	22511	22351.0	69.4	22511	22379.6	47.0	22511	22369.7	40.3	22438	22362.4	51.6	22511	22352.2	62.5	22511
10	22702	22591.0	88.5	22702	22616.0	102.9	22702	22600.1	99.8	22702	22576.5	91.0	22702	22572.9	88.8	22702
1	41395	41329.0	48.5	41395	41324.1	47.4	41395	41329.1	49.8	41395	41312.4	51.8	41393	41309.0	48.7	41395
2	42344	42214.0	49.5	42344	42233.5	47.0	42344	42232.0	60.4	42344	42210.2	45.5	42344	42221.4	54.9	42344
3	42401	42300.0	58.1	42401	42309.0	38.4	42401	42311.5	41.7	42401	42316.1	47.2	42401	42313.6	43.5	42401
4	45624	45461.0	73.6	45624	45484.2	69.4	45624	45450.2	70.9	45585	45462.3	71.6	45585	45474.4	64.2	45598
5	41884	41739.0	57.3	41884	41770.0	53.0	41884	41769.9	52.0	41884	41758.6	53.0	41884	41750.4	50.2	41884
6	42995	42909.0	76.3	42995	42910.6	76.5	42995	42898.4	72.7	42995	42891.3	78.1	42995	42923.4	69.8	42995
7	43574	43464.0	71.7	43553	43466.9	50.0	43553	43470.0	43.0	43553	43479.0	47.6	43553	43463.7	46.5	43552
8	42970	42903.0	47.7	42970	42904.7	39.4	42970	42901.5	48.1	42970	42924.6	35.3	42970	42915.2	40.1	42970
9	42212	42146.0	48.0	42212	42167.3	39.8	42212	42165.7	39.7	42212	42160.6	38.4	42212	42162.5	42.2	42212
10	41207	41067.0	89.7	41207	41098.7	36.9	41207	41085.9	39.0	41134	41093.5	38.3	41207	41077.7	44.8	41207
1	57375	57318.0	59.5	57375	57295.9	66.1	57375	57307.8	68.1	57375	57311.7	74.1	57375	57321.9	61.3	57375
2	58978	58889.0	40.2	58978	58914.2	32.4	58978	58899.4	54.6	58978	58886.4	43.3	58934	58898.1	24.1	58978
3	58391	58333.0	29.5	58391	58337.7	26.3	58391	58321.2	47.8	58391	58326.8	32.5	58391	58335.4	27.2	58391
4	61966	61885.0	42.4	61966	61891.2	36.4	61966	61876.0	47.9	61966	61873.9	40.6	61966	61882.7	36.6	61966
5	60803	60798.0	5.0	60803	60800.6	3.0	60803	60799.9	3.2	60803	60800.5	3.1	60803	60800.0	5.2	60803
6	61437	61293.0	52.7	61437	61295.3	55.6	61437	61294.1	52.6	61437	61288.3	48.6	61437	61297.5	52.5	61437
7	56377	56324.0	35.7	56377	56319.0	35.4	56377	56311.0	47.0	56377	56313.9	49.1	56377	56328.3	33.8	56377
8	59391	59339.0	53.3	59391	59340.7	42.6	59391	59331.5	51.4	59391	59331.2	53.0	59391	59341.3	37.5	59391
9	60205	60146.0	62.6	60205	60167.7	50.7	60205	60123.1	73.5	60205	60096.8	70.9	60205	60155.8	56.9	60205
10	60633	60605.0	36.1	60633	60613.9	32.2	60633	60589.4	47.6	60633	60571.7	48.4	60633	60613.5	30.7	60633

budget B. The obtained parameter values after 3500 evaluations of EVOCA are: (1) for COISA are $N_{SOL} = 16$, $N_{HOL} = 11$, $N_{WOL} = 8$ and $B = 0.211$, (2) for SOL are $N_{SOL} = 20$ and $B = 0.241$, (3) for HOL are $N_{HOL} = 2$ and $B = 0.422$, (4) for WOL are $N_{HOL} = 16$ and $B = 0.408$. Table 1 shows the results obtained for the 10×100 set and Table 2 shows the results for the 5×100 set. We considered 50 independent runs per instance, each with 60000 evaluations ($maxRes$). Light grey cells show the best average quality (AVG) of the 50 seeds and dark grey cells show the Best quality solution obtained. Also, the standard deviation (SDV) is shown for each instance and algorithm. First, results show that AK could find most of the best known solutions for the instances from both sets (51 of the 60 instances). Moreover, $COISA$-AK outperformed AK obtaining the best known solution in 53 of the 60 instances. This shows that the collaboration between sub-colonies is better than each method on their own. Regarding the average quality, results show that AK obtained better results in the 5×100 set and $COISA$-AK was better for the 10×100 set. Finally, considering the independent and the cooperative approaches, all the best known solutions can be found using opposite information. The non-parametric Wilcoxon test was applied to assess that these algorithms are statistically different ($pvalue = 0.01$). About the Speedup obtained by $COISA$-AK, the average was 1.8, with a maximum of 4.9 and a minimum of 1.4. As the FS only consumes 20% of the evaluations, these metrics show the positive effect of using a parallel architecture.

Table 2. Results for set 5×100 from OR library

#	BK	AK AVG	SDV	BEST	COISA-AK AVG	SDV	BEST	SOL-AK AVG	SDV	BEST	HOL-AK AVG	SDV	BEST	WOL-AK AVG	SDV	BEST
1	24381	**24342.0**	29.3	24381	24340.6	29.0	24381	24329.4	38.2	24381	24335.1	35.3	24381	24330.4	31.4	24381
2	24274	**24247.0**	38.5	24274	24241.2	35.1	24274	24234.9	42.8	24274	24246.0	33.5	24274	24229.2	38.8	24274
3	23551	**23529.0**	8.0	23551	23527.2	9.2	23551	23526.3	13.6	23551	23525.6	14.1	23551	23527.1	11.9	23551
4	23534	**23462.0**	32.6	23534	23460.1	32.7	23527	23453.3	44.4	23534	23458.4	34.8	23527	23457.8	30.2	23511
5	23991	23946.0	31.8	23991	23942.5	26.8	23991	23934.2	33.8	23991	23940.1	35.0	23991	**23950.5**	29.0	23991
6	24613	**24587.0**	31.3	24613	24585.3	25.8	24613	24583.0	28.9	24613	24573.3	34.2	24613	24579.2	28.8	24613
7	25591	25512.0	43.8	25591	25521.8	41.3	25591	**25524.2**	47.8	25591	25506.2	39.7	25591	25509.5	45.9	25591
8	23410	23371.0	30.3	23410	23375.1	33.5	23410	23378.2	29.6	23410	23378.8	29.4	23410	**23381.5**	31.9	23410
9	24216	24172.0	32.9	24216	**24177.0**	31.1	24216	24171.7	32.7	24216	24163.1	38.0	24216	24164.5	39.3	24216
10	24411	**24356.0**	44.3	24411	24346.1	45.8	24411	24340.5	44.5	24411	24342.9	47.2	24411	24346.7	44.8	24411
1	42757	42704.0	14.3	42757	42706.5	25.4	42757	**42709.8**	21.0	42757	42700.6	11.4	42757	42701.6	14.8	42757
2	42545	42456.0	15.8	42510	42458.4	14.6	42471	**42459.5**	12.9	42494	42455.0	21.3	42545	42458.8	25.4	42545
3	41968	41934.0	22.3	41967	**41939.8**	15.9	41968	41935.2	23.0	41967	41930.9	26.3	41967	41930.8	27.7	41967
4	45090	45056.0	24.0	45071	45056.1	24.1	45071	**45058.3**	23.1	45071	45041.0	29.4	45071	45049.4	31.7	45071
5	42218	42194.0	33.2	42218	42201.9	31.4	42218	42196.0	31.2	42218	42189.6	41.8	42218	**42202.2**	28.1	42218
6	42927	42911.0	33.3	42927	**42913.5**	32.6	42927	42903.5	40.8	42927	42913.0	34.0	42927	42908.0	39.5	42927
7	42009	41977.0	45.2	42009	**41985.2**	40.9	42009	41978.9	42.9	42009	41984.6	40.6	42009	41978.0	49.0	42009
8	45020	44971.0	32.5	45010	**44988.8**	22.1	45020	44984.4	29.2	45020	44969.9	35.5	45010	44979.7	31.9	45010
9	43441	**43356.0**	38.5	43441	43349.1	42.9	43441	43353.6	49.7	43441	43345.1	40.9	43441	43347.1	40.7	43441
10	44554	44506.0	25.2	44554	44512.8	23.5	44554	44513.9	25.5	44554	44510.4	28.2	44554	**44515.8**	25.2	44554
1	59822	59821.0	3.2	59822	**59822.0**	0.0	59822	59815.1	20.2	59822	**59822.0**	0.0	59822	**59822.0**	0.0	59822
2	62081	62010.0	47.1	62081	62010.7	44.1	62081	62003.6	44.8	62081	61994.7	31.0	62081	**62011.0**	49.0	62081
3	59802	59759.0	21.7	59802	59757.9	16.1	59802	59745.8	24.7	59802	59750.2	20.2	59802	**59760.1**	22.2	59802
4	60479	60428.0	21.8	60479	**60444.8**	27.3	60479	60417.2	30.0	60479	60438.6	24.2	60479	60435.8	23.6	60479
5	61091	61072.0	20.0	61091	61075.5	18.7	61091	61066.8	35.9	61091	**61078.4**	21.2	61091	61077.6	18.2	61091
6	58959	**58945.0**	14.5	58959	58940.6	12.3	58959	58929.4	35.5	58959	58943.9	19.4	58959	58943.3	14.1	58959
7	61538	**61514.0**	24.0	61538	61511.7	26.6	61538	61508.9	27.2	61538	61498.9	36.9	61538	61513.6	25.9	61538
8	61520	61492.0	25.6	61520	61494.0	22.7	61520	61473.7	34.7	61520	61475.4	32.7	61520	**61496.5**	23.0	61520
9	59453	**59436.0**	40.5	59453	59434.8	43.1	59453	59413.9	59.1	59453	59427.4	53.4	59453	59435.2	39.7	59453
10	59965	59958.0	8.4	59965	59956.0	11.2	59965	59944.9	26.9	59965	59946.0	25.8	59960	**59959.1**	5.2	59965

6   Conclusions

In this work, we proposed a Cooperative Opposite-Inspired Strategy for ants-based algorithms. The objective of this approach is to obtain information about some $u\mathcal{D}$-characteristic that could bias the search process to poor quality solutions. We proposed to divide the search process into two steps: a *First Step* for learning about an $u\mathcal{D}$-characteristic and, a *Second Step* performed by a target ant-based algorithm. During the *First Step*, three sub-colonies of ants cooperate to define an initial pheromone matrix. Each sub-colony is guided by one *Method*: SOL, HOL and WOL. To evaluate our strategy, we used the well-known Ant Knapsack algorithm for solving the MKP. Our preliminary results show that the inclusion of *COISA* in Ant Knapsack improves its robustness and helps to obtain better quality solutions. Additionally, we were able to show that the cooperation between the three methods adopted is better than using only one of them in isolation. As part of our future work, we are interested in evaluating *COISA* in other ant-based algorithms for solving other combinatorial optimization problems. Also, we are interested in comparing *COISA* with other existing pre-processing schemes for ant-based algorithms.

Acknowledgements. Second author is partially supported by the Centro Científico Tecnológico de Valparaíso (CCTVal) Project No. FB0821. The first author is supported by CONICYT-PCHA/National Doctorate/2015-21150696. Third author acknowledges support from CONACyT project no. 221551.

References

1. Alaya, I., Solnon, C., Ghedira, K.: Ant algorithm for the multi-dimensional knapsack problem. In: International Conference on Bioinspired Optimization Methods and their Applications (BIOMA 2004). Citeseer (2004)
2. Cordon, O., de Viana, I., Herrera, F., Moreno, L.: A new ACO model integrating evolutionary computation concepts: the best-worst ant system (2000)
3. Dorigo, M., Stützle, T.: Ant Colony Optimization. Bradford Company, Scituate (2004)
4. Malisia, A.R.: Investigating the application of opposition-based ideas to ant algorithms. Master's thesis, University of Waterloo (2007)
5. Malisia, A.R.: Improving the exploration ability of ant-based algorithms. In: Tizhoosh, H.R., Ventresca, M. (eds.) Oppositional Concepts in Computational Intelligence. SCI, vol. 155, pp. 121–142. Springer, Heidelberg (2008). https://doi.org/10.1007/978-3-540-70829-2_7
6. Malisia, A., Tizhoosh, H.: Applying opposition-based ideas to the ant colony system. In: 2007 IEEE Swarm Intelligence Symposium, SIS 2007, pp. 182–189 (2007)
7. Montgomery, J., Randall, M.: Anti-pheromone as a tool for better exploration of search space. In: Dorigo, M., Di Caro, G., Sampels, M. (eds.) ANTS 2002. LNCS, vol. 2463, pp. 100–110. Springer, Heidelberg (2002). https://doi.org/10.1007/3-540-45724-0_9
8. Puchinger, J., Raidl, G.R., Pferschy, U.: The core concept for the multidimensional knapsack problem. In: Gottlieb, J., Raidl, G.R. (eds.) EvoCOP 2006. LNCS, vol. 3906, pp. 195–208. Springer, Heidelberg (2006). https://doi.org/10.1007/11730095_17
9. Riff, M.C., Montero, E.: A new algorithm for reducing metaheuristic design effort. In: Proceedings of the IEEE Congress on Evolutionary Computation, CEC 2013, Cancun, Mexico, 20–23 June 2013, pp. 3283–3290. IEEE (2013). http://ieeexplore.ieee.org/xpl/mostRecentIssue.jsp?punumber=6552460
10. Rojas-Morales, N., Riff, M.C., Montero, E.: Ants can learn from the opposite. In: Friedrich, T., Neumann, F., Sutton, A.M. (eds.) Proceedings of the 2016 on Genetic and Evolutionary Computation Conference, Denver, CO, USA, 20–24 July 2016, pp. 389–396. ACM (2016). https://doi.org/10.1145/2908812
11. Rojas-Morales, N., Riff, M.C., Montero, E.: Learning from the opposite: strategies for Ants that solve Multidimensional Knapsack problem. In: IEEE Congress on Evolutionary Computation, CEC 2016, Vancouver, BC, Canada, 24–29 July 2016, pp. 193–200. IEEE (2016)
12. Rojas-Morales, N., Riff, M.C., Montero, E.: A survey and classification of opposition-based metaheuristics. Comput. Ind. Eng. 110, 424–435 (2017)
13. Schoonderwoerd, R., Bruten, J.L., Holland, O.E., Rothkrantz, L.J.M.: Ant-based load balancing in telecommunications networks. Adapt. Behav. 5(2), 169–207 (1996)
14. Stützle, T., Hoos, H.H.: MAX-MIN ant system. Future Gener. Comput. Syst. 16(8), 889–914 (2000)

A Solution for the Team Selection Problem Using ACO

Lázaro Lugo[1], Marilyn Bello[1,2(✉)], Ann Nowe[3], and Rafael Bello[1]

[1] Department of Computer Science, Universidad Central "Marta Abreu" de Las Villas, Santa Clara, Cuba
{ljplugo,mbgarcia}@uclv.cu, rbellop@uclv.edu.cu
[2] Faculty of Business Economics, Hasselt University, Hasselt, Belgium
marilyn.bellogarcia@uhasselt.be
[3] Artificial Intelligence Lab, Vrije Universiteit Brussel, Brussels, Belgium
ann.nowe@vub.ac.be

Abstract. The team selection problem is usually solved by ranking candidates based on the preferences of decision-makers and allowing the decision-makers to take turns selecting candidates. While this solution method is simple and might seem fair it usually results in an unfair allocation of candidates to the different teams, i.e. the quality of the teams might be quite different according to the rankings articulated by the decision-makers. In this paper, we propose a new method based on Ant Colony Optimization (ACO), where the selection process is performed in a new context, with more than two decision-makers selecting from a common set of candidates. Furthermore, a plugin implementing this method for the KNIME platform was developed.

1 Introduction

Personnel selection is the process by which one or more people are chosen for a job, depending on how suitable their characteristics are. It is one of the main processes of any company or organization, and it is expected to take on the right employee for the right job at the right time [13].

Today, many tools and techniques are used in this specific decision-making problem [5]. One of the first works where the problem was presented from the perspective of intelligent systems was reported in [11]. Extensions based on Multi-Criteria Decision-Making (MCDM) have also been proposed [14] in which the decision-maker seeks to optimize a combination of criteria associated with the candidates [15]. According to this goal, they can be used to rank alternatives (to build a ranking).

To obtain a ranking of candidates is especially interesting when the management of human resources is directed to organize, manage and lead a team instead of selecting an employee for a simple vacant; this contributes to the success of the project and creates a competitive advantage for the organization. A variety of approaches are proposed for the selection of the members of a team,

© Springer Nature Switzerland AG 2018
M. Dorigo et al. (Eds.): ANTS 2018, LNCS 11172, pp. 325–332, 2018.
https://doi.org/10.1007/978-3-030-00533-7_26

most of them aimed at forming teams in the field of business, industry and sport [1,6,10,19,21].

In this research, this problem is tried in a framework different to the classic, due to the process of selection is realized in a competitive environment, that is, when two or more decision-makers should to form their teams by choosing the personnel from the same set of candidates; and at the same time, it is necessary to form teams as similar as possible to the preferences established by each decision-maker on the candidates.

An example that illustrates this problem is the draft process used in the United States, Canada, Australia and Mexico to assign certain players to sports teams. In a draft, the teams take turns selecting a group of eligible players. When a team chooses a player, the team receives exclusive rights to sign a contract with him/her, and no other team in the league can choose that player. A draft avoids expensive bidding wars for young talents and ensures that no team can monopolize all the best young players and make the leagues uncompetitive [7].

Recently, the Artificial Intelligence Laboratory from UCLV has been working on this topic [2,3], proposing the development of a method for the formation of teams in a competitive environment, where two decision-makers try to shape their team from a list of candidates.

In this work we propose a generalization of this method for cases with more than two decision-makers. Furthermore, we developed a plugin for the KNIME platform [4], where a method based on Ant Colony Optimization (ACO) is implemented for the formation of teams in a competitive environment.

2   Formulation of Personnel Selection Problem in a Competitive Environment

Given N candidates $C = \{C_1, C_2, C_3, \ldots, C_N\}$ and Q decision-makers $D = \{D_1, D_2, D_3, \ldots, D_Q\}$, the objective is to form N teams. For that, each decision-makers order the N candidates according to their preference, obtaining a set of rankings for each decision-maker $R = \{R_1, R_2, R_3, \ldots, R_Q\}$. From these, each decision-maker forms his team selecting the candidate that first appears in his preference ranking, as long as, this candidate has not already been selected by another decision-maker. This process is repeated while there are still candidates available to be selected. As a result of this process, we obtain a set of Q teams $R* = \{R_1*, R_2*, R_3*, \ldots, R_Q*\}$, that is, one team per decision-maker. See example 1 below.

Example 1. Suppose that $N = 10$, $Q = 5$, $C = \{0, 1, 2, 3, 4, 5, 6, 7, 8, 9\}$ and $R = \{\{7, 4, 3, 2, 6, 9, 0, 5, 8, 1\}, \{6, 3, 5, 8, 1, 2, 4, 7, 0, 9\}, \{7, 2, 3, 9, 8, 5, 4, 0, 1, 6\}, \{6, 3, 1, 0, 7, 8, 9, 4, 5, 2\}, \{3, 8, 7, 9, 2, 0, 4, 1, 6, 5\}\}$ and the order in which the decision-makers will select is the following: $\{D_3, D_2, D_1, D_4, D_5\}$. Using these values, the resulting teams will be the set $R* = \{R_1* = \{4, 9\}, R_2* = \{6, 5\}, R_3* = \{7, 2\}, R_4* = \{3, 1\}, R_5* = \{8, 0\}\}$.

The problem is to build the set of Q teams $R*$ which minimize the differences between R_i and R_i*, for all i, that is, $min \sum_{i=1}^{Q} d(R_i, R_i*)$. This is a discrete optimization problem.

3 A Method for Solving the Team Selection Problem in a Competitive Environment Using Ant Colony Optimization

ACO is inspired by the behavior governing ants looking to find the shortest paths between food sources and their anthill; it has been extensively used to solve combinatorial optimization problems. Several models of the ACO metaheuristic have been proposed in the literature [8,9,16,18,20].

In the context of this work, a hybrid between the Max-Min Ants System (MMAS) [18] and the Multi-type Ants [16,20] is used. The steps of the ASMMTS algorithm (Ant System Max Min Teams Selection) for the Team Selection are described below:

1. A complete graph with N nodes is constructed, where each node represents a candidate for selection.
2. Q types of ants are generated, one for each decision-maker.
3. Ants are grouped into groups $(h_{1k}, h_{2k}, \ldots, h_{Qk})$, which always include one ant of each type, so that all decision-makers are represented. There are m groups $(h_{11}, h_{21}, \ldots, h_{Q1}), (h_{12}, h_{22}, \ldots, h_{Q2}), \ldots, (h_{1m}, h_{2m}, \ldots, h_{Qm})$, each working to build a solution to the problem together.
4. In each iteration, the ants are randomly distributed with the constraint that no other ant from the same group can be on the same node. This is to enforce that decision-makers cannot select the same candidate.
5. The initial values of pheromones associated with the link between the nodes i and j for each type of ant take the value $\tau_{ijp} = \tau_{max\_initial}$. In this way, there is a high initial exploration of the search space.
6. The heuristic value used to evaluate the quality of each possible successor node is denoted as η_{ijp}; and it is calculated by the Eq. (1).

$$\eta_{ijp} = 1/O(j) \tag{1}$$

Where $O(j)$ is the position in which node j is located in the preference ranking p of the set R.

7. The neighborhood in node i, denoted by V_{ipk} of the kth ant of type p (h_{ipk}) is the set of all the nodes that have not yet been selected by it or any other ant in its group.
8. The probabilistic rule that decides the new node to choose is defined by the Eq. (2), being α, β and θ constant parameters of this rule.

$$p_{ij}^{pk} = \left(\frac{([\tau_{ij}^p]^\alpha * [\eta_{ij}^p]^\beta) / [\sum_{n=1, n \neq p}^{Q} (\tau_{ij}^n)]^\theta}{(\sum_{j \in V_i^{pk}} [\tau_{ij}^p]^\alpha * [\eta_{ij}^p]^\beta) / [\sum_{n=1, n \neq m} (\tau_{ij}^n)]^\theta} \right) \tag{2}$$

9. Each ant ends when there are no nodes to select. At the end of each cycle each group of ants $(h_{1k}, h_{2k}, \ldots, h_{Qk})$ will generate a solution that together form the set $R* = \{R_1*, R_2*, R_3*, \ldots, R_p*, \ldots, R_Q*\}$, where the node-path of ant h_{pk} represents the team selected by decision-maker D_p (R_p*).

10. The solutions found for each group of ants are evaluated at the end of each cycle using Eqs. (3), (4) and (5). In Eq. (3), the first term measures how similar the teams are to the resulting ranking and the second term measures how similar is the level of satisfaction of the decision-makers.

$$min \to Eval^k = A + B \tag{3}$$

$$A = (\sum_{p=1}^{Q} eval(s(h_{pk}), R_p)/Q) \tag{4}$$

$$B = |eval(s(h_{pk}), R_p) - \sum_{n=1, n \neq p}^{Q} eval(s(h_{nk}), R_n)| \tag{5}$$

Where $s(h_{pk})$ is the solution found by ant h_{pk}, and the value of $eval(s(h_{pk}), R_p)$ is obtained according to Eq. (6).

$$eval(R_p*, R_p) = \sum_{\forall c \in R_p*} \pi(c) \tag{6}$$

Where $\pi(c)$ is the value of the candidate $c \in R_p*$ according to its position in the R_p ranking; the function π assigns the value 1 to the first place in the ranking, 2 to the second place and so on until the last place in the ranking is assigned the value N.

Example 2. Given a set of 6 candidates $C = \{1, 2, 3, 4, 5, 6\}$ and 2 decision-makers; the rankings $R1 = \{2, 0, 1, 3, 5, 4\}$ and $R2 = \{3, 2, 1, 5, 0, 4\}$; and the resulting subsets $R1* = \{0, 2, 4\}$ and $R2* = \{3, 5, 1\}$. Applying Eq. (4) we have $eval(R_1*, R_1) = 2 + 1 + 6 = 9$ and $eval(R_2*, R_2) = 1 + 4 + 3 = 8$.

11. Whenever a cycle is finished, that is, when all the ants have traversed all the nodes, the evaporation of pheromones happens, and new pheromones are deposited. The pheromone values are decremented using Eq. (7), where ρ is the evaporation constant, which is a value between 0 and 1.

$$\tau_{ij}(t+1) = \rho * \tau_{ij}(t) \tag{7}$$

As the model proposed in this section is based on the Max-Min Ants System, the pheromones are deposited on the arcs ij that appear in the solutions found by the best group of ants during the current iteration and those appearing in the best global solution (the group of ants that has obtained the best solution since the beginning of execution). The pheromone deposit is calculated using Eq. (8).

$$\tau_{ijp}(t+1) = \tau_{ijp}(t) + (1/eval(s(h_{pk}), R_p)) \tag{8}$$

12. At the end of the search process the solutions $s(h_{1k}), s(h_{2k}), \ldots, s(h_{Qk})$ associated with the group $(h_{1k}, h_{2k}, \ldots, h_{Qk})$ that have the lowest (and thus best) value of $Eval^k$ will be the set $R* = R_1*, R_2*, R_3*, \ldots, R_Q*$ resulting for the Q decision-makers.

13. The pheromone levels are bounded each time a cycle ends, using a maximum and a minimum level, so that no trace is less than a minimum level τ_{min} or greater than a maximum level τ_{max}. If any trail of pheromone is smaller than the minimum level, it is re-initialized to τ_{min}. In the same way, all pheromone values that exceed the maximum level are re-initialized to τ_{max}. The max and min values are calculated using Eqs. (9) and (10) respectively [17].

$$\tau_{max} = (1/(1 - \rho)) * 1/Eval^{better\_global\_group} \tag{9}$$

$$\tau_{min} = \tau_{max}/(10 * N) \tag{10}$$

Where ρ is the value of the evaporation constant and $Eval^{better\_global\_group}$ represents the quality of the best overall solution found by the colony throughout the search process, and is calculated using Eq. (3). In Eq. (10) N represents the number of candidates.

4   Experimental Study

The purpose of this study is to illustrate the effectiveness of the ASMMTS algorithm using some examples. To do this, we make a simulation of the process of selecting candidates in order to form the teams, using R sets of preference rankings with N elements for each decision-maker that are randomly generated. In the study, values between 10 and 20 were considered for N with 5 decision-makers. Different values were evaluated as input parameters of the algorithm, the results shown in the following table were obtained with the values: $\alpha = \beta = \theta = 1$, $\rho = 0.75$, $NmaxC = 10$ and $m = N$ (i.e. the number of ants is the same as the number of candidates).

In the experimental study, a new measure is used to compute the degree of hardness of the set of rankings used to build the team; that is, while greater is the correlation between the rankings greater is the complexity of applying the methods for building the teams selection. The Eq. (11) defines a measure of hardness, where the correlation Kendall is used [22], denoted by t, and $U = \frac{1}{2} * Q(Q - 1)$.

$$H(R) = \sum_{i=1}^{U} t(R_i, R_j) \tag{11}$$

Table 1 shows the results achieved using the proposed algorithm. This table contains three rows that correspond to different examples generated for that number of candidates; the second column shows the result reached by each of the decision-makers if the selection of the candidates is made according to the order established in the rankings; and the third column lists the results obtained after applying the ASMMTS algorithm. Both columns have the value of Eq. (3),

Table 1. 10 candidates, 5 decisions-makers.

Rankings	Order in the rankings	ASMMTS
$R1 = \{7,4,3,2,6,9,0,5,8,1\}$	$R1^* = \{4,9\}$	$R1^* = \{4,6\}$
$R2 = \{6,3,5,8,1,2,4,7,0,9\}$	$R2^* = \{5,0\}$	$R2^* = \{5,2\}$
$R3 = \{7,2,3,9,8,5,4,0,1,6\}$	$R3^* = \{7,2\}$	$R3^* = \{8,7\}$
$R4 = \{6,3,1,0,7,8,9,4,5,2\}$	$R4^* = \{6,1\}$	$R4^* = \{1,3\}$
$R5 = \{3,8,7,9,2,0,4,1,6,5\}$	$R5^* = \{3,8\}$	$R5^* = \{0,9\}$
$H(R) = -1.21$	$Eval(R^*, R) = 20.0$	$\mathbf{Eval(R^*,R) = 12.4}$
$R1 = \{5,9,1,4,2,6,0,3,7,8\}$	$R1^* = \{9,2\}$	$R1^* = \{9,6\}$
$R2 = \{7,8,9,1,5,6,3,0,2,4\}$	$R2^* = \{7,8\}$	$R2^* = \{1,7\}$
$R3 = \{5,7,1,8,0,6,2,3,9,4\}$	$R3^* = \{5,1\}$	$R3^* = \{5,2\}$
$R4 = \{4,0,9,8,6,7,3,5,2,1\}$	$R4^* = \{4,0\}$	$R4^* = \{4,8\}$
$R5 = \{3,8,7,0,5,1,2,4,9,6\}$	$R5^* = \{3,6\}$	$R5^* = \{3,0\}$
$H(R) = -0.13$	$Eval(R^*, R) = 19.6$	$\mathbf{Eval(R^*,R) = 9.2}$
$R1 = \{0,8,7,6,1,3,9,4,2,5\}$	$R1^* = \{0,1\}$	$R1^* = \{0,6\}$
$R2 = \{4,2,7,1,8,9,6,5,3,0\}$	$R2^* = \{2,7\}$	$R2^* = \{1,7\}$
$R3 = \{5,4,8,1,2,7,0,6,3,9\}$	$R3^* = \{5,3\}$	$R3^* = \{2,4\}$
$R4 = \{4,6,1,5,2,9,7,0,8,3\}$	$R4^* = \{4,6\}$	$R4^* = \{9,5\}$
$R5 = \{8,5,7,0,9,2,1,3,6,4\}$	$R5^* = \{8,9\}$	$R5^* = \{3,8\}$
$H(R) = -0.49$	$Eval(R^*, R) = 24.0$	$\mathbf{Eval(R^*,R) = 12.6}$

lower values means that better teams was formed according to previously ranking made by the decision-makers.

Analyzing the results reported in the experimental study, we can conclude that it is possible to obtain higher quality results when using the ASMMTS method instead of selecting strictly following the order established in the rankings; it allows to form teams closer to the preferences of each decision-maker, achieving a comparable satisfaction levels for all decision-makers. An important conclusion is that the greater the number of candidates, the more effective the behavior of the proposed method.

5   Plugin ASMMTS for KNIME

KNIME (Konstanz Information Miner) is an open source platform with more than 1000 modules, hundreds of ready-to-run examples, a wide range of integrated tools and a wide variety of advanced algorithms available [4,12]. It can be easily extended because it is based on the Eclipse Enriched Client Platform [12]. In order to add new plugins based on the Eclipse platform, the tool proposes a class structure based on a Node class, which presents all the functionalities that will allow the implementation of the NodeModel class, optionally a NodeDialog and one or more instances of the NodeView class. In addition, each node contains the number of data entry and exit ports, where the information will transit.

The ASMMTS node has an entry, which corresponds to the rankings of preferences of each of the decision-makers, obtained from the File Reader node, this node is responsible for transforming the read data into an ordered list representing the preferences of the decision-makers.

Before proceeding to the execution of the ASMMTS node, the configuration dialog of the node is accessed and the necessary data are introduced in order to execute the algorithm. After executing the node, the view of the resulting data can be analyzed using the ViewData option, which shows the team formed by each decision-maker after having executed the algorithm.

6 Conclusions

In this paper, we propose a new method to solve the team selection problem based on the Multi-type Ant Colony Optimization. In this case, the decision-makers select from a common set of candidates, each decision-maker defines a ranking expressing his preference on the candidates and they want to form the team that is closest to his ranking. But as the preferences of decision-makers may be similar, the rankings established by them may be similar.

A common approach for developing the selection of candidates is to allow the decision-makers to alternatively select one candidate according to their ranking. While this seems a fair approach, it does not necessarily in a fair set of teams as shown in the experimental study.

The conducted experimental study shows that the proposed method allows to form teams that are equally close to the preferences of both employers, yet are fair, and that efficiency is more noticeable for larger numbers of candidates.

A plugin related to the team selection is developed for the KNIME platform.

References

1. Ahmed, F., Deb, K., Jindal, A.: Multi-objective optimization and decision making approaches to cricket team selection. Appl. Soft Comput. **13**(1), 402–414 (2013)
2. Bello, M., Bello, R., Nowé, A., García-Lorenzo, M.M.: A method for the team selection problem between two decision-makers using the ant colony optimization. In: Collan, M., Kacprzyk, J. (eds.) Soft Computing Applications for Group Decision-making and Consensus Modeling. SFSC, vol. 357, pp. 391–410. Springer, Cham (2018). https://doi.org/10.1007/978-3-319-60207-3_23
3. Bello, M., Lugo, L., García, M.M., Bello, R.: Un método para la generación de rankings en la selección de equipos de trabajo en ambiente competitivo basado en algoritmos genéticos. Revista Cubana de Ciencias Informáticas **10**(2), 196–210 (2016)
4. Berthold, M.R., et al.: KNIME - the Konstanz information miner: version 2.0 and beyond. ACM SIGKDD Explor. Newsl. **11**(1), 26–31 (2009)
5. Canós, L., Casasús, T., Liern, V., Pérez, J.C.: Soft computing methods for personnel selection based on the valuation of competences. Int. J. Intell. Syst. **29**(12), 1079–1099 (2014)

6. Dadelo, S., Turskis, Z., Zavadskas, E.K., Dadeliene, R.: Multi-criteria assessment and ranking system of sport team formation based on objective-measured values of criteria set. Expert Syst. Appl. **41**(14), 6106–6113 (2014)

7. Diario, A.: NFL draft 2016: todas las elecciones de 1ª y 2ª ronda (2016)

8. Dorigo, M., Gambardella, L.M.: Ant colony system: a cooperative learning approach to the traveling salesman problem. IEEE Trans. Evol. Comput. **1**(1), 53–66 (1997)

9. Dorigo, M., Maniezzo, V., Colorni, A.: Ant system: optimization by a colony of cooperating agents. IEEE Trans. Syst. Man. Cybern. Part B (Cybern.) **26**(1), 29–41 (1996)

10. Hayano, M., Hamada, D., Sugawara, T.: Role and member selection in team formation using resource estimation for large-scale multi-agent systems. Neurocomputing **146**, 164–172 (2014)

11. Hooper, R.S., Galvin, T.P., Kilmer, R.A., Liebowitz, J.: Use of an expert system in a personnel selection process1. Expert Syst. Appl. **14**(4), 425–432 (1998)

12. Iglesias, A.I., Ilisástigui, L.B., Cordovéz, T.C., Rodríguez, D.M.: Nuevos plugins para la herramienta knime para el uso de sus flujos de trabajo desde otras aplicaciones. Ciencias de la Información **46**(1), 47–52 (2015)

13. Kulik, C.T., Roberson, L., Perry, E.L.: The multiple-category problem: category activation and inhibition in the hiring process. Acad. Manag. Rev. **32**(2), 529–548 (2007)

14. Lai, Y.J.: IMOST: interactive multiple objective system technique. J. Oper. Res. Soc. **46**(8), 958–976 (1995)

15. Mohamed, F., Ahmed, A.: Personnel training selection problem based on SDV-MOORA. Life Sci. J. **10**(1) (2013)

16. Nowé, A., Verbeeck, K., Vrancx, P.: Multi-type ant colony: the edge disjoint paths problem. In: Dorigo, M., Birattari, M., Blum, C., Gambardella, L.M., Mondada, F., Stützle, T. (eds.) ANTS 2004. LNCS, vol. 3172, pp. 202–213. Springer, Heidelberg (2004). https://doi.org/10.1007/978-3-540-28646-2_18

17. Puris, A., Bello, R., Herrera, F.: Analysis of the efficacy of a two-stage methodology for ant colony optimization: case of study with TSP and QAP. Expert Syst. Appl. **37**(7), 5443–5453 (2010)

18. Stützle, T., Hoos, H.H.: Max-min ant system. Future Gener. Comput. Syst. **16**(8), 889–914 (2000)

19. Tavana, M., Azizi, F., Azizi, F., Behzadian, M.: A fuzzy inference system with application to player selection and team formation in multi-player sports. Sport Manag. Rev. **16**(1), 97–110 (2013)

20. Vrancx, P., Nowé, A., Steenhaut, K.: Multi-type ACO for light path protection. In: Tuyls, K., Hoen, P.J., Verbeeck, K., Sen, S. (eds.) LAMAS 2005. LNCS (LNAI), vol. 3898, pp. 207–215. Springer, Heidelberg (2006). https://doi.org/10.1007/11691839_13

21. Wang, J., Zhang, J.: A win-win team formation problem based on the negotiation. Eng. Appl. Artif. Intell. **44**, 137–152 (2015)

22. Webber, W., Moffat, A., Zobel, J.: A similarity measure for indefinite rankings. ACM Trans. Inform. Syst. (TOIS) **28**(4), 20 (2010)

Boundary Constraint Handling Techniques for Particle Swarm Optimization in High Dimensional Problem Spaces

Elre T. Oldewage[1,2]([⊠]), Andries P. Engelbrecht[1,3],
and Christopher W. Cleghorn[1]

[1] Department of Computer Science, University of Pretoria, Pretoria, South Africa
vze.ezv@gmail.com, {engel,ccleghorn}@cs.up.ac.za
[2] Council for Scientific and Industrial Research, Pretoria, South Africa
[3] Institute for Big Data and Data Science, Pretoria, South Africa

Abstract. This paper investigates the use of boundary constraint handling mechanisms to prevent unwanted particle roaming behaviour in high dimensional spaces. The paper tests a range of strategies on a benchmark for large scale optimization. The empirical analysis shows that the hyperbolic strategy, which scales down a particle's velocity as it approaches the boundary, performs statistically significantly better than the other methods considered in terms of the best objective function value achieved. The hyperbolic strategy directly addresses the velocity explosion, thereby preventing unwanted roaming.

1 Introduction

Particle swarm optimization (PSO) is a stochastic, population-based optimization algorithm [9]. A swarm consists of a number of particles. Each particle's position in the search space represents a possible solution to an optimization problem. The particles move through the search space, guided by local and global information. This paper considers PSO with inertia weight [20].

Previous studies in literature have emphasized the importance of boundary handling techniques for PSO, especially in high dimensional spaces [12,15]. As problem dimensionality increases, the particles become increasingly likely to leave the search space and exhibit unwanted roaming behaviour [12]. Application of boundary constraint handling techniques may mitigate particles' roaming behaviour and allow the search to continue even in high dimensional spaces.

A number of constraint handling strategies are examined that may be employed to mitigate particle roaming behaviour. Classical boundary constraint handling methods have been suggested [8,17], but the methods utilize information about the optima locations and/or gradient information. For black-box optimization problems, such information is typically not available. The boundary constraint handling techniques considered in this paper do not make use of additional information about the optimization function.

© Springer Nature Switzerland AG 2018
M. Dorigo et al. (Eds.): ANTS 2018, LNCS 11172, pp. 333–341, 2018.
https://doi.org/10.1007/978-3-030-00533-7_27

Boundary constraint handling techniques bias the particles towards certain parts of the search space [10,11]. Thus, the best choice in technique usually depends on the location of the optima. However, additional information about the optima locations may not be known. This paper considers minimization problems that have been shifted by a random vector, distributed uniformly throughout the search space. There is thus no clear pattern regarding the optima locations, i.e. they are not near the boundaries or near the center of the search space in all dimensions. The paper discusses the performance of a selection of boundary handling techniques so that practitioners are guided in choosing a strategy when the locations of the optima are unknown or differ widely among dimensions.

The paper proceeds as follows: Sect. 2 discusses the boundary constraint handling techniques being considered. Section 3 describes the experimental method. Section 4 presents the empirical results and Sect. 5 concludes the paper.

2   Background

This section discusses a number of the most common boundary constraint handling techniques. Section 2.1 discusses position repair methods. Section 2.2 introduces velocity repair strategies that can be combined with the position repair strategies. Section 2.3 discusses techniques that do not fall into either category.

2.1   Position Repair Strategies

This section lists position repair strategies, which modify a particle's position so that it no longer violates the boundary.

Infinity: The first strategy, suggested by [3] only modifies the PSO algorithm by constraining the personal and global best positions to be within the search space. Thus, the particles may leave the search space, but their local and global attractors will always be within bounds, thereby encouraging the particles to return to the search space. There have been suggestions in literature to make this approach standard practice [3]. This method is also known as the "invisible wall" [18]. The *infinity* method has the advantage of not modifying the velocity or position vectors directly, thereby preventing the algorithm from becoming biased as a side effect of the boundary handling technique.

Random: The *random* method [4,11,16] re-initializes any invalid position component uniformly within the search space. A possible side effect of this method is to inject diversity into the swarm by randomly selecting position components that particles would have been unlikely to encounter otherwise. Another variant, *random-half* [19], re-initializes any invalid position components within the half of the search space nearest the violated boundary.

Absorb: The *absorb* strategy repairs a particle's position by moving it back onto the boundary in every violated dimension. This approach biases particles towards solutions that are on the boundary of the search space. The *absorb* strategy is also known as *truncate* [1], *nearest* [10], or *boundary* [16].

Exponential: The *exponential* method as originally proposed [1] repairs a particle's position in every dimension by moving the particle to a point between its previous position and the violated boundary. The new position is sampled from a truncated exponential distribution, oriented so that there is a higher probability of sampling a position near the boundary. An alternative method [16] samples from a truncated exponential distribution spread across the entire search space in the violated dimension (oriented so that positions near the violated boundary are more likely). The original method is referred to as *exponential-confined* and the latter as *exponential-spread*. The *exponential* method introduces less artificial diversity than the *random* and *random-half* methods, and also preserves search information about good solutions near the boundaries by sampling from a biased distribution.

2.2   Velocity Repair Methods

A particle's velocity vector contains information about favourable search directions in relation to its local and global attractors. If the particle is relocated, then the relative direction of its attractors change. Modifying a particle's position without also adjusting its velocity may cause the particle to move in directions that are unfavourable due to its outdated momentum component. Additionally, if the particle left the search space due to high velocity in a given dimension, then the particle is likely to leave the search space again after being moved inside the search space (because it still has a large, outward momentum component). A variety of velocity repair methods are discussed below:

Zero: Set the velocity to zero in violated dimensions.

Adjust: The *adjust* strategy [10, 11] performs a backward calculation to obtain the repaired velocity after applying a position repair strategy, $\mathbf{v}_i^{t+1} = \mathbf{x}_i^{t+1} - \mathbf{x}_i^t$, where $\mathbf{v}_i^{t+1}$ denotes the velocity and $\mathbf{x}_i^{t+1}$ denotes the position of the i-th particle at iteration $t + 1$. This strategy records a particle's movement to a feasible location. For certain position repair strategies, this may help to prevent the particle from leaving the search space again or from moving in unfavourable directions due to an outdated momentum component.

Reflect: The *reflect* strategy reflects the particle's velocity in the violated dimension. Reflection ensures that particles do not have large, outward momentum components after their positions have been repaired, with the aim of reducing their propensity to leave in the following iteration.

Random Damping: *Random damping.*[13] is a hybrid between reflection and absorption. If a particle exceeds the boundary in a given dimension, its velocity is partially reflected and partially absorbed. This forces the particle back into valid space and decreases its velocity. The fraction of the velocity to be reflected or absorbed is uniform random.

Damping: *Damping* is a deterministic version of the *random damping* method in [13]. The parameter λ is used to determine how much of the velocity is reflected or absorbed. λ may be a constant value or, as in [14], may depend on the particle's distance from the boundary. In this paper, $\lambda = 0.5$.

2.3 Other Strategies

This section lists strategies that prescribe how a particle's position and velocity should be repaired or interpreted when a boundary is violated.

PBest: The *pBest* method, as proposed in [14], re-positions a particle to its personal best position and sets its velocity to **0** if it leaves the search space (in any dimension). Since particles frequently leave the search space in high dimensions, this may lead to most of the swarm being relocated frequently. Relocating a particle encourages highly exploitative behaviour: in the following iteration, the velocity's cognitive component will be zero and the momentum component is zero, since the velocity was zeroed. Thus, the particle's movement depends only on its social component, causing the particles to move towards the global best position. In this paper, another version of the *pBest* strategy is suggested in which only violated dimensions are reset. This method is referred to as *pBest-dim*. Resetting only violated dimensions will reduce the chances of premature convergence since fewer dimensions will rely on social-only velocity updates.

Hyperbolic: The *hyperbolic* strategy [6] prevents a particle from ever reaching the boundary by scaling its velocity. The closer a particle is to the boundary, the smaller its scaled velocity is. Scaling is performed as follows:

$$v_{i,j}^t = \begin{cases} \frac{v_{i,j}^t}{1+|v_{i,j}^t/(U_j - x_{i,j}^t)|} & \text{if } v_{ij}^t > \frac{U_j + L_j}{2} \\ \frac{v_{i,j}^t}{1+|v_{i,j}^t/(x_{i,j}^t - L_j)|} & \text{if } v_{ij}^t \leq \frac{U_j + L_j}{2} \end{cases} \tag{1}$$

where U_j and L_j denote the upper and lower boundaries in the j-th dimension.

Resampling Stochastic Scalars: The *RES* or *resampling* method [1] resamples the stochastic scalars $r_{1,j}$ and $r_{2,j}$ in every dimension until the resulting velocity does not cause the particle to leave the search space. This strategy will have a non-deterministic run time. Particles close to the boundary may have to draw many random numbers to obtain satisfactory values for $r_{1,j}$ and $r_{2,j}$.

Periodic PSO: The *periodic* strategy [22] does not modify particle positions or velocities. Instead, the search space is extended with infinitely many copies that the particles can traverse without further consideration to the boundaries. A particle's position is mapped back to the original search space for evaluation, where each dimension is mapped by M_j as described below:

$$x_{i,j}^t \overset{M_j}{\mapsto} L_j + (x_{i,j}^t \% (U_j - L_j)) \tag{2}$$

where $\%$ is the modulo operator. A particle's score or fitness is given by $f(M(\mathbf{x}_i^t))$, where f denotes the objective function. Although the strategy works well on some search spaces, adjoining copies of the search space may introduce sharp discontinuities, which make the search space more difficult to traverse.

3   Experimental Method

This section describes the empirical method. The experiments used PSO with inertia weight [20] with the global best topology. The selected inertia weight, $w = 0.7298$ and the acceleration coefficients $c_1 = c_2 = 1.49618$ are known good values suggested by Clerc [7] that guarantee convergence of the swarm (in terms of expectation and variance of particle positions [5]). Each swarm consisted of 50 particles. The different boundary-handling techniques were tested on problems from the CEC 2010 Benchmark Suite for Large Scale Optimization [21] with dimensionality of 1000 ($n = 1000$). The suite consists of minimization problems that includes separable, non-separable, and partially separable problems. The degree of separability is controlled by a parameter m which was set to 10 for these experiments. The original definition of the benchmark suite uses a vector of random numbers distributed normally throughout the search space (in each dimension). However, this will bias the location of the optima to be near the center of the search space. In order to prevent such bias, the shift vectors used in this paper are distributed uniformly throughout the search space. Every boundary-handling technique was run on each of the 20 benchmark problems 30 times for statistical significance. Every simulation was allowed 5000 iterations.

4   Results

Every boundary handling technique was assigned a rank score that depends on the best score achieved over all simulations as proposed in [2]. These scores were normalized so that the best rank score is 1 and the worst is 0 (as shown in Fig. 1). A score is related to the number of statistically significant "wins" when a strategy is compared in a pairwise manner to all the other strategies across all benchmark functions (in terms of solution accuracy). Comparisons are performed using a Mann-Whitney U test with $p = 0.05$. Additionally, an average normalized fitness was calculated for each technique according to:

$$\frac{1}{|\mathcal{F}|} \sum_{f \in \mathcal{F}} \frac{\overline{f(\mathbf{y})}}{\widetilde{f(\mathbf{y})}} \tag{3}$$

where $\mathcal{F}$ denotes the set of benchmark functions, $|.|$ denotes set cardinality, $\overline{f(\mathbf{y})}$ denotes the best fitness attained by the strategy on function f, averaged across all runs and $\widetilde{f(\mathbf{y})}$ denotes the worst average fitness attained by any strategy on function f. Thus, if a strategy always performed the worst on all functions, it would receive an average normalized fitness of 1.

Figure 1, which plots the rank score and average normalized fitness, shows that the *hyperbolic* strategy exhibits the best performance in terms of both measures. *Hyperbolic* also performed statistically significantly better than the other four best strategies on 16 out of the 20 benchmark functions. It is known from literature [12,15] that a large factor in PSO's poor performance in high dimensional problem spaces is the initial velocity explosion and the consequent roaming

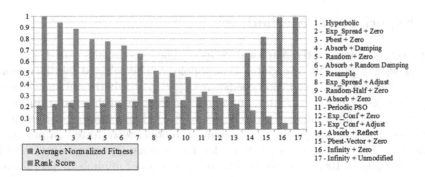

Fig. 1. Average normalized fitness and rank scores

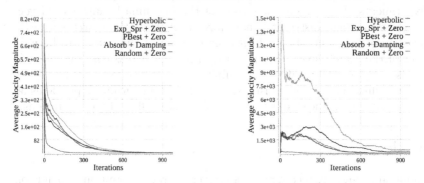

Fig. 2. Velocity magnitude (top 5)   **Fig. 3.** Velocity variance (top 5)

behaviour. The *hyperbolic* strategy completely mitigates the effects of the velocity explosion, since the velocity is scaled down to ensure that the particles remain in valid space. Although this prevents particles from attaining optima that are on the boundaries, the strategy does not affect the particles' search direction or artificially introduce or inhibit swarm diversity. The reduction in the velocity explosion is apparent in Figs. 2 and 3 which plot the average velocity magnitude and the average variance in velocity for the five best-performing strategies.

For 15 out of the 19 strategies, the average normalized fitness was between 0.21 and 0.32. Therefore, although the difference in performance among the strategies were statistically significant, the actual difference in fitness between most of the strategies was not very large. This is likely due to the problem-dependent nature of boundary handling strategies, which is known in literature.

Due to space limitations, only a few of the strategies' behaviour are discussed in detail. Although it may be expected that the *pBest* strategy will converge prematurely, Fig. 4 shows that the *pBest* strategy failed to converge. Instead, the swarm's diversity oscillates with every iteration, as the particles attempt to explore, leave the search space immediately and are reset. Since no searching could take place, the personal bests were almost never updated and thus the global best was almost never updated. In contrast, the *pBest-dim* strategy per-

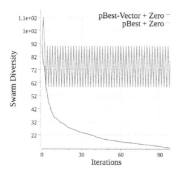

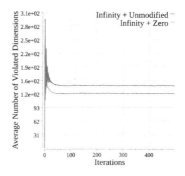

Fig. 4. Swarm diversity of PBest strategies on F10 (first 100 iterations)

Fig. 5. Average number of violated dimensions on F11 (first 500 iterations)

formed quite well and achieved the 3-rd best score. *pBest-dim* also performed better than randomly re-initializing, since resetting the position to a known good location encourages the search to exploit within a known good region.

The two strategies that performed the worst were *infinity + zero* and *infinity + unmodified*. All of the particles left the search space and remained out of bounds for the remainder of the search. The average number of violated boundaries was fewer for *infinity + zero* than for *infinity + unmodified* (see Fig. 5). Thus, zeroing the velocity component when a particle is out of bounds does improve the particle's ability to return to the search space to some extent.

In all cases where comparison was possible, zeroing the velocity performed better than adjusting or reflection. *Damping* and *random damping* performed better than *zero*. However, *damping* and *random-damping* are not applicable for many of the position repair strategies. All three of the best-performing velocity repair strategies reduce the velocity in some manner, thereby reducing the velocity explosion and the consequent roaming behaviour.

5   Conclusion

This paper tested PSO with a variety of boundary constraint handling techniques on high dimensional problems with optima that were distributed uniformly throughout the search space. The best-performing strategy was *hyperbolic*, which rescales a particle's velocity so that the particle can never reach the boundary, thereby preventing the velocity explosion and subsequent unwanted roaming. The five best performing strategies were *hyperbolic*, *exp_spread + zero*, *pBest + zero*, *absorb + damping* and *random + zero*. Although the difference in performance of these strategies are statistically significant, their performance is highly problem-dependent and their average normalized fitness values were similar. In general, velocity repair strategies such as *zero* and *damping* performed better than the other velocity repair strategies. The worst strategies were *infinity+zero* and *infinity+unmodified*, which were also the least restrictive.

Acknowledgments. This work is based on the research supported by the National Research Foundation (NRF) of South Africa (Grant Number 46712). The opinions, findings and conclusions or recommendations expressed in this article is that of the author(s) alone, and not that of the NRF. The NRF accepts no liability whatsoever in this regard.

References

1. Alvarez-Benitez, J.E., Everson, R.M., Fieldsend, J.E.: A MOPSO algorithm based exclusively on pareto dominance concepts. In: Coello Coello, C.A., Hernández Aguirre, A., Zitzler, E. (eds.) EMO 2005. LNCS, vol. 3410, pp. 459–473. Springer, Heidelberg (2005). https://doi.org/10.1007/978-3-540-31880-4_32
2. Bonyadi, M.R., Michalewicz, Z.: Impacts of coefficients on movement patterns in the particle swarm optimization algorithm. IEEE Trans. Evol. Comput. **21**(3), 378–390 (2017). https://doi.org/10.1109/TEVC.2016.2605668
3. Bratton, D., Kennedy, J.: Defining a standard for particle swarm optimization. In: Proceedings of the IEEE Swarm Intelligence Symposium, pp. 120–127. IEEE Computer Society (2007). https://doi.org/10.1109/SIS.2007.368035
4. Chu, W., Gao, X., Sorooshian, S.: Handling boundary constraints for particle swarm optimization in high-dimensional search space. Inf. Sci. **181**(20), 4569–4581 (2011). https://doi.org/10.1016/j.ins.2010.11.030
5. Cleghorn, C.W., Engelbrecht, A.P.: Particle swarm stability: a theoretical extension using the non-stagnate distribution assumption. Swarm Intell. **12**, 1–22 (2017). https://doi.org/10.1007/s11721-017-0141-x
6. Clerc, M.: Confinements and biases in particle swarm optimization, March 2006. http://clerc.maurice.free.fr/pso/. Accessed 12 Mar 2006
7. Clerc, M., Kennedy, J.: The particle swarm - explosion, stability, and convergence in a multidimensional complex space. IEEE Trans. Evol. Comput. **6**(1), 58–73 (2002). https://doi.org/10.1109/4235.985692
8. Deb, K.: Optimization for Engineering Design: Algorithms and Examples. Prentice-Hall, Upper Saddle River (1995)
9. Eberhart, R., Kennedy, J.: A new optimizer using particle swarm theory. In: Proceedings of the Sixth International Symposium on Micro Machine and Human Science, pp. 39–43 (Oct 1995). https://doi.org/10.1109/MHS.1995.494215
10. Helwig, S., Branke, J., Mostaghim, S.: Experimental analysis of bound handling techniques in particle swarm optimization. IEEE Trans. Evol. Comput. **17**(2), 259–271 (2013). https://doi.org/10.1109/TEVC.2012.2189404
11. Helwig, S., Wanka, R.: Particle swarm optimization in high-dimensional bounded search spaces. In: Proceedings of the IEEE Swarm Intelligence Symposium, pp. 198–205. IEEE Computer Society, April 2007. https://doi.org/10.1109/SIS.2007.368046
12. Helwig, S., Wanka, R.: Theoretical analysis of initial particle swarm behavior. In: Rudolph, G., Jansen, T., Beume, N., Lucas, S., Poloni, C. (eds.) PPSN 2008. LNCS, vol. 5199, pp. 889–898. Springer, Heidelberg (2008). https://doi.org/10.1007/978-3-540-87700-4_88
13. Huang, T., Mohan, A.S.: A hybrid boundary condition for robust particle swarm optimization. IEEE Antennas Wirel. Propag. Lett. **4**, 112–117 (2005). https://doi.org/10.1109/LAWP.2005.846166

14. Mostaghim, S., Mostaghim, S., Halter, W., Wille, A.: Linear multi-objective particle swarm optimization. In: Ajith, A., Crina, G., Vitorino, R. (eds.) Stigmergic Optimization, vol. 31, pp. 209–238. Springer, Heidelberg (2006). https://doi.org/10.1007/978-3-540-34690-6_9

15. Oldewage, E.: The perils of particle swarm optimisation in high dimensional problem spaces. Master's thesis, University of Pretoria, Pretoria, South Africa (2018)

16. Padhye, N., Deb, K., Mittal, P.: Boundary handling approaches in particle swarm optimization. In: Bansal, J.C., Singh, P.K., Deep, K., Pant, M., Nagar, A.K. (eds.) Proceedings of Seventh International Conference on Bio-Inspired Computing: Theories and Applications, vol. 1, pp. 287–298. Springer, India (2013). https://doi.org/10.1007/978-81-322-1038-2_25

17. Reklaitis, G., Ravindran, A., Ragsdell, K.: Engineering Optimization Methods and Applications. Wiley, Hoboken (1983)

18. Robinson, J., Rahmat-Samii, Y.: Particle swarm optimization in electromagnetics. IEEE Trans. Antennas Propag. $52(2)$, 397–407 (2004). https://doi.org/10.1109/TAP.2004.823969

19. Shi, Y., Cheng, S., Qin, Q.: Experimental study on boundary constraints handling in particle swarm optimization: From population diversity perspective. Int. J. Swarm Intell. Res. $2(3)$, 43–69 (2011). https://doi.org/10.4018/jsir.2011070104

20. Shi, Y., Eberhart, R.: A modified particle swarm optimizer. In: Proceedings of the IEEE International Conference on Evolutionary Computation, pp. 69–73, May 1998. https://doi.org/10.1109/ICEC.1998.699146

21. Tang, K., Li, X., Suganthan, P.N., Yang, Z., Weise, T.: Benchmark functions for the CEC'2010 special session and competition on large-scale global optimization. Technical report, Nature Inspired Computation and Applications Laboratory (2009)

22. Zhang, W., Xie, X., Bi, D.: Handling boundary constraints for numerical optimization by particle swarm flying in periodic search space. CoRR abs/cs/0505069 (2005). http://arxiv.org/abs/cs/0505069

Does the ACO$_\mathbb{R}$ Algorithm Benefit from the Use of Crossover?

Ashraf M. Abdelbar$^{1(\boxtimes)}$ (ID) and Khalid M. Salama^2 (ID)

^{1} Department of Mathematics and Computer Science, Brandon University,
Brandon, Canada
abdelbara@brandonu.ca
^{2} School of Computing, University of Kent, Canterbury, UK
kms39@kent.ac.uk

Abstract. The ACO$_\mathbb{R}$ algorithm is based on the Ant Colony Optimization (ACO) metaphor, and a crossover operator does not naturally within this metaphor. In spite of this, we investigate in this paper whether the performance of ACO$_\mathbb{R}$ would benefit from the deployment, with a fixed probability, of a crossover operator. Our extensive experimental evaluation uses two applications: (1) training feedforward neural networks for classification using 65 benchmark datasets from the UCI repository; and (2) optimizing several popular synthetic benchmark continuous-domain functions with the number of dimensions varying from 10 up to 10,000. Our experimental results confirm that the use of crossover does improve performance on both applications to a statistically significant extent.

1 Overview

ACO$_\mathbb{R}$ [8,19] is an established ant colony optimization (ACO) [4] algorithm for continuous-domain optimization. Even though a crossover operator does not naturally fit within the ACO metaphor, this paper investigates whether the performance of ACO$_\mathbb{R}$ would benefit from the periodic deployment of such an operator.

We propose a variation of ACO$_\mathbb{R}$, called ACO$_\mathbb{R}$-**R**, that includes a crossover operator (specifically uniform crossover in the present work), that is used for solution construction with a fixed probability P_r. Although crossover does not fit naturally within the ACO metaphor, it is nonetheless interesting to explore whether the use of crossover would improve the performance of ACO$_\mathbb{R}$. In previous work [3], we investigated the use of crossover (recombination) within the iMOACO$_\mathbb{R}$ algorithm [6] for multi-objective optimization, which is based on the ACO$_\mathbb{R}$ algorithm, and found performance to significantly improve.

Our experimental evaluation is based on two applications: (1) the problem of training feedforward neural networks for classification, using 65 benchmark datasets from the University of California Irvine (UCI) repository; (2) optimizing several popular synthetic benchmark continuous-domain functions with the number of dimensions varying from 10 up to 10,000.

© Springer Nature Switzerland AG 2018
M. Dorigo et al. (Eds.): ANTS 2018, LNCS 11172, pp. 342–350, 2018.
https://doi.org/10.1007/978-3-030-00533-7_28

2 Review of the ACO$_\mathbb{R}$ Algorithm

ACO has been applied to a wide variety of domains [2,10,11,13–17] with the majority of research focusing on discrete (combinatorial) optimization problems [5]. However, ACO methods for continuous problem domains have also been investigated [18–20]. In this paper, we focus on the ACO$_\mathbb{R}$ algorithm [8,19], which has been applied to a number of continuous optimization problems, including neural network training [1,12,18].

Suppose the ACO$_\mathbb{R}$ algorithm is to be applied to an optimization problem over n real-valued variables $V_1, V_2, \ldots, V_n$. The central data structure, analogous to pheromone information in natural ants, that is maintained by ACO$_\mathbb{R}$ is an archive A of R previously-generated candidate solutions. Each element s_a in the archive, for $a = 1, 2, \ldots, R$, is an n-dimensional real-valued vector, $s_a = (s_{a,1}, s_{a,2}, \ldots, s_{a,n})$. The archive is sorted by solution quality, so that $Q(s_1) \geq Q(s_2) \geq \ldots \geq Q(s_R)$. Each solution s_a in the archive has an associated weight ω_a that is related to $Q(s_a)$, so that $\omega_1 \geq \omega_2 \geq \ldots \geq \omega_R$.

In each iteration of the ACO$_\mathbb{R}$ algorithm, there are two phases: solution construction and pheromone update. In the solution construction phase, each ant probabilistically constructs a solution based on the solution archive A (representing pheromone information). The solution archive A is initialized with R randomly-generated solutions, where the size R is a user-supplied parameter of the ACO$_\mathbb{R}$ algorithm. Then, in the pheromone update phase, the m constructed solutions (where m is the number of ants) are added to A, resulting in the size of A temporarily being $R + m$. The archive A is then sorted by solution quality, and the m worst solutions are discarded, so that the size of A returns to being R.

In the solution construction phase, each ant i generates a candidate solution s_i and is influenced by one of the R solutions in the archive A. The ant first probabilistically selects one of the R solutions in the archive according to:

$$\Pr(\text{select } s_a) = \frac{\omega_a}{\sum_{r=1}^{R} \omega_r} \tag{1}$$

The weights ω_a that are used in Eq. (1) are constructed in each iteration as:

$$\omega_a = g(a; 1, qR) \tag{2}$$

where g is the Gaussian function:

$$g(y; \mu, \sigma) = \frac{1}{\sigma\sqrt{2\pi}} e^{-\frac{(y-\mu)^2}{2\sigma^2}} \tag{3}$$

Thus, Eq. (2) assigns the weight ω_a to be the value of the Gaussian function with argument a, mean 1.0, and standard deviation (qR), where q is another user-supplied parameter.

Let s_a be the solution of A that is selected by ant i according to Eq. (1) in a given iteration. Ant i then generates each solution element $s_{i,a}$ by sampling the Gaussian probability density function (PDF):

$$s_{i,j} \sim N(s_{a,j}, \sigma_{a,j}) \tag{4}$$

where $N(\mu, \sigma)$ represents the Gaussian PDF with mean μ and standard deviation σ. The standard deviation $\sigma_{a,j}$ is computed according to:

$$\sigma_{a,j} = \xi \sum_{r=1}^{R} \frac{|s_{a,j} - s_{r,j}|}{R-1} \tag{5}$$

where ξ is a user-supplied parameter of the algorithm. Once each ant constructs its solution, the archive A is updated as described above, and the process repeats.

If the top solution in the archive remains unchanged for I_{stag} iterations, then the archive is re-initialized with random solutions. The algorithm terminates when the total number of iterations reaches I_{max}.

3 $ACO_{\mathbb{R}}$ with Crossover

Crossover (recombination) is a standard component of Evolutionary Algorithms. Hybrid approaches that combine discrete ACO models with crossover have been explored [7]. In previous work [3], we have proposed a recombination-based variation of the iMOACO$_{\mathbb{R}}$ algorithm [6] for multi-objective optimization (which is built on the $ACO_{\mathbb{R}}$ algorithm).

There are multiple ways that recombination can potentially be incorporated within $ACO_{\mathbb{R}}$. We propose a baseline approach called $ACO_{\mathbb{R}}$-**R** which differs from $ACO_{\mathbb{R}}$ in the following. When an ant i starts to generate a candidate solution s_i, its first step is to decide with a probability P_r to apply crossover to two archived solutions, or with the inverse probability, to apply the standard $ACO_{\mathbb{R}}$ solution generation mechanism (Eqs. 1–5).

If it decides to apply crossover, then a crossover operator is used to construct a candidate solution instead of $ACO_{\mathbb{R}}$'s usual solution construction mechanism. The present work uses uniform crossover, although other crossover operators can potentially be applied. Two parents are selected from the $ACO_{\mathbb{R}}$ solution archive. One parent, s_a, is selected by applying Eq. (1) in the usual $ACO_{\mathbb{R}}$ way (i.e. rank-proportionate selection). The second parent, s_b, is selected from the archive probabilistically with uniform distribution. A single offspring, s_i, is then generated by uniform crossover: for $j = 1, \ldots, N$, each solution element s_{ij} is set equal to s_{aj} with 50% probability, or to s_{bj} with 50% probability. The generated offspring s_i then becomes one of the m constructed solutions that compete with each other and with the existing R solutions for a place in the archive.

It is worthwhile to emphasize that $ACO_{\mathbb{R}}$-**R** carries out the same number of fitness function evaluations per iteration as $ACO_{\mathbb{R}}$. In each iteration, m solutions are constructed and evaluated; each solution may be constructed either by crossover with a randomly-selected population element or by $ACO_{\mathbb{R}}$'s usual solution construction mechanism (Eqs. 1–5). The former is slightly less computationally expensive than the latter, because it does not require the application of Eq. (5) to compute the standard deviation for each solution dimension. Thus, $ACO_{\mathbb{R}}$-**R** does not carry any additional computational burden relative to $ACO_{\mathbb{R}}$.

4   Experimental Methodology and Results

Our experimental comparison is in the context of ACO$_\mathbb{R}$ applied to two problem domains: training feedforward neural networks for classification, and optimizing several popular synthetic continuous-domain benchmark functions. We adopt the ACO$_\mathbb{R}$ parameter settings of [8]. Specifically, we use:

$$m = 5, \quad R = 90, \quad q = 0.05, \quad I_{stag} = 650, \quad I_{max} = 5000, \quad \xi = 0.68 \qquad (6)$$

ACO$_\mathbb{R}$ has previously been applied to the training of three-layer feedforward neural networks [1,18]. In this application of ACO$_\mathbb{R}$, a candidate solution consists of a value of the neural network's weight vector, and the fitness function consists of initializing a neural network with the weight vector under evaluation, applying the training set, pattern by pattern, and computing the training set classification accuracy. The test set is used only once. After the ACO$_\mathbb{R}$ algorithm terminates, a network is initialized with the top solution in the archive, and the test set classification accuracy is computed.

Ideally, the number of neurons in the hidden layer should be tuned for each dataset individually. However, for convenience and standardization, we set the number of hidden neurons to be the sum of the number of input neurons and output neurons. We use the standard sigmoid activation function.

Table 1. Synthetic benchmark functions used in experimental evaluation.

Name	Mathematical representation	Search range	Init. range
Sphere	$f(x) = \sum_{i=1}^{D} x_i^2$	$(-100, 100)^D$	$(50, 100)^D$
Rosenbrock	$f(x) = \sum_{i=1}^{D-1} \left[100 \left(x_{i+1} - x_i^2 \right)^2 + (x_i - 1)^2 \right]$	$(-100, 100)^D$	$(15, 30)^D$
Rastrigin	$f(x) = \sum_{i=1}^{D} \left[x_i^2 - 10 \cos (2\pi x_i) + 10 \right]$	$(-10, 10)^D$	$(2.56, 5.12)^D$
Griewank	$f(x) = \frac{1}{4000} \sum_{i=1}^{D} x_i^2 - \prod_{i=1}^{D} \cos \left(\frac{x_i}{\sqrt{i}} \right) + 1$	$(-600, 600)^D$	$(300, 600)^D$
Ellipsoid	$f(x) = \sum_{i=1}^{D} \left(10^6 \right)^{\frac{i-1}{D-1}} x_i^2$	$(-100, 100)^D$	$(-100, 100)^D$
Ackley	$f(x) = -20 \exp \left(-0.2 \sqrt{\frac{1}{D} \sum_{i=1}^{D} x + i^2} \right)$ $- \exp \left(\frac{1}{D} \sum_{i=1}^{D} \cos (2\pi x_i) \right) + 20 + e$	$(-32, 32)^D$	$(-32, 32)^D$

Before being presented to the network, the dataset undergoes some preprocessing. Any duplicate instances are removed from the dataset before the partitioning into cross-validation folds (see below). Continuous attributes are scaled to the range $[0, 1]$ and any missing values are set to the mean value for that attribute. Any missing values for a categorical attribute are set to the mode for that attribute. Then, each categorical attribute, with c category labels, is converted to c numeric attributes, where one of the numeric attributes has a value of 1, and each of the other $(c - 1)$ attributes has a value of 0. If the dataset has m possible classes, then the network will have m output neurons. We use 65 datasets from the University of California Irvine (UCI) Repository.

Table 2. Neural network test set classification accuracy (%) results.

Dataset	ACO_R	ACO_R-R	Dataset	ACO_R	ACO_R-R
1. abalone	13.65	**19.63**	34. letter-r	9.30	**21.60**
2. adult	83.01	**85.17**	35. libras	23.40	**56.19**
3. annealing	71.73	**72.19**	36. liver-disorders	67.65	**69.24**
4. audiology	28.82	**60.37**	37. lung-cancer	**47.09**	41.02
5. automobile	59.14	**65.38**	38. lymphography	83.05	**86.88**
6. balance	91.11	**91.84**	39. mammographic	**51.42**	50.93
7. bcancer	**73.20**	73.05	40. monks	**76.69**	75.97
8. bcancer-wisc-diag	**88.37**	86.98	41. mushrooms	99.58	**99.93**
9. bcancer-wisc-orig	90.41	**91.22**	42. musk	72.26	**80.86**
10. breast-wisc-prog	71.98	**72.25**	43. nursery	93.11	**93.54**
11. biology	83.01	**86.06**	44. ozone	93.63	**93.66**
12. breast-tissue	**62.17**	61.77	45. page-blocks	**92.42**	91.25
13. car	**92.99**	92.87	46. parkinsons	84.03	**84.26**
14. chess	94.05	**96.46**	47. pima	74.11	**74.41**
15. cmc	53.81	**55.47**	48. pop	54.99	**56.41**
16. credit-a	86.29	**86.58**	49. s-heart	**82.26**	81.71
17. credit-g	77.54	**77.94**	50. seeds	**93.39**	93.12
18. cylinder	68.80	**72.30**	51. segmentation	83.10	**88.45**
19. dermatology	88.31	**96.09**	52. sensorless	14.88	**22.16**
20. ecoli	81.66	**85.93**	53. sonar	77.47	**81.18**
21. EEG	51.28	**51.75**	54. soybean	42.62	**74.99**
22. gesture	40.90	**45.79**	55. spam	89.75	**92.11**
23. glass	54.61	**57.00**	56. thyroid	95.29	**95.43**
24. GTC	88.83	**89.70**	57. transfusion	71.48	**72.34**
25. haberman	70.33	**70.37**	58. ttt	92.52	**95.02**
26. hay	76.78	**78.84**	59. vehicle	62.56	**67.07**
27. heart-c	62.39	**62.50**	60. vertebral-column-2c	74.35	**74.93**
28. heart-h	**64.92**	62.11	61. vertebral-column-3c	**57.92**	57.60
29. hepatitis	**85.79**	85.46	62. voting	95.14	**95.32**
30. horse	**79.50**	78.48	63. wave	76.55	**84.44**
31. ionosphere	90.98	**92.94**	64. wine	96.74	**97.48**
32. iris	**94.41**	94.38	65. zoo	86.90	**90.59**
33. lenses	76.29	**78.40**			

Our experiments were carried out using the well-known stratified 4-fold cross-validation procedure. This means that a dataset is divided into four mutually exclusive partitions (folds), with approximately the same number of instances and class distribution in each fold. Each algorithm is run four times, with a different fold acting as test set each time. The entire process is then repeated 10 times, with different random seeds. Average test set classification accuracy is then computed over the 40 runs.

In addition to neural network training, our experimental evaluation also uses six popular synthetic benchmark continuous-domain functions. Table 1 lists these

Table 3. Results for the synthetic continuous-domain benchmark functions.

D	Sphere		Rosenbrock		Rastrigin		Griewank		Ellipsoid		Ackley	
	ACO_R	ACO_R-R	ACO_R	ACO_R-R	ACO_R	ACO_R-R	ACO_R	ACO_R-R	ACO_R	ACO_R-R	ACO_R	ACO_R-R
10	4.3E3	3.1E-97	1.2E7	3.8E1	4.0E1	5.0E1	4.0E1	2.9E-2	1.8E7	3.0E-95	2.0E0	1.2E-4
20	9.4E3	3.6E-44	3.4E7	7.1E1	1.0E2	1.4E2	8.6E1	1.2E-2	9.5E7	2.7E-40	2.1E0	1.2E0
30	1.5E4	1.5E-22	6.0E7	1.1E2	1.8E2	2.4E2	1.3E2	7.6E-2	1.7E8	1.6E-19	2.4E0	2.9E0
40	2.0E4	1.4E-12	8.2E7	2.0E2	2.6E2	3.7E2	1.8E2	5.8E-1	3.2E8	2.6E-9	3.4E0	6.0E0
50	2.5E4	4.0E-6	1.1E8	3.4E2	3.4E2	4.8E2	2.3E2	4.3E-1	4.6E8	5.4E2	4.1E0	8.8E0
60	3.1E4	5.7E-2	1.3E8	8.8E2	4.3E2	6.0E2	2.8E2	3.7E-1	5.0E8	8.7E4	5.1E0	1.0E1
70	3.6E4	1.0E2	1.6E8	1.5E4	5.1E2	7.2E2	3.3E2	4.2E0	7.5E8	3.4E4	6.7E0	1.2E1
80	4.2E4	3.8E2	1.8E8	2.6E5	6.0E2	8.6E2	3.8E2	5.2E0	9.7E8	6.1E5	7.2E0	1.4E1
90	4.8E4	8.4E2	2.1E8	9.2E5	7.0E2	9.8E2	4.4E2	1.8E1	1.1E9	1.4E6	9.2E0	1.5E1
100	5.6E4	1.9E3	2.4E8	2.9E6	8.0E2	1.1E3	5.0E2	2.6E1	1.2E9	1.4E6	1.1E1	1.5E1
200	1.8E5	7.9E4	6.1E8	2.4E8	2.1E3	2.6E3	1.6E3	7.1E2	3.3E9	1.2E8	1.9E1	1.9E1
300	4.2E5	2.6E5	1.5E9	1.0E9	4.0E3	4.4E3	3.7E3	2.4E3	5.9E9	5.3E8	2.1E1	1.9E1
400	5.8E5	5.3E5	2.3E9	2.2E9	5.9E3	6.3E3	5.2E3	4.8E3	2.3E9	1.6E9	2.1E1	2.0E1
500	1.1E6	8.4E5	5.1E9	3.6E9	8.9E3	8.3E3	1.0E4	7.5E3	1.3E10	3.4E9	2.1E1	2.0E1
600	1.4E6	1.2E6	6.6E9	5.1E9	1.2E4	1.0E4	1.2E4	1.1E4	7.8E9	5.9E9	2.1E1	2.0E1
700	1.8E6	1.6E6	1.0E10	6.9E9	1.5E4	1.2E4	1.6E4	1.4E4	1.2E10	8.7E9	2.1E1	2.0E1
800	2.2E6	2.0E6	1.3E10	8.9E9	1.8E4	1.5E4	2.0E4	1.8E4	1.7E10	1.2E10	2.1E1	2.0E1
900	2.7E6	2.4E6	1.8E10	1.1E10	2.1E4	1.7E4	2.4E4	2.2E4	2.3E10	1.7E10	2.1E1	2.1E1
1,000	3.4E6	2.8E6	2.4E10	1.3E10	2.4E4	1.9E4	3.1E4	2.5E4	4.7E10	2.2E10	2.1E1	2.1E1
2,000	8.3E6	7.5E6	5.5E10	3.6E10	4.8E4	4.1E4	7.5E4	6.7E4	1.5E11	9.8E10	2.1E1	2.1E1
3,000	1.4E7	1.3E7	8.4E10	6.1E10	7.2E4	6.4E4	1.2E5	1.1E5	3.2E11	2.1E11	2.1E1	2.1E1
4,000	1.9E7	1.8E7	1.1E11	8.7E10	9.7E4	8.8E4	1.7E5	1.6E5	5.1E11	3.4E11	2.1E1	2.1E1
5,000	2.5E7	2.3E7	1.4E11	1.1E11	1.2E5	1.1E5	2.2E5	2.1E5	7.4E11	4.8E11	2.1E1	2.1E1
6,000	3.0E7	2.8E7	1.7E11	1.4E11	1.5E5	1.4E5	2.7E5	2.5E5	1.1E12	6.4E11	2.1E1	2.1E1
7,000	3.6E7	3.4E7	2.0E11	1.7E11	1.7E5	1.6E5	3.2E5	3.0E5	1.4E12	8.1E11	2.1E1	2.1E1
8,000	4.2E7	3.9E7	2.3E11	2.0E11	2.0E5	1.8E5	3.7E5	3.5E5	1.7E12	9.8E11	2.1E1	2.1E1
9,000	4.7E7	4.4E7	2.6E11	2.2E11	2.2E5	2.1E5	4.2E5	4.0E5	2.0E12	1.2E12	2.1E1	2.1E1
10,000	5.3E7	5.0E7	2.8E11	2.5E11	2.4E5	2.3E5	4.7E5	4.5E5	2.2E12	1.3E12	2.1E1	2.1E1

Table 4. Results of Wilcoxon signed-rank tests comparing $ACO_\mathbb{R}$-**R** to $ACO_\mathbb{R}$.

Domain	Comparison	N	W	z	p	Sig.?
NN	$ACO_\mathbb{R}$-**R** vs. $ACO_\mathbb{R}$	65	336.0	−4.8130	**1.5E-06**	Yes
Bench	$ACO_\mathbb{R}$-**R** vs. $ACO_\mathbb{R}$	168	702.0	−10.1298	**8.9E-16**	Yes

functions, along with their initialization and search ranges. These ranges follow [9]. In our experiments, we use 28 different settings for the number of dimensions D, varying from 10 to 10000. Specifically, we use the following settings:

$$D \in \{10, 20, \ldots, 100, 200, \ldots, 1000, 2000, \ldots, 10000\} \qquad (7)$$

Each algorithm under evaluation is applied to each of the six functions for each of the 28 dimensionalities. In each case, the algorithm is run 100 times (with different seeds), and the average over the 100 runs is computed.

Initial experimentation indicated 0.4 to be a good "default" value for the P_r parameter. The results for neural network classification are reported in Table 2 for $ACO_\mathbb{R}$ and $ACO_\mathbb{R}$-**R** (with $P_r = 0.4$). Each row shows the average test set classification accuracy for each of the two algorithms, with the better performance for each dataset shown in boldface. Table 3 shows the analogous results for the benchmark functions.

The tables indicate that $ACO_\mathbb{R}$-**R** performed better on 50 out of 65 datasets (i.e. 77% of the datasets) for neural network training. For the synthetic benchmark functions, $ACO_\mathbb{R}$-**R** performed better on 147 out of 168 cases overall (i.e. 88%), and on 100% of the cases where D is greater than 400. Table 4 reports the results of applying a (non-parameteric) Wilcoxon signed-rank test to the results for each of the two applications, and indicates that there is a statistically significant difference, with $p < 0.001$ for both application domains.

5   Concluding Remarks

This paper is concerned with determining whether there is the potential for the performance of $ACO_\mathbb{R}$ to be improved by the inclusion of a crossover operator. We proposed and evaluated a simple variation of $ACO_\mathbb{R}$, called $ACO_\mathbb{R}$-**R**, that applies uniform crossover, with a fixed user-supplied probability P_r, in place of $ACO_\mathbb{R}$'s usual solution construction mechanism. Our results indicate that there is indeed the potential to significantly improve the performance of $ACO_\mathbb{R}$ with the inclusion of a crossover operator. Our previous work [3], which considered a crossover-based variation of $iMOACO_\mathbb{R}$[6], reached a similar conclusion.

In future work, we would like to consider self-adaptive approaches which would eliminate the need for a user-supplied parameter P_r, by allowing the frequency of applying crossover to be dynamically adapted during the course of the algorithm's execution, perhaps based on the relative quality of solutions obtained by crossover.

Acknowledgments. Partial support of a grant from the Brandon University Research Council (BURC) is gratefully acknowledged.

References

1. Abdelbar, A.M., Salama, K.M.: A gradient-guided ACO algorithm for neural network learning. In: Proceedings IEEE Swarm Intelligence Symposium (SIS-2015), pp. 1133–1140 (2015)
2. Abdelbar, A.M., Salama, K.M.: An extension of the ACO$_\mathbb{R}$ algorithm with time-decaying search width, with application to neural network training. In: Proceedings IEEE Congress on Evolutionary Computation (CEC-2016), pp. 2360–2366 (2016)
3. Abdelbar, A.M., Salama, K.M.: Solution recombination in an indicator-based many-objective ant colony optimizer for continuous search spaces. In: Proceedings IEEE Swarm Intelligence Symposium (SIS-2017), pp. 1–8 (2017)
4. Dorigo, M., Stützle, T.: Ant Colony Optimization. MIT Press, Cambridge (2004)
5. Dorigo, M., Stützle, T.: Ant colony optimization: overview and recent advances. In: Gendreau, M., Potvin, Y. (eds.) Handbook of Metaheuristics, pp. 227–263. Springer, New York, NY, USA (2010). https://doi.org/10.1007/978-1-4419-1665-5_8
6. Falcón-Cardona, J.G., Coello Coello, C.A.: A new indicator-based many-objective ant colony optimizer for continuous search spaces. Swarm Intell. **11**, 71–100 (2017)
7. Kalinli, A., Sarikoc, F.: A parallel ant colony optimization algorithm based on crossover operation. In: Siarry, P., Michalewicz, Z. (eds.) Advances in Metaheuristics for Hard Optimization, pp. 87–110. Springer, Berlin Heidelberg (2008). https://doi.org/10.1007/978-3-540-72960-0_5
8. Liao, T., Socha, K., Montes de Oca, M., Stützle, T., Dorigo, M.: Ant colony optimization for mixed-variable optimization problems. IEEE Trans. Evol. Comput. **18**(4), 503–518 (2014)
9. Ratnaweera, A., Halgamuge, S., Watson, H.: Self-organizing hierarchical particle swarm optimizer with time-varying acceleration coefficients. IEEE Trans. Evol. Comput. **8**(3), 240–255 (2004)
10. Salama, K.M., Abdelbar, A.M.: Extensions to the Ant-Miner classification rule discovery algorithm. In: Dorigo, M. (ed.) ANTS 2010. LNCS, vol. 6234, pp. 167–178. Springer, Heidelberg (2010). https://doi.org/10.1007/978-3-642-15461-4_15
11. Salama, K.M., Abdelbar, A.M.: Exploring different rule quality evaluation functions in ACO-based classification algorithms. In: IEEE Swarm Intelligence Symposium, pp. 1–8 (2011)
12. Salama, K.M., Abdelbar, A.M.: Learning neural network structures with ant colony algorithms. Swarm Intell. **9**(4), 229–265 (2015)
13. Salama, K.M., Abdelbar, A.M.: Instance-based classification with ant colony optimization. Intell. Data Anal. **21**(4), 913–944 (2017)
14. Salama, K.M., Abdelbar, A.M.: Learning cluster-based classification systems with ant colony optimization algorithms. Swarm Intell. **11**(2–3), 211–242 (2017)
15. Salama, K.M., Abdelbar, A.M., Anwar, I.: Data reduction for classification with ant colony algorithms. Intell. Data Anal. **20**(5), 1021–1059 (2016)
16. Salama, K.M., Abdelbar, A.M., Freitas, A.: Multiple pheromone types and other extensions to the Ant-Miner classification rule discovery algorithm. Swarm Intell. **5**(3–4), 149–182 (2011)

17. Salama, K.M., Abdelbar, A.M., Otero, F., Freitas, A.: Utilizing multiple pheromones in an ant-based algorithm for continuous-attribute classification rule discovery. Appl. Soft Comput. **13**(1), 667–675 (2013)
18. Socha, K., Blum, C.: An ant colony optimization algorithm for continuous optimization: application to feed-forward neural network training. Neural Comput. Appl. **16**, 235–247 (2007)
19. Socha, K., Dorigo, M.: Ant colony optimization for continuous domains. Eur. J. Oper. Res. **185**, 1155–1173 (2008)
20. Tsutsui, S.: Ant colony optimisation for continuous domains with aggregation pheromones metaphor. In: Proceedings International Conference on Recent Advances in Soft Computing (RASC-2004), pp. 207–212 (2004)

Embodied Evolution of Self-organised Aggregation by Cultural Propagation

Nicolas Cambier[1(✉)], Vincent Frémont[1], Vito Trianni[2],
and Eliseo Ferrante[3,4(✉)]

[1] Sorbonne Universités, Université de Technologie de Compiègne,
UMR CNRS 7253 Heudiasyc, Compiègne, France
{nicolas.cambier,vincent.fremont}@hds.utc.fr
[2] Institute of Cognitive Sciences and Technologies,
National Research Council, Rome, Italy
vito.trianni@istc.cnr.it
[3] Laboratory of Socio-ecology and Social Evolution, KU Leuven,
Leuven, Belgium
eliseo.ferrante@kuleuven.be
[4] School of Computer Science, University of Birmingham,
Dubai, United Arab Emirates
e.ferrante@bham.ac.uk

Abstract. Probabilistic aggregation is a self-organised behaviour studied in swarm robotics. It aims at gathering a population of robots in the same place, in order to favour the execution of other more complex collective behaviours or tasks. However, probabilistic aggregation is extremely sensitive to experimental conditions, and thus requires specific parameter tuning for different conditions such as population size or density. To tackle this challenge, in this paper, we present a novel embodied evolution approach for swarm robotics based on social dynamics. This idea hinges on the cultural evolution metaphor, which postulates that good ideas spread widely in a population. Thus, we propose that good parameter settings can spread following a social dynamics process. Testing this idea on probabilistic aggregation and using the minimal naming game to emulate social dynamics, we observe a significant improvement in the scalability of the aggregation process.

1 Introduction

Aggregation processes are extremely common in Nature [10], and consist in animals aggregating in common areas in the environment. They have been studied in a variety of biological systems [14,20] and also implemented on distributed robotic systems [17–19], according to the swarm robotics ethos of getting inspiration from natural phenomena to solve engineering problems [25], while relying only on local interactions [7]. Furthermore, aggregation is a prerequisite for other cooperative behaviours [16].

There are many approaches to designing self-organised aggregation, from evolutionary solutions [31] to minimal deterministic models [19]. Probabilistic

© Springer Nature Switzerland AG 2018
M. Dorigo et al. (Eds.): ANTS 2018, LNCS 11172, pp. 351–359, 2018.
https://doi.org/10.1007/978-3-030-00533-7_29

approaches are most commonly used [3] because of their simple implementation and direct natural inspiration [5,18,20]. Nevertheless, the aggregation quality obtained using probabilistic approaches is extremely sensitive to experimental conditions such as population size [3] or agent's capabilities (e.g., speed, communication range) [13], and therefore requires supervised tuning of internal model parameters in order to be effective in a specific setting [3,27]. Obviously, this necessity to tune collective behaviours to specific environmental conditions hinders both scalability and flexibility, both desired features of swarm robotics [7].

Evolutionary swarm robotics is the approach typically used for synthesising collective behaviours for specific experimental settings via off-line parameter tuning [21,31,32]. An approach of on-line parameter tuning is instead represented by embodied evolution [8], whereby parameters controlling the robot behaviour are continuously adapted as a result of the interaction among robots and between robots and environment. Here, the evolutionary process is not driven by an explicit fitness measure that evaluates the quality of the collective behaviour, but emerges implicitly from the dynamics of interaction among the agents [4]. This approach with implicit fitness characterises algorithms like Environment-driven Distributed Evolutionary Adaptation (EDEA) [9] and subsequent extensions such as MONEE [22]. These algorithms are inspired by natural evolution, and require (i) a high mobility of the agents to spread their genomes, and (ii) a survival criteria to decide whether individuals live (and reproduce) or die, usually implemented by means of a foraging metaphor. Both these criteria are hard to meet in self-organised aggregation, as agents aggregated on different clusters hardly communicate between each other, and a new survival criteria would thus need to be devised. Hence, different evolutionary dynamics must be devised.

In this paper, we propose a novel embodied evolutionary process inspired by the cultural evolution metaphor [29] that postulates that good ideas spread widely in a population as a result of social dynamics [12]. In our model, social dynamics are coupled with self-organised aggregation dynamics: parameters beneficial for aggregation spread widely in the robot swarm.

As a model of cultural evolution, we use the Naming Game (NG), which was developed to study the evolution of human language through statistical physics [2] and artificial life experiments [28]. The NG has actually already been studied within robotic swarms [11,30], whereby the swarm dynamics (e.g., random walk and aggregation) and the NG had a mutual effect on each other. However, to the best of our knowledge, there has been no attempt to use the NG as an embodied evolution approach. In this paper, we move towards this direction by studying the dynamics of self-organised aggregation when coupled with the naming game framed as a cultural evolutionary process. The model is introduced in Sect. 2, while, in Sect. 3, we present and discuss experimental observations. Section 4 concludes the paper with an outlook of future research.

2   Model

The baseline aggregation controller we used is described in Sect. 2.1, while in Sect. 2.2, we describe the NG protocol and how it can be coupled with aggregation in our cultural evolution framework.

2.1   Self-organised Aggregation Controller

One of the first models of probabilistic aggregation in swarm robotics takes inspiration from cockroaches collective behaviour model [14,18,20]. Here, agents move within an arena following a random walk [15]. They decide to stop according to some probability dependent on the number n of perceived neighbours, as observed in cockroaches [20]. To prevent the creation of a large number of small clusters, exploration is introduced by allowing agents to leave their cluster with a probability inversely proportional to n.

As in previous models [13], we implemented a PFSA (see Fig. 1b) with two states: WALK (random walk as in [23]) and STAY. To gain a higher control over the aggregation process, we replaced the transitions probabilities in [13] (Fig. 1a) with probability functions with parametrisable steepness, allowing to strengthen or weaken the agents' alignment and dispersion at will (see Fig. 1c). For both P_{Join} and P_{Leave}, we used exponential decay functions which, for our baseline, were tuned to fit the values in Fig. 1a:

$$P_{Join} = 0.03 + 0.48 * (1 - e^{-an}) \tag{1}$$

In this equation, a is a parameter that handles the strength of the alignment. Indeed P_{Join} becomes steeper as a increases.

P_{Leave}, on the other hand, is a straightforward exponential decay function:

$$P_{Leave} = e^{-bn} \tag{2}$$

Here, b handles the strength of the dispersion as P_{Leave} becomes steeper when b increases, and thus dispersion weakens. Figure 1c shows the fitted values of these functions to the baseline values from Fig. 1a.

We can understand the effect of parameters a and b as strengthening (increase a or b) or weakening (decrease a or b) the cohesion of a cluster. A trade-off between these two forces must be found: If cohesion is too weak (low a and b), no durable cluster will form; if cohesion is too strong (high a and b), the robots aggregate in several sparse and static clusters that never break. Following the above observations, we formulate the following premise:

Premise 1. *Robots with (near) optimal parameters a and b have more neighbours, on average, than robots with suboptimal parameters.*

Based on this premise we build an implicit fitness for embodied evolution.

2.2   Cultural Evolution

Self-organised aggregation and the minimal naming game (MNG)[1,2] have been coupled as in [11]. Each agent possesses an individual lexicon (i.e., a list of words) and can be both a speaker (only when in the STAY state) and hearer (at all times). A speaker communicates a word from its lexicon to its neighbours, or generates a new one when the lexicon is empty. The neighbours in the speaker's

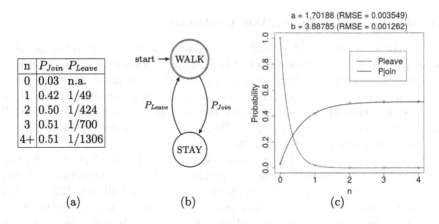

Fig. 1. (a) Averaged probability to join and leave a cluster as computed by [13] from observation of gregarious arthropods [20]. (b) PFSA of our aggregation controller. (c) Plotting of P_{join} and P_{Leave} according to (resp.) Eqs. 1 and 2 and parametrised with the values for a and b that best fit the values from Table 1a [13] (circles).

communication range (i.e. the hearers) receive this word. If the word is already known to the hearer, the game is a success and the hearer deletes all words from its lexicon except for the one it just received. Otherwise, the game is a fail and the hearer adds the word to its lexicon. In our implementation, the amount of a hearer's successful games in a time-step is used as n for the computation of the transition probabilities in the PFSA.

Moreover, the MNG presents one interesting feature: words promoted by agents with more neighbours have a higher chance to eventually become the chosen name for the predetermined topic, at least in static networks [1]. Therefore, we formulate a second premise:

Premise 2. *Words promoted by robots with more neighbours spread more on average.*

From Premises 1 and 2, we conclude that words promoted by robots with (near) optimal parameters a and b should propagate more on average. Therefore, by using an encoding of the values of a and b as the words used in the MNG, we close a positive feedback loop whereby better parameters propagate more and, as they are shared by new robots, propagate even more.

In our experiments, we used the concatenation of the encoded values of a and b as the "meme" of our cultural evolution process. We assumed a minimal message size of one byte, and each parameter was therefore expressed with four bits. With these four bits we coded the $[1.25, 5]$ interval with steps of 0.25.

The above described MNG can converge only to a word generated by speakers at the beginning of an experiments. To allow for novelty, we added noise to the messages following the noisy channel model [26]. Noise has two effects in our implementations. Firstly, it acts as mutation operator and allows to explore the solution landscape [33]. Secondly, it impacts the quantity of (un)successful games as it can mutate a word known to the hearer to an unknown one, thus slowing down convergence time of the MNG and making it more compatible with the time-scale of self-organised aggregation. However, a high mutation rate m may completely prevent the MNG from converging.

3   Experimental Results

In the experiments, we used MarXbots [6] simulated within the ARGoS simulator [24]. They moved at a speed of 10 cm/s and with a communication range of 70 cm within a circular arena of constant radius $r = 10$ m. We studied three different population sizes $N = \{25, 50, 100\}$ and evaluated the aggregation behaviour using the cluster metric in [19], which is the ratio between the size of the biggest cluster and the swarm size N.

   To highlight the dynamics produced by embodied evolution, we contrasted it with selected non-evolving instances, namely the *baseline* controller obtained by fitting the parameters a and b to the probability table from [13], and the *optimal* controller obtained with the parameter settings that maximise the cluster metric separately for each swarm size through brute-force search. Therefore we performed:

- 20 runs of embodied evolution with mutation rate $m = 0.001$
- 20 runs of the baseline controller featuring fixed parameters $(a, b) = (1.70188, 3.88785)$, set to fit [13]
- 10 runs of the optimal controller with fixed parameters:
    - $N = 25 : (a, b) = (2.25, 3.5)$
    - $N = 50 : (a, b) = (1.25, 2.0)$
    - $N = 100 : (a, b) = (1.25, 1.25)$

In addition to the cluster metric, we recorded the variation over time of the number of clusters formed and the number of free agents. The average figures are shown in Fig. 2. It is possible to notice that the evolutionary model fails to produce stable aggregates when $N = 25$. This is because the MNG is particularly slow at low densities, because interactions among agents happen with very low probability [11,30]. As a consequence, the number of successful games is small—also due to mutations disturbing the language dynamics—and clusters quickly disband. However, as N increases, we can see that embodied evolution presents dynamics that are very close to the baseline aggregation behaviour [13], i.e. a short phase of building aggregates followed by stagnation. For $N = 100$, embodied evolution attains values for the cluster metric that are larger than the baseline controller. Additionally, evolution has different dynamics from those of the baseline: Almost all agents with the baseline behaviour stay in clusters after the build-up phase, whilst the evolutionary model continues to explore for a longer time and never entirely stops. This is partly the consequence of the MNG and its mutation-induced failures (see Sect. 2.2), but also demonstrates a better handling of the cohesion trade-off (see Sect. 2.1), which explains our model's higher scalability, especially for large N. The optimal controllers, instead, slowly and constantly build up a large aggregate, maintaining at the same time a large fraction of exploring robots. This slow process represents the only means to increase the size of the largest cluster at the expenses of small clusters, when parameters are fixed and the system is homogeneous. However, a very specific parameterisation is necessary to observe this behaviour, especially for large N.

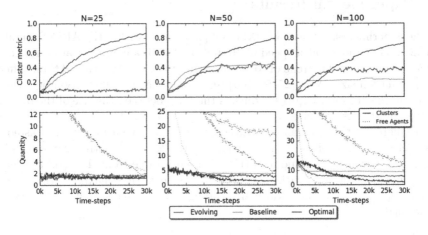

Fig. 2. Averaged time evolution over i runs of the three variations: embodied evolution ($i = 20$), *baseline* controller with parameters fixed to match the probability table from [13], $i = 20$) and *optimal* ($i = 10$).

We speculate that the efficiency of the MNG (and, thus, of embodied evolution) depends on the amount of interactions and, thus, can not work in low density settings. To test this, we confront three experimental conditions in which density of robots is maintained constant: $N = 25$ with $r = 5\,\mathrm{m}$, $N = 50$ with $r = 7\,\mathrm{m}$, and $N = 100$ with $r = 10\,\mathrm{m}$. We contrast embodied evolution with the baseline controller and with the *optimal* controller obtained with brute force search for $N = 25$ and $r = 5\,\mathrm{m}$. The latter is tested also in the other conditions, to assess whether fixed parameterisation optimal on a given setting also perform well in other settings. The results are presented in Fig. 3. We can see that, with sufficient robot density, the evolutionary model initially performs as well as the baseline behaviour [13] and not too distant from the optimal behaviour. However, embodied evolution scales up better than either of the fixed-parameters alternatives. We conclude that embodied evolution represents a promising solution for scalable behaviour rather than optimal probabilistic aggregation, provided that the density of robots remains sufficiently high.

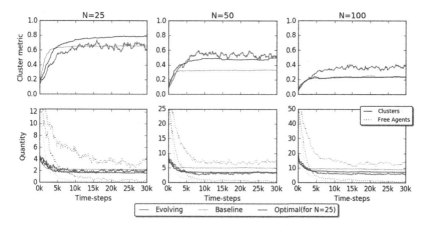

Fig. 3. Averaged time evolution over 20 runs of the three variations with constant density: embodied evolution, baseline behaviour and optimal controller for $N = 25$ and $r = 5\,\mathrm{m}$ (i.e. $(a, b) = (1.5, 2.75)$).

4   Conclusions

In this paper, we presented a novel embodied evolution approach for swarm robotics based on social dynamics. The main underlying idea is coupling opinion spreading in the population with the self-organising aggregation process. Inspired by the cultural evolution metaphor, which postulates that good ideas spread widely in a population, we propose that good parameters of a self-organising behaviour can spread following a social dynamics process, leading to a swarm capable of adapting its behaviour to the current environmental conditions. This is possible to the extent that these coupled dynamics and the opinion spreading are self-sustaining. To test this idea we considered probabilistic aggregation as the self-organising swarm behaviour and the minimal naming game as a model of social dynamics. Experimental results show that the proposed embodied evolution process autonomously selects performing parameters at different scales provided that an adequate robot density is present.

Our future work will be to (i) evaluate our model's flexibility and adaptivity by varying the size N of the population as well as the environment's features within runs, (ii) study the criticality of the mutation rate for our algorithm's performances and (iii) assessing its efficiency as a natural communication noise in an embodied setting.

Acknowledgments. This work was funded in the framework of the Labex MS2T. It was supported by the French Government, through the program "Investments for the future" managed by the National Agency for Research (Reference ANR-11-IDEX-0004-02). Vito Trianni acknowledges support from the project DICE (FP7 Marie Curie Career Integration Grant, ID: 631297).

References

1. Baronchelli, A.: Role of feedback and broadcasting in the naming game. Phys. Rev. E **83**(4), 046103 (2011)
2. Baronchelli, A., Felici, M., Loreto, V., Caglioti, E., Steels, L.: Sharp transition towards shared vocabularies in multi-agent systems. J. Stat. Mech.: Theory Exp. **2006**(06), P06014 (2006)
3. Bayindir, L., Sahin, E.: Modeling self-organized aggregation in swarm robotic systems. In: IEEE Swarm Intelligence Symposium, SIS 2009, pp. 88–95. IEEE (2009)
4. Bianco, R., Nolfi, S.: Toward open-ended evolutionary robotics: evolving elementary robotic units able to self-assemble and self-reproduce. Connect. Sci. **16**(4), 227–248 (2004)
5. Bodi, M., Thenius, R., Szopek, M., Schmickl, T., Crailsheim, K.: Interaction of robot swarms using the honeybee-inspired control algorithm beeclust. Math. Comput. Model. Dyn. Syst. **18**(1), 87–100 (2012)
6. Bonani, M., et al.: The marXbot, a miniature mobile robot opening new perspectives for the collective-robotic research. In: 2010 IEEE/RSJ International Conference on Intelligent Robots and Systems (IROS), pp. 4187–4193. IEEE (2010)
7. Brambilla, M., Ferrante, E., Birattari, M., Dorigo, M.: Swarm robotics: a review from the swarm engineering perspective. Swarm Intell. **7**(1), 1–41 (2013)
8. Bredeche, N., Haasdijk, E., Prieto, A.: Embodied evolution in collective robotics: a review. Front. Rob. AI **5**, 12 (2018)
9. Bredeche, N., Montanier, J.-M.: Environment-driven embodied evolution in a population of autonomous agents. In: Schaefer, R., Cotta, C., Kołodziej, J., Rudolph, G. (eds.) PPSN 2010. LNCS, vol. 6239, pp. 290–299. Springer, Heidelberg (2010). https://doi.org/10.1007/978-3-642-15871-1_30
10. Camazine, S.: Self-organization in Biological Systems. Princeton University Press, Princeton (2003)
11. Cambier, N., Frémont, V., Ferrante, E.: Group-size regulation in self-organised aggregation through the naming game. In: International Symposium on Swarm Behavior and Bio-Inspired Robotics (SWARM 2017), Kyoto, Japan, October 2017. https://hal.archives-ouvertes.fr/hal-01679600
12. Castellano, C., Fortunato, S., Loreto, V.: Statistical physics of social dynamics. Rev. Mod. Phys. **81**(2), 591–646 (2009)
13. Correll, N., Martinoli, A.: Modeling and designing self-organized aggregation in a swarm of miniature robots. Int. J. Rob. Res. **30**(5), 615–626 (2011)
14. Deneubourg, J.L., Lioni, A., Detrain, C.: Dynamics of aggregation and emergence of cooperation. Biol. Bull. **202**(3), 262–267 (2002)
15. Dimidov, C., Oriolo, G., Trianni, V.: Random walks in swarm robotics: an experiment with kilobots. In: Dorigo, M., et al. (eds.) ANTS 2016. LNCS, vol. 9882, pp. 185–196. Springer, Cham (2016). https://doi.org/10.1007/978-3-319-44427-7_16
16. Dorigo, M., et al.: Evolving self-organizing behaviors for a swarm-bot. Auton. Rob. **17**(2), 223–245 (2004)
17. Garnier, S., et al.: The embodiment of cockroach aggregation behavior in a group of micro-robots. Artif. Life **14**(4), 387–408 (2008)
18. Garnier, S., et al.: Aggregation behaviour as a source of collective decision in a group of cockroach-like-robots. In: Capcarrère, M.S., Freitas, A.A., Bentley, P.J., Johnson, C.G., Timmis, J. (eds.) ECAL 2005. LNCS (LNAI), vol. 3630, pp. 169–178. Springer, Heidelberg (2005). https://doi.org/10.1007/11553090_18

19. Gauci, M., Chen, J., Li, W., Dodd, T.J., Groß, R.: Self-organized aggregation without computation. Int. J. Rob. Res. **33**(8), 1145–1161 (2014)
20. Jeanson, R., et al.: Self-organized aggregation in cockroaches. Anim. Behav. **69**(1), 169–180 (2005)
21. Nolfi, S., Floreano, D.: Evolutionary Robotics: The Biology, Intelligence, and Technology of Self-organizing Machines. MIT Press, Cambridge (2000)
22. Noskov, N., Haasdijk, E., Weel, B., Eiben, A.E.: MONEE: using parental investment to combine open-ended and task-driven evolution. In: Esparcia-Alcázar, A.I. (ed.) EvoApplications 2013. LNCS, vol. 7835, pp. 569–578. Springer, Heidelberg (2013). https://doi.org/10.1007/978-3-642-37192-9_57
23. Nouyan, S., Dorigo, M.: Chain based path formation in swarms of robots. In: Dorigo, M., Gambardella, L.M., Birattari, M., Martinoli, A., Poli, R., Stützle, T. (eds.) ANTS 2006. LNCS, vol. 4150, pp. 120–131. Springer, Heidelberg (2006). https://doi.org/10.1007/11839088_11
24. Pinciroli, C., et al.: ARGoS: a modular, parallel, multi-engine simulator for multi-robot systems. Swarm Intell. **6**(4), 271–295 (2012)
25. Şahin, E.: Swarm robotics: from sources of inspiration to domains of application. In: Şahin, E., Spears, W.M. (eds.) SR 2004. LNCS, vol. 3342, pp. 10–20. Springer, Heidelberg (2005). https://doi.org/10.1007/978-3-540-30552-1_2
26. Shannon, C.: A mathematical theory of communication. Bell Syst. Tech. J. **27**, 379–423, 623–656 (1948)
27. Soysal, O., Sahin, E.: Probabilistic aggregation strategies in swarm robotic systems. In: Proceedings of 2005 IEEE Swarm Intelligence Symposium, SIS 2005, pp. 325–332. IEEE (2005)
28. Steels, L.: A self-organizing spatial vocabulary. Artif. Life **2**(3), 319–332 (1995)
29. Steels, L.: Modeling the cultural evolution of language. Phys. Life Rev. **8**(4), 339–356 (2011)
30. Trianni, V., De Simone, D., Reina, A., Baronchelli, A.: Emergence of consensus in a multi-robot network: from abstract models to empirical validation. IEEE Rob. Autom. Lett. **1**(1), 348–353 (2016)
31. Trianni, V., Groß, R., Labella, T.H., Şahin, E., Dorigo, M.: Evolving aggregation behaviors in a swarm of robots. In: Banzhaf, W., Ziegler, J., Christaller, T., Dittrich, P., Kim, J.T. (eds.) ECAL 2003. LNCS (LNAI), vol. 2801, pp. 865–874. Springer, Heidelberg (2003). https://doi.org/10.1007/978-3-540-39432-7_93
32. Trianni, V., Nolfi, S., Dorigo, M.: Evolution, self-organization and swarm robotics. In: Blum, C., Merkle, D. (eds.) Swarm Intelligence. NCS, pp. 163–191. Springer, Heidelberg (2008). https://doi.org/10.1007/978-3-540-74089-6_5
33. Winfield, A.F., Erbas, M.D.: On embodied memetic evolution and the emergence of behavioural traditions in robots. Memet. Comput. **3**(4), 261–270 (2011)

Experimental Evaluation of ACO for Continuous Domains to Solve Function Optimization Problems

Ryouei Takahashi$^{(\boxtimes)}$, Yukihiro Nakamura, and Toshihide Ibaraki

The Kyoto College of Graduate Studies for Informatics, Sakyo-ku, Kyoto, Japan
{r_takahashi,y_nakamura,t_ibaraki}@kcg.ac.jp

Abstract. A new Ant Colony Optimization ACO_B to solve function optimization problems (FOP) is evaluated experimentally by using ten standard multimodal test functions such as Michaelwicz's function. In ACO_B, ants search for solutions in binary search space and can improve the accuracy of solutions by the stepwise localization of search space. Experiments show that ACO_B can keep the balance between accuracy and efficiency to search for optimum solutions, and that it can reduce the population size of ACO_R, which is a preceding ACO based on real search space. It is also shown that Covariance Matrix Adaptation-Evolution Strategy (CMA-ES) is superior in computational time but lacks the accuracy of solutions, and that Genetic Algorithm (GA) is superior in the ratio of getting the optimum solutions but weak in the performance.

1 Introduction

The purpose of our study is to reveal evolutionary process of optimization methods experimentally to invent a new optimization method. In this study, we evaluate the capability of ACO [1,2] for continuous domains such as for solving the function optimization problem (FOP) [3,4] with n real variables xi $(i = 1, 2, \ldots, n)$. ACO is developed as a general-purpose optimization method for solving combinatorial optimization problems (COP) [5,6] based on indirect communication with pheromone called stigmergy. And a variant ACO called ACO_R [7–9] is developed to solve FOP over continuous domains. In our earlier paper [10], we proposed a new ACO called ACO_B for solving FOP, by utilizing the ideas of Ant System (AS), Elitist Ant System (EAS) and improved-EAS (i-EAS), which are originally designed to solve the travelling salesman problem (TSP). ACO_B searches solutions by approximating continuous values of each real variable xi with discrete values. It improves the accuracy of solutions through stepwise localization of the search space. Among existing algorithms for solving FOP, we consider Matrix Adaptation Evolution Strategy CMA-ES [11], and Genetic Algorithms GA [12] with Blend Crossover BLX-α [13], Simplex Crossover SPX, Changing Crossover Operators CXO (SPX→BLX) [14] and Two-point Crossover 2X [12]. In the paper [10], ACO_B is computationally compared with GAs such as BLX-α, SPX and CXO (SPX→BLX), but not with other optimization methods.

© Springer Nature Switzerland AG 2018
M. Dorigo et al. (Eds.): ANTS 2018, LNCS 11172, pp. 360–367, 2018.
https://doi.org/10.1007/978-3-030-00533-7_30

Hence, in this paper, ACO_B is compared with four optimization methods, CMA-ES, ACO_R, and GAs with BLX-α and 2X to complete the comparisons. Thirteen test cases using ten well known multimodal test functions are used for the evaluation. It is found that ACO_B can keep good balance between accuracy and efficiency for searching optimum solutions.

2   Function Optimization Problem (FOP)

2.1   Definition of FOP

We suppose that the function is denoted by $y = f(x_1, \ldots, x_n)$, where each x_i $(i = 1, 2, \ldots, n)$ is an independent variable in real data space R. The function optimization problem (FOP) is to find the optimal real value $y_{opt} = f_{opt}$, which is the minimum value f_{min} or the maximum value f_{max}, together with its optimal solution $x_o = (x_{1o}, \ldots, x_{no})$ in a certain closed domain $D_i = [-a_i, a_i]$ for each x_i, where a_i is an initial positive constant given corresponding to the test function f. However, in this paper, we study only minimization problems, and therefore optimal solutions are usually referred to as minimum solutions.

2.2   Structures of Solutions

(1) Real coded solutions. To represent solutions with real coded algorithm, each variable x_i is represented as a real number of double precision. Optimization methods CMA-ES, ACO_R and BLX-α which is one of the real coded GA belong to this category.

(2) Binary coded solutions. When we solve the function value $y = f(x_1, \ldots, x_n)$ minimization problem with ACO_B or 2X, candidates for solutions $(x_1, \ldots, x_n)^s$ to the problem are designed by gene arrays of binary data taking each of their values of 0 or 1, and they represent a sequence of values of variables x_i. Each x_i can only represent 2^l discrete real data on a certain closed continuous domain $D_i = [-a_i, a_i]$ $(i = a_i)$ $(i = 1, 2, \ldots, n)$, where l is length of its genes. In the structures of solutions of Binary Coded Solutions of ACO_B [10], ant k probabilistically determines whether x_{ij} takes the value of zero or one, referring the pheromone $\tau_{ij\nu}$, where x_{ij} is the gene on the j-th locus $(j = 1, 2, \ldots, l)$ on the i-th variable x_i and ν is the above determination which takes its value of zero or one. 2X has the same structures of solutions as ACO_B, but the solution is generated stochastically by two-point crossover operation on a pair of parent genes called chromosome.

3   Optimization Methods

We briefly explain five optimization methods tested in this paper.

3.1   Covariance Matrix Adaptation-Evolution Strategy (CMA-ES)

CMA-ES is one of (μ, λ)-ES which searches for solutions with sequences of mutation operations [15]. CMA-ES generates λ individuals on the first stage in the
$(g+1)$-th generation. They are randomly created on the assumption that minimum solutions may exist probabilistically according to multivariate normal distribution $N(m(g), \sigma(g)^2 \times \Sigma'(g))$. Here, $m(g)$ is the gravity vector on the g-th
generation. Covariance matrix $\Sigma(g)$ is recalculated as $\Sigma'(g)$ by considering the
amount of changes of the above gravity vector $m(g)$ through the time elapses.
$\sigma(g)$ is a parameter called "step size", and it harmonizes the convergence speed
to the minimum solution.

3.2   Ant Colony Optimization for Continuous Domains (ACO$_R$)

ACO is developed as a general purpose optimization method to solve combinatorial optimization problems (COP), which is indirect pheromone communication
to search minimum ones among discrete N^M candidate solutions probabilistically, where N and M are given integers [5,6]. ACO$_R$ is an extension of the
idea of pheromone communication between ants in ACO, in order to search
for solutions in real data space. ACO$_R$ starts to generate k individuals $X[i]$
$(i = 1, 2, \ldots, k)$ randomly in the domain on the initial generation. Then, it
determines the individual $X[k_0]$ stochastically, which is the center of the search
of the solutions on the next generation, according to the amount of pheromone
$w(X[i]) = \frac{1}{qk\sqrt{2\pi}} \, e^{-\frac{(i-1)^2}{2(qk)^2}}$ deposited on individual $X[i]$, where $w(X[i])$ is proportional to the inverse of the function value $f(X[i])$, where q is a constant. On
each generation, m individuals are updated.

3.3   Blend Crossover with Stepwise Localization (BLX-α)

Genetic Algorithm (GA) is developed as optimization methods to solve combinatorial optimization problems based on Darwin's natural selection principle.
Among those algorithms based on GA, BLX-α solves FOP by extending the
search area from discrete space to real space by considering the continuity of functions. We assume that two individuals denoted by $X_1 = (x_{11}, \ldots, x_{1i}, \ldots, x_{1n})$
and $X_2 = (x_{21}, \ldots, x_{2i}, \ldots, x_{2n})$ are selected with roulette wheel selection [12].
BLX-α randomly searches for solutions $X = (x_1, \ldots, x_i, \ldots, x_n)$ in the intervals $(S_1, \ldots, S_i, \ldots, S_n)$ which include the above parental chromosomes X_1 and
X_2 with extension ratio of α of 0.5. Our BLX-α implemented in our previous paper [10] has mechanism of stepwise localization of the search space. Let
g be the number of layers of the localization. The best individual $X(g) =
(x_1(g), x_2(g), \ldots, x_n(g))$ found in the g-th layer is used as the center of the
search on the next $(g+1)$-th layer. One layer is composed of several generations.
One generation produces individuals of the population size. On the g-th layer, we
reduce the search space $D_i(g)$ so narrow as $[x_i(g) - R_i(g), x_i(g) + R_i(g)]$, where
$R_i(g)$ satisfies the exponential order of equation $R_i(g) = (1/2)^{g \times \log(g)} R_i(0)$. By

searching solutions in the above narrowed domain, we can improve the accuracy of solutions.

3.4   ACO$_B$ with Stepwise Localization of Binary Search Space

ACO$_B$ implemented in our previous paper [10] is an adaptation of elitist ant system (EAS) which was originally considered to solve TSP. In ACO$_B$, the value of each independent variable xi, which approximately realizes the solution $X = (x_1, \ldots, x_i, \ldots, x_n)$ in real data space, is expressed as a genetic string of binary data (0 or 1) of length l. ACO$_B$ has functions of stepwise localization of search space similar to that of BLX-α described in above 3.3. We assume that we can reduce the search space $D_i(g) = [x_i(g) - R_i(g), x_i(g) + R_i(g)]$ on the g-th layer. Although the domain $D_i(g)$ is reduced, the number of observed points can still be expressed as a single-digit binary number (2^l), so the accuracy of the solution is improved as $d_i(g) = R_i(0) \times (1/2)^{(g \times log(g)) + l - 1}$, where $d_i(g)$ is the interval of observable data and candidate solutions are expressed by $x_i = x_i(g) - R_i(g) + k \times d_i(g)(k = 0, 1, \ldots, 2^l - 1)$. As $d_i(g) \to 0$ when $g \to \infty$, so for any positive number γ, we can get solutions approximated to the order of $(1/2)^\gamma$ using the properties of continuous functions and the continuity of real numbers.

3.5   Two Points Crossover with Stepwise Localiztion (2X)

The method 2X is the same as that De Jong uses to solve FOP in GA. Similar to ACO$_B$ described in Sect. 3.4, the solutions are represented by gene arrays of binary data. 2X generates offspring by using well known two-point crossover operations [12] on a pair of selected parents. Our 2X has also functions of stepwise localization of the search space just like ACO$_B$.

4   Test Functions Investigated in This Study

Test functions $f(x_1, x_2, \ldots, x_n)$ and their domains investigated in this study are defined as follows shown in the papers [3]. Minimum values found through our experiments are also illustrated.

(1) Schwefel's test function: $- \sum\limits_{i=1}^{10} x_i \times \sin(\sqrt{|x_i|})$,
$-500 \leq x_i \leq 500, -4189.828872724338$

(2) Griewank's test function: $\frac{1}{4000} \sum\limits_{i=1}^{10} x_i{}^2 - \prod\limits_{i=1}^{10} \cos(\frac{x_i}{\sqrt{i}}) + 1$,
$-600 \leq x_i \leq 600, 0.0$

(3) Ackley's test function:
$-20 \times \exp(-0.2 \times \sqrt{\frac{1}{10} \times \sum\limits_{i=1}^{10} x_i^2}) - \exp(\frac{1}{10} \times \sum\limits_{i=1}^{10} \cos(2\pi x_i)) + 20 + e$,
$-32 \leq x_i \leq 32, 0.44409 \times 10^{-15}$

(4) Shubert's test function:

$$\left[\sum_{i=1}^{5} i \times \cos(i + (i+1)x_1)\right] \times \left[\sum_{i=1}^{5} i \times \cos(i + (i+1)x_2)\right],$$

$-10 \leq x_i \leq 10, -186.73090883102398$

(5) Six-hump camel back function:

$(4 - 2.1x_1^2 + \frac{1}{3}x_1^4)x_1^2 + x_1x_2 + 4(x_2^2 - 1)x_2^2,$

$-3 \leq x_1 \leq 3, -2 \leq x_2 \leq 2, -1.03162845348987741723$

(6) Easom's function: $-(-1)^n(\prod_{i=1}^{n} \cos(x_i)) \times \exp\left[-\sum_{i=1}^{n}(x_i - \pi)^2\right],$

$-100 \leq x_i \leq 100, -1.0 \ (n = 2); -2\pi \leq x_i \leq 2\pi, -1.0 \ (n = 10)$

(7) Michaelwicz's function: $-\sum_{i=1}^{n} \sin(x_i) \times \left[\sin(\frac{ix_i^2}{\pi})\right]^{2m}, m = 10,$

$0 \leq x_i \leq \pi, -1.80130341009855365897(n = 2)$
and $-9.660151715641434966786(n = 10)$

(8) Perm function type (D, BETA):

$\sum_{j=1}^{n}\{\sum_{i=1}^{n}(i^j + \beta)[(\frac{x_i}{i})^j - 1]\}^2, \beta = 0.5,$

$-n \leq x_i \leq n, 0.0(n = 2)$ and $0.00258877904512366374(n = 5)$

(9) Perm functions type (0, D, BETA):

$\sum_{j=1}^{10}\{\sum_{i=1}^{10}(i + \beta)[(x_i)^j - (\frac{1}{i})^j]\}^2, \beta = 10,$

$-1 \leq x_i \leq 1, 0.1374139896037 \times 10^{-7}$

(10) Xin-She Yang's functions: $(\sum_{i=1}^{10}|x_i|)\exp[-\sum_{i=1}^{10}\sin(x_i^2)],$

$-2\pi \leq x_i \leq 2\pi, 0.56606799135206409 \times 10^{-3}$

5 Experiments

Five optimization methods such as CMA-ES, ACO_R, BLX-α, 2X, ACO_B are evaluated. All of the optimization methods are coded in C language. C compiler is Vidual C++ 2017. Test runs are executed on the machine DELL VOSTRO, Intel Core i7-7500 CPU, 2.7 GHz, 8.0 GB RAM, two cores, and four threads.

5.1 Measures to Evaluate Methods

Five methods are compared with each other from the following five viewpoints to measure accuracy and performance.

Accuracy measures. For each method k, the following ratios are defined.

(a) The ratio of attaining minimum values: $Ratio(\#OPT)$. $\#r(=15$ in this experiments) is the number of independent runs from different initial uniform random numbers called seed_ids for each test case. $\#test\_f(=13)$ is the number of test cases using ten test functions defined in Sect. 4. $\#OPT(f_i)$ is the number of trials that attain the minimum values f_{min} in $\#r$ independent runs for each test function f_i.

$$Ratio(\#OPT) = (\sum_{i=1}^{\#test\_f} \#OPT(f_i))/(\#test\_f \times \#r) \tag{1}$$

(b) The ratio of times method k is judged to be good as $METHOD$
$_{BEST}(f)$ **through T-test, averaged over all f:** $Ratio(METH_{ALMOST\_BEST})$.
$METH_{A\_BEST}(f_i)$ in the formula takes its value of one if this method is selected as one of almost best minimization models $METH_{ALMOST\_BEST}(f_i)$, otherwise it takes its value of zero. $METHOD_{BEST}(f)$ is the best minimum method among five tested methods for each function f, if it attains the most minimum function value $BEST(f)$, where ties are broken firstly by preferring smaller average function value $AVG(f)$ searched for $\#r$ independent runs, secondly smaller sample standard deviation $STD(f)$, and lastly smaller average number of individuals required for searching for the minimum solutions $AVG_{\#IND}(f)$.

$$Ratio(METH_{ALMOST\_BEST}) = \sum_{i=1}^{\#test\_f} METH_{A\_BEST}(f_i) \ / \ \#test\_f \tag{2}$$

Performance measures
(c) The expected value of the number $AVG_{\#IND}$ of generations averaged overall f: $E.AVG_{\#IND}$

$$E.AVG_{\#IND} = \sum_{i=1}^{\#test\_f} AVG_{\#IND}(f_i) \ / \ \#test\_f \tag{3}$$

(d) The expected value of computational time: $E.COMP$. In the formula, $AVG_{COMP}(f)$ is the average computational time required for searching for the minimum solutions for each function f.

$$E.COMP = \sum_{i=1}^{\#test\_f} AVG_{COMP}(f_i) \ / \ \#test\_f \tag{4}$$

(e) Population size (Pop_size). The number of individuals in a population is called population size in GA (BLX-α, 2X) While alternating generations, each generation generates the population size individuals with sequences of genetic operations. Hence the population size is a metric to measure the required memory space to generate individuals for each generation. It is λ in (μ, λ)-ES, and it is m defined in Sect. 3.2 in ACO_R, and it is the pheromone update cycle m in ACO_B.

5.2 Experimental Results

Our experimental results from the above five viewpoints are illustrated in Table 1. The table shows that as ratio of the number of minimum trials ($Ratio(\#OPT)$)

improves, computational time ($E.COMP$) takes longer. Results show that among five optimization methods ACO_B, which has the function of stepwise localization of search space, can keep the balance between the accuracy of solutions and that of efficiency to search solutions. Minimum solutions found through our tests are shown to have higher or equivalent accuracy compared with other websites such as http://www.sfu.ca/~ssurjano/optimization.html.

Table 1. Evaluation of optimization methods

Optimization method		Accuracy		Performance		
		Ratio(#OPT)	Ratio (METH$_{ALMOST\_BEST}$)	E. AVG$_{\#IND}$	E.COMP (second)	Pop_size
Real coded	CMA-ES	0(=0/195)	0.23(=3/13)	153,030	2.8	100
	ACO$_R$	0.38(=74/ 195)	0.62(=8/13)	85,120	40.5	100
	BLX-α	0.44(=86/195)	0.85(=11/13)	2,197,684	484.8	300
Binary coded	2X	0.21(=40/195)	0.62(=8/13)	94,569	18.2	300
	ACO$_B$	0.27(=53/195)	0.54(=7/13)	1,115	43.8	10

6   Conclusions and Future Work

In solving FOP, the accuracy of computed solutions and performance of ACO_B are compared with other methods such as CMA-ES, ACO_R, BLX-α, and 2X by using thirteen test cases of ten well known test functions. We observe the following results. (1) The best method $METHOD_{BEST}(f)$ varies corresponding to tested functions f. In the experiment, ACO_R and ACO_B have the high $Ratio(METOD_{BEST})$ of 0.46 (=6/13) and 0.3 (=4/13) respectively. (2)BLX-α has the highest $Ratio(\#OPT)$ of 0.44(=86/195) and $Ratio(METH_{ALMOST\_BEST})$ of 0.85(=11/13), showing its high accuracy, but poor from the viewpoint of performance. (3) ACO_R has the second highest $Ratio(\#OPT)$ and $Ratio(METH_{ALMOST\_BEST})$. It has less $AVG_{\#IND}$ and less computational time than s.t.-BLX-α. (4) Even though ACO_B is the third measure of $Ratio(\#OPT)$, it has the least $AVG_{\#IND}$, and the least population size. (5) CMA-ES has the least computational time $E.COMP$ (=2.8 sec.) but behaves poorly in the accuracy measure. However, the error between the function value found through CMA-ES and that found through other optimization methods is very small (i.e. below $10^{-9} \sim 10^{-11}$). (6) It is also verified that stepwise localization of search space can keep the balance between accuracy and efficiency by experiments with ACO_B, BLX-α and 2X. (7) It is also observed that no method has both high accuracy and high performance. Our future work is to verify it by extending search space.

Acknowledgements. This work was supported by JSPS KAKENHI Grant Number JP15K00347, Grant-in-Aid for Scientific Research (C). We would like to thank them for supporting our work.

References

1. Dorigo, M., Stützle, T.: Ant Colony Optimization. The MIT Press, Cambridge (2004)
2. Dorigo, M., Blum, C.: Ant colony optimization theory: a survey. Theor. Comput. Sci. **344**, 243–278 (2005)
3. Yang, X.-S.: Test problems in optimization. In: Yang, X.-S. (ed.) Engineering Optimization: An Introduction with Metaheuristic Applications. Wiley (2010)
4. Borhani, R.: Machine Learning Refined: Foundations, Algorithms, and Applications. Cambridge University Press, Cambridge (2016)
5. Aarts, E., Lenstra, J.K.: Local Search in Combinatorial Optimization. Princeton University Press, Princeton (2003)
6. Sait, S.M., Youssef, H.: Iterative Computer Algorithms with Applications in Engineering, (translated into Japanese by Y. Shiraishi), Maruzen Co., Ltd (2002)
7. Socha, K., Dorigo, M.: Ant colony optimization for continuous domains. Eur. J. Oper. Res. **185**, 1155–1173 (2008)
8. Socha, K.: ACO for continuous and mixed-variable optimization. In: Dorigo, M., Birattari, M., Blum, C., Gambardella, L.M., Mondada, F., Stützle, T. (eds.) ANTS 2004. LNCS, vol. 3172, pp. 25–36. Springer, Heidelberg (2004). https://doi.org/10.1007/978-3-540-28646-2_3
9. Ojha, V.K., Abraham, A., Snásel, V.: ACO for continuous function optimization: a performance analysis. In: Proceeding of 14th International Conference on Intelligent Systems Design and Applications (ISDA), pp. 145–150. IEEE (2014)
10. Takahashi, R., Nakamura, Y.: Ant colony optimization with stepwise localization of the discrete search space to solve function optimization problems. In: Proceedings of 16th IEEE International Conference on Machine Learning and Applications (ICMLA17), pp. 701–706 (2017)
11. Hansen, N.: The CMA evolution strategy: a comparing review. In: Lozano, J.A., Larrañaga, P., Inza, I., Bengoetxea, E. (eds.) Towards a New Evolutionary Computation. STUDFUZZ, pp. 75–102. Springer, Heidelberg (2006). https://doi.org/10.1007/3-540-32494-1_4
12. Goldberg, D.E.: Genetic Algorithms in Search, Optimization, and Machine Learning. Addison-Wesley Publishing Company Inc., Boston (1989)
13. Eshelman, L.J., Schaffer, J.D.: Real coded genetic algorithms and interval-schemata. Found. Genet. Algorithms **2**, 187–202 (1993)
14. Takahashi, R.: Empirical evaluation of changing crossover operators to solve function optimization problems. In: Proceedings of the 2016 IEEE Symposium Series on Computational Intelligence (IEEE SSCI 2016), pp. 1–10 (2016). https://doi.org/10.1109/SSCI.2016.7850141
15. Schwefel, H.P., Wegener, I., Weinert, K.: Advances in Computational Intelligence. Springer, Heiderberg (2003). https://doi.org/10.1007/978-3-662-05609-7

Gaussian-Valued Particle Swarm Optimization

Kyle Robert Harrison[1(✉)], Beatrice M. Ombuki-Berman[2],
and Andries P. Engelbrecht[1]

[1] Department of Computer Science, University of Pretoria, Pretoria, South Africa
kharrison@outlook.com, engel@cs.up.ac.za
[2] Department of Computer Science, Brock University, St. Catharines, Canada
bombuki@brocku.ca

Abstract. This paper examines the position update equation of the particle swarm optimization (PSO) algorithm, leading to the proposal of a simplified position update based upon a Gaussian distribution. The proposed algorithm, Gaussian-valued particle swarm optimization (GVPSO), generates probabilistic positions by retaining key elements of the canonical update procedure while also removing the need to specify values for the traditional PSO control parameters. Experimental results across a set of 60 benchmark problems indicate that GVPSO outperforms both the standard PSO and the bare bones particle swarm optimization (BBPSO) algorithm, which also employs a Gaussian distribution to generate particle positions.

1 Introduction

The particle swarm optimization (PSO) algorithm [22] is a stochastic optimization technique based upon the social dynamics of a flock of birds. The PSO algorithm generates new positions stochastically based upon the position of two key attractors in the search space, namely the personal and neighbourhood best positions. The step sizes, and therefore the degree of exploration and exploitation, are then controlled via the values of three control parameters [1,4,20,28]. The values of the control parameters directly influence the particle movement patterns [1,3]. However, the best control parameter values are problem dependent and effective tuning is needed to improve PSO performance [2,4,37].

While parameter tuning is clearly warranted in the PSO algorithm, it is typically a time-consuming process whereby a large number of candidate parameter configurations must be analysed. Fortunately, there have been a number of studies that have suggested general-purpose PSO parameters based on empirical evidence [3–5,8,18,19,28,37,41]. While these studies have made use of the implicit assumption that *a priori* tuning of control parameters is sufficient to optimize performance, recent evidence suggests that the best PSO parameters to employ change over time [18]. Similar results have also been found for heterogeneous PSOs [26,29,30,38] and for dynamic PSOs [24].

M. Dorigo et al. (Eds.): ANTS 2018, LNCS 11172, pp. 368–377, 2018.
https://doi.org/10.1007/978-3-030-00533-7_31

An alternative approach is to use PSO variants that do not rely on *a priori* control parameter values. A prominent example of this approach lies in the development of self-adaptive PSO (SAPSO) techniques, which continuously adapt the values of their control parameters throughout the search process [25,27,31,32,34–36,39,40]. However, many of the SAPSO algorithms have been shown to exhibit either premature convergence or rapid divergence, thereby leading to poor performance [14,15,17,42]. An additional example is the bare bones PSO (BBPSO) [21], which updates particle positions probabilistically using a Gaussian distribution. However, the manner in which particle positions are created via BBPSO is strikingly dissimilar to how the conventional PSO determines updated particle positions.

In this paper, a new PSO variant is proposed by formulating a new probabilistic approach to generating particle positions. The new approach is inspired by the BBPSO algorithm, but differs significantly in the manner by which particle positions are generated. Notably, the proposed algorithm generates particle positions using a model that more closely resembles the canonical PSO, which as this paper will demonstrate, provides a clear performance advantage over BBPSO and other PSO configurations.

The remainder of this paper is structured as follows. Section 2 provides the necessary background information about PSO and BBPSO. The proposed algorithm is described in Sect. 3, while Sect. 4 details the empirical analysis and results. Finally, concluding remarks and avenues of future work are provided in Sect. 5.

2   Background

This section provides the necessary background information about the PSO and BBPSO algorithms. The PSO algorithm is described in Sect. 2.1 while the BBPSO algorithm is outlined in Sect. 2.2.

2.1   Particle Swarm Optimization

The PSO algorithm [22] consists of a collection of agents, referred to as particles, which each represent a candidate solution to an optimization problem. Each particle retains three pieces of information, namely its current position, velocity, and its (personal) best position found within the search space. Particle positions are updated each iteration via the calculation and subsequent addition of a velocity to the particle's current position. A particle's velocity is based on its attraction towards two (promising) locations in the search space, namely the best position found by the particle itself and the best position found by any particle within the particle's immediate neighbourhood [23]. The neighbourhood of a particle is defined as the other particles within the swarm from which it may take influence, which is most commonly the entire swarm or, alternatively, the immediate neighbours when arranged in a ring [23].

To facilitate movement in the PSO algorithm, the velocity is calculated for particle i according to the inertia weight model [34] as

$$v_{ij}(t+1) = \omega v_{ij}(t) + c_1 r_{1ij}(t)(y_{ij}(t) - x_{ij}(t)) + c_2 r_{2ij}(t)(\hat{y}_{ij}(t) - x_{ij}(t)), \tag{1}$$

where $v_{ij}(t)$ and $x_{ij}(t)$ are the velocity and position in dimension j at time t, respectively. The inertia weight is given by ω while c_1 and c_2 represent the cognitive and social acceleration coefficients, respectively. The stochastic component of the algorithm is provided by the random values $r_{1ij}(t), r_{2ij}(t) \sim U(0,1)$, which are independently sampled each iteration for all components of each particle's velocity. Finally, $y_{ij}(t)$ and $\hat{y}_{ij}(t)$ denote the personal and neighbourhood best positions in dimension j, respectively. Particle positions are then updated according to

$$x_{ij}(t+1) = x_{ij}(t) + v_{ij}(t+1). \tag{2}$$

2.2 Barebones Particle Swarm Optimization

When examining particle movement patterns, Kennedy noted that particle positions formed a bell curve centered around the midpoint between the global and personal best positions [21]. Based on this result, the BBPSO algorithm [21] eliminates the velocity component of PSO and rather updates particle positions probabilistically according to

$$x_{ij}(t+1) = \begin{cases} y_{ij}(t) & \text{if } U(0,1) < e \\ \mathcal{N}\left(\frac{c_1 y_{ij}(t) + c_2 \hat{y}_{ij}(t)}{c_1 + c_2}, |y_{ij}(t) - \hat{y}_{ij}(t)|\right) & \text{otherwise} \end{cases}, \tag{3}$$

where $\mathcal{N}(\mu, \sigma)$ denotes a normal distribution with mean μ and standard deviation σ and e is a parameter representing the per-dimension chance of selecting the personal best position. In the original formulation of BBPSO, the control parameters were set as $c_1 = c_2 = 1$ [21]. Later theoretical results supported the observation of Kennedy by showing that, using the stagnation and deterministic assumptions, each particle will converge to the point $\frac{c_1 y_i + c_2 \hat{y}_i}{c_1 + c_2}$ [2,37].

3 Gaussian Valued Particle Swarm Optimization

To provide the motivation for the proposed algorithm, consider the PSO velocity equation given in Eq. (1) when $\omega = 0$. With no inertia, the velocity calculation simplifies to

$$v_{ij}(t+1) = c_1 r_{1ij}(t)(y_{ij}(t) - x_{ij}(t)) + c_2 r_{2ij}(t)(\hat{y}_{ij}(t) - x_{ij}(t)). \tag{4}$$

Note that because $r_{1ij}(t), r_{2ij}(t) \sim U(0,1)$, Eq. (4) can be reformulated as

$$v_{ij}(t+1) = v_{1ij}(t) + v_{2ij}(t) \tag{5}$$

where $v_{1ij}(t) \sim U(0, c_1(y_{ij}(t) - x_{ij}(t)))$ and $v_{2ij}(t) \sim U(0, c_2(\hat{y}_{ij}(t) - x_{ij}(t)))$[1]. It can be easily observed that the position update becomes a sum of two uniform distributions, thereby leading to a trapezoidal distribution. The shape of the resulting trapezoidal distribution is then governed by the distance between the current particle's position and the position of the personal and neighbourhood best, respectively. Even with the reintroduction of the inertia component, the same general observation can be made; the particle position update depends heavily upon not only the personal and neighbourhood best positions, but rather the distance between the current particle and these two attractors.

The position update mechanism for GVPSO is formulated by employing a Gaussian distribution centered at a random point taken from the aforementioned trapezoidal distribution. The Gaussian distribution is used to modulate the particle step sizes based upon the distance between the current position and the personal and neighbourhood best positions. Specifically, an ancillary position, $\Delta_{ij}(t)$, is calculated for each particle in every dimension using Eqs. (1) and (2) with $\omega = 0$ and $c_1 = c_2 = 1$. This effectively retains the core movement pattern of PSO without the reliance on control parameter values. The particle's new position is then determined using a Gaussian distribution centered between the current position and $\Delta_{ij}(t)$ with a standard deviation based on the magnitude of the distance between the current position and $\Delta_{ij}(t)$ according to

$$
x_{ij}(t+1) = \begin{cases} y_{ij}(t) & \text{if } U(0,1) < e \\ \mathcal{N}\left(\frac{x_{ij}(t) + \Delta_{ij}(t)}{2}, |\Delta_{ij}(t) - x_{ij}(t)|\right) & \text{otherwise} \end{cases}, \quad (6)
$$

where e is the exploitation parameter, as seen in Eq. (3). Note that GVPSO, in the same manner as BBPSO, eliminates the need for the conventional PSO parameters ω, c_1, and c_2. However, the GVPSO algorithm differs from BBPSO by creating particle positions that more closely mimic the canonical position update of PSO through the use of distance information and thus the two attractors remain to have a strong influence. Furthermore, the step sizes in the GVPSO are implicitly controlled by the distances between the current particle and the two attractors, thereby leading to diminishing step sizes as the positions and attractors inevitably converge. Thus, the GVPSO is expected to exhibit both initial exploration and exploitation in the later phase of the search process.

4   Experimental Results and Discussion

This section presents the experimental design regarding the empirical examination of GVPSO. Section 4.1 describes the parameterization, benchmark suite, and statistical analysis. Section 4.2 presents a sensitivity analysis on the exploitation parameter while Sect. 4.3 presents a comparison of GVPSO to other PSO variants.

---

[1] Without loss of generality, this assumes that $c_1(y_{ij}(t) - x_{ij}(t)) > 0$ and $c_2(\hat{y}_{ij}(t) - x_{ij}(t)) > 0$, otherwise the bounds must be flipped, i.e., 0 becomes the upper bound.

4.1 Experimental Setup

To first examine the effect of the exploitation probability parameter e, 10 values of e were examined for GVPSO and BBPSO, namely values between 0.0 and 0.9 in increments of 0.1. Linearly decreasing variants (GVPSO-LD and BBPSO-LD), whereby the value of e was linearly decreased from 0.9 to 0.0, were also examined. The performance of GVPSO was then compared against the following PSO strategies:

- BBPSO
- Three static PSO parameter configurations: PSO-1 ($\omega = 0.7298, c_1 = c_2 = 1.49618$) [7], PSO-2 ($\omega = 0.729, c_1 = 2.0412, c_2 = 0.9477$) [4], and PSO-7 ($\omega = 0.785, c_1 = c_2 = 1.331$) [41], which were found to be the best performing of 14 commonly recommended PSO parametrizations [16]
- PSO with time-varying acceleration coefficients (PSO-TVAC) [32]
- PSO with improved random constants (PSO-iRC) [16]

All examined variants consisted of 30 particles arranged in a star neighbourhood and used a synchronous update strategy. To prevent invalid attractors, a particle's personal best position was only updated if a new position had a better objective function value and was within the feasible bounds of the search space. For the BBPSO algorithm, the original parametrization of $c_1 = c_2 = 1$ was used. Where applicable, particle velocities were initialized to zero [9]. For PSO-TVAC, the social acceleration coefficient was linearly increased from 0.5 to 2.5 while the values of the cognitive and inertia control parameters were linearly decreased from 2.5 to 0.5 and 0.9 to 0.4, respectively. For PSO-iRC, parameter configurations were re-sampled every 5 iterations (i.e., according to PSO-iRC-p5 [16]). The value of the objective function (i.e., the fitness), averaged over 50 independent runs each consisting of 5000 iterations, was taken as the measure of performance for each algorithm.

Benchmark Problems. A suite of 60 minimization problems, originally used by [10], were used in this study. The suite has been demonstrated to include a range of different landscape characteristics [13]. All functions were optimized in 30 dimensions. Further information about the benchmark suite can be found in [10] and [14].

Statistical Analysis. Statistical analysis of results was done by way of Friedman's test for multiple comparisons among all methods [11,12], as recommended by Derrac *et al.* [6]. Furthermore, Shaffer's post-hoc procedure [33] was performed as a means to identify the pairwise comparisons that produced significant differences. Finally, the statistical results are visualized via critical difference plots, whereby algorithms to the left of the plot (i.e., those with lower average ranks) demonstrated superior performance. The critical difference (CD) denotes the difference in average rank that was found to be statistically significant. Therefore, algorithms that are grouped by a line (i.e., those with a difference in rank less than CD) were found to have statistically insignificant differences in performance.

4.2   Examining the Exploitation Probability

Figures 1 and 2 show the critical difference plots for the examined values of e for both the GVPSO and BBPSO algorithms over the entire set of problems. While the exact values that lead to the best performance were different among the two algorithms, the general trends were the same. In general, mid-range values of e (i.e., 0.4–0.7) tended to perform the best, showing that both exploration and exploitation were beneficial to the GVPSO algorithm. Based upon these results, GVPSO and BBPSO with values of e set to 0.5, 0.6, and 0.7 were compared against other PSO techniques in the next section.

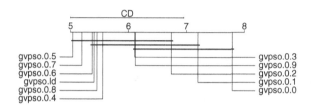

Fig. 1. Comparison of GVPSO exploit probabilities over all 60 benchmark problems.

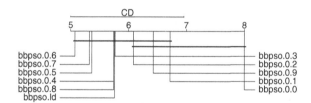

Fig. 2. Comparison of BBPSO exploit probabilities over all 60 benchmark problems.

4.3   Comparison with Other Particle Swarm Optimization Techniques

This section presents the results from comparing GVPSO (with $e = \{0.5, 0.6, 0.7\}$) against the other PSO variants. Figure 3 shows the results across all benchmark problems. It was first observed that the best average ranks across all benchmark problems were attained by the three configurations of GVPSO, clearly indicating the merit of this approach. Despite the better average rank attained by GVPSO, the critical difference plot indicates there was no significant difference in performance between the different GVPSO and BBPSO configurations as well as PSO-2. However, it was also observed from Fig. 3 that PSO-2 attained a notably worse average rank than each of the GVPSO and BBPSO configurations. The remaining PSO variants, namely PSO-1, PSO-7, PSO-TVAC, and PSO-iRC-p5 all performed significantly worse than GVPSO.

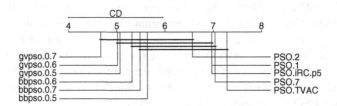

Fig. 3. Comparison of GVPSO with other PSO variants over all 60 benchmark problems.

5   Conclusions and Future Work

This paper proposed a new particle swarm optimization (PSO) variant, entitled Gaussian-valued PSO (GVPSO), which generates particle positions probabilistically according a Gaussian distribution. The GVPSO algorithm is loosely inspired by the bare bones PSO (BBPSO) but differs significantly from the BBPSO algorithm by generating particles according to a distribution that more closely resembles the conventional PSO position update. An analysis of the single parameter of GVPSO was first performed, followed by a comparison of GVPSO to BBPSO and five additional PSO configurations. Results indicate that GVPSO generally outperforms the other strategies.

An immediate avenue of future work lies in the self-adaptation of the single GVPSO parameter, resulting in a parameter-free algorithm. Further work will also examine the proposed algorithm in different dimensionalities and compare its performance against additional PSO variants, including improved implementations of BBPSO.

Acknowledgments. This work is based on the research supported by the National Research Foundation (NRF) of South Africa (Grant Number 46712). The opinions, findings and conclusions or recommendations expressed in this article is that of the author(s) alone, and not that of the NRF. The NRF accepts no liability whatsoever in this regard. This work is also supported by the Natural Sciences and Engineering Research Council of Canada (NSERC).

References

1. van den Bergh, F.: An analysis of particle swarm optimizers. Ph.D. thesis, University of Pretoria (2001)
2. van den Bergh, F., Engelbrecht, A.P.: A study of particle swarm optimization particle trajectories. Inf. Sci. **176**(8), 937–971 (2006). https://doi.org/10.1016/j.ins.2005.02.003
3. Bonyadi, M., Michalewicz, Z.: Impacts of coefficients on movement patterns in the particle swarm optimization algorithm. IEEE Trans. Evol. Comput. **21**(3), 1 (2016). https://doi.org/10.1109/TEVC.2016.2605668
4. Carlisle, A., Dozier, G.: An off-the-shelf PSO. In: Proceedings of the Workshop on Particle Swarm Optimization, vol. 1, pp. 1–6. Purdue School of Engineering and Technology (2001)
5. Clerc, M.: Stagnation analysis in particle swarm optimisation or what happens when nothing happens. Technical report 1, HAL (2006)
6. Derrac, J., García, S., Molina, D., Herrera, F.: A practical tutorial on the use of nonparametric statistical tests as a methodology for comparing evolutionary and swarm intelligence algorithms. Swarm Evol. Comput. **1**(1), 3–18 (2011). https://doi.org/10.1016/j.swevo.2011.02.002
7. Eberhart, R., Kennedy, J.: A new optimizer using particle swarm theory. In: Proceedings of the Sixth International Symposium on Micro Machine and Human Science, vol. 12, pp. 39–43. IEEE (2008). https://doi.org/10.1109/MHS.1995.494215
8. Eberhart, R., Shi, Y.: Comparing inertia weights and constriction factors in particle swarm optimization. In: Proceedings of the 2000 IEEE Congress on Evolutionary Computation, pp. 84–88. IEEE (2000). https://doi.org/10.1109/CEC.2000.870279
9. Engelbrecht, A.: Particle swarm optimization: velocity initialization. In: Proceedings of the 2012 IEEE Congress on Evolutionary Computation, pp. 1–8. IEEE (2012). https://doi.org/10.1109/CEC.2012.6256112
10. Engelbrecht, A.: Particle swarm optimization: global best or local best? In: Proceedings of the 2013 BRICS Congress on Computational Intelligence and 11th Brazilian Congress on Computational Intelligence, pp. 124–135. IEEE (2013). https://doi.org/10.1109/BRICS-CCI-CBIC.2013.31
11. Friedman, M.: A comparison of alternative tests of significance for the problem of m rankings. Ann. Math. Stat. **11**(1), 86–92 (1940). https://doi.org/10.1214/aoms/1177731944
12. Friedman, M.: The use of ranks to avoid the assumption of normality implicit in the analysis of variance. J. Am. Stat. Assoc. **32**(32), 675–701 (1937). https://doi.org/10.1080/01621459.1937.10503522
13. Garden, R.W., Engelbrecht, A.P.: Analysis and classification of optimisation benchmark functions and benchmark suites. In: Proceedings of the 2014 IEEE Congress on Evolutionary Computation, pp. 1641–1649. IEEE (2014). https://doi.org/10.1109/CEC.2014.6900240
14. Harrison, K.R., Engelbrecht, A.P., Ombuki-Berman, B.M.: Inertia weight control strategies for particle swarm optimization. Swarm Intell. **10**(4), 267–305 (2016). https://doi.org/10.1007/s11721-016-0128-z
15. Harrison, K.R., Engelbrecht, A.P., Ombuki-Berman, B.M.: The sad state of self-adaptive particle swarm optimizers. In: Proceedings of the 2016 IEEE Congress on Evolutionary Computation, pp. 431–439. IEEE (2016). https://doi.org/10.1109/CEC.2016.7743826

16. Harrison, K.R., Engelbrecht, A.P., Ombuki-Berman, B.M.: An adaptive particle swarm optimization algorithm based on optimal parameter regions. In: Proceedings of the 2017 IEEE Symposium Series on Computational Intelligence, pp. 1606–1613. IEEE (2017). https://doi.org/10.1109/SSCI.2017.8285342

17. Harrison, K.R., Engelbrecht, A.P., Ombuki-Berman, B.M.: Self-adaptive particle swarm optimization: a review and analysis of convergence. Swarm Intell. 12, 187–226 (2017). https://doi.org/10.1007/s11721-017-0150-9

18. Harrison, K.R., Engelbrecht, A.P., Ombuki-Berman, B.M.: Optimal parameter regions and the time-dependence of control parameter values for the particle swarm optimization algorithm. Swarm Evol. Comput. (2018). https://doi.org/10.1016/j.swevo.2018.01.006

19. Jiang, M., Luo, Y., Yang, S.: Particle swarm optimization - stochastic trajectory analysis and parameter selection. In: Swarm Intelligence, Focus on Ant and Particle Swarm Optimization, pp. 179–198. I-Tech Education and Publishing, December 2007. https://doi.org/10.5772/5104

20. Jiang, M., Luo, Y., Yang, S.: Stochastic convergence analysis and parameter selection of the standard particle swarm optimization algorithm. Inf. Process. Lett. 102(1), 8–16 (2007). https://doi.org/10.1016/j.ipl.2006.10.005

21. Kennedy, J.: Bare bones particle swarms. In: Proceedings of the 2003 IEEE Swarm Intelligence Symposium, pp. 80–87. IEEE (2003). https://doi.org/10.1109/SIS.2003.1202251

22. Kennedy, J., Eberhart, R.: Particle swarm optimization. In: Proceedings of the International Conference on Neural Networks, vol. 4, pp. 1942–1948. IEEE (1995). https://doi.org/10.1109/ICNN.1995.488968

23. Kennedy, J., Mendes, R.: Population structure and particle swarm performance. In: Proceedings of the 2002 Congress on Evolutionary Computation, vol. 2, pp. 1671–1676. IEEE (2002). https://doi.org/10.1109/CEC.2002.1004493

24. Leonard, B.J., Engelbrecht, A.P.: On the optimality of particle swarm parameters in dynamic environments. In: Proceedings of the 2013 IEEE Congress on Evolutionary Computation, pp. 1564–1569. IEEE (2013). https://doi.org/10.1109/CEC.2013.6557748

25. Leu, M.S., Yeh, M.F.: Grey particle swarm optimization. Appl. Soft Comput. 12(9), 2985–2996 (2012). https://doi.org/10.1016/j.asoc.2012.04.030

26. Li, C., Yang, S., Nguyen, T.T.: A self-learning particle swarm optimizer for global optimization problems. IEEE Trans. Syst. Man Cybern. B (Cybern.) 42(3), 627–646 (2012). https://doi.org/10.1109/TSMCB.2011.2171946

27. Li, X., Fu, H., Zhang, C.: A self-adaptive particle swarm optimization algorithm. In: Proceedings of the 2008 International Conference on Computer Science and Software Engineering, vol. 5, pp. 186–189. IEEE (2008). https://doi.org/10.1109/CSSE.2008.142

28. Liu, Q.: Order-2 stability analysis of particle swarm optimization. Evol. Comput. 23(2), 187–216 (2015). https://doi.org/10.1162/EVCO_a_00129

29. Montes de Oca, M.A., Peña, J., Stützle, T., Pinciroli, C., Dorigo, M.: Heterogeneous particle swarm optimizers. In: Proceedings of the 2009 IEEE Congress on Evolutionary Computation, pp. 698–705. IEEE (2009). https://doi.org/10.1109/CEC.2009.4983013

30. Nepomuceno, F.V., Engelbrecht, A.P.: A self-adaptive heterogeneous PSO for real-parameter optimization. In: Proceedings of the 2013 IEEE Congress on Evolutionary Computation, pp. 361–368. IEEE (2013). https://doi.org/10.1109/CEC.2013.6557592

31. Nickabadi, A., Ebadzadeh, M.M., Safabakhsh, R.: A novel particle swarm optimization algorithm with adaptive inertia weight. Appl. Soft Comput. **11**(4), 3658–3670 (2011). https://doi.org/10.1016/j.asoc.2011.01.037

32. Ratnaweera, A., Halgamuge, S., Watson, H.: Self-organizing hierarchical particle swarm optimizer with time-varying acceleration coefficients. IEEE Trans. Evol. Comput. **8**(3), 240–255 (2004). https://doi.org/10.1109/TEVC.2004.826071

33. Shaffer, J.P.: Modified sequentially rejective multiple test procedures. J. Am. Stat. Assoc. **81**(395), 826–831 (1986). https://doi.org/10.1080/01621459.1986.10478341

34. Shi, Y., Eberhart, R.: A modified particle swarm optimizer. In: Proceedings of the 1998 IEEE International Conference on Evolutionary Computation, pp. 69–73. IEEE (1998). https://doi.org/10.1109/ICEC.1998.699146

35. Shi, Y., Eberhart, R.: Empirical study of particle swarm optimization. In: Proceedings of the 1999 Congress on Evolutionary Computation, vol. 3, pp. 1945–1950. IEEE (1999). https://doi.org/10.1109/CEC.1999.785511

36. Tanweer, M., Suresh, S., Sundararajan, N.: Self regulating particle swarm optimization algorithm. Inf. Sci. **294**, 182–202 (2015). https://doi.org/10.1016/j.ins.2014.09.053

37. Trelea, I.C.: The particle swarm optimization algorithm: convergence analysis and parameter selection. Inf. Process. Lett. **85**(6), 317–325 (2003). https://doi.org/10.1016/S0020-0190(02)00447-7

38. Wang, Y., Li, B., Weise, T., Wang, J., Yuan, B., Tian, Q.: Self-adaptive learning based particle swarm optimization. Inf. Sci. **181**(20), 4515–4538 (2011). https://doi.org/10.1016/j.ins.2010.07.013

39. Xu, G.: An adaptive parameter tuning of particle swarm optimization algorithm. Appl. Math. Comput. **219**(9), 4560–4569 (2013). https://doi.org/10.1016/j.amc.2012.10.067

40. Yang, X., Yuan, J., Yuan, J., Mao, H.: A modified particle swarm optimizer with dynamic adaptation. Appl. Math. Comput. **189**(2), 1205–1213 (2007). https://doi.org/10.1016/j.amc.2006.12.045

41. Zhang, W., Ma, D., Wei, J.J., Liang, H.F.: A parameter selection strategy for particle swarm optimization based on particle positions. Expert Syst. Appl. **41**(7), 3576–3584 (2014). https://doi.org/10.1016/j.eswa.2013.10.061

42. van Zyl, E., Engelbrecht, A.: Comparison of self-adaptive particle swarm optimizers. In: Proceedings of the 2014 IEEE Symposium on Swarm Intelligence, pp. 1–9. IEEE (2014). https://doi.org/10.1109/SIS.2014.7011775

Individual Activity Level and Mobility Patterns of Ants Within Nest Site

Kazutaka Shoji[✉]

Tokyo Metropolitan University, Minami-Osawa 1-1, Hachioji-shi, Tokyo, Japan
kazutakashoji.ants@gmail.com

Abstract. Augmented reality (AR) tracking method allowed us not only to obtain entire interaction data but also entire behavioral big data, of ants, at the same time. Individual behavioral data may provide us a way to analyze individual personality behavioral responses to environmental condition, and an automatic way to detect their task allocation in a colony. In this study, individual behavioral differences were assessed by comparing individual behavior under normal and harsh environments, to evaluate individual responses. Individuals were classified based on their behavior; mobility patterns were analyzed to understand their relationship with respective network structure. These results show that individual behaviors are regulated as responses to environmental conditions and reactions differ depending on colony size. Individual classification and mobility patterns show that this method can be used to distinguish individuals solely by their behavioral and mobility patterns, which may have important roles in network structure pattern.

1 Introduction

Individual behavioral differences such as speed, spatial distribution, and personality [4] may have an important role in emergence of network structure in ants [7,9,11]. Environmental factors almost certainly play an important role in defining individual and group-level behavioral patterns [6]. In this study, influence of environmental factors on behavior and individual mobility patterns of ants among various colony sizes were investigated. To reveal how environmental conditions affect individual behavior within nest sites, colonies were exposed to harsh environments such as strong light, which can induce migratory behavior in ants [2–4,6]. Previous studies [6] only observed arenas because of the difficulty of keeping their ID numbers for ordinary tracking system in each individual, continuously. Therefore, it is unknown whether individual ants at their nest sites respond the same way as on the arena. In a previous study [8], individual ants were classified, based on observed task allocation from manually recorded video, which suggested that classification and special distribution are correlated. Observation of numerous ants and lengthy video are time consuming. To reduce costs and to obtain classification information automatically, behavioral parameters such as speed, heading angle, and distance obtained from AR tracking are useful.

© Springer Nature Switzerland AG 2018
M. Dorigo et al. (Eds.): ANTS 2018, LNCS 11172, pp. 378–384, 2018.
https://doi.org/10.1007/978-3-030-00533-7_32

The AR tracking system provides XY positional data [1] and it can be utilized for detection of individual interaction and network analysis [10]. Network analysis only focused on structure and ignored individual (node) differences in colonies [8,11]. To provide individual differences data on nodes and to facilitate network analysis, behavioral analysis was conducted. There are two approaches to behavioral analysis in this study; (1) analyzing individual speed, heading angle and total walking distance; (2) by using individual data from (1), classifying individuals and analyzing their mobility patterns, which may have influence on network structure. In this study, (1) and (2) behavioral analyses were conducted to show their functionality and usefulness as follows: (a) Individual speed, heading angle, and total working distance were compared between normal and harsh environment within the nest sites to reveal whether they work hard only on the trail [6] or also within nest sites; (b) individuals were classified based on obtained XY positional data, from AR tracking process, within nest sites and their mobility patterns were analyzed.

2 Methods and Materials

2.1 Ants and AR Tracking

Entire colonies of *Myrmecina nipponica* were collected from broadleaf forest near Chitose City in Hokkaido, northern Japan (N42 470' E141 340', altitude approximately 100 m) in September 2016 and maintained in artificial laboratory nests, 4.8 cm × 2.8 cm in area and 3 mm high with a 6 mm wide entrance, using standard protocols [2, 10]. Nest walls were made of styrene plastic. Top of the nest was covered with slide glass and a redcolored plastic sheet. Nine colonies of varying size were selected for observation.

The AR tags contain XY position, ID, heading direction rotation, pitch, and yaw. AR barcode tags were made from those presented with BEEtag [1]. BEEtag is a source code which can be operated on MATLAB. Barcodes of 0.8 mm were printed on paper. These tags were glued (Henkel, LOCTITE pin pointer jelly type) on thorax of ants in colonies. A high-resolution mirrorless camera (Panasonic, DMC-GH4H) was used to take digital photographs of ants every second. Six LED light sources (Style+, PAR38-12*3W) were used to illuminate experimental chamber. Before taking photographs, ants were allowed to acclimate to the arena for 60 min. In total, 14,400 images were recorded in 4 h.

2.2 Individual Behavior Under Harsh Environment

Individual behavior within nest sites under normal (red colored roof of nest site) and harsh environment (clear roof of nest site), which can induce migratory behavior [2,3,6], was monitored with AR tracking. Medians of mean speed, relative heading angle, and total distance were compared between normal and harsh environmental condition. Wilcoxon signed rank test with Bonferroni adjustment was used throughout the analysis. Differences of mean speed between normal and harsh condition were compared in each individual activity-level such as, high, middle, and low-activity level; classification of individual activity-levels is described in the following section with ANCOVA (Analysis of covariance).

2.3 Individual Classification Based on Their Behavior and Mobility Patterns

To detect individual differences such as task allocation, automatically, ants were classified based on individual activity-level. Individual speed, total travel distance, and relative heading angle were used to distinguish individuals based on movement. From this data, individuals in a colony were categorized by using Byesian information criterion (BIC) for parameterized Gaussian mixture models fitted by expectation-maximization (EM) algorithm initialized by model-based hierarchical clustering [10]. Number of classes was determined based on a previous study [8], which assumed three classes such as nurse, cleaner, and forager, beforehand. Based on mixed distribution model, they were classified as high, middle, and low-activity ants, based on mean speed. To identify whether mobility patterns in the previous study [8] can be detected by behavioral data base individual classification, individual mobility patterns were visualized using Kernel density estimation method [12] within nest sites. The ant's silhouette was obtained by overlapping tracked image on respective heat maps.

3   Results

3.1   Individual Behavior Under Harsh Environment

AR tracking precision was around 60–80%. Individual speed, heading angle, and distance were compared between normal rearing condition and harsh condition. Each value was compared using median at colony level. Wilcoxon signed rank test was used for statistical analysis throughout the comparison; there were significant differences in speed (Fig. 1a, Wilcoxon signed rank test; median difference $= -34.68$, $V = 4$, $P < 0.05$) and distance (Fig. 1b, Wilcoxon signed rank test; median difference $= -34.68$, $V = 4$, $P < 0.05$). However, there were no significant differences in relative heading angle between red light and strong light conditions (Fig. 1c, Wilcoxon signed rank test; median difference $= 0.375$, $V = 18$, $P = 0.375$). The increases in speed and distance indicate that individuals became more active in response to harsh environment.

The relationship between colony size and individual mean speed per frame within nest sites in a colony were analyzed between the environmental conditions. In ANCOVA, there were no significant differences in mean speed between environments in high activity-level (Fig. 2. ANCOVA, Sum Sq $= 9908$, Df $= 1$, F value $= 1.44$, $P = 0.23$); there were interactions between colony size and environment in middle activity-level (Fig. 2. ANCOVA, $P = 0.01$); and there were significant differences in mean speed between environment in low activity-individual

(Fig. 2. ANCOVA, Sum Sq = 12132, Df = 1, F value = 0.2, P = 0.0001). There-fore, activity-level in high-activity individuals did not differ between environments. Middle-activity level individuals responded differently to the environmental condition with colony size. Therefore, middle-activity level individual became more active in small colonies under harsh environment.

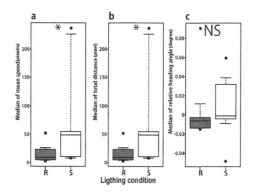

Fig. 1. (a) Median of mean speed under red light condition. Wilcoxon signed rank test: median difference = −34.68458, V = 4, P < 0.05. The circle is outliners. X-axis is lighting condition (R, red filter; S, strong light). Y-axis is median of mean speed of each colony. (b) Median of total distance under red light condition. Wilcoxon signed rank test: median difference = −34.68458, V = 4, P < 0.05. X-axis is lighting condition (R, red filter; S, strong light). Y-axis is median of the mean total distance of each colony. (c) Median of relative heading angle under red light condition. Wilcoxon signed rank test: median difference = 0.375, V = 18, P = 0.375. X-axis is lighting condition (R, red filter; S, strong light). Y-axis is median of the mean relative heading angle of each colony. (Color figure online)

3.2 Individual Classification and Mobility Patterns

Kernel density in individual spatial distribution was estimated (evaluation points = 4000) using the results calculated in individual classification results. Individual trajectories in each class and estimated densities were visualized as heat map as shown in Fig. 3. Mobility patterns seem to contain layers, such as center (around the brood), middle, and outside, based on heat maps. It corresponds to their activity. Low-activity individuals were distributed in the center layer, middle activity individuals were in the middle layer, and high activity individuals were in the outside layer.

4 Discussion

4.1 Individual Behavior Under Harsh Versus Normal Environments

Not only social environmental factors such as interaction or task allocation; but also abiotic environmental factor such as temperature, light, and humidity are

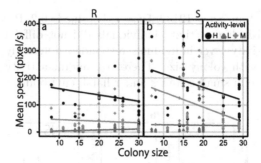

Fig. 2. Colony size versus mean speed under normal condition (R) and harsh condition (S). Colony size is obtained from R condition (normal condition, number of detected tag IDs). The X-axis is colony size. Y-axis is mean speed. H (black circle) is high, M (green cross) is middle and L (red triangle) is low activity-level. The line indicates regression line calculated by least squares method. (a) Normal condition (R) and (b) harsh condition (S) (Color figure online)

some of the most important factors influencing the entire colony function [5,6]. In this study, individual behaviors between normal and harsh condition were compared. Individual activity under harsh environment was higher than that of baseline in middle and low-activity level (Fig. 2). The relationship between colony size and activity shows the same tendency in the nest site as in a previous study, which observed individual on trail [6] in middle activity-level individuals. In middle activity-level, individuals responded differently with colony size under harsh environment. Middle activity-level individuals became more active in smaller colonies than in larger colonies under harsh environment. Under harsh environment, ants tried to find another nest site. High activity-level ants go out of their nest site and try to find a new one. All the ants in a colony move to the new nest site by themselves without carrying behavior. Therefore, middle activity-level individuals, in smaller colonies under harsh environment, are ready to engage in searching behavior to gather information or share information about new nest sites, where migration behavior follows the pheromone trail.

4.2   Individual Classification and Spatial Distribution

To reduce the costs and obtain information on classes, behavioral parameters are useful. Only by using three behavioral parameters such as speed, heading angle and distance, results similar to that of special distribution in Mersch et al. [8] can be reproduced. In the analysis, automatic detection of individual classes may provide a way to categorize individuals while avoiding any arbitrariness during classification or observation of videos. Spatial distribution in each class was visualized using Kernel density estimation. Individuals that mostly stayed around the brood, showed the smallest speed and total working distance among the three classes. Ants distributed around the center individuals showed middle speed and distance. The outlier individuals showed the highest speed and

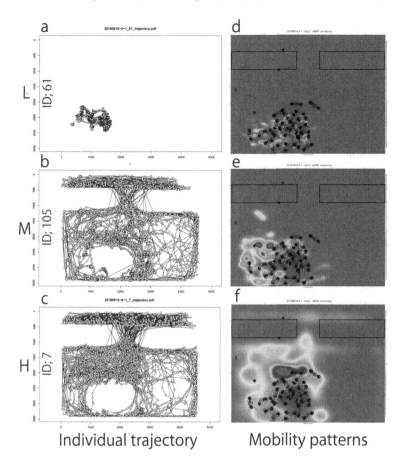

Fig. 3. (a) Class L (center layer) individual trajectory. Colony ID is 20160918-9-1. Individual ID is 61. Circles are obtained positions. Top of the picture is entrance side. Individual classes are indicated as L, low-mean speed; M, middle-mean speed; H, high-mean speed. (b) Class M (middle layer) individual trajectory. Individual ID is 105. (c) Class H (outside layer) individual trajectory. Individual ID is 7. (d) the mobility patterns of center layer (class L). Colony ID is 20160918-9-1. Top of the fig is entrance side. The black line indicates the wall of the nest site. Red color (dark area) indicates low-density and the region around purple color (bright area) is the high-density area. The silhouettes of ants were overlapped. Center layer individuals were distributed around their brood. (e) the mobility patterns of middle layer (class M). Middle layer individuals were distributed around the center layer individuals. (f) Mobility patterns of outside layer (class H). Outside layer individuals were distributed around or overlapped with center or middle layer individuals, working around the nest site and frequently went out of the nest sites so the entrance can seem clearly. (Color figure online)

distance among the classes. Therefore, individuals around the broods can be presumed to be nurses or cleaners, ants around the center individuals can be presumed as cleaner, nurse, scout, or forager and outliner can be presumed as scout, forager, or cleaner. Middle activity-level individuals became more active in small colonies under the harsh environment; therefore, the individuals can be presumed as buffering the number of tasks within a colony by becoming active and engaging in searching behavior as scouts to gather information about new nest sites. Therefore, network structure and information spreading process among individuals can be interesting points of view to reveal how mobility patterns and individual activity-level contribute to the performance of collective behavior in ants.

Acknowledgments. This study was supported by JSPS KAKENHI Grant Number JP18J20064.

References

1. Crall, J.D., Gravish, N., Mountcastle, A.M., Combes, S.A.: BEEtag: a low-cost, image-based tracking system for the study of animal behavior and locomotion. PloS One **10**(9), e0136487 (2015)
2. Cronin, A.L.: Consensus decision making in the ant Myrmecina nipponica: house-hunters combine pheromone trails with quorum responses. Anim. Behav. **84**(5), 1243–1251 (2012)
3. Cronin, A.L.: Conditional use of social and private information guides house-hunting ants. PLoS One **8**(5), e64668 (2013)
4. Cronin, A.L.: Individual and group personalities characterise consensus decision-making in an ant. Ethology **121**(7), 703–713 (2015)
5. Cronin, A.L.: Group size advantages to decision making are environmentally contingent in house-hunting Myrmecina ants. Anim. Behav. **118**, 171–179 (2016)
6. Cronin, A.L., Stumpe, M.C.: Ants work harder during consensus decision-making in small groups. J. Roy. Soc. Interf. **11**(98) (2014)
7. Jeanson, R.: Long-term dynamics in proximity networks in ants. Anim. Behav. **83**(4), 915–923 (2012)
8. Mersch, D.P., Crespi, A., Keller, L.: Tracking individuals shows spatial fidelity is a key regulator of ant social organization. Science **340**(6136), 1090–1093 (2013)
9. Moreau, M., Arrufat, P., Latil, G., Jeanson, R.: Use of radio-tagging to map spatial organization and social interactions in insects. J. Exp. Biol. **214**(1), 17–21 (2011)
10. Scrucca, L., Fop, M., Murphy, T.B., Raftery, A.E.: mclust 5: clustering, classification and density estimation using Gaussian finite mixture models. R J. **8**(1), 289 (2016)
11. Shoji, K.: Interaction networks in ants. Master's thesis. Tokyo Metropolitan University (2018)
12. Venables, W.N., Ripley, B.D.: Modern Applied Statistics with S-PLUS. Springer, New York (2013)

Learning Based Leadership in Swarm Navigation

Ovunc Tuzel, Gilberto Marcon dos Santos, Chloë Fleming[✉],
and Julie A. Adams[✉]

Collaborative Robotics and Intelligent Systems Institute, Oregon State University,
Corvallis, OR, USA
{flemichl,julie.a.adams}@oregonstate.edu

Abstract. Collective migration in biological species is often guided by distributed leaders that modulate their peers' motion behaviors. Distributed leadership is important for artificial swarms, but designing the leaders' controllers is difficult. A swarm control strategy that leverages trained leaders to influence the collective's trajectory in spatial navigation tasks was formulated. The neuro-evolutionary learning based control method was used to train a few leaders to influence motion behaviors. The leadership control strategy is applied to a rally task with varying swarm sizes and leadership percentages. Increasing the leadership representation improved task performance. Leaders moved quickly when the swarm had a higher percentage of leaders and slowly when the percentage was small.

1 Introduction

Biologically-inspired swarm robotic systems exhibit emergent behavior based on local interactions. However, coordinating swarms is challenging. Swarms' distributed, localized communication networks hinder access to global information [7], including navigation goals.

Collective behaviors in fish, birds, and bees suggests that motion coordination can be achieved by a decentralized system without global control or communication mechanisms [13]. Navigation tasks are often facilitated by distributed leaders responding to environmental stimuli [5, 17–20, 26, 29]. The leaders are typically anonymous in large homogeneous collectives [11] and only directly influence individuals within their localized interaction neighborhoods; however, their actions propagate, creating a collective response [8, 34].

Leaders play an important role in biological swarm coordination, but it is unclear how they tune their behaviors to maximize their influence over the swarm's behavior. A neuro-evolutionary learning method to train leaders is developed and evaluated in order to explore leadership mechanisms for artificial swarms. A key contribution is a learning based leadership strategy, where a simulated swarm can be influenced with leadership percentages as low as 4%.

M. Dorigo et al. (Eds.): ANTS 2018, LNCS 11172, pp. 385–394, 2018.
https://doi.org/10.1007/978-3-030-00533-7_33

2   Related Work

Collective navigation is critical for large groups and is often guided by individuals assuming leadership roles. Leadership in biological swarms can be a transient role assumed by any individual [17,21,29]. Fish trained to forage based on environmental features were inserted into a shoal of naive fish and led the shoal to the food source, even though the majority were uninformed [26]. Leadership that emerges based on internal and environmental conditions allows any swarm member to assume a leadership role, which makes them anonymous and the swarm robust to leader loss. The leader percentage in biological swarms is often small: 5% for swarming honeybees [30], and 9% for fish [26].

The mechanisms biological leaders use to guide a swarm are not well-characterized, but they can assume frontal positions [10,24]. Honeybees moving to a new hive have a few fast-flying members [5]. Frontal fish, often faster swimming food deprived individuals, have greater influence on the shoal's direction [19]. However, leaders that move too aggressively tend to leave the swarm members behind, suggesting that leaders must remain aware of their followers [16].

Learning based methods for deriving controllers are often applied to robotic swarms. Attributing global performance to individual agents' behaviors is difficult for large scale systems [33], particularly when the global state is unobservable by the individuals [4]. Several related efforts mitigate this multi-agent credit assignment problem using *team learning* [22] to train the swarm using identical controllers and reward signals for all agents [1–3,14,25].

Neuro-evolutionary team learning methods are common when generating swarm agents' controllers using a fitness function representative of collective behavior. The design paradigm emphasizes simple agent control policies based on locally observable information [7], where directly mapping sensor inputs to control actions is often suitable [2,6,23,31,32]. Several efforts [2,31,32] demonstrated that swarm aggregation tasks can be accomplished with neural network (NN) controllers using this mapping approach. Similar methods successfully generated controllers for more complex tasks (i.e., predator avoidance [28] and collaborative foraging [12]). However, influencing the swarm via distributed leaders is a complex task that has not been investigated with learning based methods.

3   Approach

A swarm of n nonholonomic homogeneous agents, $R = \{r_1, r_2, \cdots, r_n\}$, navigates a continuous 2D environment to perform a rally task, in which leaders, $L \subset R$, know the goal location and are to lead the swarm to the goal. All agents in $S = R - L$ are oblivious to the goal. Each agent controls its velocity $v \in [0, v_{max}]$, where v_{max} is a constant upper limit, and desired heading angle $\psi \in [-\pi, \pi]$.

All agents in S interact based on Reynolds's rules [27]: repulsion (r_{rep}), orientation (r_{ori}), and attraction (r_{att}) that delineate 2D zones around each agent, where $r_{rep} < r_{ori} < r_{att}$. These agents: (1) Veer away from all neighbors within

r_{rep}, (2) Align with neighbors at distances between r_{rep} and r_{ori}, and (3) Move towards neighbors between r_{ori} and r_{att}.

Leaders exert influence by moving among the swarm, using a policy determined by a NN to apply repulsion-orientation-attraction forces on the other agents. All leaders use the same NN, with identical weights and train using team learning.

3.1 Neuro-Evolutionary Learning Method

A set of m two-layer NNs $A = \{a_1, a_2, \cdots, a_m\}$ is initialized with random weights generated by a Gaussian distribution centered at zero, with a standard deviation of 1 ($\mu = 0, \sigma = 1$). The NNs' hidden and output layers consist of units with arc tangent activation functions to provide symmetry, and bounded the output to $[-\pi, \pi]$, consistent with the leader robots' desired heading angle, ψ.

The NN's four sensory inputs represent the polar coordinates of two points, relative to the leader's reference frame. The first pair of inputs are the polar coordinates, distance (d) and heading (θ), between the robot and the goal position, $\langle d_g, \theta_g \rangle$. The second pair of inputs are the polar coordinates from the robot to the centroid of the swarm within the leader's perceptual range, $\langle d_s, \theta_s \rangle$. All possible inputs to the NN are defined by the input vector $NN_x = \langle d_g, \theta_g, d_s, \theta_s \rangle$. The NN's outputs are the leader's desired velocity and heading $NN_o = \langle v, \psi \rangle$.

Each epoch simulates a rally task for all m NNs in the population A. An episode loads a NN a_i into the leaders L, positions the leaders and non-leaders S randomly within a starting location, and simulates a fixed number of steps τ. The NN's performance is evaluated using the cost function E (Eq. 1) upon episode completion. All m NNs are evaluated and the top-performing λ networks, called parents, are retained. A new set of $m - \lambda$ NNs are generated by randomly sampling (with replacement) from the λ parents. These $m - \lambda$ NNs are mutated by applying zero-mean Gaussian noise with a fixed standard deviation NN_{mut} ($\mu = 0, \sigma = NN_{mut}$) to every NN weight. The mutated NNs are incorporated into the evolutionary population; thus, returning the population size back to m.

At each simulation step t, the temporal factor, $\frac{t}{\tau}$ in $[0, 1]$, represents time progress over the episode's total steps τ. The temporal weight w_t is a function of the temporal factor, $w_t = 1 - cos(\pi \cdot \frac{t}{\tau})$, over its valid input range, $\frac{t}{\tau} \in [0, 1]$. The area under the curve, $w_t(\frac{t}{\tau})$, is 1.

The Euclidean distance between each non-leader agent i and the goal at step t, d_t^i, is evaluated and weighted by the temporal weight. The weighted accumulated distance is summed over every step, providing an average weighted distance $d_{avg}^i = \frac{1}{\tau} \cdot \sum_{t=0}^{\tau} d_t^i \cdot w_t$ representing the cost associated with agent i for the entire episode. The temporal weighting increases the influence of agents' deviations from the goal late in the episode. The weighting rewards NNs that consistently converge towards the goal, rather than those that initially drive the swarm towards the goal, but later disperse or lose control of the swarm.

The average accumulated weighted distance d_{avg}^i is averaged across all agents in S at the end of each episode, defining the NN fitness function:

$$E = \frac{1}{|S|} \cdot \sum_{i \in S} d_{avg}^i. \tag{1}$$

4 Experimental Design and Results

The primary research question is whether a small percentage of leaders using the neuro-evolutionary learning algorithm can influence a robot swarm to significantly outperform a baseline model that does not incorporate learning.

4.1 Experimental Design

The independent variables are the leadership model, swarm size and the leadership percentage. The leadership model is either the learning based model described in Sect. 3, or a baseline model. Baseline leaders do not learn and always align their heading towards the goal. Swarms of 50 and 100 agents were evaluated with leadership percentages ranging from 4% to 24%, as shown in Table 1. These percentages reflect observations on leadership in biological swarms [11, 26, 30].

The swarm begins each rally trial gathered at a starting point d_{init} distance units (u) from, and at a random angle to the goal. $d_{init} = 400$ distance units (u) to minimize locating the goal by chance, while also completing the trial within a reasonable number of time steps. A zero-mean Gaussian noise with variance σ_{init} was added to the starting positions of each agent, in order to avoid collisions. Swarm agents are initialized with uniform random orientations, and the swarm's starting speed is set to v_{init}, which is 50% of the swarm's maximum speed, v_{max}. This stochasticity encourages the learning of generalized behaviors. The r_{rep}, r_{ori}, and r_{att} radii govern the non-leader agents' motion, and were selected to be 20 u, 30 u, and 50 u, respectively, based on biological swarms [15]. The total number of NNs, m, and the number of parents, λ, were set to be 15 and 5, respectively. The training session lasted 400 epochs, ensuring convergence of all training errors.

The percentage of non-leader agents within a radius of the goal location, r_{goal}, is calculated at trial completion, and averaged over all trials to calculate the percent reached (PR). The test error (E) represents the accumulated distance to the goal, and is calculated using Eq. 1.

Leaders were trained using all independent variable configurations. After 400 training epochs, each NN was evaluated over 100 trials without any mutations, and the NN with the minimum root-mean-squared error deemed the *champion*. The process was repeated 10 times, resulting in 10 champions. The champion NNs' performance metrics are reported in all results.

Table 1. Experimental parameters and independent variables.

Parameters	Values	Parameters	Values
Swarm size	50, 100	r_{rep}	20 u
Leader percentage	4%, 8%, 12%, 16%, 20%, 24%	r_{ori}	30 u
τ	20000 steps	r_{atr}	50 u
v_{init}	1 u/step	r_{goal}	150 u
σ_{init}	50	v_{max}	2 u/step
λ	5	m	15

4.2   Results

The performance, percent reached (PR), improved with increasing leadership percentage, as shown in Table 2 and Fig. 1. The learning based agents successfully guided the swarm, with both 50 and 100 agents, even with the 4% leadership. However, the baseline leaders generally failed to lead any swarm members to the goal with the 4% and 8% leadership. The baseline model matched or exceeded the learning model when leaders composed 20% and 24% of the swarm, but pairwise T-tests (degrees of freedom [dof] $= 999$ in all tests) found no significant differences between the models. The learning based method significantly outperformed the baseline for all other cases ($p < 0.01$). The PR was generally better with a swarm size of 50 and pairwise T-tests comparing PR by swarm size found significant differences only at the 4% and 8% leadership percentages ($p < 0.01$).

Table 2. The percent reached (PR) descriptive statistics (mean - (μ), standard error - SE) by swarm size, leadership percentage, and leadership model.

Leader %	50 agents				100 agents			
	Baseline		Learning		Baseline		Learning	
	μ	SE	μ	SE	μ	SE	μ	SE
4%	0.04	0.06	32.01	2.75	0.01	0.02	20.31	2.38
8%	0.02	0.04	54.24	3.01	0.01	0.02	40.33	2.94
12%	21.21	8.01	69.73	2.81	5.05	4.29	69.52	2.79
16%	64.67	9.36	84.77	2.21	43.43	9.71	86.34	2.08
20%	89.92	5.89	87.48	2.02	75.76	8.40	90.00	1.82
24%	97.98	2.76	93.71	1.49	93.94	4.68	85.42	2.14

The test error (E) results, presented in Table 3, were grouped into bins (size $= 10$), as shown in Fig. 2. The E for a majority of trials ($>80\%$) was less than 400, and trials with $E \geq 400$ were deemed unsuccessful, and are grouped into the final bin. E is impacted by the time required to reach the goal, but there is

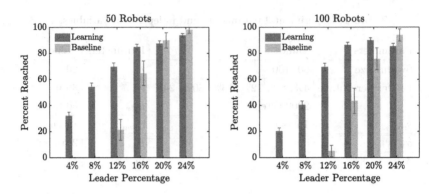

Fig. 1. Percent reached by leadership model, leadership percentage and swarm size.

no effect on the PR if the agents reach the goal by trial completion. Thus, slow moving swarms have higher Es. The PR metric suggests that agents occasionally reach the goal area, even with 4% and 8% leadership, but with higher minimum errors than swarms with higher leadership percentages. The leaders move slower when their percentage is low, and faster when their percentage is higher.

Table 3. The test error (E) descriptive statistics by leadership model, swarm size and leadership percentage ($L\%$). Median is reported due to a large number of outliers.

L%	50 agents						100 agents					
	Baseline			Learning			Baseline			Learning		
	Med	Min	Max	Med	Min	Max	Med	Min	Max	Med	Min	Max
4%	308	261	1033	193	67	20992	313	262	993	376	76	19742
8%	302	255	923	140	38	7361	307	274	766	197	43	20247
12%	282	40	320	81	22	6268	304	42	324	83	32	6236
16%	38	36	313	31	20	6540	271	37	321	30	21	6528
20%	36	35	307	29	19	7721	37	36	317	23	20	2681
24%	35	34	317	21	18	9092	35	34	308	25	19	8893

Median and minimum Es of the learning model were lower than the baseline with leadership percentages higher than 16%, despite the PR not being statistically significant, suggesting that learning agents move faster than the baseline agents. Trials where the fast moving learning leaders fail to guide the swarm explain the high maximum E, as the swarms travel farther away from the goal. The median and minimum Es decreased with increasing leadership percentage. The largest reduction in the median occurred when the percentage increased from 4% to 12%. The change in the median and minimum Es was minimal past the 16% leadership percentage.

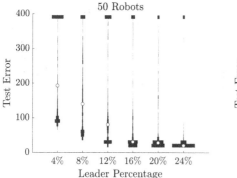
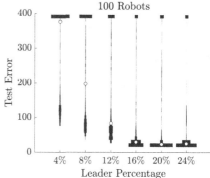

Fig. 2. Learning model test errors by leadership percentage and swarm size. Errors are packed into bins of size 10, and errors greater than 400 are grouped together into the top bin (400). The circles represent the median error.

5   Discussion

The learning model resulted in leaders that successfully influenced the swarm to achieve the rally task with very small leadership percentages, which validates the proposed learning based controller and answers the primary research question. Generally, the learning model leaders outperformed the baseline model and were able to perform significantly better at the lowest leadership percentages. While swarms led by small sets of leaders took longer and were less likely to reach the goal, they were able to achieve the task at leadership percentages representative of biological species, which can be as low as 5% [30] and 9% [26].

Lower leadership percentages (i.e., 4%) require the leaders to learn more nuanced behaviors in order to be effective, which explains the slow convergence of the training errors relative to higher percentages. The leaders' movements are fast, aggressive and goal driven when their influence over the swarm is high, and resemble the baseline model. Thus, the baseline method is only viable if the leader influence is guaranteed to be high throughout the duration of the task.

The biological literature demonstrates that leaders must balance goal-oriented actions with socially-oriented ones in order to be effective [16]. Leaders following the learning model act based on both the goal and their followers, while the baseline leaders are indifferent to their followers. The significant performance differences between the two models emphasize the importance of spatial awareness, and confirms that the learning based model successfully combines goal-oriented and socially-oriented actions.

Biological swarms rely only on local interactions, and typically use an implicit leadership mechanism [9]. The learning based strategy draws inspiration from biological swarms in that it is based on implicit communication and local decision making. Further, no agent knows whether another agent is a leader or not.

The leadership percentage strongly affects the characteristics of the learned behaviors. The learned behaviors are more aggressive, and the leaders travel straight towards the goal when their overall influence is higher. However, the leaders follow more complex movement patterns when the leadership percentage is low. Leaders must be aware of their followers when the leaders' influence is low, and use these complex movement patterns in order to ensure they are being followed, otherwise the leaders lose track of the swarm. The proposed neuro-controller solves this problem by integrating both the follower positions and the goal position into the decision making process, which enables reliable swarm control with only a small percentage of informed leaders.

6 Conclusion

A learning based leadership strategy was developed that allowed small percentages of leaders to drastically improve its task performance over a baseline model. The leadership model incorporates implicit leadership and communication, which allows any agent to assume a leadership role at any given time. While a higher leadership percentage improved task performance, the increase was minimal with percentages >16%. The task was successful with leadership percentages as low as 4%, but the consistency of success increased with higher percentages.

Acknowledgments. This work was partially supported by NSF Grant #1723924 and DARPA award W31P4Q18C0034.

References

1. Ampatzis, C., Tuci, E., Trianni, V., Dorigo, M.: Evolution of signaling in a multi-robot system: categorization and communication. Adapt. Behav. **16**(1), 5–26 (2008)
2. Baldassarre, G., Nolfi, S., Parisi, D.: Evolving mobile robots able to display collective behaviors. Artif. Life **9**(3), 255–267 (2003)
3. Baldassarre, G., Trianni, V., Bonani, M., Mondada, F., Dorigo, M., Nolfi, S.: Self-organized coordinated motion in groups of physically connected robots. IEEE Trans. Syst. Man Cybern. Part B (Cybern.) **37**(1), 224–239 (2007)
4. Barca, J.C., Sekercioglu, Y.A.: Swarm robotics reviewed. Robotica **31**(3), 345–359 (2013)
5. Beekman, M., Fathke, R.L., Seeley, T.D.: How does an informed minority of scouts guide a honeybee swarm as it flies to its new home? Anim. Behav. **71**(1), 161–171 (2006)
6. Beer, R.D., Gallagher, J.C.: Evolving dynamical neural networks for adaptive behavior. Adapt. Behav. **1**(1), 91–122 (1992)
7. Brambilla, M., Ferrante, E., Birattari, M., Dorigo, M.: Swarm robotics: a review from the swarm engineering perspective. Swarm Intell. **7**(1), 1–41 (2013)
8. Cavagna, A., et al.: Scale-free correlations in starling flocks. Proc. Natl. Acad. Sci. **107**(26), 11865–11870 (2010)
9. Colby, M.K., Chung, J.J., Tumer, K.: Implicit adaptive multi-robot coordination in dynamic environments. In: IEEE/RSJ International Conference on Intelligent Robots and Systems, pp. 5168–5173 (2015)

10. Couzin, I.D., Krause, J.: Collective memory and spatial sorting in animal groups. Theoret. Biol. **218**, 1–11 (2002)
11. Couzin, I.D., Krause, J., Franks, N.R., Levin, S.A.: Effective leadership and decision-making in animal groups on the move. Nature **433**(7025), 513–516 (2005)
12. Pérez, I.F., Boumaza, A., Charpillet, F.: Learning collaborative foraging in a swarm of robots using embodied evolution. In: The European Conference on Artificial Life, pp. 162–161 (2017)
13. Garnier, S., Gautrais, J., Theraulaz, G.: The biological principles of swarm intelligence. Swarm Intell. **1**(1), 3–31 (2007)
14. Gross, R., Dorigo, M.: Towards group transport by swarms of robots. Int. J. Bio-Inspir. Comput. **1**(1–2), 1–13 (2009)
15. Haque, M., Ren, C., Baker, E., Kirkpatrick, D., Adams, J.A.: Analysis of swarm communication models. In: International Workshop on Combinations of Intelligent Methods and Applications, p. 29 (2016)
16. Ioannou, C.C., Singh, M., Couzin, I.D.: Potential leaders trade off goal-oriented and socially oriented behavior in mobile animal groups. Am. Nat. **186**(2), 284–293 (2015)
17. Kilgour, R., Scott, T.: Leadership in a herd of dairy cows. Proc. N. Z. Soc. Anim. Prod. **19**, 36–43 (1959)
18. King, A.J., Johnson, D.D., Van Vugt, M.: The origins and evolution of leadership. Curr. Biol. **19**(19), R911–R916 (2009)
19. Krause, J., Hoare, D., Krause, S., Hemelrijk, C., Rubenstein, D.: Leadership in fish shoals. Fish Fish. **1**(1), 82–89 (2000)
20. Leca, J.B., Gunst, N., Thierry, B., Petit, O.: Distributed leadership in semifree-ranging white-faced capuchin monkeys. Anim. Behav. **66**(6), 1045–1052 (2003)
21. Meese, G., Ewbank, R.: Exploratory behaviour and leadership in the domesticated pig. Br. Vet. J. **129**(3), 251–259 (1973)
22. Panait, L., Luke, S.: Cooperative multi-agent learning: the state of the art. Auton. Agents Multi-Agent Syst. **11**(3), 387–434 (2005)
23. Pini, G., Tuci, E.: On the design of neuro-controllers for individual and social learning behaviour in autonomous robots: an evolutionary approach. Connect. Sci. **20**(2–3), 211–230 (2008)
24. Portugal, S.J., et al.: Upwash exploitation and downwash avoidance by flap phasing in ibis formation flight. Nature **505**(7483), 399–402 (2014)
25. Pugh, J., Martinoli, A.: Parallel learning in heterogeneous multi-robot swarms. In: IEEE Congress on Evolutionary Computation, pp. 3839–3846 (2007)
26. Reebs, S.G.: Can a minority of informed leaders determine the foraging movements of a fish shoal? Anim. Behav. **59**(2), 403–409 (2000)
27. Reynolds, C.W.: Flocks, herds and schools: a distributed behavioral model. Comput. Graph. **21**(4), 25–34 (1987)
28. Ripon, K.S.N., Jakobsen, E., Tannum, C., Montanier, J.M.: Assessing the effect of self-assembly ports in evolutionary swarm robotics. In: IEEE Symposium Series on Computational Intelligence, pp. 1–8 (2016)
29. Sato, S.: Leadership during actual grazing in a small herd of cattle. Appl. Anim. Ethol. **8**(1–2), 53–65 (1982)
30. Seeley, T.D.: The Wisdom of the Hive: The Social Physiology of Honey Bee Colonies. Harvard University Press, Cambridge (2009)
31. Soysal, O., Bahçeci, E., Şahİn, E.: Aggregation in swarm robotic systems: evolution and probabilistic control. Turk. J. Electr. Eng. Comput. Sci. **15**(2), 199–225 (2007)

32. Trianni, V., Groß, R., Labella, T.H., Şahin, E., Dorigo, M.: Evolving aggregation behaviors in a swarm of robots. In: Banzhaf, W., Ziegler, J., Christaller, T., Dittrich, P., Kim, J.T. (eds.) ECAL 2003. LNCS, vol. 2801, pp. 865–874. Springer, Heidelberg (2003). https://doi.org/10.1007/978-3-540-39432-7_93
33. Wolpert, D.H., Tumer, K.: An introduction to collective intelligence. arXiv preprint arXiv:cs/9908014 (1999)
34. Xu, X.K., Kattas, G.D., Small, M.: Reciprocal relationships in collective flights of homing pigeons. Phys. Rev. E **85**(2), 026120 (2012)

Maintaining Diversity in Robot Swarms with Distributed Embodied Evolution

Iñaki Fernández Pérez[1,2(✉)], Amine Boumaza[2], and François Charpillet[3]

[1] University of Toulouse, IRIT, UMR 5505, Toulouse, France
inaki.fernandez-perez@irit.fr
[2] Université de Lorraine, LORIA, Nancy, France
[3] Inria Nancy Grand-Est, Villers-lès-Nancy, France

Abstract. In this paper, we investigate how behavioral diversity can be maintained in evolving robot swarms by using distributed Embodied Evolution. In these approaches, each robot in the swarm runs a separate evolutionary algorithm, and populations on each robot are built through local communication when robots meet; therefore, genome survival results not only from fitness-based selection but also from spatial spread. To better understand how diversity is maintained in distributed EE, we propose a postanalysis diversity measure, that we take from two perspectives, global diversity (over the swarm), and local diversity (on each robot), on two swarm robotic tasks (navigation and item collection), with different intensities of selection pressure, and compare the results of distributed EE to a centralized case. We conclude that distributed evolution intrinsically maintains a larger behavioral diversity when compared to centralized evolution, which allows for the search algorithm to reach higher performances, especially in the more challenging collection task.

1 Introduction

Diversity in an evolving population, as a measure of how different its individuals are, is crucial for effective evolutionary adaptation. In artificial evolution and evolutionary robotics, diversity has been investigated either to analyze the dynamics of the evolutionary process, or to explicitly promote the search for diverse or novel individuals [12,17]. An adequate level of diversity through evolution allows to better search, balancing between exploration, to find promising areas, and exploitation, to refine good solutions. This is even more necessary when the search space is deceptive, *i.e.* it is rugged, with valleys and many local optima, which corresponds to difficult optimization problems. A very active research topic in Evolutionary Computation concerns the explicit promotion of diversity, where diversity measures are used as an auxiliary objective to be maximized: searching for diverse solutions to the problem [5,12]. Diversity measures can also be used to monitor and analyze the evolutionary process, better understand its dynamics, and trigger specific events depending on the diversity in the population (*e.g.* restarting an evolutionary process to enhance exploration, or stop evolution when the diversity gets too low). Typically, work on diversity

© Springer Nature Switzerland AG 2018
M. Dorigo et al. (Eds.): ANTS 2018, LNCS 11172, pp. 395–402, 2018.
https://doi.org/10.1007/978-3-030-00533-7_34

in evolutionary robotics is restricted to evolving single-robot behaviors with a centralized evolutionary algorithm. The work by Gomes [10] is an exception, where the authors evolve behaviors for multirobot and swarm robotic systems using a novelty-based centralized algorithm. On the other hand, in distributed Embodied Evolution (dEE), [1,18] robots in a swarm locally communicate with each other to build their respective local populations. This entails different evolutionary dynamics to the global process, compared to centralized algorithms, due to local interactions between robots. Here, we analyze the influence on the diversity of the evolved behaviors of the distributed nature of dEE algorithms and the intensity of local selection pressure. Our experiments aim at answering the following questions: (a) does distributed Embodied Evolution for robot swarms intrinsically maintains more diversity than centralized evolution?, and (b) does local selection pressure influence diversity in distributed EE as it does in centralized algorithms? We first describe related work on dEE, and approaches to measure diversity in single and multirobot systems. Then, we describe the distributed EE algorithm used in our experiments, and our proposed generic diversity metric, that we compute at two levels, *i.e.* *global* (over the swarm) and *local* (on each local population). Finally, we detail our experiments, discuss the results, conclude and provide further research questions.

2   Related Work

A particularity of dEE is that selection is decentralized, with each robot of the swarm selecting over its local population, which is progressively built over the evaluation of controllers: robots exchange their active controllers and their respective fitness value when meeting. As such, local populations on different robots are different, and selection pressure applied over such subpopulations has different dynamics as compared to more classical centralized EAs. In [2], the authors investigate the influence of the environment on the behaviors evolved by mEDEA, a dEE algorithm that does not use a fitness measure to perform selection: selection is performed at random inside the local population of each robot. As such, the algorithm does not apply any task-driven selection pressure: it is rather the environmental selection pressure to reproduce and spread their genes that pushes evolution toward behaviors adapted to the environment that maximize the opportunities to meet other robots and mate. In [6], the authors evaluate the impact on the performance of the swarm of the intensity of selection pressure of the local selection operator in a dEE algorithm. The authors evolve neurocontrollers in a swarm of robots using different intensities of selection pressure, and conclude that the higher the selection pressure, the higher the performance, as opposed to classical centralized evolutionary algorithms, in which a lower intensity of selection pressure is usually preferred to maintain diversity in the population. This could indicate that distributed EE algorithms maintain such a diversity, necessary for the search to escape local minima.

Measuring diversity has been a topic of interest in the literature, and typically aims at two non-exclusive goals: understanding the dynamics of an evolutionary algorithm (diversity *analysis, e.g.* [13]), and reinjecting diversity measures into the EA, *e.g.* for diversity *promotion* (*e.g.* Novelty Search [12]), to evolve a diverse set of individuals (*e.g.* Quality-Diversity algorithms [16]), to restart the algorithm [8], or to maintain a population able to adapt to unforeseen changes [14]. Generally, when investigating diversity in Evolutionary Robotics it is measured based on *behaviors*, instead of genotypic or phenotypic diversity. A behavioral descriptor must be defined (task-specific or task-agnostic, *i.e.* generic, based on sensorimotor values) to capture adequate features of the behavior resulting from a controller. These are then used by distance functions to compute diversity metrics. In [4], the authors propose four different behavioral diversity measures as a auxiliary objectives to evolve single-robot behaviors, which help circumvent the deceptiveness of the chosen task. In [9], the authors propose two diversity measures specifically designed for swarms of robots by capturing features of the joint behavior of a swarm, instead of features of single-robot behaviors. In their paper, the authors use these measures as novelty objective, linearized with fitness values into a single objective, for a centralized novelty-based EA to evolve diverse behaviors for robot swarms. In this paper, we measure behavioral diversity as a postanalysis measure to provide insights on the internal dynamics of distributed evolution. Specifically, we propose a generic behavioral diversity metric for distributed Embodied Evolution, taken at two levels (*global*, over the swarm, and *local* diversity, on the local population of each robot). While the algorithm on each robot can only rely on local information, since the diversity measures are not used by the robots, but used to analyze how diverse the behaviors are, this does not contradict the decentralized nature of the approach. Since we focus on characterizing diversity between individual robot behaviors, either among local populations or in the swarm, and not joint swarm behaviors, we chose to use mono-robot behavioral diversity measures, closer to [4], instead of basing our study on the diversity measures for swarm robotics in [9].

3   Methods and Experiments

The algorithm used in our experiments corresponds mEDEA with task-driven selection pressure [2, 6]. Each robot in the swarm runs an independent instance of the algorithm. At every moment, a robot carries an active genome corresponding to its current neurocontroller, which is randomly initialized at the beginning of each experiment. A robot executes its controller for some time T_e, while estimating its fitness and continuously broadcasting the active genome and its current fitness estimate to other nearby robots (and vice versa). Once T_e timesteps are elapsed, the robot stops and selects a parent genome using a given selection operator. The selected genome is mutated and replaces the active genome (no crossover is used), the local population l is emptied, and a new generation begins. We designed a parameterized tournament selection operator, that, given a parameter $\theta_{sp} \in [0, 1]$ and a local population, selects the genome with the best fitness in a random θ_{sp} fraction of the population. The parameter θsp influences selection pressure by determining the actual tournament size, and the higher the tournament size, the stronger the selection pressure. If $\theta_{sp} = 0$, the fitness is disregarded and selection is random, while if $\theta_{sp} = 1$, the best genome in the population is selected (maximal selection pressure). Each experiment consists in running this algorithm for a given task, with a given θ_{sp}, and either with selection operating on local populations (*distributed*), or on the global one (*centralized*), *i.e.* the set of all active genomes in the swarm. At each generation, in addition to measuring the swarm's average fitness, we measure behavioral diversity using our proposed metric of dispersion among a set of behaviors $\mathbf{b}$:

$$Div(\mathbf{b}) = \frac{2}{|\mathbf{b}| \cdot (|\mathbf{b}| - 1)} \sum_{i=0}^{|\mathbf{b}-1|} \sum_{j=i+1}^{|\mathbf{b}|} d(b_i, b_j), \tag{1}$$

where $\mathbf{b}$ is a set of behavioral descriptors b_i, and $d(\cdot, \cdot)$ is a distance function between two behavioral descriptors. We aim at defining a diversity measure as generic as possible while still capturing differences in functional features of the corresponding neurocontrollers. In our approach, a behavioral descriptor for a given robot controller is defined as the list of motor outputs corresponding to an input dataset I, sampled at the beginning of each run, $I = [in^1, in^2, \ldots, in^N]$. Each in^k is a random vector of the size of the inputs of the controllers, uniformly sampled in the corresponding value range. To compute the behavioral descriptor of a controller c_i, the entries in the input dataset are fed to the controller, and the corresponding outputs $(o_i^k = c_i(in^k)$ are recorded, serving as the behavioral descriptor for c_i, *i.e.* $b_i = [o_i^1, o_i^2, \ldots, o_i^N]$. The distance between two behaviors, b_i and b_j, is computed as the average Euclidian distance between all their paired elements from b_i and b_j. In other words, the distance measures how different are the motor outputs computed by two neurocontrollers when confronted with the same set of inputs, and the global diversity, $Div(\cdot)$, is then computed as the average functional distance between each pair of behaviors in $\mathbf{b}$. We use our proposed diversity metric to evaluate at each generation how diverse are the behaviors at the *global* level of the swarm ($Div(\mathbf{b}^g_{swarm})$, where $\mathbf{b}^g_{swarm}$ is the set of behavioral descriptors of the active robot controllers in the swarm at generation g), and at the *local* level of the local populations (for each robot r, $Div(\mathbf{b}^g_r)$, where $\mathbf{b}^g_r$ is the set of behavioral descriptors of the local population of r at generation g; we report the average over the swarm).

We measure the fitness and behavioral diversity over time when a swarm of robots uses this algorithm to adapt to two classical benchmark tasks for swarm robotics: navigation and item collection. For each task, we perform 10 variants, with 5 levels of selection pressure, $\theta_{\mathrm{sp}} \in \{0, 0.25, 0.5, 0.75, 1\}$, with either robots locally exchanging genomes (distributed), or selecting on the global population (centralized). The experiments with selection on the global population do not comply with the distributed nature of swarm systems, and are used as control experiments to test if dEE intrinsically maintain more diversity than when selection is performed on the global population. In each experiment, a swarm of robotic agents is deployed in a simulated environment (Fig. 1), containing food items in the collection task. Our experiments are run using the RoboRobo simulator [3], which is a fast simulator for collective robotics. For the navigation task,

Fig. 1. Simulated environment: enclosed square arena containing a swarm of robots and items (black and blue circles). (Color figure online)

Table 1. T_e and σ are the evaluation time and std. dev. of the Gaussian mutation.

# Robots	80
# Items	80
Envir. size	$1000 \times 1000px$
Sensor range	$30px$
# runs	30
Generations	200
T_e	800 *steps*
σ	0.1

each robot has 8 proximity sensors evenly spaced around the robot, which detect walls and other robots, with 8 additional item proximity sensors in the collection task. Each robot is controlled by a fully-connected perceptron with a bias neuron and no hidden layers, and maps sensory inputs to motor outputs (left and right wheel speed). The genome corresponds to a real-valued vector containing the weights of the controller (18 for navigation, and 34 for collection), adapted by either the distributed algorithm, or the centralized version (Table 1).

The fitness for navigation rewards moving fast, straight and avoiding obstacles [15], while in item collection it is the number of items collected by a robot. To evaluate the impact of distributed evolution on swarm performance and diversity, at every generation of each experiment, we measure the swarm fitness (average fitness over all the robots), and the global and local diversity. We compare the results (swarm fitness and diversity) of distributed evolution to centralized evolution, and the impact of the intensity of selection pressure in both cases.

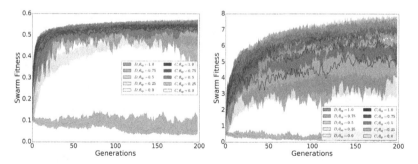

Fig. 2. Swarm fitness over generations for navigation (left) and item collection (right). Blue curves represent centralized evolution (C), while orange curves represent distributed evolution (D). θ_{sp} is the intensity of selection pressure. (Color figure online)

4   Results and Conclusion

To compare diversity (either global or local) between centralized and distributed evolution, we use 2D histograms represented as heatmaps, where the x-axis and the y-axis correspond to the diversity in the distributed variant and in the centralized variant, respectively. Each datapoint is then the pair of diversity values corresponding to the same generation g in a distributed and a centralized run (randomly paired), i.e. (Div_D^g, Div_C^g) for each pair of runs. The density of each bin in the histogram corresponds to the number of generations across all the runs when the pair of diversity values from the distributed variant and the centralized falls into that bin. If a plot is denser under the diagonal, it means that, overall, distributed evolution maintains more diversity, and vice versa. When comparisons are made between swarm fitness values, difference is reported *iff* Mann-Whitney tests yield $p < 0.05$. Figure 2 (resp. Figure 3) show the fitness of the swarm over generations for the navigation and the collection task (resp. the global and local behavioral diversity heatmaps). In both tasks, robots adapt solve the task, reaching high fitness in all the experiments except for the centralized experiment with $\theta_{sp} = 0.0$, which corresponds to random search in the entire population. The distributed variants with $\theta_{sp} \neq 0.0$ reach slightly higher values with lower variance than the centralized variants, especially in the more challenging collection task. Regarding item collection, the intensity of selection pressure seems to have little impact on the fitness in the distributed case, while in the centralized case, the highest performance is obtained when $\theta_{sp} = 0.25$ or $\theta_{sp} = 0.5$. On the other hand, when $\theta_{sp} = 0.75$, and especially when $\theta_{sp} = 1$, the swarm fitness is lower. This could be due to a possible loss of diversity when selection pressure is strong in the centralized case. Search could stagnate in local minima, being unable to escape, and thus yielding lower fitness, especially since item collection is arguably more difficult to evolve than navigation: the search space is bigger, and information from sensors of different nature needs to be integrated. In the case of distributed evolution with $\theta_{sp} = 0.0$, which corresponds to mEDEA algorithm, there is also an improvement, although slower, even in the absence of task-driven selection pressure. This is due to environmental selection pressure pushing toward behaviors that maximize mating chances by navigating the environment, and collecting items by chance in the item collection task. Figure 3 show that, when there is selection pressure ($\theta_{sp} \neq 0.0$), distributed evolution maintains more diversity, both local and global (denser areas under the diagonal). In the case of $\theta_{sp} = 0.0$, centralized evolution yields higher diversity than distributed evolution: the centralized case corresponds to random search, and, even if a diversity of behaviors is maintained, those behaviors do not provide any fitness, as shown before.

In this paper, our main hypothesis is that such algorithms intrinsically maintain diversity, since the genomes on the local repositories of the robots are built through local exchanges between robots when meeting, and are therefore different. To test such a hypothesis, we perform a set of experiments where a swarm of robots adapts to given tasks using a distributed EE algorithm. We test 5 intensities of selection pressure, in the distributed algorithm and in a control experiment

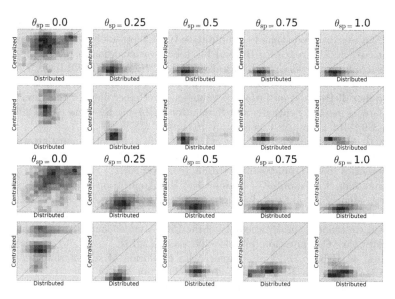

Fig. 3. Heatmap for comparing global and local diversity between centralized and distributed experiments in navigation (top 2 rows) and collection (bottom 2 rows).

with selection on the global population. We measure both the performance on the tasks and a proposed diversity measure designed for distributed evolution in robot swarms, both from local and global perspectives, and we conclude that, when there is selection pressure in our experiments, this approach systematically maintains more diversity, compared to centralized evolution, allowing to reach slightly higher performances, especially in the item collection task. This work opens questions on how to exploit such diversity measures: they could help regulating evolutionary operators, including the *mating* operator that defines genome migration between robots in distributed evolution [1]: mating could be restricted to robots with similar behaviors, a form of reproductive isolation, which might favor the evolution of specialized niches. On the other hand, diversity measures could be used as novelty objectives. Searching for novelty in distributed evolution has recently received attention [7,11], and we believe that our proposed diversity measures could be used to guide search in robot swarms.

References

1. Bredèche, N., Haasdijk, E., Prieto, A.: Embodied evolution in collective robotics: a review. Front. Robot. AI **5**, 12 (2018)
2. Bredèche, N., Montanier, J.-M.: Environment-driven embodied evolution in a population of autonomous agents. In: Schaefer, R., Cotta, C., Kołodziej, J., Rudolph, G. (eds.) PPSN 2010. LNCS, vol. 6239, pp. 290–299. Springer, Heidelberg (2010). https://doi.org/10.1007/978-3-642-15871-1_30
3. Bredèche, N., Montanier, J.M., Weel, B., Haasdijk, E.: Roborobo! a fast robot simulator for swarm and collective robotics. CoRR abs/1304.2888 (2013)

4. Doncieux, S., Mouret, J.B.: Behavioral diversity measures for evolutionary robotics. In: Congress on Evolutionary Computation (CEC), pp. 1303–1310. Espagne (2010)

5. Doncieux, S., Mouret, J.B.: Beyond black-box optimization: a review of selective pressures for evolutionary robotics. Evol. Intell. **7**(2), 71–93 (2014)

6. Fernández Pérez, I., Boumaza, A., Charpillet, F.: Comparison of selection methods in on-line distributed evolutionary robotics. In: Proceedings of the International Conference on the Synthesis and Simulation of Living Systems (Alife 2014), pp. 282–289. MIT Press, New York, July 2014

7. Galassi, M., Capodieci, N., Cabri, G., Leonardi, L.: Evolutionary strategies for novelty-based online neuroevolution in swarm robotics. In: Systems, Man, and Cybernetics (SMC), pp. 002026–002032. IEEE (2016)

8. Ghannadian, F., Alford, C., Shonkwiler, R.: Application of random restart to genetic algorithms. Inf. Sci. **95**(1–2), 81–102 (1996)

9. Gomes, J., Christensen, A.L.: Generic behaviour similarity measures for evolutionary swarm robotics. In: Proceedings of the 15th Annual Conference on Genetic and Evolutionary Computation, pp. 199–206. ACM (2013)

10. Gomes, J., Urbano, P., Christensen, A.L.: Evolution of swarm robotics systems with novelty search. Swarm Intell. **7**(2–3), 115–144 (2013)

11. Hart, E., Steyven, A.S., Paechter, B.: Evolution of a functionally diverse swarm via a novel decentralised quality-diversity algorithm (2018)

12. Lehman, J., Stanley, K.O.: Abandoning objectives: evolution through the search for novelty alone. Evol. Comput. **19**(2), 189–223 (2011)

13. Morrison, R.W., De Jong, K.A.: Measurement of population diversity. In: Collet, P., Fonlupt, C., Hao, J.-K., Lutton, E., Schoenauer, M. (eds.) EA 2001. LNCS, vol. 2310, pp. 31–41. Springer, Heidelberg (2002). https://doi.org/10.1007/3-540-46033-0_3

14. Nguyen, T.T., Yang, S., Branke, J.: Evolutionary dynamic optimization: a survey of the state of the art. Swarm Evol. Comput. **6**, 1–24 (2012)

15. Nolfi, S., Floreano, D.: Evolutionary Robotics. MIT Press, Cambridge (2000)

16. Pugh, J.K., Soros, L.B., Stanley, K.O.: Quality diversity: a new frontier for evolutionary computation. Front. Robot. AI **3**, 40 (2016)

17. Ursem, R.K.: Diversity-guided evolutionary algorithms. In: Guervós, J.J.M., Adamidis, P., Beyer, H.-G., Schwefel, H.-P., Fernández-Villacañas, J.-L. (eds.) PPSN 2002. LNCS, vol. 2439, pp. 462–471. Springer, Heidelberg (2002). https://doi.org/10.1007/3-540-45712-7_45

18. Watson, R.A., Ficici, S.G., Pollack, J.B.: Embodied evolution: distributing an evolutionary algorithm in a population of robots. Robot. Auton. Syst. **39**, 1–18 (2002)

On Steering Swarms

Ariel Barel$^{(\boxtimes)}$ (ID), Rotem Manor (ID), and Alfred M. Bruckstein

Technion - Israel Institute of Technology, Technion City, Haifa, Israel
arielbarel@gmail.com

Abstract. The main contribution of this paper is a novel method allowing an external observer/controller to steer and guide swarms of identical and indistinguishable agents, in spite of the agents' lack of information on absolute location and orientation. Importantly, this is done via simple global broadcast signals, based on the observed average swarm location, with no need to send control signals to any specific agent in the swarm.

1 Introduction

This paper deals with steering multi-agent systems, based on decentralized gathering laws, using an external broadcast control signal. Agents move according to local information provided by their sensors. The agents are assumed to be identical and indistinguishable, memoryless (oblivious), with no explicit communication between them. The agents do not share a common frame of reference i.e. agents are not equipped with either GPS systems or compasses. By assumption, agents sense the distance and/or bearing to their neighbours, within a finite or infinite range of visibility. An external observer/controller continuously monitors the swarm's location and broadcasts the same control signal, based on the centroid of the agents' constellation. We present a simple yet practical method to steer the swarm and guide it to a given destination.

Note that unlike the simple agents that are anonymous, unaware of their position, lack memory, and do not use explicit communication to maintain the swarm cohesion, the external controller does need the ability to continuously monitor the trajectory of the swarm location. Due to these capabilities, the controller is able to influence the movement of the swarm, with a very simple global control signal broadcast simultaneously to all agents.

The inspiration to this control method came from the following observation: some of the gathering algorithms, while they ensure the convergence of agents to a bounded area, do not imply that the centroid of the agents' location remains stationary in the plane [1, 4, 6, 9, 12–14, 16, 17]. In fact, some gathering algorithms exhibits random walk like behaviour of the centroid of the agents' constellation after gathering as discussed in [3]. The method to steer the swarm to a target point, presented herein, exploits the movements of the system's center of gravity due to the agents' compliance with the distributed convergence algorithm.

© Springer Nature Switzerland AG 2018
M. Dorigo et al. (Eds.): ANTS 2018, LNCS 11172, pp. 403–410, 2018.
https://doi.org/10.1007/978-3-030-00533-7_35

2   How to Control a Single Agent

We first describe the basic idea in conjunction with a single agent performing a random walk in the plane, and then extend the discussion to multi-agent systems carrying out various cohesion ensuring gathering algorithms. Assume a drunkard agent is moving in the plane in the following random way: at discrete times $k = 1, 2, 3, \ldots$ he selects a new destination for time $k + 1$. The destination location $\tilde{p}(k + 1)$ is randomly and homogeneously distributed in a unit disc centered at its current position $p(k)$, so that $\tilde{p}(k+1) = p(k) + \tilde{\Delta}(k)$, where $\tilde{\Delta}(k)$ is a random vector uniformly distributed in a unit disc. After selecting $\tilde{p}(k + 1)$ the agent starts going there from $p(k)$ in a straight path. By monitoring his motion, one can steer him in any direction with the following control rule: if the projection of his current movement on the required direction is positive - allow the drunkard to finish his step. Otherwise, stop him after a fraction of the unit interval $\mu < 1$, by broadcasting (shouting) a startling "stop!" signal.

This process will cause the drunkard to perform a biased walk, making, in expectation, bigger steps in the desired direction. To bring the drunkard toward a region near a precise target point in the plane, one may define the desired direction to always point from the current location of the drunkard to the goal. Assume first, for simplicity, that the desired direction is fixed. Let $p(k)$ be the current position of the agent and let $d \in \mathbb{R}^2$ be a unit vector in the direction in which we require the agent to move. Denote by $\tilde{\Delta}(k)$ the *planned* travel vector of the agent for the current time period $[k, k+1)$, from $p(k)$, its position at time k, to a homogeneously distributed random point in a unit disc centered at $p(k)$, and by $\Delta(k)$ its *actual* travel vector. The relation between $\tilde{\Delta}(k)$ and $\Delta(k)$ is as follows: at time k the agent starts traveling from its existing position $p(k)$ to its planned position $\tilde{p}(k+1) = p(k) + \tilde{\Delta}(k)$ in a piecewise constant velocity equal to $\tilde{\Delta}(k)/1$. If $\tilde{\Delta}(k)^T d \leq 0$, the external controller stops the agent at a fraction μ of the time-step, i.e. $\Delta(k) = \mu\tilde{\Delta}(k)$, otherwise the controller does not interrupt its motion during the current time period, hence $\Delta(k) = \tilde{\Delta}(k)$. Therefore we have

$$p(k + 1) = p(k) + c(k)\tilde{\Delta}(k)$$

$$c(k) = \begin{cases} \mu & \tilde{\Delta}(k)^T d < 0 \\ 1 & o.w. \end{cases} \tag{1}$$

where $\tilde{\Delta}(k)$ is a vector from $p(k)$ to the homogeneously distributed random point in a unit disc centered at $p(k)$. By symmetry of the random distribution function, for any direction x, we have that the expectation of a planned step is $\mathbf{E}\{\tilde{\Delta}x(k)\} = 0$. The required direction of movement d is, without loss of generality, towards the positive x axis, i.e. to the right. Clearly, by the symmetry of the distribution function, we have that the probabilities that the drunkard moves right and left are same and equal 0.5. Hence, the expected actual travel of the agent, given external controller's (possible) interruptions, is (omitting the time index (k) for simplicity):

$$\mathbf{E}\{\Delta x\} = 0.5\mathbf{E}\{\Delta x \mid \tilde{\Delta}x \geq 0\} + 0.5\mathbf{E}\{\Delta x \mid \tilde{\Delta}x < 0\} = 0.5(1-\mu)\mathbf{E}(\tilde{\Delta}x \mid \tilde{\Delta}x \geq 0) \tag{2}$$

In order to guide an agent to a target point, the controller can set the required direction at each time-step, from the current position of the agent to the target point. Let us find the expected position of the agent at time $(k+1)$ given $p(k)$, i.e. $\mathbf{E}\{\|p(k+1)\|^2 \mid p(k)\}$. By the law of cosines in a triangle [5] we obtain that

$$\mathbf{E}\{\|p(k+1)\|^2 \mid p(k)\} = p(k)^2 - A(\frac{1-\mu}{2})\|p(k)\| + B(1+\mu^2) \tag{3}$$

where $A = \mathbf{E}\left\{ \frac{\tilde{\Delta}(k)^T p(k)}{\|p(k)\|} \, \mathrm{sgn}\left\{ \frac{\tilde{\Delta}(k)^T p(k)}{\|p(k)\|} \right\} \right\}$ is positive and depends only on the direction vector $d(k) = \frac{p(k)}{\|p(k)\|}$, and for a rotationally symmetric $\tilde{\Delta}(k)$ it is independent of $d(k)$ (and on $p(k)$ of course), and $B = \mathbf{E}\{\|\tilde{\Delta}(k)\|^2\}$ is positive and obviously independent on $p(k)$. From this result it follows that

$$\mathbf{E}\{\|p(k+1)\|^2\} = \mathbf{E}\{\|p(k)\|^2\} - \left(A\left(\frac{1-\mu}{2}\right)\mathbf{E}\{\|p(k)\|\} - B(1+\mu^2) \right) \tag{4}$$

We have that if the right expression in big parentheses in (4) is bigger than δ, $\mathbf{E}\{\|p(k)\|^2\}$ decreases by δ, and while this inequality persists, it will decrease until $\mathbf{E}\{\|p(k)\|\} \leq \left(\frac{B(1+\mu^2)+\delta}{A(\frac{1-\mu}{2})} \right)$. Returning to (4) we have that after $k(\delta)$ steps, given by

$$k(\delta) = \frac{D^2(0) - \left(\frac{B(1+\mu^2)+\delta}{A(\frac{1-\mu}{2})} \right)^2}{\delta} \tag{5}$$

the process will necessarily stop and the agent will be "near" the target. Simulated results of k vs. δ for some different initial values of $D(0)$ and the graph of Eq. (5) plotted in Fig. 1 shows that the theoretical $k(\delta)$ is indeed a rather loose upper bound on the number of steps needed to reach the target's neigbourhood.

3   Controlling Multi-agent Systems - the Idea

Let us adopt this steering method to a multi-agent system. Suppose there is a multi-agent system which converges to a bounded area. The lack of a global orientation of the agents prevents the viewer from simply broadcasting the desired direction of movement as suggested by Azuma et al. [2] and others, since the agents are unable to obey global-direction-based commands. Research methods that draw inspiration from animal behaviour in herds in nature e.g. [7] are based on the fact that part of the group moves in a certain direction and indirectly influences the group's behaviour, but in this article we assume that even leaders do not know how to orient themselves and find the desired direction of movement. Additionally, recall that our agents are anonymous and indistinguishable,

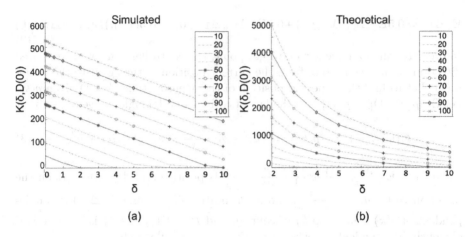

Fig. 1. Plot of k vs. δ for some $D(0)$ values from 10 to 100 units. (a) Simulation results, and (b) The theoretical bound. Here $\mu = 0.1$ and number of simulation runs is $10,000$.

hence an external observer wishing to lead the system in a required direction can not steer individual agents separately by transmitting control commands to each one of them. We show here that an external observer can lead a multi-agent system in a required direction (while the agents also converge to a bounded region), by only sensing the motion of the system's centroid. This information represents for the external controller the location of the group, and it is feasible to measure or estimate in real life multi-agent scenarios, especially for large numbers of agents, such as swarms of drones. Let $p_{cm}(t) = \frac{1}{n} \sum_{j=1}^{n} p_i(t)$ be the system's centroid. The velocity of the centroid is the average velocities of the agents $\dot{p}_{cm}(t) = \frac{1}{n} \sum_{j=1}^{n} \dot{p}_i(t)$ and we have that while all agent velocities are constant the centroid velocity is constant as well. We assume that during each time interval $k = 1, 2, 3, \ldots$ each agent's velocity is constant, therefore we have that $\hat{p}_{cm}(t)$, the direction of the centroid movement is piecewise constant (i.e. does not change during time intervals hence moves in straight lines). Similar to our discussion in Sect. 2, here, the external controller tracks the motion of the *centroid* of the system. If the projection of its movement is on the required direction $(\tilde{\Delta}_{cm}(k)^T d \geq 0)$ - it allows *all the agents* to finish their planned travels. Otherwise, it stops them all after a fraction μ of the time-step, i.e. when they complete a fraction μ their planned travel. We discuss in detail different types of such systems, and bound the expected "velocity" of the swarm's centroid due to this control mechanism.

3.1   Steering a System of Agents with Infinite Visibility and Full Sensing

We begin with a simple linear multi-agent gathering process in discrete time for the infinite visibility and full sensing case. Each agent i moves according to the decentralized dynamic law: $p_i(k+1) = p_i(k) - \sigma \sum_{j=1}^{n}(p_i(k) - p_j(k))$, where $0 < \sigma < \frac{2}{n}$ is a constant gain factor, i.e. at each time-step, each agent jumps proportionally to the sum of relative position vectors to all the other agents (recall system $\mathcal{S}_2$, in [3]). As proved by Gazi, Passino et al. [8], since the dynamics of such system is governed by an antisymmetric pairwise interaction function, the average position of the agents is invariant. To steer this system in some desired direction, we would like to bias the motion of the system centroid by measuring its trend, hence we assume some additive "noise" that breaks symmetry and causes the center of the system to move. We hence assume that each agent, in addition to obeying the distributed control law above, also moves to a randomly selected point at each time step:

$$p_i(k+1) = p_i(k) - \sigma \sum_{j=1}^{n}(p_i(k) - p_j(k)) + \tilde{\Delta}_i(k) \tag{6}$$

where $\tilde{\Delta}_i(k)$ is a randomly selected point in a unit disc. Here too, at time k the agents start traveling from their existing positions $p_i(k)$ towards their next planned positions $\tilde{p}_i(k+1)$ in piecewise constant velocities equal to their distance from it $[-\sigma \sum_{j=1}^{n}(p_i(k) - p_j(k)) + \tilde{\Delta}_i(k)]/1$, so that if an external controller does not intervene, all the agents arrive at their destinations simultaneously at time $k+1$. Hence we may denote the planned motion of the centroid to be $\tilde{\Delta}_{cm}(k) = \bar{\tilde{p}}(k+1) - \bar{p}(k) = \frac{1}{n}\sum_{i=1}^{n}\tilde{\Delta}_i(k)$, and the control mechanism for system (6) is:

$$p_i(k+1) = p_i(k) + c(k)[-\sigma \sum_{j=1}^{n}(p_i(k) - p_j(k)) + \tilde{\Delta}_i(k)]$$

$$c(k) = \begin{cases} \mu & \tilde{\Delta}_{cm}(k)^T d < 0 \\ 1 & o.w. \end{cases} \tag{7}$$

Here $c(k)$ represents the optional "stop" signal received simultaneously at fraction μ of the time-step by all agents, $\tilde{\Delta}_{cm}(k) = \frac{1}{n}\sum_{i=1}^{n}\tilde{\Delta}_i(k)$ is the planned travel of the centroid of the agents, and d is the required direction of movement of the system. Since the projection on x of the second moment of a disc of radius r is $\frac{1}{4}\pi r^4$, we have in this system [5] that $\mathbf{E}\{\Delta x_{cm}\} \geq 0.5(1-\mu)\frac{1}{8n}$ i.e. the bound on the expected step of the centroid is inversely proportional to the number of agents. To guide a system to a goal point, the observer controller should set the desired direction at every time interval so $d(k)$ is a unit vector from the centroid of the system to the goal point. Figure 2 presents a typical simulation result of this system with full visibility and complete sensing, with some evenly distributed noise jump to a unit disc of each agent, as presented in Eq. (7).

3.2 Steering a System of Agents with Limited Visibility and Bearing Only Sensing

Here we assume that the agents are able to sense the direction to their neighbours (i.e. bearing only sensing), and their motions being determined by the set of unit vectors pointing from their current location to their neighbours. The neighbours are defined for each agent i at time-step k as the set of agents located within a given visibility range V form its position $p_i(k)$. Manor et al. [15] modified Gordon's et al. motion laws [10,11], and proved that the new law gathers the agents of the system to a disc with a radius equal to the agents' maximal step size σ within a finite expected number of time steps, and that the distribution of the agents' average position converges in probability to the distribution of a random-walk. As in Sect. 3.1, we assume here piecewise continuous dynamics (where agents continuously move towards their new locations), so that the formal steering algorithm for this system is:

$$p_i(k+1) = \begin{cases} p_i(k) & \psi_i(k) \geq \pi \text{ or } \chi_i(k) = 0 \\ p_i(k) + c(k)\tilde{\Delta}_i(k) & o.w. \end{cases}$$

$$\chi_i(k) = \begin{cases} 1 & \text{w.p. } \delta \\ 0 & \text{w.p. } 1 - \delta \end{cases} \tag{8}$$

$$c(k) = \begin{cases} \mu & \tilde{\Delta}_{cm}(k)^T d < 0 \\ 1 & o.w. \end{cases}$$

$$\tilde{\Delta}_i(k) = \text{vector from } p_i(k) \text{ to a random point in } ar_i(k)$$

where $\tilde{\Delta}_{cm}(k) = \sum_{i=1}^{n} \tilde{\Delta}_i(k)$ is the planned jump of the centroid of the system, and d is a unit vector in the required moving direction of the system. It was proved in [15] that the original model, given no external control, satisfies $\mathbf{E}\{\Delta_{cm}(k)\} = 0$, and that

$$\mathbf{E}\{\Delta x_{cm}\} \geq 0.25(1 - \mu)\frac{1}{n^2}Var^* \tag{9}$$

$$Var^* = \delta^2\left(\frac{\sigma}{2}\right)^2 \frac{1 - \cos^4\left(\frac{\pi - \psi_*}{2}\right)}{\frac{\pi - \psi_*}{2} - \frac{1}{2}\sin(\pi - \psi_*)}$$

Figure 2 presents simulations result of this system (8). The system gathers and moves to a goal, and the trace of the travel of the system's centroid is plotted.

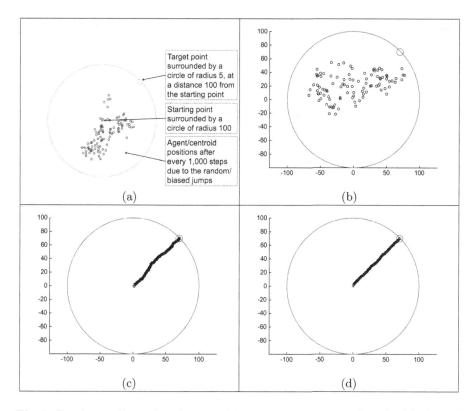

Fig. 2. Random walk vs. Steering a multi-agent system to a goal point: (a) General legend of the simulation settings. (b) Typical 100, 000 random unit steps of a drunkard agent with no bias (agents' position was plotted every 1, 000 steps for enhanced readability). (c) Typical simulation run of the system in Sect. 3.1 with $n = 10$ and $\mu = 0.01$. The system centroid first entered the goal area in less than 1, 600 time steps. (d) Typical simulation run of the system in Sect. 3.2 with $n = 10$ and $\mu = 0.01$. The system centroid first entered the goal area in less than 9, 000 time steps.

4 Conclusions

A method has been introduced here that allows an external observer to control a multi-agent system and guide it to a desired destination even when the agents are very primitive. According to our paradigm all the agents are identical (anonymous), therefore the external observer can not send a separate command to each agent, but can broadcast the same command to all the agents. The viewer controls the swarm by means of an identical command sent simultaneously to all agents. The method was tested for different cases: the control of a single moving agent performing random-walk, steering of a system with infinite visibility and relative distance and bearing measurement, and control of a system with partial information (limited visibility and bearing only measurement).

Acknowledgments. This research was partly supported by Technion Autonomous Systems Program (TASP).

References

1. Ando, H., Oasa, Y., Suzuki, I., Yamashita, M.: Distributed memoryless point convergence algorithm for mobile robots with limited visibility. IEEE Trans. Robot. Autom. **15**(5), 818–828 (1999)
2. Azuma, S.I., Yoshimura, R., Sugie, T.: Broadcast control of multi-agent systems. Automatica **49**(8), 2307–2316 (2013)
3. Barel, A., Manor, R., Bruckstein, A.M.: Come together: multi-agent geometric consensus (gathering, rendezvous, clustering, aggregation). Technical report, CIS Technical Report, TASP (2016)
4. Barel, A., Manor, R., Bruckstein, A.M.: Probabilistic gathering of agents with simple sensors. Technical report, CIS Technical Report, TASP (2017)
5. Barel, A., Manor, R., Bruckstein, A.M.: On steering swarms. Technical report, CIS Technical Report, TASP (2018)
6. Bellaiche, L.I., Bruckstein, A.M.: Continuous time gathering of agents with limited visibility and bearing-only sensing. Technical report, CIS Technical Report, TASP (2015)
7. Couzin, I.D., Krause, J., Franks, N.R., Levin, S.A.: Effective leadership and decision-making in animal groups on the move. Nature **433**(7025), 513 (2005)
8. Gazi, V., Passino, K.M.: Stability analysis of social foraging swarms. IEEE Trans. Syst. Man Cybern. Part B: Cybern. **34**(1), 539–557 (2004)
9. Gordon, N., Elor, Y., Bruckstein, A.M.: Gathering multiple robotic agents with crude distance sensing capabilities. In: Dorigo, M., Birattari, M., Blum, C., Clerc, M., Stützle, T., Winfield, A.F.T. (eds.) ANTS 2008. LNCS, vol. 5217, pp. 72–83. Springer, Heidelberg (2008). https://doi.org/10.1007/978-3-540-87527-7_7
10. Gordon, N., Wagner, I.A., Bruckstein, A.M.: Gathering multiple robotic a(ge)nts with limited sensing capabilities. In: Dorigo, M., Birattari, M., Blum, C., Gambardella, L.M., Mondada, F., Stützle, T. (eds.) ANTS 2004. LNCS, vol. 3172, pp. 142–153. Springer, Heidelberg (2004). https://doi.org/10.1007/978-3-540-28646-2_13
11. Gordon, N., Wagner, I.A., Bruckstein, A.M.: A randomized gathering algorithm for multiple robots with limited sensing capabilities. In: Proceedings of MARS 2005 Workshop at ICINCO Barcelona (2005)
12. Jadbabaie, A., Lin, J., Morse, A.S.: Coordination of groups of mobile autonomous agents using nearest neighbor rules. IEEE Trans. Autom. Control. **48**(6), 988–1001 (2003)
13. Ji, M., Egerstedt, M.B.: Distributed coordination control of multi-agent systems while preserving connectedness. IEEE Trans. Robot. **23**(4), 693–703 (2007)
14. Manor, R., Bruckstein, A.M.: Chase your farthest neighbour: a simple gathering algorithm for anonymous, oblivious and non-communicating agents. Technical report, CIS Technical Report, TASP (2016)
15. Manor, R., Bruckstein, A.M.: Discrete time gathering of agents with bearing only and limited visibility range sensors. Technical report, CIS Technical Report, TASP (2017)
16. Olfati-Saber, R.: Flocking for multi-agent dynamic systems: algorithms and theory. IEEE Trans. Autom. Control. **51**(3), 401–420 (2006)
17. Olfati-Saber, R., Fax, J.A., Murray, R.M.: Consensus and cooperation in networked multi-agent systems. Proc. IEEE **95**(1), 215–233 (2007)

Vector Field Benchmark for Collective Search in Unknown Dynamic Environments

Palina Bartashevich[✉][iD], Welf Knors, and Sanaz Mostaghim

Faculty of Computer Science, University of Magdeburg, Magdeburg, Germany
{palina.bartashevich,sanaz.mostaghim}@ovgu.de,welf.knors@st.ovgu.de

Abstract. This paper presents a Vector Field Benchmark (VFB) generator to study and evaluate the performance of collective search algorithms under the influence of unknown external dynamic environments. The VFB generator is inspired by nature (simulating wind or flow) and constructs artificially dynamic environments based on time-dependent vector fields with moving singularities (vortices). Some experiments using the Particle Swarm Optimization (PSO) algorithm, along with two specially developed updating mechanisms for the global knowledge about the external environment, are conducted to investigate the performance of the proposed benchmarks.

1 Introduction

Swarm Intelligence algorithms such as Particle Swarm Optimization (PSO) [13] and Ant Colony Optimization [6] are shown to be very effective in solving optimization problems. Due to their distributed nature, they can be easily used in swarm robotic search scenarios [8]. In the past years, PSO has been successfully used in this context [1,5,11,16]. However, there are only a few existing methods that have addressed the influence of the external environments on the collective search algorithms [2,3,9,12,17]. The main goal of this paper is to provide new benchmark problems, simulating the influence of the dynamic external environment (such as wind, flow, etc.), which will serve as a baseline testbed for the development of new collective search mechanisms, that are robust to the unknown perturbations and can be further employed in real-world applications. In [2], the authors have introduced vector fields to simulate the external dynamics in a PSO-based collective search scenario designed for a swarm of aerial micro-robots. This approach only considered static vector fields, which are rather rare in nature, as the external dynamics change over time (i.e. unsteady flows). In this paper, we consider time-dependent vector fields and propose a unified method, called Vector Field Benchmark (VFB), to construct such dynamic environments using singular points [4]. In order to test the proposed VFB system, experiments are made using VFM-PSO [2] under the composition of changing vector fields and moving singular points. The results show that the VFB system can give different properties by simply setting the environmental types. The

© Springer Nature Switzerland AG 2018
M. Dorigo et al. (Eds.): ANTS 2018, LNCS 11172, pp. 411–419, 2018.
https://doi.org/10.1007/978-3-030-00533-7_36

paper is organized as follows. We define VFB generator in Sect. 2. In Sect. 3, we describe the generalized "VFB-Map Exploration Framework"in the context of which we test the proposed benchmark. Section 4 contains several experiments to test the performance of VFB. The paper is concluded in Sect. 5.

2   Vector Fields Benchmark (VFB)

The time-varying dynamics of the environment are modeled by the unsteady vector fields with or without vortices, which are described below. In general, most vortex definitions are characterized by means of differential properties of the observed vector fields. For simplicity, our VFB functions are limited to a two-dimensional space (as horizontal wind).

Definition 1. *A* **Vector Field** $\boldsymbol{VF}$ *on a planar domain* $D \subset \mathbb{R}^2$ *is a function assigning to each point* $(x, y) \in D$ *a 2-dimensional vector* $\boldsymbol{VF}(x, y) = (u(x, y), v(x, y))$.

Definition 2. *A point* $(x_0, y_0) \in D$ *is singular for* $\boldsymbol{VF}$ *if* $\boldsymbol{VF}(x_0, y_0) = (0, 0)$.

The values at any point $\boldsymbol{p} \in D$ of local vector field $\boldsymbol{SP}$ defined by corresponding singular point can be calculated as follows [14]:

$$SP(p) = e^{-d\|p - p_0\|^2} JV(p - p_0),   \tag{1}$$

where JV is the Jacobian matrix of the desired Singular Point, $p_0 = (x_0, y_0)$ is the center of the Singular Point and d is a decay constant limiting the intensity of the Singular Point influence with increasing distance to its center $\boldsymbol{p_0}$.

Definition 3. *The* **spatial Jacobian** *JV is an $n \times n$ matrix that contains a first-order description of how the flow $\boldsymbol{VF}$ behaves locally around a given location.* Following Hartman-Grobman theorem [10], singular points can be partly classified by looking to the eigenvalues of the Jacobian matrix at that point (see Fig. 1, where k denotes the spread of $\boldsymbol{SP}$).

However, in measured data, the vector field $\boldsymbol{VF}$ is not given as a differentiable function. Following that, we discretize a domain $D \subset \mathbb{R}^2$ of the vector field and assume that we have the values of $\boldsymbol{VF}$ at the points (x_i, y_j) of a regular grid of size $M \times N$ cells. We will denote the unit cell $c_{i,j}$ of the grid by sample point (x_i, y_j) as follows $c_{i,j} = (u_{i,j}, v_{i,j}) = \boldsymbol{VF}(x_i, y_j)$.

Definition 4. *An* **Unsteady Vector Field** *varies over time and is given as a time-dependent map* $\boldsymbol{VF}(\boldsymbol{x}, t) = \boldsymbol{VF}(x, y, t) : D \times T \to D$. *It can be also written as an $(2 + 1)$-dimensional steady field* [7]: $\boldsymbol{VF}(x, y, t) = (u(x, y, t), v(x, y, t), t)$.

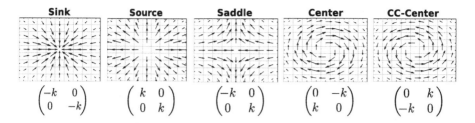

Fig. 1. Five types of singular points shown on grids with 9×9 cells with $k = 5$ and corresponding values of JV matrix below.

VFB Generator. When there are 2 or more Singular Points in a Vector Field, the velocities are calculated by simply adding the vector fields $\boldsymbol{SP}$ of each Singular Point. The overall VFB is defined as a sum of the underlying VF and the Singular Points influences:

$$VFB := \boldsymbol{VF}(\boldsymbol{p}, t) + \sum \boldsymbol{SP}(\boldsymbol{p}; JV(k), center : \boldsymbol{p_0}, decay : d, MoveType), \quad (2)$$

where singular points $\boldsymbol{p_0}$ are characterized by their type and strength k performed with Jacobian $JV(k)$, and movement types $MoveType$. The dynamics of SPs are organized mainly by the moving center point $\boldsymbol{p_0}$ of the defined SP according to some law of movement (denoted by $MoveType$). Additionally, SPs movements are also affected by the underlying VF (if any), which means that the velocity vectors of the underlying $\boldsymbol{VF}$ at the SPs current centers positions, i.e. $\boldsymbol{VF}(\boldsymbol{p_0})$, will be added to the $MoveType$ movement. The dynamics of underlying VFs, as well as of SPs, can also be complicated through multiplication on the Rotation Matrix $Rot(\alpha, t)$, thereby making their vectors rotating by some angle α at each time step t.

3   VFB-Map Exploration Framework

In order to test the proposed VFB functions, we take the same concept of Information Map (IM) approach for steady vector fields which was proposed in [2]. We adapt the concept for time-dependent flows with new ways of saving information in IM, which are supposed to catch the main features of unknown unsteady external dynamics. To estimate the external dynamics, the explorer population is used with simple movements based on the value of the VFB at their own positions. The explorers are coming from one and the same initial positions every Δt time steps and save the information about VFBs magnitude and direction at their positions in the IM, which is a global and central archive accessible by all individuals. The optimizers follow the rules of PSO and retrieve the information about VFB from the IM to organize a better collective search process by full compensation of negative factors at already explored regions. The type of the optimizers decision, based on the IM, is not limited to full compensation (see for example, [3]), however, in this paper we consider only this type of action. As the

VFB (i.e. $\boldsymbol{VF}(\boldsymbol{x},t)$) changes over time, its values at the same positions might be different at different time steps t. So it is the question of how to store the measured values in IM, in order to take the relation between the past and the present into account. We present two update mechanisms of the IM for storing collected data:

(1) **Recent (Rec)** saves only the most recent measured values of the cells and left them unchangeable in the IM until their next visit. When a cell is visited a second time, all information from the first visit is replaced.

(2) **Evaporating Mean (EM)** computes the mean of measured values, if the cell $c_{i,j}$ is visited several times, and applies an evaporation operator ρ_t to it, which linearly decreases the saved value for the $c_{i,j}$ by subtracting from its initially saved value the evaporation rate ρ_t^0 constantly at each time step t. The evaporation continues until the value in $c_{i,j}$ reaches the minimum limit of ρ^{min}, after which it is stopped, in order to preserve the information in already information-starved environment. The value of evaporation rate itself is constant $\rho_t^0 = \rho^0$ unless the cell is visited only once, otherwise its value is decreased by dividing on the number of cell visits N_{vis} for each cell $c_{i,j}$ individually, as a means of 'confidence' in the saved value. In other words, the more times the cell is visited the less it is evaporated, i.e. $\rho_t^0(c_{i,j})$ is a function, which takes a value $(\sum N_{vis}(c_{i,j}))^{-1}\rho_{t-1}^0(c_{i,j})$ if $N_{vis}(c_{i,j}) > 1$ and $\rho_{t-1}^0(c_{i,j})$ otherwise.

Both of the above approaches consider the cells separately, therefore we refer to them as discrete methods denoted by Rec-D and EM-D. We also consider their continuous variants (denoted by Rec-C and EM-C) by using interpolation and extrapolation for the rest of the cells inside and outside the convex hull, defined by the cells with already saved information in IM. The *Nearest* interpolation [15] is used, as the fastest interpolation method among the others known in the literature (what is sufficient for time-dependent changes).

4   Experimental Study

The goal of the experiments is to demonstrate the usage of the introduced VFBs and to estimate the performance of the proposed updating mechanisms.

Parameter Settings. Similar to [2], we use a VFM-PSO algorithm with 20 optimizers initialized randomly over the search space $S : [-15, 15] \times [-15, 15]$. The velocity limit v_{max} is set to 2, inertia weight w is selected to be 0.6 along with acceleration coefficients $C_1, C_2 = 1$. The number of explorers is set to 10 with frequency of update each $\Delta t = 10$ iterations. The total number of iterations is 150. The algorithm has been run on Sphere, Ackley and Rosenbrock over proposed further VFBs. We compare the proposed update mechanisms for the IM both for discrete (*Rec-D* and *EM-D*) and continuous (*Rec-C* and *EM-C*) variants. In the experiments, *"None"* indicates the approach without IM (i.e., without explorers). According to the preliminary experiments, the evaporation rate ρ^0 is set to 0.3 and ρ^{min} is 0.5. Each experiment is repeated 30 times with different random initializations for both optimizers and explorers. Table 1 provides the function description of considered VFBs without SPs, i.e. VFB1-VFB3. Table 2 describes the VFBs containing moving SPs, i.e. VFB4-VFB7. For each VFB4-VFB7 one considers 9 singular points of at most two types with given coordinates (x_0, y_0). Each scenario can have an underlying vector field, indicated as VF in the last row of Table 2 and equations for which can be taken from Table 1. For all used in VFB4-VFB7 SPs, spread k is set to 15 and decay d is 0.4. $MoveType$ is defined by sinus law in horizontal direction. The only exception is VFB7, where SPs move according to the velocities of the underlying VF, i.e. Waves.

Table 1. Function descriptions for VFBs without singularities: VFB1-VFB3.

VBF1	CrossRot	$VF(x_1, x_2) = Rot(5, t) * (x_2, x_1)$
VBF2	Waves	$VF(x_1, x_2) = (10, cos(x_1 - 0.5 * t) * 3)$
VBF3	UniformRot	$VF(x_1, x_2) = Rot(5, t) * (3, 3)$
	Sheared	$VF(x_1, x_2) = (x_1 + x_2, x_2)$

Table 2. Parameters descriptions for VFBs with singularities: VFB4-VFB7.

	VBF4	VBF5	VBF6	VBF7	Coordinates (x_0, y_0)
Type	Source	Source	Center	Center	$(-10, -8)$ $(-10, 10)$ $(0, -1)$ $(0, -6)$ $(10, 12)$
	Saddle	Saddle			$(-10, 1)$ $(0, -10)$ $(0, 8)$ $(10, 3)$
VF	Cross	Sheared	None	Waves	

Results. Figure 2 shows a comparison of median fitness values obtained using *None, Rec-D, EM-D, Rec-C* and *EM-C* (from left to right) within considered VFB (i.e. VFB1-VFB7 indicated by columns) on the corresponding objective function (indicated by rows). Since the main objective of the experiments is to demonstrate the usage of the introduced VFBs and to estimate the performance of the proposed exploration techniques, we have made multiple pairwise statistical comparison tests to identify which of the approaches are specifically different. Pairwise Mood's median tests were performed for VFBs, which have indicated statistical differences in at least one of the medians, i.e. VFB2-VFB5 and VFB7. For visual representation of the statistical differences between approaches the reader is referred to Fig. 2. The boxes, which do not share any letter in common within one and the same VFB over certain objective function, indicate statistical differences between compared types of updating mechanisms. From this we can see that the obtained results reveal our hypothesis as on the most of the considered VFBs, regardless of the objective function, discrete mechanisms (*Rec-D, EM-D*) are not statistically different from each other and *None*. While almost in all of the cases, continuous updating mechanisms (*Rec-C, EM-C*) are statistically different from *None* and their discrete analogies (i.e. *Rec-D, EM-D*). *Rec-C* seems to be the most successful among the presented approaches, as its median

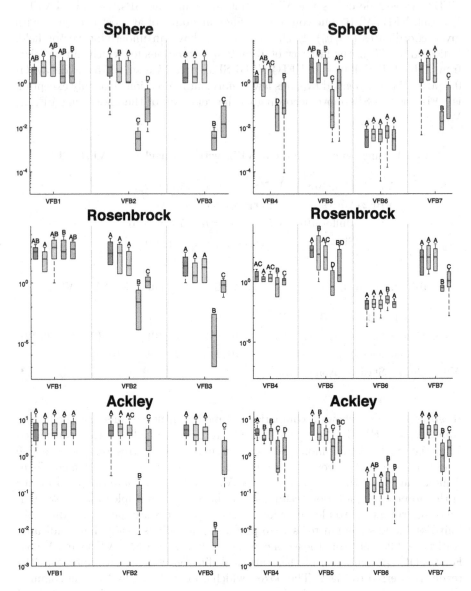

Fig. 2. Boxplots of the median fitness values obtained by *None*, *Rec-D*, *EM-D*, *Rec-C* and *EM-C* (from left to right) within VFB1 to VFB7. The central mark on each boxplot indicates the median. Boxplots which share at least one common letter within one and the same VFB indicate not statistical difference in median fitness values with significance level $\alpha = 0.05$ according to Pairwise Mood's Test.

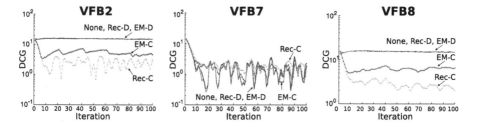

Fig. 3. Distances between the center of the swarm and the global best over the iterations obtained on Ackley function over VFB2 (full covering VF without SPs), VFB7 (consists only of SPs) and VFB8 (composition of VFB2 and VFB7).

fitness values are significantly lower than those using *EM-C* on VFB2, VFB3, VFB4 and VFB7. Although, *EM* approach was supposed to be a compromise between taking changes into account and compensating for the cases with partial covering by the VF at a time (as on VFB7), *EM-C* is statistically worse than *Rec-C* on all considered VFBs, including VFB7.

In the following we also report the convergence behavior of the proposed algorithms. Figure 3 illustrates Euclidean distances between the center of the swarm and the obtained global best over the iterations for VFB2, VFB7 and VFB8, which is a composition of VFB2 and VFB7. It can be observed that *None*, *Rec-D* and *EM-D* variants have similar behavior and do not change over iterations on VFB2 and VFB8, while *EM-C* and *Rec-C* reproduce an oscillation behavior, indicating that they have found the equilibrium point and it does not change with time anymore. The results differ for VFB7, as in this case the SPs only partially influence the movements of the optimizers, so we expect that the continuous update mechanisms do not help as they disturb the movements themselves at the positions where there is no VF influence. This can be observed on the performance of *EM-C* and *Rec-C* in VFB7, while the other update approaches help to converge until a change in the environment occurs. The changes in the environment can be depicted by the oscillating behaviors in the convergence plots. The results reported in Fig. 3 are obtained on the Ackley search landscape. However, our experiments show that the observed movement patterns (i.e. oscillations) are the same for all the other considered objective functions on the same VFBs. The only difference is in the value of the drift (i.e. vertical shift on the plots in Fig. 3), as it is defined by the found global best solution. In comparison to standard PSO problems (i.e. without VFBs), where distance between the swarm center and the global best is constantly decreasing as the particles converge to the best, acting under VFBs the swarm can not really converge. Therefore, in order to improve its performance, we observe oscillations in certain limited area around the best so-far obtained solution.

5   Conclusions and Future Work

This paper presents new benchmark functions for simulating and modeling the external dynamics for swarm robotics applications. We propose to use time-dependent vector fields with moving singularities and analyze their influence on the existing PSO-based methods in the context of "VFB-Map Exploration Framework". The results illustrate the strong influence of the environment on the collective search. One feature, imposed by moving singularities, concerns the oscillating behavior in the convergence plots. We have tested two various schemes based on continuous and discrete updating mechanisms for storing the global information about the unknown environment. The results show the advantage of the continuous variant over discrete in unsteady environments without singularities, while this degrades for environments which contain ones. Our experiments illustrate that the VFB can be used as a good base for developing search algorithms and is not limited to the proposed exploration framework and PSO. In future, we aim to work on other swarm based collective search mechanisms on the presented VFB.

References

1. Atyabi, A., Phon-Amnuaisuk, S., Ho, C.K.: Navigating a robotic swarm in an uncharted 2D landscape. Appl. Soft Comput. **10**(1), 149–169 (2010)
2. Bartashevich, P., Grimaldi, L., Mostaghim, S.: PSO-based search mechanism in dynamic environments: swarms in vector fields. In: 2017 IEEE Congress on Evolutionary Computation, pp. 1263–1270 (2017)
3. Bartashevich, P., Koerte, D., Mostaghim, S.: Energy-saving decision making for aerial swarms: PSO-based navigation in vector fields. In: 2017 IEEE Symposium Series on Computational Intelligence, pp. 1–8 (2017)
4. Demazure, M.: Singular points of vector fields. In: Demazure, M. (ed.) Bifurcations and Catastrophes, pp. 219–247. Springer, Berlin (2000). https://doi.org/10.1007/978-3-642-57134-3_9
5. Doctor, S., Venayagamoorthy, G.K., Gudise, V.G.: Optimal PSO for collective robotic search applications. In: Proceedings of the 2004 Congress on Evolutionary Computation, Vol. 2, pp. 1390–1395 (2004)
6. Dorigo, M., Stützle, T.: Ant Colony Optimization. Bradford Company, Cambridge (2004)
7. Günther, T., Theisel, H.: The state of the art in vortex extraction. Computer Graphics Forum, To appear (2018)
8. Hamann, H.: Scenarios of swarm robotics. In: Hamann, H. (ed.) Swarm Robotics: A Formal Approach, pp. 65–93. Springer, Cham (2018). https://doi.org/10.1007/978-3-319-74528-2_4
9. Meng, Q.-H., Yang, W.-X., Wang, Y., Zeng, M.: Collective odor source estimation and search in time-variant airflow environments using mobile robots. Sensors **11**(11), 10415–10443 (2011). https://doi.org/10.3390/s111110415
10. Helman, J., Hesselink, L.: Representation and display of vector field topology in fluid flow data sets. Computer **22**(8), 27–36 (1989)
11. Hereford, J.M., Siebold, M., Nichols, S.: Using the particle swarm optimization algorithm for robotic search applications. In: 2007 IEEE Swarm Intelligence Symposium, pp. 53–59 (2007)

12. Jatmiko, W., Sekiyama, K., Fukuda, T.: A mobile robots PSO-based for odor source localization in dynamic advection-diffusion environment. In: 2006 IEEE/RSJ International Conference on Intelligent Robots and Systems (2006)
13. Kennedy, J., Eberhart, R.: Particle swarm optimization. In: Proceedings of IEEE International Conference on Neural Networks, vol. 4, pp. 1942–1948 (1995)
14. Marin, R.D.C.: Vector field design notes (2008)
15. Rukundo, O., Hanqiang, C.: Nearest neighbor value interpolation. In: International Journal of Advanced Computer Science and Applications (2012)
16. Sheetal, Venayagamoorthy, G.K.: Unmanned vehicle navigation using swarm intelligence. In: Proceedings of International Conference on Intelligent Sensing and Information Processing (2004)
17. Wang, X., Yi, P., Hong, Y.: Dynamic optimization for multi-agent systems with external disturbances. Control. Theory Technol. **12**(2), 132–138 (2014)

Extended Abstracts

A Honey Bees Mating Optimization Algorithm with Path Relinking for the Vehicle Routing Problem with Stochastic Demands

Yannis Marinakis[✉] and Magdalene Marinaki

School of Production Engineering and Management,
Technical University of Crete, Chania, Greece
marinakis@ergasya.tuc.gr, magda@dssl.tuc.gr

One of the most known nature inspired algorithms based on the marriage behaviour of the bees is the Honey Bees Mating Optimization (HBMO) algorithm which simulates the mating process of the queen of the hive [1]. In this paper, as there are not any competitive nature inspired methods based on HBMO algorithm for the solution of the Vehicle Routing Problem with Stochastic Demands (VRPSD), at least to our knowledge, we would like to propose such an algorithm and to test its efficiency compared to other nature inspired and classic metaheuristic algorithms. The proposed algorithm adopts the basic characteristics of the initially proposed HBMO algorithm and, simultaneously, uses a number of characteristics of the HBMO based algorithms that were used for the solution of other Vehicle Routing Problem variants. A novelty of the proposed algorithm is the replacement of the crossover operator with a Path Relinking (PR) procedure in the mating phase in order to produce more efficient broods. Finally, a Variable Neighborhood Search (VNS) algorithm is used for the local search phase of the algorithm. The algorithm is compared with a number of algorithms from the literature and with two versions of the HBMO algorithm, the one presented by Abbass in [1] (HBMO1) and the other presented by Marinakis et al. in [2] (HBMO2). The two versions of the HBMO have been modified accordingly by the authors in order to be suitable for their application in the VRPSD. The VRPSD is a NP-hard problem, where a vehicle with finite capacity leaves from the depot with full load and has to serve a set of customers whose demands are known only when the vehicle arrives to them. As in the most VRP variants, the vehicle begins from the depot and visits each customer exactly once and returns to the depot. This is called an a priori tour [3], which is a template for the visiting sequence of all customers. In most of the algorithms used for the solution of the problem, a *preventive restocking* strategy [3] is used where although the expected demand of the customer is less than the load of the vehicle, it is chosen the return of the vehicle to the depot for replenishment. This happens in order to avoid the risk of the vehicle to go to the next customer without having enough load to satisfy him (route failure). For analytical formulation of the VRPSD please see [3]. The results of the algorithm with and without the preventive restocking strategy and comparisons with other algorithms from the literature are presented in the following Table.

© Springer Nature Switzerland AG 2018
M. Dorigo et al. (Eds.): ANTS 2018, LNCS 11172, pp. 423–424, 2018.
https://doi.org/10.1007/978-3-030-00533-7

Computational results **with** the preventive restocking strategy

Instance	BKS1	PSO		HBMO1		HBMO2		HBMOPR	
A-n32-k5	820.5	821.65	0.14	842.02	2.62	841.36	2.54	841.05	2.50
A-n33-k5	684.2	687.04	0.42	688.74	0.66	687.77	0.52	687.32	0.46
A-n33-k6	762.4	769.62	0.95	768.32	0.78	767.32	0.65	766.76	0.57
A-n37-k6	999.72	999.72	0.00	1009.37	0.96	1008.27	0.86	1007.1	0.74
A-n38-k5	752.2	756.56	0.58	758.49	0.84	757.93	0.76	757.51	0.71
A-n39-k5	853.08	853.08	0.00	869.02	1.87	868.50	1.81	868.15	1.77
A-n44-k6	978.83	978.83	0.00	998.82	2.04	997.00	1.86	996.8	1.84
A-n45-k6	996.86	997.41	0.06	998.87	0.20	998.86	0.20	998.73	0.19
A-n53-k7	1096.6	1096.6	0.00	1098.27	0.15	1098.07	0.13	1096.8	0.02
A-n55-k9	1124.3	1124.3	0.00	1126.90	0.23	1124.89	0.05	1124.5	0.02
E-n33-k4	847.38	847.38	0.00	849.72	0.28	849.71	0.27	849.5	0.25
E-n51-k5	544.86	544.86	0.00	547.54	0.49	547.50	0.48	546.63	0.32
P-n55-k15	1002.6	1008.6	0.60	996.56	−0.60	995.99	−0.66	995.98	−0.66
P-n60-k10	772.86	772.86	0.00	776.22	0.44	774.86	0.26	774.81	0.25
P-n60-k15	1021.58	1021.58	0.00	1044.84	2.28	1043.79	2.17	1043.2	2.12

Computational results **without** the preventive restocking strategy

Instance	BKS2	PSO		HBMO1		HBMO2		HBMOPR	
A-n32-k5	853.6	853.6	0.00	855.2	0.19	854.1	0.06	853.6	0.00
A-n33-k5	704.2	704.5	0.04	705.8	0.23	704.6	0.06	704.2	0.00
A-n33-k6	793.9	794.4	0.06	796.8	0.37	794.5	0.08	793.9	0.00
A-n37-k6	1030.73	1031.21	0.05	1032.14	0.14	1031.92	0.12	1030.87	0.01
A-n38-k5	775.14	778.24	0.40	777.35	0.29	776.88	0.22	776.15	0.13
A-n39-k5	869.18	869.18	0.00	871.25	0.24	869.18	0.00	869.18	0.00
A-n44-k6	1025.48	1026.42	0.09	1027.55	0.20	1026.31	0.08	1025.69	0.02
A-n45-k6	1026.73	1027.58	0.08	1027.71	0.10	1027.31	0.06	1026.85	0.01
A-n53-k7	1124.27	1126.95	0.24	1128.42	0.37	1125.52	0.11	1124.71	0.04
A-n55-k9	1179.11	1182.41	0.28	1184.37	0.45	1180.25	0.10	1179.11	0.00
E-n33-k4	850.27	850.27	0.00	850.27	0.00	850.27	0.00	850.27	0.00
E-n51-k5	552.26	554.11	0.33	556.28	0.73	554.12	0.34	553.17	0.16
P-n55-k15	1068.05	1068.05	0.00	1073.28	0.49	1068.05	0.00	1068.05	0.00
P-n60-k10	804.24	807.04	0.35	811.31	0.88	807.15	0.36	805.41	0.15
P-n60-k15	1085.49	1089.12	0.33	1095.42	0.91	1091.18	0.52	1088.51	0.28

References

1. Abbass, H.A.: A monogenous MBO approach to satisfiability. In: Proceeding of the International Conference on Computational Intelligence for Modelling, Control and Automation, CIMCA 2001, Las Vegas, NV, USA (2001)
2. Marinaki, M., Marinakis, Y., Zopounidis, C.: Honey bees mating optimization algorithm for financial classification problems. Appl. Soft Comput. **10**, 806–812 (2010)
3. Marinakis, Y., Marinaki, M.: Combinatorial neighborhood topology bumble bees mating optimization for the vehicle routing problem with stochastic demands. Soft Comput. **19**, 353–373 (2015)

Blockchain Technology for Robot Swarms: A Shared Knowledge and Reputation Management System for Collective Estimation

Volker Strobel$^{(\boxtimes)}$ and Marco Dorigo

IRIDIA, Université Libre de Bruxelles, Brussels, Belgium
{vstrobel,mdorigo}@ulb.ac.be

In swarm robotics research, it is often assumed that robots do not have access to shared knowledge. By sharing knowledge, however, it may become easier to determine whether the robots agree on an outcome or to aggregate the information of the individual robots. We argue that having a medium of shared knowledge can possibly facilitate the implementation of several swarm robotics algorithms and can pave the way for novel swarm robotics applications (e.g., computationally lightweight machine learning algorithms). Sharing knowledge, however, introduces several challenges, such as detecting whether someone has tampered with the shared data or how to reach consensus in case of conflicting information.

We propose a blockchain as shared knowledge medium, computing platform, and reputation management system for robot swarms. A blockchain is a tamper-proof decentralized system used as database and computing platform. Using the *Ethereum* framework [1], arbitrary applications can be executed in a decentralized and secure way via blockchain technology. These applications—called blockchain-based smart contracts—are containers that encapsulate variables and functions executed via blockchain technology. The participants (robots in this work) of a blockchain network locally keep a copy of the blockchain. They create transactions and distribute them among their peers, whenever they are in communication range. The data of the transactions is then used as input to the functions of smart contracts.

The idea of using blockchain technology in combination with robot swarms was first proposed in [2]. In previous work [4], we provided the first proof-of-concept by adding a security layer on top of an existing collective-decision making approach via blockchain technology. In the scope of the present work, we developed a new algorithm exploiting the blockchain's possibilities: the robots use a blockchain-based smart contract to collectively estimate the relative frequency of black tiles (a value between 0.0 and 1.0) in an environment where the floor is covered with black and white tiles.

We conducted three experiments using the robot swarm simulator ARGoS [3], showing (i) the feasibility of the approach, (ii) the trade-off between blockchain size and accuracy, and (iii) the suitability of the blockchain as a reputation management system. In each time-step, a robot determines if it is above a black

© Springer Nature Switzerland AG 2018
M. Dorigo et al. (Eds.): ANTS 2018, LNCS 11172, pp. 425–426, 2018.
https://doi.org/10.1007/978-3-030-00533-7

or a white tile and, every 30 sec, it creates a blockchain transaction with its quality estimate. The smart contract aggregates the estimates of the individual robots to obtain a collective estimate and tells the robots to stop exploring as soon as the uncertainty in the estimate is below a threshold. Therefore, consensus in the swarm is achieved in a fully decentralized way without the need of an external observer. To be able to identify robots with malfunctioning sensors, the reputation of the individual robots is stored and managed via a blockchain-based smart contract. Robots with properly functioning sensors are likely to increase their reputation, while a malfunctioning robot's reputation is likely to decrease.

In case of conflicting blockchain versions, Ethereum achieves consensus by agreeing on the longest blockchain, i.e., the one that required the highest Proof-of-Work (PoW). PoW requires the participants to solve a computational puzzle. This ensures that writing information into the blockchain is computationally expensive. The complexity of the puzzle depends on the computational power of the blockchain participants, i.e., it can adapt to the limited power of robots. Other consensus protocols, such as Proof-of-Stake or using permissioned blockchains might be suitable alternatives to PoW.

Blockchain technology introduces additional computational and memory requirements for the robot swarm. These requirements depend on the number of participants in the network (the blockchain size scales linearly with the number of robots) and the amount of information that is sent to the blockchain. Therefore, it is important to determine which information is security-relevant and should be stored on the blockchain, and which information can be locally processed by the robots.

In future work, we will transfer the blockchain system to physical e-puck robots to study the energy impact of blockchain technology and the possibilities of consensus protocols tailored to robot swarms.

Acknowledgments. Volker Strobel and Marco Dorigo acknowledge support from the Belgian F.R.S.-FNRS and from the FLAG-ERA project RoboCom++.

References

1. Buterin, V.: A next-generation smart contract and decentralized application platform. Ethereum project white paper. (2014). https://github.com/ethereum/wiki/wiki/White-Paper
2. Castelló Ferrer, E.: The blockchain: a new framework for robotic swarm systems. pre-print (2016). arXiv:1608.00695v3
3. Pinciroli, C., Trianni, V., O'Grady, R., Pini, G., Brutschy, A., Brambilla, M., Mathews, N., Ferrante, E., Di Caro, G., Ducatelle, F., Birattari, M., Gambardella, L.M., Dorigo, M.: ARGoS: a modular, parallel, multi-engine simulator for multi-robot systems. Swarm Intell. **6**(4), 271–295 (2012)
4. Strobel, V., Castelló Ferrer, E., Dorigo, M.: Managing Byzantine robots via blockchain technology in a swarm robotics collective decision making scenario. In: Dastani, M., Sukthankar, G., André, E., Koenig, S. (eds.) Proceedings of 17th International Conference on Autonomous Agents and Multiagent Systems (AAMAS 2018). IFAAMAS (2018, in press)

Declarative Physicomimetics for Tangible Swarm Application Development

Ayberk Özgür[1]([⊠]), Wafa Johal[1,2], Arzu Guneysu Ozgur[1], Francesco Mondada[2], and Pierre Dillenbourg[1]

[1] CHILI, EPFL, Lausanne, Switzerland
{ayberk.ozgur,wafa.johal,arzu.guneysu,pierre.dillenbourg}@epfl.ch
[2] LSRO, EPFL, Lausanne, Switzerland
francesco.mondada@epfl.ch

Emerging interest in exploring proximal Human-Swarm Interaction has started approaching the Human-Computer Interaction (HCI) vision of tangible, bidirectional interaction with intelligent swarms made up of "radical atoms" [2]. The recent years have witnessed significant progress towards these once-hypothetical materials in the form of Tangible Swarm Robots, for which the focus shifted towards developing *applications* for the user to interact with, rather than *controllers* for the robots to solve tasks. Still, how and with which tools HCI designers could build such applications in a swift and reusable manner is an open issue.

We propose here such an application development framework which combines two existing approaches: First, we program swarms with *virtual forces* that describe and create robot motion (*i.e.* physicomimetics [4]) instead of coding individual or collective actions over time. Second, we use the Qt Modeling Language (QML) [1], a declarative programming language originally designed to develop graphical user interfaces by *declaring objects* and *binding* their *properties* and *events* to create the program's structure and flow. Our core idea is to define the swarm of robots and their behaviors (forces, tangible input detectors *etc.*) in terms of these modular and reusable constructs. Below, we provide the program for a rudimentary "bubble shooter" game that illustrates this (details such as calibration values and game logic omitted), see Fig. 1 for its operation.

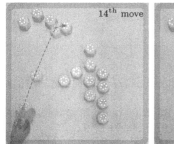

Fig. 1. Bubble shooter game with Cellulo platform [3] where the goal is *e.g.* to build the largest group of the same color. Launched robots collide elastically with existing robots (all 3 figures) and the walls (middle and right) before stopping.

M. Dorigo et al. (Eds.): ANTS 2018, LNCS 11172, pp. 427–428, 2018.
https://doi.org/10.1007/978-3-030-00533-7

```
Swarm{           //Runs physics simulation and other periodic updates of Robots
  Repeater{
    count: 20
    Robot{     //Imaginary robot (a point mass) that the physical robot follows
      color: "white"      //Property that updates LEDs of robot when changed
      GraspDetector{ id: graspDetector
        onGrasped: {      //Event that fires when the robot is touched by user
          var rand = Math.random();
          if(rand < 0.3333)       parent.color = "red";
          else if(rand < 0.6667) parent.color = "green";
          else                    parent.color = "blue";
        }
      }
      FixedToPhysicalRobot{     //Follows physical robot instead when enabled
        enabled: graspDetector.isGrasped
      }
      LaunchDetector{
        onLaunched: {                      //Fires when grasp is released by user
          parent.vel = launchVel;    //launchVel is mean vel. of last N frames
          dampingTimer.timedDisable();
        }
      }
      ViscousDamping{     //Force against and proportional to velocity, F = -cV
        coeff: 4.0                              //Coefficient c in F = -cV
        Timer{ id: dampingTimer                     //Standard QML object
          interval: 5000                              //In milliseconds
          onTriggered: parent.coeff = 4.0        //Fires when timer elapses
          function timedDisable(){
            parent.coeff = 0.0;
            restart();         //Starts timer, restarts if already running
          }
        }
      }
    }
  }            //For forces below, physical robot width is assumed to be 75 mm
  AlignmentAttraction{      //Aligns Robot pairs towards axes with force that
    dist: 150               //is orthogonal to pair, when closer than dist mm
    anglePeriod: Math.PI/3  //Aligns to 0, 60, 120, 180, 240 and 300 degrees
  }
  Attraction{ dist: 150 } //Pulls Robot pairs closer when closer than dist mm
  Repulsion{ dist: 100 }  //Pushes Robot pairs away when closer than dist mm
  BouncyContainer{          //Applies inwards force when container is exited
    rect: { x: 100, y: 100, w: 800, h: 800 }    //Whole arena is 1000x1000 mm
  }
}
```

The resulting programs are concise and contain no robotic implementation details, hiding what is uninteresting from the HCI perspective and exposing what is essential. However, our approach is centralized and cannot be readily applied to many existing platforms who do not guarantee global awareness. Moreover, developers must design and/or tune the desired forces, which may not be trivial.

Acknowledgments. Supported by the Swiss National Science Foundation through the National Centre of Competence in Research (NCCR) Robotics.

References

1. QML Applications. https://doc.qt.io/qt-5/qmlapplications.html. Accessed 28 June 2018
2. Ishii, H., Lakatos, D., Bonanni, L., Labrune, J.B.: Radical atoms: beyond tangible bits, toward transformable materials. Interactions **19**(1), 38–51 (2012)
3. Özgür, A., Lemaignan, S., Johal, W., Beltran, M., Briod, M., Pereyre, L., Mondada, F., Dillenbourg, P.: Cellulo: versatile handheld robots for education. In: ACM/IEEE International Conference on Human-Robot Interaction (2017)
4. Spears, W.M., Spears, D.F., Heil, R., Kerr, W., Hettiarachchi, S.: An overview of physicomimetics. In: Şahin, E., Spears, W.M. (eds.) SR 2004. LNCS, vol. 3342, pp. 84–97. Springer, Heidelberg (2005). https://doi.org/10.1007/978-3-540-30552-1_8

Influence of Leaders and Predators on Steering a Large-Scale Robot Swarm

John D. Lewis[1]([✉]), Himanshi Jain[2], and Sujit P. Baliyarasimhuni[1]

[1] IIIT Delhi, New Delhi, India
{john16095,sujit}@iiitd.ac.in
[2] Delhi Technological University, New Delhi, India
himshijain.hj@gmail.com

A common task involving robotic swarm is to navigate from the current location to a goal location. In order to perform the task, each agent must have some higher level control. There are several ways of imparting this higher knowledge to the agents. One way to control large-scale robotic swarm is by introducing few influential agents (leaders or predators), which can influence the swarm to steer them towards the goal. The leader-based swarms have cohesiveness property by which they can steer a group, however, they do not have the responsibility or role in ensuring transition of the complete swarm to the goal leading to loss of agents. Contrary to the leader-based swarm control, the predator-based swarm control enables the transfer of all the agents to the goal without losing any agents. In this paper, we study, the influence of leaders and predators on three different type of swarm models, namely, shepherding model [3], Couzins model [1], and physicomimetics model [2]. We select these three models because of different underlying principles for swarm behavior. We evaluate the performance of these models under the time to steer and the minimum number of influential agents required for herding metrics.

Predator Model: The predator model utilizes the *"fear"* to influence the agents. We use the sheep/sheep-dog model [3]. Multiple predators are incorporated into the algorithm by dividing the swarm and assigning each predator a section of the swarm. If the agents move out of the flock, then it changes its role to collecting, brings back the stray agents and reverts its role to herding.

In Couzins highly parallel model, the predator(s) influence the swarm only to correct the swarm's heading angle towards the goal, the motion of the predator is controlled such that the resultant movement of the swarm is towards the goal. In Physicomimetics model, the herd repels to the presence of the predator, we allow the swarm to recover and influence again.

Leader Models: Affinity towards the leader propels the remaining agents to follow the leader. We have used leader-based approach [4] and extended it to physicomimetic and shepherding models.

Monte-Carlo simulation results in terms of time taken by the predators/leaders in steering the swarm to the goal are shown in Fig. 1. If the agents do not reach the goal by 1000 s, then the simulation is deemed failure. The results show that the

© Springer Nature Switzerland AG 2018
M. Dorigo et al. (Eds.): ANTS 2018, LNCS 11172, pp. 429–430, 2018.
https://doi.org/10.1007/978-3-030-00533-7

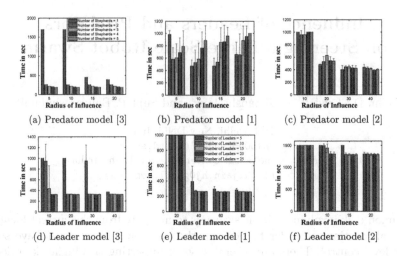

Fig. 1. Average time taken for different number of influential agents with varying radii of influence. Swarm size is 200. The agents are normally distributed $\mathcal{N}(0; 30)$ around the origin $(0; 0)$. The goal area is of $20m$ radius at $(125; 125)$. The parameters of the sheep model and predators are given in [3]. The parameters for the physicomimetic model are $R = 5$ and $C = 50$. For the Couzin's model, the parameters are $r_r = 1$, $r_o = 20$, $r_a = 60$, $\theta = 40°$, and $\alpha = 270°$. For (f), the speed of the leader is 0.25 times that of the agents.

predator-based steering given in [3] outperforms the predator-based swarm models of [1, 2]. In the leaders-based models, Couzin's model in highly parallel motion outperforms models from [2, 3]. However, from Fig. 1(a) and (e), we can see that predator model given in [1] performs better than the leader-based couzin's model. Further analysis on effect of noise and partial information must be investigated for better understanding of the swarm model performances.

Acknowledgments. This work is partially funded by EPSRC GCRF grant EP/P02839X/1.

References

1. Couzin, I.D., Krause, J., Franks, N.R., Levin, S.A.: Effective leadership and decision-making in animal groups on the move. Nature **433**(7025), 513–516 (2005)
2. Spears, W.M., Spears, D.F., Heil, R., Kerr, W., Hettiarachchi, S.: An overview of physicomimetics. In: Şahin, E., Spears, W.M. (eds.) SR 2004. LNCS, vol. 3342, pp. 84–97. Springer, Heidelberg (2005). https://doi.org/10.1007/978-3-540-30552-1_8
3. Strömbom, D., Mann, R.P., Wilson, A.M., Hailes, S., Morton, A.J., Sumpter, D.J., King, A.J.: Solving the shepherding problem: heuristics for herding autonomous, interacting agents. J. R. Soc. Interface **11**(100), 20140719 (2014)
4. Tiwari, R., Jain, P., Butail, S., Baliyarasimhuni, S.P., Goodrich, M.A.: Effect of leader placement on robotic swarm control. In: Proceedings of the Conference on Autonomous Agents and MultiAgent Systems, pp. 1387–1394 (2017)

Movement-Based Localisation for PSO-Inspired Search Behaviour of Robotic Swarms

Sebastian Mai[(✉)], Christoph Steup, and Sanaz Mostaghim[iD]

Faculty of Computer Science, Otto von Guericke University, Magdeburg, Germany
{sebastian.mai,steup,sanaz.mostaghim}@ovgu.de

Particle Swarm Optimization (PSO) is a popular algorithm in swarm robotic search applications. However, a typical PSO method such as Standard PSO [2] usually assumes that the location of all particles are known. Obtaining an estimate for the location of other robots' is necessary to perform the search.

In this paper, we developed an algorithm called Movement-Based Localisation for Robotic Search (MoBaLoRS) that combines a variation of the GDL [1] localisation algorithm with the swarm movement. We have used the SPSO 2011 [2] as the main PSO for search. Nevertheless, other PSO algorithms can be adapted to generate the movement commands. Our experiments show that the movement-based localisation works well with the movements generated by PSO. At the same time, the PSO algorithm continues to work with the position estimates. In each time step our algorithm executes the following computations. First, velocity vectors and particle distances are measured. From those measurements location estimates for the particles are computed. Then PSO is used to compute the particle velocities for the next time step. Similar to GDL [1], our localisation uses measured distances and velocity vectors to compute a position estimate. We assume the particle A to be immovable while particle B moves with the combined speed of both particles $v = v_{B,t} - v_{A,t}$. This is mathematically equivalent to GDL [1] but results in equations that are less complicated and easier to work with.

Given two particles A and B, their relative movement in one time step can be modelled as shown in Fig. 1. The angle α from the triangle formed by particle A, old position of B_{t-1} and new position B_t, is computed with $cos(\alpha) = \frac{d_t^2 + |v|^2 - d_{t-1}^2}{2 \cdot |v| \cdot d_{t-1}}$. The angle ϕ is known from the direction of the combined movement v. Finally, the position can be computed as $\{\tilde{x}, \tilde{x}'\} = (d_t \; sin(\phi \pm \alpha), d_t \; cos(\phi \pm \alpha))$. Because of symmetry there are two solutions $\tilde{x}$ and $\tilde{x}'$, from which the solution that is more likely as position is selected by one of the following two methods: In the first method, we compute $(\hat{x} = x_{t-1} + v)$, where $\hat{x}$ is the old position updated with the velocity vector. Then $\hat{x}$ is compared to the two solutions $\{\tilde{x}, \tilde{x}'\}$. Of the two solutions the one closer to $\hat{x}$ is chosen as the estimate for the current time step. We call this method Closest Solution Selection (CSS).

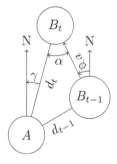

Fig. 1. Geometrical Configuration.

© Springer Nature Switzerland AG 2018
M. Dorigo et al. (Eds.): ANTS 2018, LNCS 11172, pp. 431–432, 2018.
https://doi.org/10.1007/978-3-030-00533-7

The second method uses an error metric for the solution based on a three way round trip. Therefore, we call this method Round Trip Selection (RTS). For points A, B, C we select one solution for each of the vectors $\boldsymbol{AB}$, $\boldsymbol{BC}$ and $\boldsymbol{CA}$ that minimises the round trip error e. e is defined as $\|\boldsymbol{AB}+\boldsymbol{BC}+\boldsymbol{CA}\| = 0 + e$. The estimates for vectors in the round trip that are not selected are deleted. Deletion starts at the round trip where e can be minimised to the lowest value and continues with the other round trips, until only one solution remains for each location estimate. The RTS method allows the algorithm to converge much quicker and has a lower localisation error than the CSS method. The CSS method requires less communication between the robots and uses less computation. The next movement of each robot is planned with a modified version of SPSO 2011 [2] with a fully connected network of particles. The PSO algorithm uses the estimated location of the global best particle L_i and the previous best position of the particle P_i. The position L_i is expressed relative to the particle computing its movement. L_i is different for each particle. Additionally, $L_i(t)$ uses the particle with the best fitness in the *current* time step.

The modified equations of PSO are shown in Eqs. 1–3. $X_i(t)$ (the location of the particle in SPSO) is always $(0,0)$. The parameters c_3 and c_4 are added to control the uniform hyper-

$$G_i(t) = \frac{1}{3}\left(c_1 U_1 \otimes P_i(t) + c_2 U_2 \otimes L_i(t)\right) \quad (1)$$

$$V_i(t+1) = \omega V_i(t) + \mathcal{H}_i(G_i, c_3 \| G_i \| + c_4) \quad (2)$$

$$X_i(t+1) = V_i(t+1) \quad (3)$$

sphere sampling. Next, the positions of the particles are updated according to the previously computed velocity. As the particles move and relative coordinates are used, the previous best needs to be updated after the movement: $P_i(t+1) = P_i(t) - V_i(t+1)$.

Our experiments show that our algorithm does not achieve a localisation error as low as GDL, but solves the behaviour of GDL to yield no solution in specific cases [1]. We found out that PSO and localisation both continue to work when used together. Moverover, higher localisation error leads to more explorative PSO behaviour. Localisation error in the experiment with PSO was lower than with random walk. We are confident that the MoBaLoRS algorithm is able to provide the localisation that is needed by swarm intelligence algorithms and hope this approach enables the use of swarm intelligence algorithms in more robotic applications than before.

References

1. Akcan, H., Kriakov, V., Brnnimann, H., Delis, A.: GPS-Free node localization in mobile wireless sensor networks. In: Proceedings of the 5th ACM International Workshop on Data Engineering for Wireless and Mobile Access, pp. 35–42. ACM (2006)
2. Zambrano-Bigiarini, M., Clerc, M., Rojas, R.: Standard particle swarm optimisation 2011 at cec-2013: a baseline for future pso improvements. In: 2013 IEEE Congress on Evolutionary Computation, pp. 2337–2344, June 2013. https://doi.org/10.1109/CEC.2013.6557848

Of Bees and Botnets

Vijay Sarvepalli[(✉)]

Software Engineering Institute, Carnegie Mellon University, Pittsburgh, USA
vssarvepalli@sei.cmu.edu

Botnets' ability to grow to large sizes, combined with our inability to exhaustively incapacitate them, has forced us to look for more effective methods to model their growth and seek ways to curtail it. Swarm behavior is a stochastic modeling technique based on the collective behavior of swarms. Botnets are like swarms of bees in several ways. For both, a successful hive grows while withstanding losses. In addition, a swarm of honeybees uses distributed decision making [2] similar to a botnet.

The Mirai botnet has been responsible for recent large-scale attacks with its elusive backend of unmanaged Internet of Things and its decentralized self-propagation of infection. This paper explores the swarming behavior of bee colonies as a model for botnets' growth. While there are several differences between this ecological behavior and botnets, the collective behavior of loosely coupled individuals exhibits some commonality that is explored here using a simplified meta-huersitic model. This model can be extended with swarm optimization techniques that can help us prepare to address the more complex botnets of the future.

My modeling uses a beehive with the broad roles of scout bees (infected bot scanners), active foragers (infected bots), and inactive bees (inactive/unreachable bots) to understand a botnet such as Mirai. This meta-heuristic uses an approximation of observed ratios of these roles that make a successful beehive [2]. This is implemented to create a Simulated Bee Colony (SBC) using a simple logic and a computer program written in Python. This logic is represented in Eq. 1, which follows the susceptible–infectious–recovered disease spread model [1].

$$\sum(\tau) = N(1 + \frac{Pscout * Psuccess}{n})^{n\tau} - \mu * N \qquad (1)$$

$\sum(\tau)$ is the total size of the botnet at any given time τ, where N is the current size of the botnet, $Psuccess$ is the probability of success for infection, $Pscout$ is the percentage of devices that are active scanners, n is the number of scan operations per time period τ, and μ represents the decrease in size of the botnet due to a simulated death or other reduction in hive size. The current model is simple in order to evaluate this meta-heuristic. After validating the stability of the simulation with 10 such simulations, one simulation representing a mid-range of botnet growth was compared to scanning data obtained from a few ISPs. The results show that scanning and growth activity can be reasonably modeled with this logic, excluding factors such as varying loss μ over time (see Fig. 1). The size of a botnet depends on scanning (exploration) and compromise

© Springer Nature Switzerland AG 2018
M. Dorigo et al. (Eds.): ANTS 2018, LNCS 11172, pp. 433–434, 2018.
https://doi.org/10.1007/978-3-030-00533-7

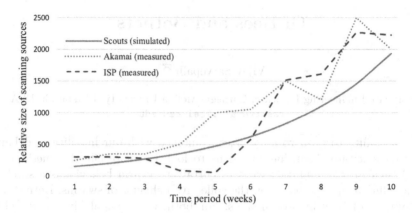

Fig. 1. Simulated bee colony vs. measured scanning activities

(exploitation) activities, which are represented by the two probabilities *Pscout* and *Psuccess* in this model.

The following recommendations for the computer security community were developed from these observations:

- ISPs should analyze their dark space for repeated scanners that appear for three or more days and target remediating those identified as "scouts."
- ISPs should monitor outgoing scanning activity and pursue modifying DHCP lease times to reduce sustained scanning activity from these devices.
- Device vendors should identify the types of devices that are becoming effective "scouts" and pursue patching and fixing their vulnerabilities.

Bio-inspired models can be effective ways to analyze and understand botnets' survival techniques that mimic characteristics of a biological system. A stochastic model such as the one proposed illuminates a botnet's strengths and weaknesses. It will allow computer security communities to begin addressing the threat of botnets that perform attacks at large scale.

Acknowledgments. I thank Soumya Moitra (Carnegie Mellon University), Angelos Stavrou and Constantinos Kolias (George Mason University), and Martin Mckeay (Akamai Technologies) for their input, support, and feedback. This work has been funded by the Department of Defense under Contract No. FA8702-15-D-0002 with Carnegie Mellon University for the operation of the Software Engineering Institute, a federally funded research and development center DM18-0329.

References

1. Diekmann, O., Heesterbeek, J.A.P.: Mathematical Epidemiology of Infectious Diseases: Model Building, Analysis and Interpretation, vol. 5. John Wiley & Sons, New York (2000)
2. Janson, S., Middendorf, M., Beekman, M.: Searching for a new homescouting behavior of honeybee swarms. Behav. Ecol. **18**(2), 384–392 (2006)

Using Particle Swarms to Build Strategies for Market Timing: A Comparative Study

Ismail Mohamed[✉] and Fernando E. B. Otero

School of Computing, University of Kent, Chatham Maritime, UK
{IM572,F.E.B.Otero}@kent.ac.uk

Market timing is the issue of identifying when to buy or sell a given asset in a financial market. A typical market timing strategy would utilize a number of signal generating components that process market related data and return a signal indicating the action to take. These components can either be technical or fundamental in nature. Each component would have a number of parameters and a weight that controlled its contribution to the overall signal. The overall signal is an aggregate of all the signals from the constituent components multiplied by their respective weights. In two recent and comprehensive studies, Hu et al. [1] and Soler-Dominguez et al. [2] investigate the use of computational intelligence techniques in finance. PSO was either used as the primary metaheuristic that optimized either the composition of components used or the parameters of a preset selection of components, or in a secondary rule that optimized another metaheuristic that performed the aforementioned task. No approach was encountered that attempted to both tune the selection of signal generating components as well as the parameter values to use for each component.

In order to adapt PSO to tackle market timing while considering both the selection of signal generating components to use as well as the parameter values to use per selected components, we introduced a number of modifications. We pushed down the implementation of the addition, subtraction and multiplication operators down to the level of the signal generating components in order to maintain parameter integrity during velocity and particle state updates. This allows us to be agnostic to the number and types of parameters used by the components, giving us greater freedom in utilizing whatever components we deem fit. In order to promote solution convergence, we also implemented both velocity scaling and Clerc's Constriction, as well as a decreasing inertia schedule. These basic modifications allowed us to apply the basic PSO algorithm to the issue of market timing, while tackling both the contribution of each component as well as the tuning of each component's parameters. Furthermore, as basic PSO does not display the best balance between exploration and exploitation, we also introduced our own model in an attempt to remedy that. This model differs from basic PSO in two ways: first, both the cognitive and social components of the particle velocity update mechanism are now done probabilistically based on the fitness of the particle's previous best and neighborhood best solutions respectively; second, we periodically prune the components that make up the candidate solutions based on their contribution to overall fitness, allowing us to shrink the candidate solutions and incur less of a computational cost while

© Springer Nature Switzerland AG 2018
M. Dorigo et al. (Eds.): ANTS 2018, LNCS 11172, pp. 435–436, 2018.
https://doi.org/10.1007/978-3-030-00533-7

not significantly sacrificing solution quality. These two modifications allow our variant to seek the least sufficing subset of components that maximizes fitness. We named this variant the Fitness Influenced Stochastic State Update with Pruning PSO (PSO-FInSSUP).

Having defined a method of applying PSO to market timing as defined above, we then assessed the performance of five PSO variants: *l*-best with velocity clamping, *l*-best Clerc's Constriction, *g*-best with velocity clamping, FInSSUP with pruning disabled and FInSSUP with pruning enabled. Both FInSSUP variants assumed a ring neighborhood and used velocity clamping. All variants also used a decreasing inertia schedule. The variants utilized the daily trading data from four stocks (MSFT, GOOG, TSLA and BP), and split the data into training (2015–2016) and testing (2017). All variants also started with six technical signal generating components: Moving Average Converge Diverge, Aroon, Relative Strength Indicator, Stochastic Oscillator, Chaikin Oscillator and On Balance Volume. The fitness of the returned solutions was evaluated using the Sharpe Ratio.

In the results obtained, we noted that models using ring-based networks fared better on average than star-based ones, with *g*-best ranking second to last. No statistical significance was observed, implying that the results from the various PSO models are competitive, suggesting the viability of the PSO-FInSSUP variants. We also noted that the standard deviation of all the PSO variants are relatively large when compared to their means, with means considerably lower than the maximum values of fitness achieved in some cases. At this point it is not clear whether this is attributable to the shape of the solution landscapes and further studies would be required to identify the reason behind the relatively large range of solutions returned. When considering the Sharpe Ratio values attained, we observed that the averages rarely reached a value of one or more. This seems to suggest that the solutions discovered by PSO so far would be considered sub par according to industry standards.

In conclusion, we were able to formalize the issue of market timing into a form that considers both the selection of signal generating components and the values of their parameters. We were able to modify PSO to tackle the issue of market timing using this formulation. One of the PSO models tested is a novel approach that attempts to find the least sufficing set of components while still maximizing Sharpe Ratio and was dubbed PSO-FInSSUP. Our results suggest that work still remains to improve the performance of this PSO approach before the metaheuristic can be considered suitable for live trading.

References

1. Hu, Y., Liu, K., Zhang, X., Su, L., Ngai, E.W.T., Liu, M.: Application of evolutionary computation for rule discovery in stock algorithmic trading: a literature review. Appl. Soft Comput. J. **36**, 534–551 (2015)
2. Soler-Dominguez, A., Juan, A.A., Kizys, R.: A survey on financial applications of metaheuristics. ACM Comput. Surv. **50**(1), 1–23 (2017)

Author Index

Abdelbar, Ashraf M. 342
Adams, Julie A. 385
Al-Hammadi, Yousof 150
Allwright, Michael 188

Baliyarasimhuni, Sujit P. 429
Barel, Ariel 403
Bartashevich, Palina 411
Bello, Marilyn 325
Bello, Rafael 325
Bhalla, Navneet 188
Birattari, Mauro 16, 30, 109
Boumaza, Amine 395
Bozhinoski, Darko 30
Bruckstein, Alfred M. 44, 403
Bullock, Seth 277

Camacho-Villalón, Christian Leonardo 302
Cambier, Nicolas 351
Charpillet, François 395
Christensen, Anders Lyhne 225
Cleghorn, Christopher W. 201, 264, 333
Coello Coello, Carlos A. 317
Coppola, Mario 123
Crowder, Richard 277

de Croon, Guido C. H. E. 123
De Masi, Giulia 239
Dillenbourg, Pierre 427
Dorigo, Marco 57, 188, 302, 425

Engelbrecht, Andries P. 163, 201, 264, 333, 368

Fernández Pérez, Iñaki 395
Ferrante, Eliseo 213, 239, 351
Fleming, Chloë 385
Font Llenas, Anna 135
Franchi, Antonio 3
Frémont, Vincent 351

Gabellieri, Chiara 3
Gomes, Jorge 225

Hamann, Heiko 290
Harrison, Kyle Robert 368
Hasselmann, Ken 16
Hofstadler, Daniel Nicolas 84
Husain, Zainab 150
Hüttenrauch, Maximilian 71

Ibaraki, Toshihide 360
Isakovic, Abdel F. 150

Johal, Wafa 427
Jain, Himanshi 429

Khaluf, Yara 252
Knors, Welf 411
Kuckling, Jonas 30

Lawry, Jonathan 97
Lee, Chanelle 97
Lewis, John D. 429
Ligot, Antoine 30, 109
Lugo, Lázaro 325

Mai, Sebastian 431
Manor, Rotem 44, 403
Marcon dos Santos, Gilberto 385
Marinaki, Magdalene 423
Marinakis, Yannis 423
Marshall, James A. R. 135, 176
Mohamed, Ismail 435
Mondada, Francesco 427
Montero, Elizabeth 317
Mostaghim, Sanaz 411, 431

Nakamura, Yukihiro 360
Neumann, Gerhard 71
Nowe, Ann 325

Oldewage, Elre T. 264, 333
Ombuki-Berman, Beatrice M. 368
Otero, Fernando E. B. 435
Ozgur, Arzu Guneysu 427
Özgür, Ayberk 427

Pallottino, Lucia   3
Pamparà, Gary   163
Pinciroli, Carlo   176, 188
Pitonakova, Lenka   277
Prasetyo, Judhi   239
Primiero, Giuseppe   213

Ranjan, Pallavi   239
Rausch, Ilja   252
Reina, Andreagiovanni   135, 176
Riff, María-Cristina   317
Robert, Frédéric   16
Rojas-Morales, Nicolás   317
Ruta, Dymitr   150

Saffre, Fabrice   150
Salama, Khalid M.   342
Sarvepalli, Vijay   433
Scheepers, Christiaan   201
Schmickl, Thomas   84
Shoji, Kazutaka   378
Simoens, Pieter   252

Šošić, Adrian   71
Steup, Christoph   431
Strobel, Volker   425
Stützle, Thomas   302

Tagliabue, Jacopo   213
Takahashi, Ryouei   360
Talamali, Mohamed S.   135, 176
Tognon, Marco   3
Trabattoni, Marco   57
Trianni, Vito   176, 351
Tuci, Elio   213
Tuzel, Ovunc   385

Valentini, Gabriele   57

Winfield, Alan   97

Xu, Xu   135

Zahadat, Payam   84

Printed in the United States
By Bookmasters